Undergraduate Lecture Notes in Physics

Undergraduate Lecture Notes in Physics (ULNP) publishes authoritative texts covering topics throughout pure and applied physics. Each title in the series is suitable as a basis for undergraduate instruction, typically containing practice problems, worked examples, chapter summaries, and suggestions for further reading.

ULNP titles must provide at least one of the following:

- An exceptionally clear and concise treatment of a standard undergraduate subject.
- A solid undergraduate-level introduction to a graduate, advanced, or non-standard subject.
- A novel perspective or an unusual approach to teaching a subject.

ULNP especially encourages new, original, and idiosyncratic approaches to physics teaching at the undergraduate level.

The purpose of ULNP is to provide intriguing, absorbing books that will continue to be the reader's preferred reference throughout their academic career.

More information about this series at https://link.springer.com/bookseries/8917

Wolfgang Demtröder

Nuclear and Particle Physics

Springer

Wolfgang Demtröder
Demtröder, Germany

ISSN 2192-4791 ISSN 2192-4805 (electronic)
Undergraduate Lecture Notes in Physics
ISBN 978-3-030-58311-8 ISBN 978-3-030-58313-2 (eBook)
https://doi.org/10.1007/978-3-030-58313-2

Translation from the German language edition: "Experimentalphysik 4" by Wolfgang Demtröder, © Springer-Verlag GmbH Deutschland 2017. Published by Springer Spektrum. All Rights Reserved.

This Springer imprint is published by the registered company Springer Nature Switzerland AG
The registered company address is: Gewerbestrasse 11, 6330 Cham, Switzerland

The original version of this chapter has been revised.
The word 'Quellen' has been changed to 'Source'.

The correction can be found in pg.283 of this book.

Preface

This textbook on the introduction to Nuclear and Particle Physics for undergraduate students, is essentially the English translation of the 4th volume of a German textbook series on Experimental Physics. It follows the foregoing three English volumes on *Mechanics and Thermodynamics*, *Electrodynamics and Optics* and *Atoms, Molecules and Photons*.

It tries to explain the essentials and basic ideas of our knowledge about atomic nuclei and elementary particles, the different ways how we got this knowledge and the physical essence of the theories based on the experimental results. The author wants to convince the readers about the importance of these fields in physics, not only because nuclei form the basic building blocks of the matter surrounding us, and they constituted the first matter at the beginning of our universe, but also because of their importance for many applications in medicine, technology, energy production and environmental sciences. It has opened the way for our understanding of stars and their radiation, about the processes at the "big bang" 14.5 billion years ago, but also about many biochemical processes in our body, the detection of diseases and the understanding of metabolic processes.

There are many other fields where the application of radioactive nuclei has brought more detailed information. One example is the age determination of fossils or minerals, which has not only increased our knowledge about the life of our ancestors several thousand years ago, but could also confirm the estimations about the age of our earth.

If the expectations about the technical realization of nuclear fusion reactors will be fulfilled, the CO_2 problem with the resultant increase of the atmospheric temperature and all its consequences can be solved.

Although the applications of particle physics are not as numerous as those of nuclear physics, its importance lies more on the intellectual side. It helps us to understand the basic foundations of our world, the questions about smallest indivisible particles and the borderline between matter and energy. Since the former belief that atoms were the smallest indivisible particles could be. meanwhile. confuted by proving that they can be divided into nuclei and electrons. Now the border to the smallest particles has been shifted farther. Now even nuclei can be split into nucleons which again consist of quarks. The question, whether quarks can be still further divided is still open, but is seems that quarks do not constitute matter in the conventional sense, but form a time-dependent mixture of energy and matter.

Such questions about the fundamental understanding of the basic elements of our world should be of great interest for every reflective student.

The author hopes that his textbook helps to improve the understanding of these fields. He would be grateful about every question and any hint to possible errors. The author tries to answer every e-mail as soon as possible.

Many thanks go to Mrs. Ute Heuser, at Springer Verlag, who was very helpful, even during the COVID-19 Pandemic, in the completion and supervision of the progress of editing this textbook.

Kaiserslautern, Germany Wolfgang Demtröder

Contents

1 Introduction

Nearly all phenomena, observed on earth can be explained by gravitational and electro-dynamical interactions. The macroscopic behavior of matter, which shows up by its mechanical, electrical or magnetic properties is essentially determined by the electron shells of atoms or molecules. The spatial distribution of electrons in atoms is governed by electro-magnetic interaction, as was outlined in Volume 3 of this series of Textbooks. Also all chemical and biological reactions, which determine life on earth are based on electro-dynamical interactions between the electron shells of atoms or molecules. Since the electron shell cloud shields largely the electric Coulomb field of the atomic nuclei, the interaction of nuclei with their surrounding (except that of the same atom) is generally completely negligible, except of the gravitational interaction of the nuclear masses.

This fact is one of the reasons, why atomic nuclei were discovered only in the twentieth century. Nuclear Physics, which deals with the properties and structures of atomic nuclei, is therefore a relatively young branch of Physics.

1.1 What is the Subject of Nuclear and Particle Physics?

Nuclear Physicists investigate the composition of nuclei, their structure and the forces that hold together the different building blocks which form the atomic nuclei. They measure the binding energy of nuclei, investigate excitation methods of nuclei, possible energy levels of excited nuclei and the different ways on which this energy is released. Interesting questions are the criteria for the stability of nuclei and under which conditions they can decompose, how nuclei react when they collide with other particles or with photons. The detailed knowledge of characteristic properties of nuclei, such as mass, charge distribution within the nucleus, electrical and magnetic moments of nuclei or their angular momentum is essential for the further investigations of the dynamics of excited nuclei.

While the atomic electron shell can be described by the well-known electro-magnetic interaction and a closed theory exists, the *Quantum-Electrodynamics,* which correctly reproduces all experimental facts quite well (although most problems can be only solved by numerical computation) the structure of atomic nuclei is governed besides by electro-magnetic forces by two new kind of forces. About these **weak and strong interactions** up to now only incomplete knowledge is available in spite of impressive progress in recent years and there exists no assured complete theory which could describe all observed phenomena. There are several phenomenological models of nuclei, which are adopted from examples of atomic shell treatment, which describe some experimental facts quite well, such as the shell model (see Sect. 5.5.2), or the liquid drop model (Sect. 2.6.3) which is taken from continuum physics.

A deeper insight into the structure of atomic nuclei has been provided by experimental achievements of high energy physics, where highly energetic particles could be used as projectiles hitting the nuclei and probing their internal structure.

It turned out, that the nucleons which form the nuclei were not elementary particles but have a substructure and can be regarded as composed of still more elementary particles called **quarks** (see Sect. 7.4). The forces between these quarks can be described by the interchange of other elementary particles, the Gluons and Vector-Bosons. The wealth of experimental information can now be explained by a closed theory called *Quantum-Chromodynamics*, which is also able to predict new particles that were indeed found later on. This theory was developed analogous to *quantum-electrodynamics* for the atomic shell.

Because of didactical reasons it therefore makes sense to study nuclear physics after atomic physics, although in the representation of the build-up principle of macroscopic bodies the quarks form to our present knowledge the smallest constituents composing the nuclei, which in turn form together with the electrons atoms and molecules as the

W. Demtröder, *Nuclear and Particle Physics*, Undergraduate Lecture Notes in Physics,
https://doi.org/10.1007/978-3-030-58313-2_1

building blocks of macroscopic matter. Starting from the smallest components in a systematic build-up theory elementary particles should be presented first followed by nuclei, atoms and molecules.

Since atomic nuclei cannot be seen by the naked eye, special detection techniques had to be developed for their "observation". These techniques are mostly based on interactions of a nucleus with the atomic shell of its own atom or with that of other atoms. Examples of the first kind are spectroscopic measurements of the hyperfine structure of atomic spectra (Sect. 2.5 and Vol. 3, Sect. 5.6). Examples of the second kind are the ionization of air-molecules in the cloud chamber (Sect. 4.3) by collisions with alpha-particles or the light emission of scintillators under bombardment by the emitted products of radioactive nuclei (Chap. 4). Such detection techniques and the interpretation of their results demand often the knowledge of atomic and solid state physics, which were not available at the beginning of the twentieth century. Therefore real knowledge about nuclear Physics became only possible after the necessary experimental techniques were available.

A further important aspect is related to the wave-nature of particles (see Vol. 3, Chap. 3). For the investigation of the internal structure of nuclei by the scattering of colliding particles they must have a sufficient spatial resolution. This implies that the de-Broglie-wavelength $\lambda_{dB} = h/(m \cdot v)$ of particles with mass m and velocity v must be smaller than the size of the nucleus. This means that their kinetic energy must be sufficiently high. If α-particles are used as probe, their velocity must be larger than 10^8 m/s i.e. their kinetic energy must be larger than 3.2×10^{-11} J = 200 MeV for a de-Broglie-wavelength of $\lambda_{dB} < 10^{-15}$ m.

Particles with such a high energy were not available at the beginning of the twentieth century because the first particle accelerators for medium energies had been built only after 1930 and for higher energies only after the Second World War, i.e. after 1945. The first investigations of the size of the atomic nucleus had been therefore performed with particles emitted by radio-active nuclei. Their kinetic energy, however, did not exceed 10 MeV. This is a further reason for the relatively late (compared to atomic physics) development of nuclear physics.

The physics of elementary particles owns its rapid progress at the end of the twentieth century mainly to the construction of giant particle accelerators (see Sect. 4.1) and the realization of sensitive detectors, but also to new ideas of theoretical physicists. The experimental and theoretical success of recent years has brought about a better understanding of elementary particles and their interactions. The nowadays widely accepted general theory is called the "*Standard Model of Particle Physics*" which explains most of the experimental facts found up to now. The interesting questions is, whether there will be new experimental results which demand an extension of the present theory beyond the Standard Model. The ultimate goal of all possible theories is the unique description of all known interactions. i.e. a unified theory of everything.

Elementary Particle Physics (also called "High energy physics") is, not only from the experimental view, but also regarding existing theories quite demanding and difficult and its detailed description would surpass the level of this introductory textbook. We will therefore treat it on a more elementary level presenting the physical essence of the experimental findings and the theoretical statements, thus explaining the essential physical insight gained from experiment and theory.

1.2 Historical Development of Nuclear and Elementary Particle Physics

Signals from atomic nuclei as radioactive radiation were first discovered by *Antoine Henri Becquerel* (1852–1910 Fig. 1.1). He found that "radiation" emitted by Uranium ore blackened photo-plates, but he did not know anything about the origin of this radiation and about atomic nuclei. Systematic investigations by *Marie Sklodowska-Curie* (1867–1908) and her husband *Pierre Curie* (1959–1906) (Fig. 1.2) lead 1898 to the discovery of two new much more intensely radiating chemical elements, the **Polonium** and **Radium**

Fig. 1.1 Antoine Henry Bequerel (*From* E. Bagge: Die Nobelpreisträger der Physik (Heinz Moos Verlag München 1964)

Fig. 1.2 Marie Sklodowska-Curie and her husband Pierre Curie (*From* E. Bagge: die Nobelpreisträger der Physik 9Heinz Moos Verlag München 1964)

(Nobel prize 1908), but it was still not clear what the source and the character of this radiation was.

Quantitative nuclear physics started at the beginning of the twentieth century (1909–1910) with the scattering experiments by *Ernest Rutherford* (Fig. 1.3). The result of these detailed experiments (see Vol. 3, Sect. 2.8.5) could be only explained when assuming that atoms are composed of a positively charged nucleus which carries nearly all of the atomic mass but has a very small volume compared to that of the atom, and a negatively charged electron shell, which determines the size of the atom but contributes a negligible fraction to the atomic mass. This result, which astonished Rutherford very much, sets the foundation of modern atomic models.

Rutherford and *Frederick Soddy* (1877–1956) could show in the following years 1902–1909 that there are three different kinds of radioactive radiation which they called α, ß and γ-radiation.

All these investigations collected a variety of facts about radioactive radiation, which were very helpful for the beginning of a better understanding of radioactivity. However, a real insight was only achieved after Rutherford published 1911 his atomic model which explained the results of his scattering experiments of α-particles by gold atoms (see Vol. 3, Sect. 2.8).The essential features of this model are also nowadays still valid. This date may be regarded as the birth of nuclear physics, because here already the characteristic properties of atomic nuclei such as size, mass density and charge had been determined.

Rutherford and *Geiger* found 1912 that α-particles were identical to twofold positively charged helium-ions He^{++} which can be produced not only as particle emission from radioactive nuclei, but also by impact ionization in a helium discharge.

Fig. 1.3 Lord Ernest Rutherford (*From* St. Weinberg, Teile des Unteilbaren. Spektrum, Heidelberg 1990)

Sir *Joseph John Thomson* (1856–1940) discovered by means of mass spectrometry (see Vol. 3, Sect. 2.7) that many chemical elements with equal number of electrons (and therefore equal chemical properties) have different masses (**isotopes**) (Nobel-prize 1906). *Francis William Aston* (1877–1945) proved later 1911 with an improved mass spectrometer that most chemical elements have several isotopes which means that their nuclei have different masses.

The time period 1918–1939 between world war 1 and 2 brought about a wealth of new discoveries and a more detailed insight into the structure of atomic nuclei., where only a few of them can be mentioned here (see time table of nuclear- and high energy- physics.

After the first detection of an artificial nuclear conversion of nitrogen nuclei by collisions with α-particles by Rutherford 1919 *Patrick Maynard Blackett* (1897–1974) could make such reactions visible by means of the **cloud chamber** developed by *Charles Thomson Rees Wilson* (1869–1959) (see Vol. 3, Fig. 2.18). With these artificial nuclear conversion the old dream of alchemists to make gold of cheaper materials can be realized, however, with an expenditure which surpasses the prize of gold by several orders of magnitude.

Analyzing the energy balance of the observed nuclear conversion Rutherford could already 1924 imply the huge binding energy of atomic nuclei which exceeds that of atomic electrons by a factor of 10^5–10^6. Rutherford concluded that very strong binding forces must exist.

While investigating nuclear reactions with an ionization chamber *Sir James Chadwick* (1891–1974) discovered 1932 the neutron (which Rutherford had already postulated 15 years ago). This allowed now realistic models of nuclei and explained the existence of isotopes.

All nuclei consist of protons and neutrons. The number Z of the positively charged protons is equal to the number of atomic electrons and therefore determines the chemical characteristics of the atom. The total mass of the nucleus is the sum of proton and neutron masses.

Different isotopes of an element have different number of neutrons but the same number of protons.

Similar to the excitation of electrons in the atomic shell the protons or neutrons can be excited into higher energy states from where they can return to the ground state by emission of radiation or particles. Such excited nuclei are the source of the observed γ-radiation. Also the emission of ß-radiation (which consists of electrons emitted by the nucleus) could be now explained. The electrons are created by the conversion of neutrons into protons. They do not stay in the nucleus but leave the nucleus immediately (see Chap. 3).

Fig. 1.4 Otto Hahn (1879–1968) and Fritz Straßmann (1902–1980)

The discovery of nuclear fission 1939 by *Otto Hahn* (1879–1968) and *Fritz Straßmann* (1902–1980) (Fig. 1.4) started very intense research activities, because the relevance of this discovery for peaceful energy production and also for possible atomic bombs were soon realized. Otto Hahn and his coworker could at first not explain their discovery. However, in the same year *Lise Meitner* (Fig. 1.5) and *Otto Robert Frisch* gave a plausible explanation of nuclear fission based on a hydrodynamic nuclear model (See Sect. 6.5). The couple *Frederic* (1900–1958) and *Irene Joliot-Curie* (1897–1956) (Fig. 1.6), daughter of *Piere* and *Marie Curie*) could detect the neutrons emitted during nuclear fission.

Fig. 1.5 Lise Meitner (1878–1768)

Important historical landmarks of the development of elementary particle physics were the discovery of the **positron** 1932 by *Carl David Anderson* (1905–1991) (Fig. 1.7) and the postulation of the neutrino by Wolfgang Pauli 1930 (Fig. 1.8), which was experimentally discovered only 1955. Pauli postulated the neutrino to explain the continuous energy spectrum of electrons emitted by ß-decay of nuclei.

Further important steps in the development of particle physics were the discoveries of the **myon μ** 1937 and the **π-meson** 1947 in the cosmic radiation and the anti-proton $\overline{p}$ 1955 in collision experiments with particle accelerators.

Essential for the understanding of symmetry principles and properties of the weak interaction was the experimental confirmation of parity violation for the ß-decay 1956 by Madam *Chien Shiung Wu* (Fig. 1.9), which had been already postulated by Lee and Yang (Fig. 1.10).

The development of the Weinberg-Salam model which includes the unification of electromagnetic and the weak interaction (electro-weak gauge field theory), and in particular the postulation of quarks as the fundamental building

Fig. 1.6 Fredric (1900–1958) and Iren Joliot-Curie (1897–1956) (Bettmann-Corvis)

Fig. 1.7 Carl David Anderson (1905–1991)

Fig. 1.8 Wolfgang Pauli (1900–1958)

blocks of nucleons by *Gellman* and *Zweig* 1996 (Fig. 1.11) has improved our understanding of elementary particles considerably. According to these theories there are only 17 really fundamental particles plus their antiparticles (12 Fermions with half integer spin, namely 6 quarks plus their antiparticles and 6 light particles (electron e^-, myon μ, tau-lepton τ^-, their neutrinos ν_e, ν_μ and ν_τ. These fundamental fermion are compiled in Table 1.1 together with the bosons (integer spin) describing the 4 kinds of interaction: The photon γ for the electro-magnetic interaction, the gluon g for the strong interaction and the Z^0 and and $W^\pm$ for the weak interaction and finally the recently discovered Higgs-boson (see Sects. 7.4–7.6).

Each column of Table 1.1 contains a family of quarks (purple) and a family of leptons (green) and the bosons describing the interactions (except the gravitation which is not included in the Standard model). The discovery of the Z^0 and $W^\pm$ bosons 1983 at the European high energy accelerator CERN close to Geneva, Switzerland have confirmed

Fig. 1.9 Chien-shiung Wu (1912–1997)

Fig. 1.10 Tsung Doa Lee and Chen Ning Yang. Nobelpreis 1957

essential statements of the Weinberg-Salam (Fig. 1.12) Theory (Standard Model). Experiments at CERN also approved that there are really only 3 families of leptons, which also suggests because of symmetry reasons the limitation of the number of quark-families to 3. This limits the total number of fundamental particles and is an important step towards a unified theory of all interactions.

Fig. 1.11 Murray Gell-Man and George Zweig (born 1937) (Bing Com. Images)

Fig. 1.12 Stephen Weinberg and Abdus Salam (By Larry D. Moore, CC BY-SA 3.0, https://commons.wikimedia.org/w/index.php?curid)

Until a "grand unification theory" can be realized and proved it will demand long and difficult experimental and theoretical efforts. By means of the recently realized large particle colliders and storage rings (see Sect. 4.1) one can hope to discover new particles which might explain several open questions such as the **dark matter** in the universe and the **dark energy** responsible for the accelerated expansion of the universe.

A good introduction into the cultural and historical development of physics can be found in in [1–8].

1.3 Relevance of Nuclear and Particle Physics; Open Questions

Mankind has always tried to enlarge the confines of perception in order to learn what lies behind the horizon of our present knowledge. Nuclear-and Particle- Physics provide an impressive example how these borders of our knowledge has been pushed into the subatomic range with dimensions smaller the that of atomic nuclei.

The development of models of elementary particles experiences the following paradoxes: If particles have a

Table 1.1 The fundamental particles according to the Standard Model (https://www.commons.wikimedia.org/wiki)

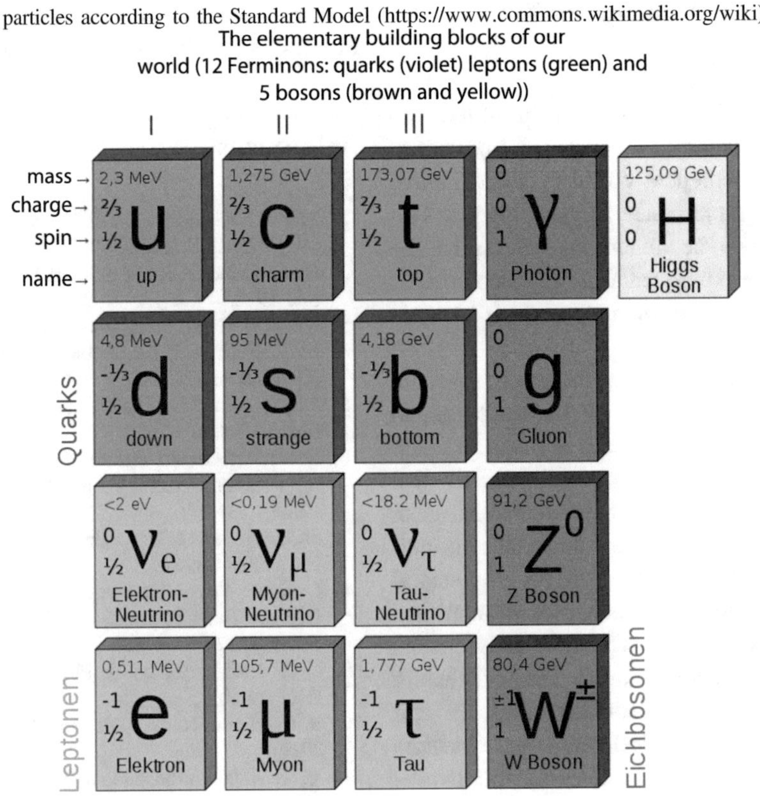

finite size, they should be in principle divisible. This means they should consist of still smaller subparticles. On the other side, if they are point-like particles they should have zero size and therefore zero mass unless we assume an infinite mass density. This shows, that we need a new concept of matter, where below a certain size or above a certain energy a distinction between matter and energy is no longer possible. Most probably symmetries and topological concepts have to be introduced, where matter and fields are described by a unified model [14–16].

The relevance of particle physics therefore extends beyond the physical concept into a frontier region between natural science and Philosophy. This becomes even more obvious, when we regard the importance of High Energy Physics for the understanding of the development of our universe. The *Big-Bang Model* und conceptions of an *inflationary universe* and the formation of the chemical elements are essentially based on information from nuclear- and particle physics.

For a long time nuclear Physics had been regarded as pure science with no applicatory prospects. Since the realization of atomic bombs it became obvious how strongly the results of nuclear physics affects mankind. Fortunately there are also many peaceful and very useful applications. Examples are the usage of radioactive isotopes in Medicine and Biology and for the solution of many technical problems (see Sect. 8.1). Although nuclear reactors are regarded as critical since the disasters of Chernobyl and Fukushima, they are still an example for CO_2-free energy sources and they are safe if maintained correctly and operated according to the rules. The still open question is the safe deposit of the nuclear waste. The hope is to realize controlled nuclear fusion in the near future, which would provide a practically inexhaustible energy source with much less radioactive waste than for fission reactors.

For hydrological investigations the tracking of underground water courses and the local distribution of groundwater using radioactive tritium has proved to be very effective and has brought more detailed insight into necessary measures for water pollution control. Radioactive methods of age determination are nowadays standard techniques for the exact age assignment of rocks, fossils and other objects of archeology.

A more extensive representation of many applications will be given in Chap. 8.

1.4 Survey About the Concept of This Textbook

This textbook tries to explain, how our concept of the subatomic structure of our word has developed and which experiments can prove theoretical concepts about atomic nuclei and elementary particles. Like in the foregoing volumes of this textbook series the author emphasizes, that the close cooperation between experimentalists and theoreticians can result in a consistent model of reality which matches with experimental facts.

Therefore in Chaps. 2 and 3 at first the most important characteristics of stable and unstable nuclei are presented together with the experimental techniques to measure them. Historically the different manifestations of radio-activity and their investigations had given the first hints to atomic nuclei (Although the discoverers did not know about nuclei, until the scattering experiments of Rutherford and his group could explain the structure of atoms consisting of a small but heavy nucleus and an electron shell. The radio-active nuclei are therefore treated extensively in Chap. 3. The next chapter introduces the experimental instrumentation of nuclear and particle physics, such as particle accelerators and detectors and explains the most important techniques for the investigation of structure and size of nuclei, namely scattering experiments and nuclear spectroscopy.

The results of such experiments have led to the models of nuclei, discussed in Chap. 5.

Analogue to chemical reactions in atomic and molecular physics which are based on collisions between atoms and molecules, nuclear reactions due to collisions of nuclei with other particles (protons, neutrons electrons, photons or other nuclei etc.) can fragmentize the nucleus under bombardment or can produce new nuclei. These new nuclei are generally unstable and emit radiation or decay into further particles. The investigation of nuclear collisions and their results are treated in Chap. 6, while nuclear reactions connected to energy production, such as nuclear fission and fusion are covered in Chap. 8.

Chapter 7 gives a compact treatment of our present knowledge about the elementary particles: quarks and leptons, their interactions and their composition to nuclei and other composed particles such as mesons and hadrons. The most important experiments which have led to our present model of particle physics are shortly discussed. In order to stress the importance of nuclear physics for technical applications, in Chap. 8 some of these applications in Medicine, Biology, environmental research and energy production are presented. In particular nuclear fission reactors are explained and the prospects of nuclear fusion reactors and their advantages are discussed.

Useful surveys and textbooks on Nuclear and Particle Physics can be found in the following reference list.

References

1. Milorad Mladjenovic: History of Early Nuclear Physics 1896 – 1931 (World Scientific 1992).
2. Bruce Cameron Reed: A short History of Nuclear Physics to the mid 1930.
3. Anton C. Capri: From Quanta to Quarks. (World Scientific 2007).
4. Jed Z. Buchwald: The Oxford Handbook of the History of Physics. (OUP Oxford 2017).
5. Stephen Hawking: Brief Answers to the Big Questions. (John Murray 2018).
6. Rutherford, Ernest (1906). "On the retardation of the α-particle from radium in passing through matter". Philosophical Magazine. 12(68): 134–146.
7. St. Krivit: Lost History: Explorations in Nuclear Research (Pacific Oaks Press 2016).
8. David N. Schwartz: The Last Man Who Knew Everything: The Life and Times of Enrico Fermi, Father of the Nuclear Age (Basic Books 2017).
9. C. A. Bertulani: Nuclear Physics in a Nutshell (Princeton Univ. Press 2015).
10. B. R. Martin, Nuclear and Particle Physics (John Wiley and Sons 2019).
11. V. K. Mittal: Introduction to Nuclear and Particle Physics, 4th edit. PHI Learning PVT. Ltd 2018.
12. S. L. Kakani: Nuclear and Particle Physics (Viva Books 2016).
13. W. N. Cottingham: An Introduction to Nuclear Physics (Cambridge Universitya Press 2001).
14. E. F. Taylor and J. A. Wheeler (1992). Spacetime Physics. San Francisco: W. H. Freeman.
15. Fernflores, Francisco (2019). "The Equivalence of Mass and Energy", The Stanford Encyclopedia of Philosophy (Fall 2019 Edition), Edward N. Zalta (ed.), https://plato.stanford.edu/archives/fall2019/entries/equivME/.
16. Matter (Wikipedia)

General Literature

1. Kenneth S. Krane: Introductory Njuclear Physics (Wiley 2008).
2. Jouni Suhonen: From Nucleons to Nuclei (Springer 2007).
3. Jean L. Basdevant, James R. Spiro: Fundamemntal Nuclear Physics (Springer 2020).
4. W. Greiner: Nuclear Physics: Present and Future (Springer Heidelberg 2016)
5. J. M. Blatt, V. Weißkopf: Theoretical Nuclear Physics (Dover, New York 1991)
6. W. N. Cottingham, Derek A. Greenwood: Introduction to Nuclear Physics (Cambridge Univ. Press 2007)
7. Reinhard Stock ed.:Encyclopedia of Nuclear Physics and its Applications (Wiley VCH, Weinheim 2013)
8. Noboru Takigawa, Kouhei Washiyama: Fundamentals of Nuclear Physics. (Springer, Heidelberg 2018)
9. J- M. Blatt, V. F. Weisskopf: Theoretical Nuclear Physics. (Dover Publ. 2012)
10. W. K. Terry: Nuclear Technology Demystified. (Canoe Tree Press Manchester 2020)

11. S. Marguet: The Physics of Nuclear Reactors (Springer, Heidelberg 2018)
12. Jose-E. Garcia-Ramos, Clara E. Alonso, Maria V. Andres, F. Perez- Bernal: (eds): Basic Concepts in Nuclear Physics. Theory, Experiments and Applications (Springer, Heidelberg 2016)
13. Anwar Kamal: Nuclear Physics (Springer, Heidelberg 2016)
14. Andrew L. Larkoski: Elementary Particle Physics. (Cambridge University Press 2019)
15. David Griffits: introduiction to Elementary Particles. (Wiley VCH 2008)
16. M. Peskin: Concept of Elementary Particles. (Oxford University Press 2019)

2 Structure of Atomic Nuclei

Before we discuss in detail in Chap. 4 the equipment and the experimental techniques used in Nuclear and High-Energy Physics we will briefly present in this chapter the basic outcomes about atomic nuclei and the resulting concept of the structure of nuclei. This helps for a better understanding of experimental details and the aims of nuclear research [1, 2].

2.1 Experimental Methods

The two essential tools for the investigation of nuclei are **scattering experiments** and **spectroscopic methods**. Our knowledge about nuclear structure and the different interactions between nuclei and other particles are mainly based on elastic, inelastic and reactive scattering at collisions of nuclei with particles, or on spectroscopic measurements of the energy states of excited nuclei and of intensity, polarization and angular distribution of radiation, which is emitted on transitions between these energy levels.

All scattering experiments use a collimated beam of particles with kinetic energy $E_0 = mv_0^2/2$, particle density n [m^{-3}] and particle flux density

$$\Phi = \boldsymbol{n} \cdot v_0 \quad \text{with} \quad [\Phi] = 1/\left(\mathrm{m}^2 \cdot \mathrm{s}\right)$$

which impinge on the target nuclei with the particle density n_T, in a defined target volume $V = F \cdot \Delta x$ with cross section F and thickness Δx (Fig. 2.1). Then $N = \Phi \cdot F$ particle pass per second through the cross section F.

For elastic scattering that fraction

$$\frac{\Delta N}{N} = f(\vartheta, E_0)\Delta\Omega \qquad (2.1)$$

with

$$f(\vartheta, E_0) = \frac{n_\mathrm{T}}{F} \cdot V \cdot \frac{\mathrm{d}\sigma}{\mathrm{d}\Omega}(\vartheta, E_0)$$

of the incident particles is measured which has been scattered by the target nuclei under the angle ϑ into the solid angle $\Delta\Omega$.

The differential scattering cross section $(\mathrm{d}\sigma/\mathrm{d}\Omega)$ gives the contribution of a single target nucleus to the fraction of particles scattered under the angle ϑ into the solid angle $\Delta\Omega = 1$ sterad. It depends on the kind of target nuclei and on the energy and the kind of incident particles. During the scattering experiments either the scattering angle ϑ or the energy E_0 of the incident particles or both are varied.

The incident particles can be

- electrons, which interact with the target nucleus only by electromagnetic or by weak interactions
- neutrons, which only probe the strong interaction
- charged heavy particles such as protons, which interact with the target nucleus by strong forces as well as by electro-magnetic interaction.

For inelastic scattering, where part of the kinetic energy is transferred into internal energy of the collision partners, in addition the energy loss of the scattered particle can be measured or the corresponding excitation energy of the target nucleus.

In reactive scattering processes the identity of the incident particle or of the target nucleus is not preserved. Either the target nucleus will be transferred into another nucleus (*artificial nuclear conversion*), or new particles are created, if the energy of the incident particle is sufficiently high. This latter process is called in high energy physics deep inelastic scattering.

The spectroscopic methods involve:

- Measurements of the atomic energy levels. Compared to the energies of the electron shell in the Coulomb field of a point like positive nuclear charge the energies are shifted because of the finite volume of the nucleus and because

W. Demtröder, *Nuclear and Particle Physics*, Undergraduate Lecture Notes in Physics,
https://doi.org/10.1007/978-3-030-58313-2_2

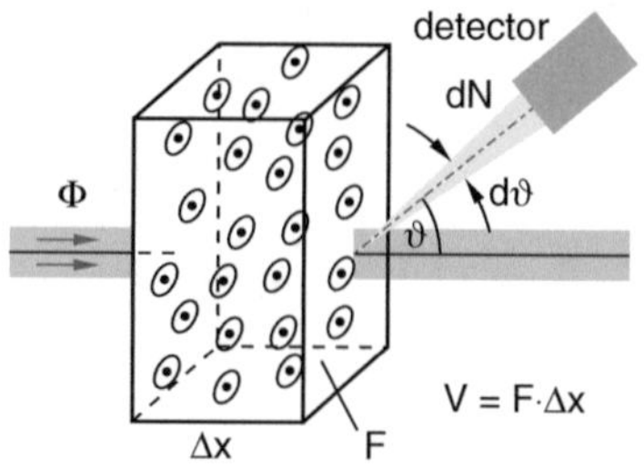

Fig. 2.1 Principle arrangement for measuring differential cross sections

of a possible nuclear quadrupole-moment or a magnetic dipole moment (*hyperfine structure* see Vol. 3, Sect. 5.6).

- Measurements of the radiative emission of excited exotic atoms which yields the energy differences between the energy levels of these atoms (one electron in the atomic shell is replaced by a muon μ^- or another heavy negatively charged particle (such as π^-, K^--meson). The Bohr-radius of these particles is smaller than that of "normal atoms" by the mass ratio m_μ/m_e. For the muon μ^- with $m_\mu = 206\ m_e$ this means a reduction by the factor 206, for the K-meson this factor is even 870. The muon and all the more the K-meson are much closer to the nucleus than the s-electron in normal atoms. In case of the K-meson it even spends most of its time inside the nucleus. The energy levels of these exotic atoms are therefore very sensitive against the charge distribution inside the nucleus (see Vol. 3, Sect. 6.7).
- Measurements of the photon energy $h \cdot \nu$ of the gamma radiation emitted by excited nuclei (γ-Fig. 2.2a)

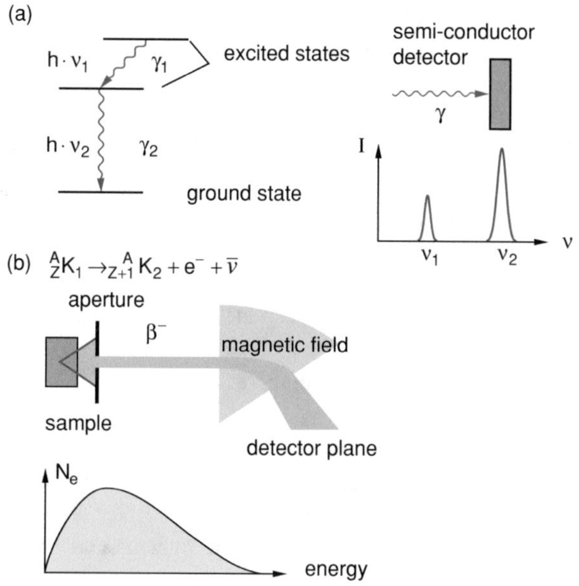

Fig. 2.2 Spectrosopy of energy levels in nuclei: **a** By measuring the energy of γ-quanta emitted by the excited nucleus. **b** By measuring the energy distribution of electrons emitted by β-radio-active nuclei

- Measurement of the energy distribution of particles emitted by unstable nuclei, for instance electrons or α-particles (see Sects. 3.3.2 and 3.3 and Fig. 2.2b)

All these methods give complimentary information about nuclear structure, excitation energies and characteristic properties of the nucleons building up the nucleus. This will be outlined later in more detail, One can find an instructive representation about many early fundamental experiments in the recommended book by Bodenstedt (Experiments on Nuclear Physics and their interpretation (B.I. Mannheim 1973) or in reference [1, 2].

2.2 Charge, Size and Mass of Nuclei

The charge of atomic nuclei could be already estimated from the scattering experiments of Rutherford (see Vol. 3, Sect. 2.8). From the differential cross section

$$\left(\frac{d\sigma}{d\Omega}\right)_\vartheta = \left(\frac{Z_1 \cdot Z_2 \cdot e^2}{4\pi\varepsilon_0 \cdot 4E_0}\right)^2 \frac{1}{\sin^4 \vartheta/2} \tag{2.2}$$

for the elastic scattering of particles with energy E_0 and charge $Z_2 \cdot e$ in the Coulomb field of a target nucleus with charge $Z_1 \cdot e$ the charge $Z_1 \cdot e$ of the target nucleus can be deduced.

A much more accurate determination of the nuclear charge was obtained by the systematic measurements of the frequencies of the X-ray K_α-lines which were performed since 1913 by *Henry Moseley* (1887–1915) for many chemical elements. As has been shown in Vol. 3, Sect. 7.6 the Moseley Law

$$\bar{\nu} = Ry(Z - S)^2\left(1/n_1^2 - 1/n_2^2\right) \tag{2.3}$$

gives the relation between the wavenumber $\bar{\nu} = 1/\lambda$ of X-ray transitions between atomic energy levels with principal quantum numbers n_1 and n_2 and the effective nuclear charge $(Z - S) \cdot e$ (this is the nuclear charge $Z \cdot e$ reduced by the quantity $S \cdot e$ due to partial screening of the nucleus by the electron cloud). Such transitions can be induced if one electron is pushed out of the inner shell by electron impact and an electron from an outer shell falls into the empty state. Ry is the Rydberg constant, which has been meanwhile measured with very high precision. For transitions into the lowest 1s-energy level is $n_1 = 1$ and the effective nuclear charge is (Z-1) i.e. the screening factor due to the remaining second electron in the 1s-state is $S \approx 1$.

Rutherford could also estimate the size of nuclei from the results of his scattering experiments [3]. He assumed that the deviation of the measured number $N(\vartheta)$ of particles scattered under large angles ϑ from that number expected from the calculated cross section for the scattering by a pure Coulomb potential was due to the finite size of the nucleus. For large scattering angles ϑ the incident particles pass the target nucleus at short distances and the influence of the nuclear force will change the interaction between the two particles and therefore change the scattering angle. However, he could investigate these deviations only for light target nuclei, because he had only as incident particles the α-particles from radioactive decay which have energies below 10 MeV. Later more accurate measurements by *Wegener* and coworkers [4] used 40 MeV α-particles from accelerators,

As long as the Coulomb law is valid the energy- and angular momentum-conservation for particles passing the target nucleus at the distance δ from the nuclear center demand

$$\frac{mv_0^2}{2} = \frac{mv^2}{2} + \frac{Z_1 Z_2 e^2}{4\pi\varepsilon_0 \cdot \delta} \quad \text{(Energy law)}, \tag{2.4a}$$

$$m \cdot b \cdot v_0 = m \cdot \delta \cdot v \quad \text{(Angular momentum conservation law)}, \tag{2.4b}$$

where v_0 is the initial velocity of the projectile, b is the impact parameter of the incident particles (Fig. 2.3) and v is the velocity of the projectile at the closest approach to the target nucleus.

For central collisions ($b = 0$) the smallest distance δ_0 to the target nucleus, where the projectile is reflected back into the incident direction can be directly obtained from the equation

$$\frac{m}{2} v_0^2 = \frac{Z_1 Z_2 e^2}{4\pi\varepsilon_0 \delta_0} \tag{2.4c}$$

Inserting (2.4c) into (2.4a) gives the equation for the general case $b \neq 0$

$$mv^2 = mv_0^2 (1 - \delta_0/\delta)$$

Squaring (2.4b) yields

$$m^2 v_0^2 b^2 = m^2 v^2 \delta^2 = m^2 v_0^2 \delta^2 (1 - \delta_0/\delta). \tag{2.4d}$$

This gives the relation between impact parameter b and smallest distance δ from the center of the target nucleus

$$b^2 = \delta^2 - \delta\delta_0. \tag{2.5}$$

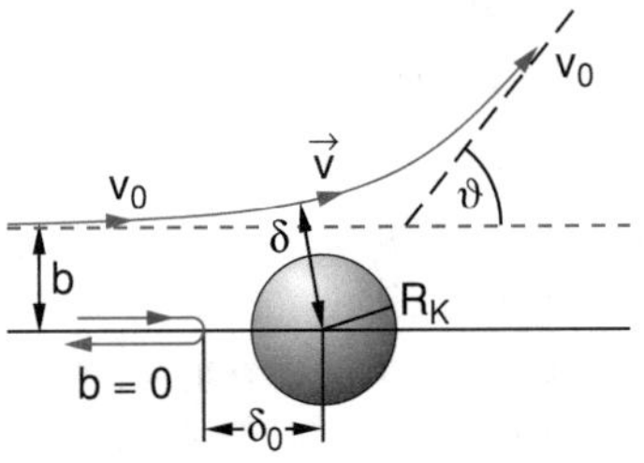

Fig. 2.3 Elastic scattering of a particle with initial velocity v_0, impact parameter b and scattering angle ϑ

In Vol. 3, Sect. 2.8.6 it was shown that for a Coulomb potential the impact parameter b can be related to the scattering angle ϑ as

$$b = \frac{Z_1 Z_2 e^2}{4\pi\varepsilon_0 m v_0^2} \cot(\vartheta/2) = \frac{\delta_0}{2} \cot(\vartheta/2)$$

Inserting this expression for b into (2.5) gives for a Coulomb potential the relation between the smallest distance δ and the scattering angle ϑ

$$\delta = \frac{\delta_0}{2}\left[1 + \frac{1}{\sin(\vartheta/2)}\right]. \tag{2.6}$$

This gives for $\vartheta = 180^\circ$ (back-scattering) for the smallest distance, the relation $\delta = \delta_0$ for central collisions.

If δ becomes smaller than a critical value δ_N the observed angular distribution $N(\vartheta)$ of the scattered particles begins to deviate from the distribution calculated for a pure Coulomb potential. This quantity δ_N can be regarded as the range of the nuclear force. In a crude model δ_N is interpreted as the nuclear radius R_N.

Note

The deviation from the Coulomb scattering is not only due to the nuclear force but also to the finite size of the nucleus. If only the Coulomb force would be present, only that fraction of the nuclear charge will contribute to the deflection of the charged projectiles which is located inside the sphere with radius $r < \delta$.

For a given kinetic energy E_0 of the projectiles the deviations of the angular distribution $N(\vartheta)$ from the pure Coulomb scattering are observed for scattering angles $\vartheta > \vartheta_k$ for which $\delta(\vartheta) \leq R_N$ (Fig. 2.4). On the other side for a given scattering angle ϑ but a variable energy E_0 deviations will be observed if E_0 becomes larger than a critical value E_k for which $\delta(E) < R_N$ (Fig. 2.5).

The analysis of such scattering experiments have been performed 1920 by *J. Chadwick*, a student of Rutherford, for a series of elements as target materials [5]. His results and those of many other scientists which later repeated the scattering experiments with greater accuracy, confirmed the nuclear radius for the element with atomic number A as

$$R_K \approx r_0 \cdot A^{1/3}, \tag{2.7}$$

which is proportional to the cubic root of the atomic number A. This means that at least for light nuclei the nuclear volume is proportional to A and implies that the mass density is independent of A and is constant for all light nuclei.

For light nuclei the quantity r_0 has the numerical value

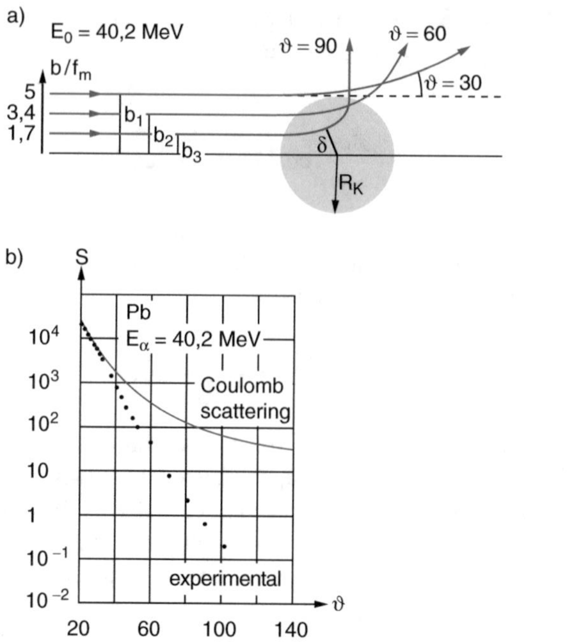

Fig. 2.4 Scattering of α-particles with given initial energy E_{kin} = 40 MeV, but different impact parameters **a** simple illustration of the dependence $\vartheta(b)$ **b** Comparison of experimental scattering rates $S(\vartheta)$ with values calculated for Coulomb scattering for the initial energy 40 MeV

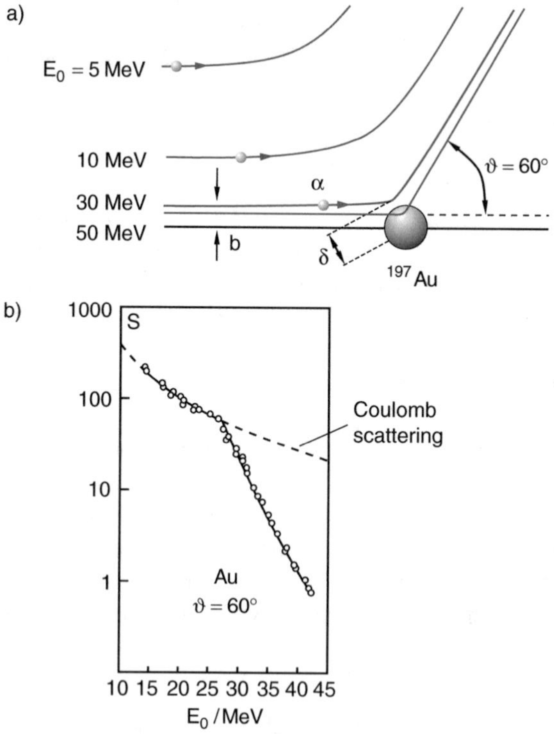

Fig. 2.5 Scattering of α-particles at gold nuclei. **a)** Relation between initial energy and impact parameter b for a given scattering angle ϑ. **b)** Scattering rate S for the scattering angel $\vartheta = 60°$ as the function of the initial energy E_0 and its comparison with Coulomb Scattering

$$r_0 = (1.3 \pm 0.1) \times 10^{-15}\,\text{m} \tag{2.8}$$

For the range of nuclei with medium mass up to very heavy nuclei r_0 varies from 0.94 fm $\leq r_0 \leq$ 1.25 fm. In nuclear physics often the length units

$$1\ \text{Fermi} = 1\ \text{Femtometer} = 1\ \text{fm} = 10^{-15}\text{m}.$$

are used.

Typical nuclear diameters are about some femtometers and are therefore about 10^5-times smaller than the atomic diameters! The nuclear volume is then 10^{15} times smaller than the atomic volume and the nucleus can be regarded for many estimations as point like mass- and charge distribution.

The nuclear masses are measured with mass spectrometers, as has been explained in Vol. 3, Sect. 2.7. They are expressed in units of the atomic mass unit AMU = $M(^{12}{}_6\text{C})/12$.

The mass spectrometers give the mass M^+ of singly ionized atoms. The nuclear mass can then be obtained form

$$M_K = M^+ - \left[(Z-1)m_e - E_B^{el}/c^2\right], \tag{2.9}$$

where m_e is the electron mass and E_B is the binding energy of the $(Z-1)$ electrons of the ion, which can be neglected for light atoms.

The construction of high resolution mass spectrometers has brought about the accurate determination of the masses of all stable nuclei (see nuclear chart of the Karlsruhe Institute in the Internet)

Another very precise measurement of nuclear masses is provided by infrared spectroscopy. The frequencies of transitions between rotational levels of diatomic molecules with rotational quantum number J and $(J + 1)$ and energies $E(J) = h \cdot c \cdot [B_v \cdot J \cdot (J + 1) - D_v \cdot J^2 \cdot (J + 1)^2]$ are given by

$$\begin{aligned} \nu(J) &= [E(J+1) - E(J)]/h \\ &= 2c \cdot \left[B_v(J+1) - D_v(J+1)^3\right] \end{aligned}$$

The molecular constants are defined as

$$B_v = \frac{\hbar}{4\pi\mu c\langle R\rangle^2}, \quad D_v = \frac{\hbar^3}{4\pi\mu^2 c \cdot k\langle R\rangle^6},$$
$$k = \frac{\partial E_{pot}/\partial R}{R - \langle R\rangle}$$

where $\mu = M_1 \cdot M_2/(M_1 + M_2)$ is the reduced mass of the diatomic molecule and $\langle R\rangle$ is the mean distance between the centers of the two atoms, averaged over the molecular vibration. The constant k is a measure of the restoring force during the elongation from the equilibrium position during the molecular vibration. It is obtained from the relation $\partial E_{pot}/\partial R = k \cdot (R - \langle R \rangle)$.

Measurements of ν(J) gives the reduced mass μ. If one of the atoms is replaced by an isotope the moment of inertia changes but not the distance between the atoms. From the change of the rotational frequencies the masses M_1 and M_2 can be determined separately.

The dependence $R_K = r_0 \cdot A^{1/3}$ shows that the nuclear mass density

$$\varrho_m = \frac{M}{V} = \frac{A \cdot \text{AME}}{\frac{4}{3}\pi R_K^3} \quad (\text{AME} = \text{atomic mass unit})$$
$$= \frac{1.66 \times 10^{-27}}{\frac{4}{3}\pi \cdot r_0^3} \approx 10^{17}\,\text{kg/m}^3 \tag{2.10}$$

is approximately independent of the atomic mass number A. It has the incredibly high value of 10^{17} kg/m^3, about 10^{14} times the density of water.

1 cm^3 of nuclear material weights 10^8 (=100 Million!) tons. Compare this with 1 cm^3 of lead which weights 11.3 g.

2.3 Mass- and Charge-Distribution within the Nucleus

In the rough model of the nucleus, presented in the foregoing chapter, the nucleus was described as a homogeneous sphere with sharp edge. The radius R_N of this sphere was obtained as 1/2 of the largest value of the minimum distance δ_{min} between the collision partners where a deviation from the Coulomb scattering was observed.

In reality the nuclear force will not suddenly vanish in spite of its short range. This implies that the nucleus will not be infinitely sharp but the mass-density ρ_m- as well as the charge-distribution ρ_e will gradually decrease monotonically to zero within a certain range of the radius r. This range is called the *boundary zone.*

Furthermore the deviation from the angular distribution of Coulomb-scattering for $\delta < R_N$ is not only caused by the nuclear force but also by diffraction effects, which appear for a projectile with de-Broglie wavelength λ_{dB} scattered by the nucleus with radius R_N. If finer details of nuclear structure should be investigated, probes with a de-Broglie wavelength $\lambda_{dB} \ll R_N$ are demanded. In this case one observes diffraction effects as maxima and minima in the scattered angular distribution

$$\Delta N(\vartheta)\mathrm{d}\vartheta = (\mathrm{d}\sigma/\mathrm{d}\Omega) n_T \cdot N_0 \cdot V \cdot \Delta\Omega$$

The angular position and the form of such diffraction maxima and minima give information about the interaction

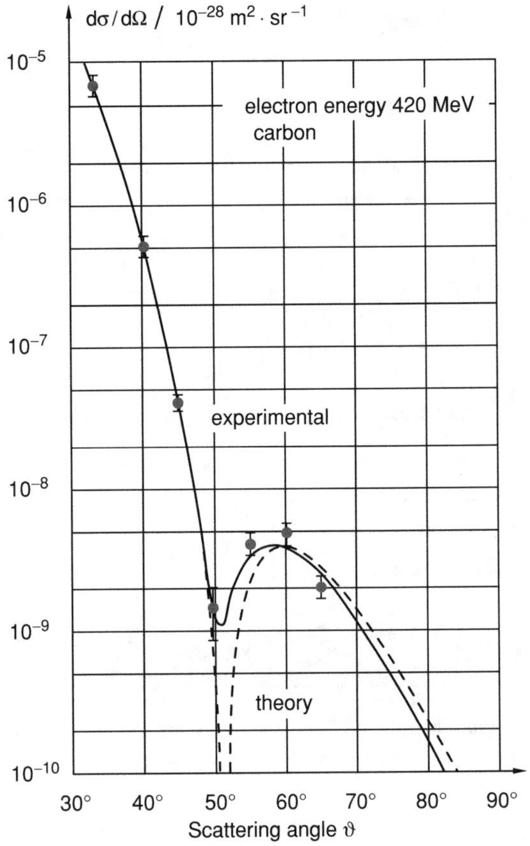

Fig. 2.6 Diffraction effect measured for the angular distribution of electrons scattered elastically at carbon nuclei. Clearly pronounced is the minimum around the scattering angle around 50° [6]

potential $E_{pot}(r)$ between the collision partners, which depends on the charge distribution in the target nucleus (see Sect. 2.3.2) (Fig. 2.6). Since the de-Broglie wavelength

$$\lambda = \frac{h}{p} \approx \frac{h}{\sqrt{2mE_{kin}}}$$

of the projectiles decreases with increasing kinetic energy, particles with sufficient energy must be used as probes.

Examples

1. α-particles with $E_{kin} = 10$ MeV have a de-Broglie wavelength $\lambda_{dB} = 1.6$ Fermi
2. For electrons with $E_{kin} = 500$ MeV $\rightarrow \lambda_{dB} = 0.4$ Fermi.

If α-particles are used as probes, Coulomb forces as well as nuclear forces contribute to the scattering and the scattering cross section depends on the charge- as well as on the mass- distribution in the target nucleus. For fast electrons only electromagnetic interaction takes place and one measures essentially the charge distribution because electrons do not feel the strong (nuclear) forces (see Sect. 7.3).

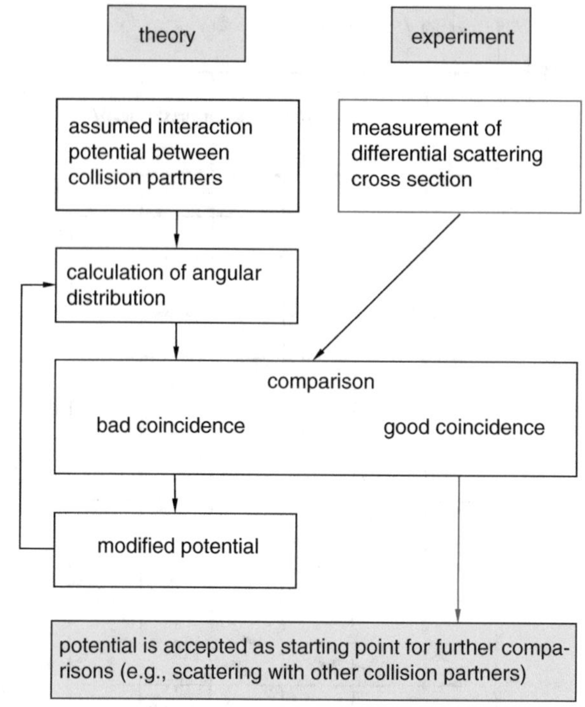

Fig. 2.7 Flowchart for the determination of interaction potentials from scattering measurements (after Mayer-Kuckuk: Kernphysik, Teubner Stuttgart 2002)

In order to measure solely the mass distribution without influence of the charge distribution neutrons are used as projectiles. This imposes, however, more experimental difficulties, because the neutrons have to be accelerated up to sufficiently high energies.

In any case projectiles with high kinetic energies have to be used, which demands particle accelerators (see Chap. 4).

As has been already discussed in Vol. 3, Sect. 2.8 it is not possible to directly derive the interaction potential, (which depends on the mass distribution $\rho_\mu(r)$, the charge distribution $\rho_e(r)$ and the kind of the interaction forces) from the results of scattering experiments. In order to determine $\rho_M(r)$ and $\rho_e(r)$ one generally assumes a model potential with free parameters. With such a model potential the scattering distribution is calculated and the free parameters are varied until the calculated distribution matches the measured one within given error limits (Fig. 2.7).

2.3.1 Mass Density Distribution within the Nucleus

Generally the mass-density distribution $\rho_m(r)$ is approximated by the function

$$\varrho(r) = \varrho_0 \frac{1}{1 + e^{\left(r - R_{\frac{1}{2}}\right)/a}} \tag{2.11}$$

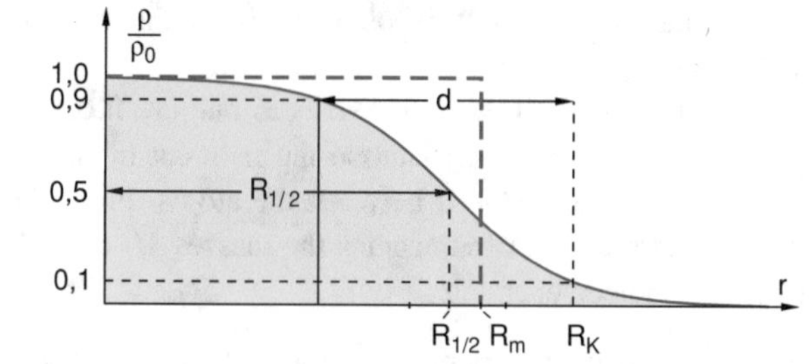

Fig. 2.8 Radial mass distribution $\rho(R)$ inside the nucleus

where the mass density has dropped at $r = R_{1/2}$ to ½ of its maximum value (Fig. 2.8).

The parameter a is a measure for the thickness of the edge range. Within the range $R_{1/2} - 2.2a < r < R_{1/2} + 2.2a$ the mass density decreases from $0.9\,\rho_0$ to $0.1\,\rho_0$. We define this range with the thickness $d = 4.4a$ as the **edge thickness**.

From measurements of scattering processes only the mean quadratic nuclear radius

$$R_m^2 = \langle r^2 \rangle = \frac{1}{M_K} \int_0^\infty r^2 \cdot \varrho(r) \cdot 4\pi r^2 \mathrm{d}r \tag{2.12}$$

of the mass density distribution can be determined.

For a homogeneous sphere with radius R and constant mass density ρ the mean quadratic radius is with $M = 4\pi\rho \int r^2 dr$

$$\langle r^2 \rangle = \frac{\int_0^{R_K} r^4 \mathrm{d}r}{\int_0^{R_K} r^2 \mathrm{d}r} = \frac{3}{5} R_K^2. \tag{2.13}$$

If the real nucleus is approximately described by a sphere with constant mass density ρ the nuclear radius can be defined as (Table 2.1).

$$R_K = \sqrt{\frac{5}{3} \langle r^2 \rangle} \tag{2.13a}$$

while the equivalent radius R_s is defined by the normalization as

$$\begin{aligned} \frac{4}{3}\pi\varrho_0 \cdot R_S^3 = M_K \Rightarrow R_S &= \left(\frac{3M_K}{4\pi\varrho_0}\right)^{1/3} \\ &= r_0 \cdot A^{1/3} \\ \text{with } r_0 &\approx 0.94 - 1.3 \text{ fm.} \end{aligned} \tag{2.13b}$$

With the experimentally obtained mass density distribution three different definitions for the nuclear radius can be derived by the following relations (Fig. 2.8):

$$R_m = \sqrt{\langle r^2 \rangle} = 0.77 \cdot R_K = 1 \cdot A^{1/3}\,(\mathrm{fm}), \tag{2.14a}$$

Table. 2.1 Mean radius $R_m = \sqrt{\langle r^2 \rangle}$, half density radius $R_{1/2}$, equivalent radius of a sphere R_K, normalized radius $R_K/A^{1/3}$ and thicknes d of surface layer for some nuclei. All numbers are given in fm, $=10^{-15}$ m (From Landolt Börnstein: Numerical Data and Functional Relationship in Science and Technology, New Series. Vol. 2: Nuclear Radii)

Nucleus	$\sqrt{\langle r^2 \rangle}$	$R_{1/2}$	R_K	$R_K/A^{1/3}$	d
^1_1H	0.80	1.03	1.03	1.03	–
^2_1D	2.17		2.80	2.22	
^4_2He	1.67	1.33	2.16	1.36	1.4
$^{12}_6C$	2.58	2.3	3.3	1.36	1.9
$^{16}_8O$	2.75	2.70	3.5	1.4	1.8
$^{24}_{12}Mg$	2.98	2.85	3.8	1.33	2.6
$^{40}_{20}Ca$	3.50	3.58	4.5	1.32	2.5
$^{197}_{79}Au$	5.32	6.38	6.87	1.18	1.3

$$R_K = R_S = \sqrt{\frac{5}{3}} R_m \approx 1.3 \cdot A^{1/3} \text{ fm}, \quad (2.14b)$$

$$R_{1/2} = (0.9 - 1.1) \cdot R_m \approx (0.9 - 1.1) \cdot A^{1/3} \text{ (fm)} \quad (2.14c)$$

where A is the atomic mass number. The three definitions have different edge thicknesses d.

This illustrates that the precise values of the nuclear radius depend on the nuclear model which is used to approximately describe the mass density distribution.

For each of the A nucleons in the atomic nucleus a volume of $(4/3)\pi R_m^3/A \approx 4.2$ fm^3 up to $(4/3)\pi R_N^3/A \approx 9$ fm^3 is available, depending on the model used for the description of the nucleus.

Fig. 2.9 Robert Hofstadter (1919–1990) (*From* Bagge; Nobelpreisträger der Physik, Heinz Moos Verlag München 1964)

2.3.2 Charge Distribution in the Nucleus

Contrary to the mass density distribution the charge distribution $\rho_e(r)$ can be measured with a much higher accuracy using high energy electrons as projectiles, because here only the well-known electro-magnetic interaction is present. The much less known nuclear forces do not act upon electrons. Therefore the scattering of the electrons is not influenced by the mass distribution but only by the charge distribution in the nucleus.

Such scattering experiments have been performed by Robert Hofstadter (Fig. 2.9) and his team at the Stanford linear accelerator [6–8]. In order to reach a sufficient spatial resolution the de-Broglie wavelength λ_{dB} of the electrons must be accordingly small. For the kinetic energy of 500 MeV for the electrons → $\lambda_{dB} = 0.4$ fm, which is small compared to the diameter of heavy nuclei.

In Fig. 2.10 the measured scattering cross sections are plotted as a function of the scattering angle ϑ for different kinetic energies of the incident electrons. One can see that with increasing energy, i.e. decreasing de Broglie wavelength the first diffraction minimum shifts towards smaller scattering angles.

We will now discuss the relation between the angular distribution $\Delta N(\vartheta)$ of the scattered electrons and the charge distribution $\rho_e(r)$ in more detail: The deflection of an electron passing through the charge distribution $\rho_e(r)$ depends on the momentum.

$$\Delta \mathbf{p} = \mathbf{p} - \mathbf{p}'$$

transferred to the electron during the scattering process, where **p** is the momentum of the incident electron and **p′** that

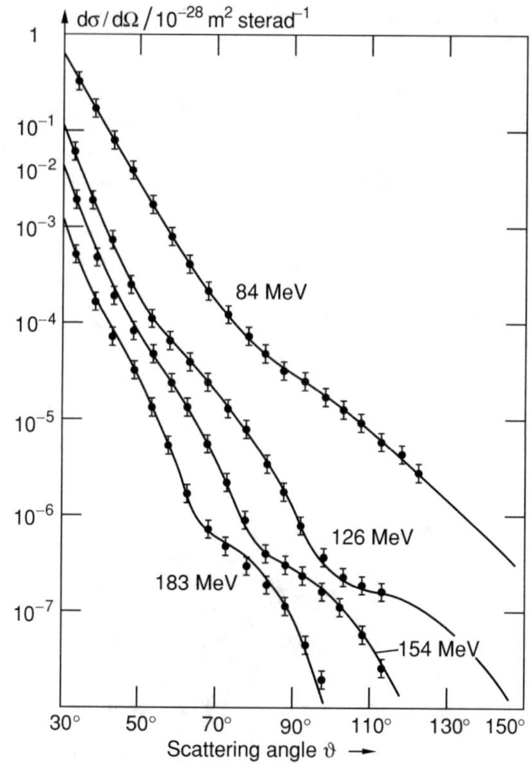

Fig. 2.10 Experimental cross sections for different electron energies as a function of the scattering angle

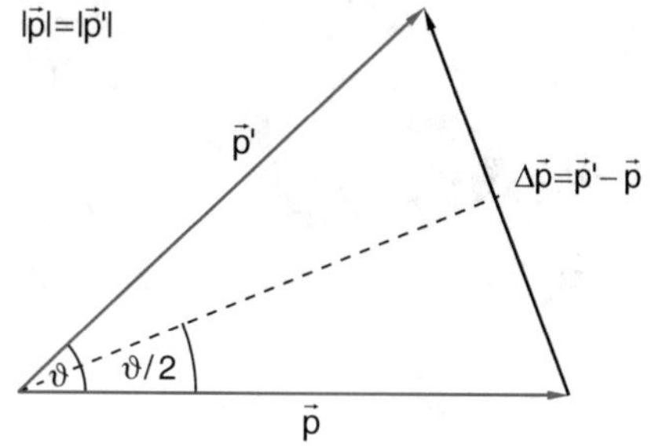

Fig. 2.11 Momentum change Δp of an elastically scattered particle

of the scattered electron. For the elastic scattering is $|\mathbf{p}| = |\mathbf{p}'|$. From Fig. 2.11 the relations

$$\sin(\vartheta/2) = \frac{1}{2}\frac{\Delta p}{p} = \frac{\Delta p}{2mv}. \qquad (2.15)$$

can be obtained. Inserting this into (2.2) one gets with $E_0 = p^2/2m$ the Rutherford scattering formula in the form

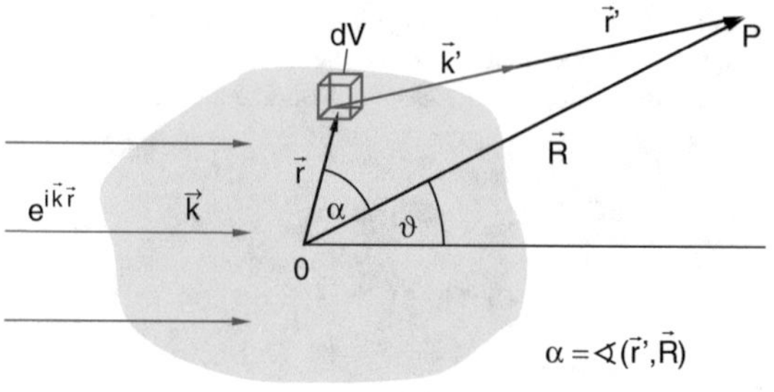

Fig. 2.12 Scattering of an incident plane wave at an extended spherically symmetric charge distribution $\rho_e(r)$

$$\left(\frac{d\sigma}{d\Omega}\right)_\vartheta = \left(\frac{Z_1 Z_2 e^2 \cdot m}{\pi \varepsilon_0}\right)^2 \cdot \frac{1}{\Delta p^4}. \qquad (2.16)$$

This equation is valid for the scattering by a point charge with mass m. How does the finite charge distribution of the nucleus alter this equation? In order to find out we describe the incident electron beam by a plane matter wave (see Vol. 3, Sect. 3.3) If this wave impinges onto the volume element dV with the charge $dq = \rho_e \cdot dV$ and the distance r from the center of the nucleus (Fig. 2.12) itscontribution to the amplitude A of the scattered spherical matter wave

$$\psi = \psi_0 \cdot e^{i(\boldsymbol{kr} - \omega t)},$$

in the observation point P at a distance $R \gg R_N$ from the center (Fig. 2.12) is

$$A = \psi_0 \cdot e^{i\boldsymbol{kr}} \cdot \frac{a}{r'} \cdot e^{i\boldsymbol{k'r'}} = \frac{a \cdot \psi_0}{r'} e^{i(\boldsymbol{kr} + \boldsymbol{k'r'})}, \qquad (2.17)$$

where the factor a depends on dq and on the scattering cross section $d\sigma/d\Omega$. According to Fig. 2.12 is

$$r = \boldsymbol{R} - \boldsymbol{r}',$$

which gives

$$\boldsymbol{k}' \cdot \boldsymbol{r}' = \boldsymbol{k}' \cdot \boldsymbol{R} - \boldsymbol{k}' \cdot \boldsymbol{r},$$

Inserting into (2.17) we obtain

$$e^{i(kr + k'r)} = e^{i\boldsymbol{k'R}} \cdot e^{i\Delta kr}$$

For $\boldsymbol{R} \gg \boldsymbol{r}$ is $\boldsymbol{r}' \approx R$ and $\boldsymbol{k}'$ is virtually parallel to $\boldsymbol{R}$. With $|\boldsymbol{k}'| = |\boldsymbol{k}| = \boldsymbol{k}$ we can replace $e^{i\boldsymbol{k'R}} \approx e^{i\boldsymbol{kr}}$. From (2.17) we then get

$$A = \frac{a \cdot \psi_0}{R} \cdot e^{ikR} \cdot e^{i(\boldsymbol{k} - \boldsymbol{k}') \cdot \boldsymbol{r}}. \qquad (2.18)$$

Integration over all volume elements dV yields the total scattering amplitude

$$A = \frac{a \cdot \psi_0 \cdot e^{ikR}}{R \cdot Q} \int \varrho_e(\boldsymbol{r}) \cdot e^{i\Delta \boldsymbol{k} \boldsymbol{r}} d\boldsymbol{r}, \qquad (2.19)$$

where $d\boldsymbol{r} = dx \cdot dy \cdot dz = dV$ ***and*** $Q = \int \rho_e dV = Z \cdot e$ is the total charge of the nucleus. The first factor describes the spherical wave scattered by a point charge $Z \cdot e$ in the center r = 0. This would be observed if only the point charge $Z \cdot e$ were present. The second factor (the integral) takes into account the finite charge distribution within the nucleus. For the differential scattering cross section which is proportional to the square of the scattering amplitude we get with the abbreviation

$$a = \frac{Z_1 \cdot Z_2 \cdot e^2}{4\pi\varepsilon_0 \cdot |\Delta p|^2} \quad \text{and} \quad \Delta p = \hbar \Delta \boldsymbol{k}$$

instead of the Rutherford scattering cross section the modified expression

$$\begin{aligned} \left(\frac{d\sigma}{d\Omega}\right)_\vartheta &= \left(\frac{d\sigma}{d\Omega}\right)_{\text{Coul}} \cdot \left| \int \varrho_e(r') \cdot e^{i\Delta k r} dr \right|^2 \\ &= \left(\frac{d\sigma}{d\Omega}\right)_{\text{Coul}} \cdot |F(\varrho, \Delta k)|^2, \end{aligned} \qquad (2.20)$$

where with electrons as projectiles $Z_1 = 1$. The differential scattering cross section for the scattering by a charge distribution $\rho_e(r)$ can therefore be written as the product of two factors:

1. the cross section (2.1) for the scattering at a Coulomb potential, i.e.by a point charge.

2. A form factor $|F(\rho_e(r)), \Delta \boldsymbol{p}|^2$ which depends on the charge distribution $\rho_e(r)$ and on the momentum change $\Delta \boldsymbol{p} = \boldsymbol{p} - \boldsymbol{p}' = \hbar \cdot \Delta \boldsymbol{k}$ transferred at the scattering to the scattered electron. One can see from (2.20) that the form factor

$$F(\varrho, \Delta \boldsymbol{k}) = \int \varrho_e(\boldsymbol{r}) \cdot e^{i\Delta \boldsymbol{k} \boldsymbol{r}} d\boldsymbol{r} \qquad (2.21)$$

is the Fourier-transform of the charge distribution $\rho_e(r)$. The form factor decreases with increasing momentum transfer $\Delta \boldsymbol{p}$. The distinct minima of the cross section at high energies of the projectiles (Figs. 2.10 and 2.13) are due to diffraction at the finite charge distribution. They do not appear for the scattering at a point charge.

The differential cross section decreases with increasing energy of the incident electrons, because

1. The first factor in (2.20) which describes the Coulomb scattering, decreases with $(1/E_0^2)$

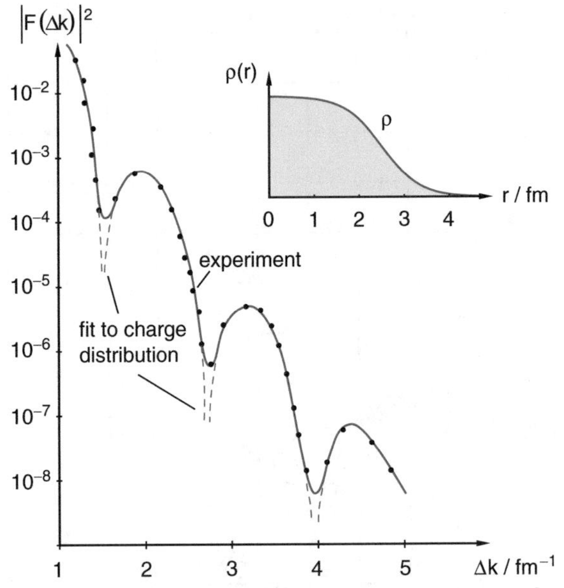

Fig. 2.13 Experimental plots of the scattering cross sections dσ/dΩ for the scattering of fast electrons at gold nuclei. (after R. Hofstadter (ed.) Electron Scattering and Nuclear Structure. Benjamin New York 1963)

2. The form factor $|F(\Delta p)|^2$ decreases with increasing momentum transfer Δp (Fig. 2.13). At a fixed scattering angle ϑ is $\Delta p \sim p = \sqrt{(2mE_0)}$.

The oscillations in the curves $d\sigma/d\Omega$ become more and more distinct with increasing energy i.e. decreasing de Broglie wavelength (Fig. 2.10). The first minimum shifts to smaller scattering angles ϑ.

When the incident particles are described by their de Broglie wavelength λ_{dB}, the first minimum appears for a diameter D of the target nucleus at the angle φ_{min} for which the condition holds $\varphi_{min} = \lambda_{dB}/D$. In this case there are always pairs of scattered particles with a path difference $\Delta s = \lambda_{dB}/2$. Their wave functions have opposite phases and they cancel each other (see Vol. 2, Sect. 10.5).

The radial charge distributions of some nuclei are illustrated in Fig. 2.14. They are based on measurements of Hofstadter and his team. These experiments have confirmed that the charge distributions generally have a slightly different radial dependence than the mass distribution. The reason for this difference is the Coulomb repulsion between the charged protons which pushes the protons more to the edge than the neutrons. The experiments further show that the charge distribution of many heavy nuclei is not spherical symmetric (see Sect. 2.5.2).

An important and accurate method of studying the charge distribution in nuclei is the spectroscopic determination of

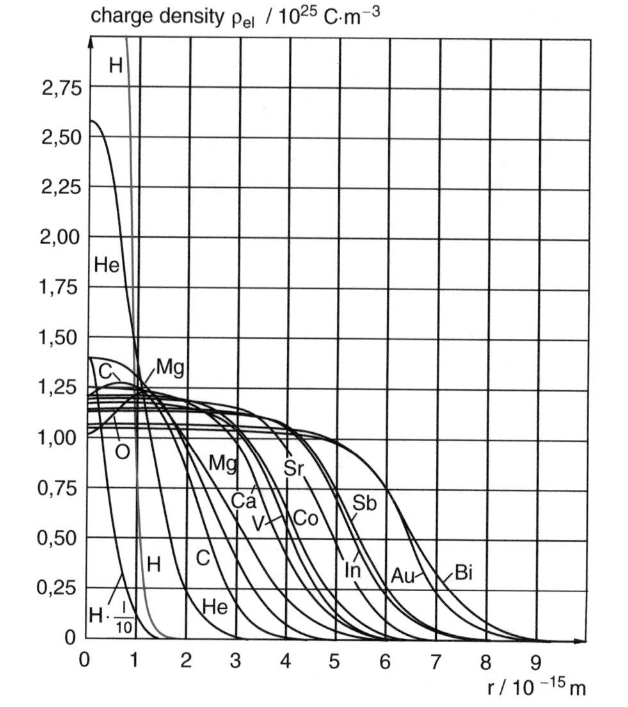

Fig. 2.14 Radial charge density distribution for some nuclei as determined by electron scattering measurements (after R. Hofstadter Ann. Rev. Nucl. Sci. 7 231, 1957)

the energy levels of "exotic atoms", where one of the inner electrons in the atomic shell is replaced by a myon μ^- or a π^--meson (see Vol. 3, Sect. 6.7).

The Bohr-radii of these particles with mass m_x is smaller by the factor m_e/m_x than that of the electron in a "conventional atom". This implies that these particles in the atomic shell are much closer to the nucleus and even can penetrate into the nucleus and therefore probe the charge distribution much more sensitive.

Accurate spectroscopic measurements of muonic hydrogen atoms where the Bohr radius of the muon is about 200 times smaller than that of the electron performed at the Paul Scherrer Institute close to Zürich in Switzerland, have brought the surprising result that the radius of the charge distribution of the proton is about 4% smaller than assumed before. According to these measurements it is 0.84184 fm instead of 0.8768 fm. This discrepancy is beyond the error limits and its cause is not known up today but intensely discussed among experts [9].

Remark For the scattering of α-particles by gold target nuclei the geometrical scattering cross section is $\sigma = \pi \cdot (r_{Au} + r_\alpha)^2$. One has to consider the radii of both particles.

2.4 Buildup of Nuclei by Nucleons; Isotopes and Isobars

When Rutherford presented his atomic model the only known elementary particles assumed as components of nuclei, were protons and electrons. Since measurements of atomic masses and charges had revealed that an atomic nucleus with the charge $Z \cdot e$ and a mass $M \approx A \cdot m_p$ which is for heavy nuclei about twice as large as Z proton masses, the nucleus could not solely be composed of protons. Therefore the hypothesis was born that nuclei were composed of A protons with charge $A \cdot e$ and $(A - Z)$ electrons with charge $-(A - Z) \cdot e$. This would give the correct mass and charge of the nuclei. This hypothesis was supported by the fact that radioactive Atoms were found which emit electrons (β-radiation). However, this hypothesis can be discounted by the following arguments and experiments :

- If electrons were enclosed into the volume of the nucleus, their spatial location could be determined within the limit $\Delta x \cdot \Delta y \cdot \Delta z = \Delta V$ of the nuclear volume. According to Heisenberg's uncertainty relation their momentum uncertainty is then $\Delta \boldsymbol{p} \geq h/\Delta \boldsymbol{r}$ and their kinetic energy is

$$E_{kin} \geq \frac{(\Delta p)^2}{2m_e} \geq \frac{h^2}{2m_e \Delta r^2}. \tag{2.22}$$

Inserting the correct numerical values, for example for the oxygen nucleus with $A = 16$ one obtains with $R_N = 3 \times 10^{-15}$ m a lower limit for the kinetic energy of the electrons $E_{kin} > 10^{11}$ eV. This is by far more than the electrostatic binding energy of protons and electrons at a mean distance of 3 fm ($|E_B| = 10^6$ eV). This means that electrons cannot be stable within the nucleus.

- A further convincing argument against the assumption that the nucleus consists of protons and electrons is provided by measurements of the nuclear angular momenta (see

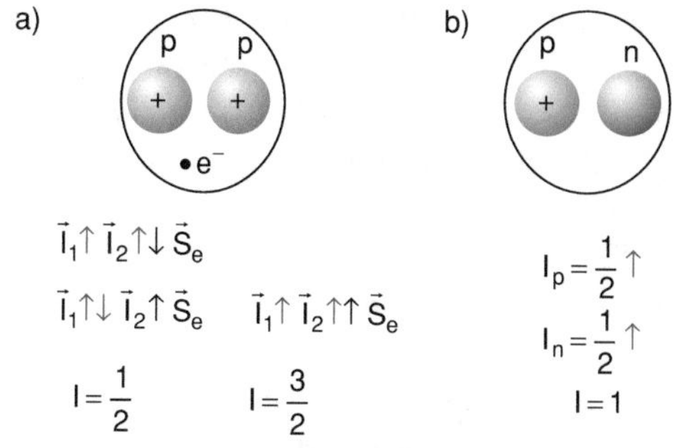

Fig. 2.15 **a)** Wrong. **b)** Correct model of the deuteron

Table. 2.2 Some designations, used in nuclear Physics

Notion	Explanation	Examples
Nucleon	Protons and neutrons	
Nuclide	Nucleus $_Z{}^AX$ with A nucleons, Z protons and $N = A - Z$ neutrons	^7_3Li, $^{238}_{92}U$
Isotopes	Nuclei with equal number of protons Z but different Neutron nubers N	$^{12}_6C$ and $^{14}_6C$ $^{235}_{92}U$ and $^{238}_{92}U$
Isobars	Nuclei with equal number A but different number Z	$^{14}_6C$ and $^{14}_7N$
Isotones	Nuclei with equal numbers N but different numbers Z	$^{14}_6C$ and $^{15}_7N$ $^{16}_8O$
Mirror nuclei	Nuclei with interchanged numbers of Z and N. $Z_1 = N_2$, $N_1 = Z_2$	^3_1H and ^3_2He $^{13}_6C$ and $^{13}_7N$
Isomers	Nuclei with equal Z and N in different energy states	$^{12}_6C(E_1)$ and $^{12}_6C(E_2)$

Sect. 2.5) (Fig. 2.15). This fact was, however, not yet known to Rutherford, because he did not know about angular momenta of nuclei). Later measurements of the hyperfine-structure of atomic spectra (see Vol. 3, Sect. 5.6) have supported the fact, that many nuclei possess a magnetic moment and an angular momentum (nuclear spin).The hyperfine structure of the H-atom revealed that the proton must have the nuclear spin $I = ½\hbar$. Since the electron spin is $½\hbar$, the nucleus of the deuterium ($Z = 1, A = 2$) should have the total spin $I = ½\hbar$ or $3/2\hbar$, if it would consist of 2 protons and one electron. Experiments prove, however, that the total spin of the deuterium nucleus is $I = 1\hbar$.

Rutherford postulated already in 1920 that nuclei must consist of charged protons and neutral particles with a similar mass as protons. When *Chadwick* discovered 1932 the neutron [10], Rutherford's assumption found a brilliant confirmation. According to the nuclear model which is still valid today a nucleus consists of Z protons and (A – Z) neutrons. These building blocks of nuclei are called **nucleons.** The masses of protons and neutrons differ only slightly. An atomic nucleus X is unambiguously defined by its number Z of protons and $N = (A - Z)$ of neutrons, i.e. by the atomic mass number $A = Z + N$ and the charge $Z \cdot e$. The internationally accepted nomenclature is $^A{}_ZX$,

Example The lithium nucleus with three protons and 4 neutrons is abbreviated as ^7_3Li.

We still have to clear two open questions:

- What is with the electrons, emitted by radioactive nuclei? How are they produced? **Answer**: They are formed in the nucleus of a radioactive atom by β-decay of one neutron according to the process (see Sect. 3.4)

$$n \rightarrow p + e^- + \bar{\nu} \tag{2.23}$$

While the proton remains in the nucleus, electron and antineutrino are emitted immediately after their generation.

- While free protons are stable (they live at least for 10^{32} years) free neutrons decay according to (2.23) with a mean lifetime of $\tau_N = 881.5 \pm 1.5$ s. The question, why neutrons, bound in a nucleus, have lifetimes of more than 10^{10} years will be answered in Sect. 3.1

Nuclei with equal numbers Z of protons but different atomic mass numbers A (i.e. different number $N = (A - Z)$ of neutrons) are called **Isotopes**. Table 2.2 compiles the nomenclature used in nuclear physics.

2.5 Nuclear Angular Momenta, Magnetic and Electric Moments

The observed hyperfine structure of energy levels in many atomic spectra prove that their nuclei possess a magnetic moment (see Vol. 3, Sect. 5.6 and [11, 12]). Analog to the explanation of the magnetic moments of the electron shell one assumes that the magnetic moment of the nucleus is related to the corresponding mechanical angular momentum of the nucleus. As shown in quantum mechanics the amount of such an angular momentum $\boldsymbol{I}$ can be written as

$$|\boldsymbol{I}| = \sqrt{I \cdot (I+1)} \cdot \hbar. \tag{2.24}$$

where $\boldsymbol{I}$ is called **nuclear spin** and I is the **nuclear spin quantum number**. In an external magnetic field the spin precesses around the field axis. Its component in the field direction is $I_z = m \cdot \hbar$.

Similar to the total angular momentum of the electron shell $\boldsymbol{J} = \Sigma(\boldsymbol{s}_i + \boldsymbol{l}_i)$ as the sum of electron spins $\boldsymbol{s}_i$ and orbital angular momenta $\boldsymbol{l}_i$ of the electrons the total nuclear angular momentum

$$\boldsymbol{I} = \sum_i (\boldsymbol{I}_i + \boldsymbol{L}_i). \tag{2.25}$$

is the vector sum of all nucleon spins and their orbital angular momenta. In Fig. 2.16 some examples are listed. For the ground state of most nuclei is $\Sigma \boldsymbol{L}_i = \boldsymbol{0}$. For the nucleus ${}^6_3\mathrm{Li}$ the two upper nuclei must occupy a higher energy level because the Pauli-Principle does not allow that more than one identical nuclei occupy the same energy level (Fig. 2.32e). Figure 2.16 illustrates that for the nucleus ${}^6_3\mathrm{Li}$ the sum of the nuclear spins of protons and neutrons is $|\Sigma(\boldsymbol{I}_\mathrm{p} + \boldsymbol{I}_\mathrm{n})| = ½\hbar$. However, the observed total nuclear spin is $3/2\hbar$. Therefore here the angular momentum cannot be zero even in the nuclear ground state but must be $|L| = |\Sigma L_i| = 1 \cdot \hbar$.

The total angular momentum $\boldsymbol{F}$ of the atom (nucleus + electron shell) is the vector sum of electronic angular momentum $\boldsymbol{J} = \boldsymbol{L} + \boldsymbol{S}$ and nuclear spin $\boldsymbol{I}$

$$\boldsymbol{F} = \boldsymbol{L} + \boldsymbol{S} + \boldsymbol{I}$$

where $\boldsymbol{L} = \Sigma \boldsymbol{l}_i$ is the vector sum of all orbital angular momenta of the electrons and $\boldsymbol{S} = \Sigma \boldsymbol{s}_i$ that of the electron spins. For atoms with only one electron is $\boldsymbol{L} = \boldsymbol{l}$ and $\boldsymbol{S} = \boldsymbol{s}$ (see Vol. 3, Sect. 6.5).

2.5.1 Nuclear Magnetic Moments

Similar to the magnetic moment of the electron shell $\boldsymbol{\mu}_J = g_J \cdot \mu_\mathrm{B} \cdot \boldsymbol{J}/\hbar$ which can be written as the product of Landé-factor g_J, Bohr magneton $\mu_\mathrm{B} = (e/2m_e) \cdot \hbar$ and total electronic angular momentum $J/\hbar$ in units of $\hbar$ (see Vol. 3, Sect. 3.5) also the nuclear magnetic moment

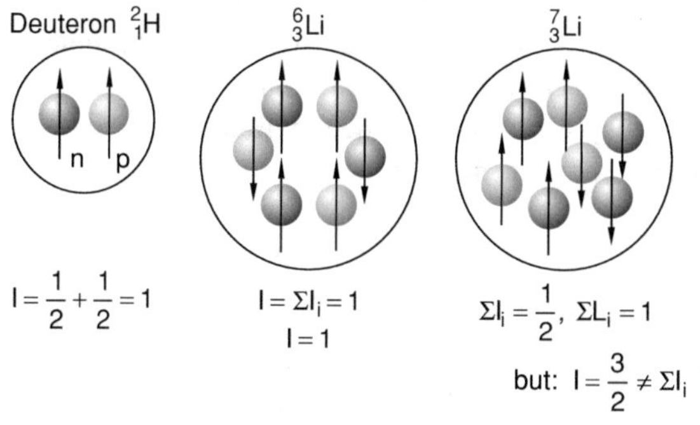

Fig. 2.16 Nuclear spin I as the vector sum of spin and orbital angular momentum of the nucleons. Without (Deuteron and Li 6) and with orbital angular momentum (Li 7)

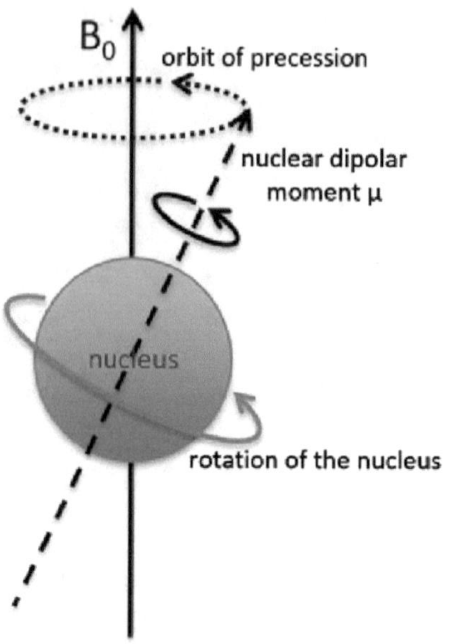

Fig. 2.17 Precession of the spin with its magnetic moment around the magnetic field axis

$$\boldsymbol{\mu}_I = g_I \cdot \mu_\mathrm{N} \cdot \boldsymbol{I}/\hbar \tag{2.26}$$

can be expressed as the product of nuclear Landé-factor g_I, nuclear magneton with the amount

$$\mu_\mathrm{N} = e/(2m_\mathrm{p}) \cdot \hbar = 5.050783 \times 10^{-27}\,\mathrm{J/T} \tag{2.27}$$

and nuclear spin $\boldsymbol{I}/\hbar$ in units of $\hbar$.

Because of the larger proton mass m_p the nuclear magneton μ_N is smaller than the Bohr magneton μ_B by a factor $m_e/m_p = 1(1836)$.

If the nuclear magnetic moment is given in units of the nuclear magneton μ_N and the nuclear spin I in units of $\hbar$, the Landé.factor

$$g_I = \frac{|\mu_I|/\mu_\mathrm{N}}{|\boldsymbol{I}|/\hbar} \tag{2.28}$$

gives the ratio of magnetic moment and mechanical spin of the nucleus. The ratio

$$\gamma = \frac{|\boldsymbol{\mu}_I|}{|\boldsymbol{I}|} \tag{2.28a}$$

is the gyro-magnetic ratio of the nucleus. The relation between nuclear Landé-factor and γ is

$$g_I = \gamma \cdot \frac{\hbar}{\mu_\mathrm{N}} = \frac{|\boldsymbol{\mu}_I|}{|\boldsymbol{\mu}_\mathrm{K}|} \cdot \frac{\hbar}{|\boldsymbol{I}|}.$$

The magnetic moment of the proton can be experimentally obtained from the hyperfine-structure of the 1S-state of the hydrogen atom (see Vol. 3, Sect. 3.5). With a total angular momentum

$$F = \ell + \mathbf{s} + \mathbf{I}$$

of electronic orbital angular momentum $\boldsymbol{l}$, electron spin $\boldsymbol{s}$ and proton spin $\boldsymbol{I}$ with the quantum numbers F, l, s, I the hyperfine splitting in the $1S$ ground state of the hydrogen atom is

$$\Delta E_{\mathrm{HFS}} = A/2[F\cdot(F+1) - s(s+1) - I(I+1)].$$

with the hyperfine coupling constant

$$A = \frac{2}{3}\mu_0 g_e \mu_B \cdot g_p \mu_K \cdot |\psi_{1s}(r=0)|^2, \qquad (2.29)$$

where $\mu_0 = 4\pi \times 10^{-7}$ (Vs/A · m) is the permeability constant, g_e the Landé-factor of the electron, μ_B the Bohr magneton and $\psi_{1s}(0)$ the wavefunction of the 1s-electron at the location of the proton.

More accurate values of the magnetic moment of the proton can be obtained from measurements of the Zeeman-splitting

$$\Delta W = \boldsymbol{\mu}\cdot\boldsymbol{B}$$

of hydrogen energy levels in an external magnetic field, which can be precisely measured by irradiation of hydrogen atoms in an atomic beams by a radio-frequency wave (Rabi-experiment [12], see Vol. 3, Sect. 10.3). The energy shift of a hyperfine-level in the magnetic field B is

$$\Delta E = -\boldsymbol{\mu}\cdot\boldsymbol{B} = g_F \mu_B B \cdot m_F \quad \text{with} \quad B = |\boldsymbol{B}|$$

and the magnetic quantum number m_F which obeys

$$-F \le m_F \le +F$$

The magnetic moment $\boldsymbol{\mu} = \boldsymbol{\mu}_J + \boldsymbol{\mu}_I$ is the total magnetic moment of the hydrogen atom in the energy level with total angular momentum $\boldsymbol{F}$. The Landé-factor

$$\begin{aligned} g_F = g_J \cdot [F(F+1) + J(J+1) - I(I+1)]/[2F(F+1)] \\ - g_I(\mu_K/\mu_B)[F(F+1) + I(I+1) \\ - J(J+1)]/[2F(F+1)] \end{aligned}$$

consists of the electronic part and the nuclear part. The two parts have a different sign because of the opposite charges of electron and proton. Since $\mu_B \gg \mu_N$ the electronic part is by three orders of magnitude larger than the nuclear part. The transition between two Zeeman levels with energies E_1 ($F = 0, m_F = 0, m_J = -½$) and E_2 ($F = 1, m_F = 0, m_J = +½$) with $J = ½$, $\Delta m_F = 0$, $\Delta m_J = +1$ (Fig. 2.18a) has the frequency $\nu = (E_2 - E_1)/h =$. Its measurement gives the total magnetic moment $\mu = \mu_{el} + \mu_N$ and therefore the nuclear magnetic moment μ_N, since the electronic magnetic moment

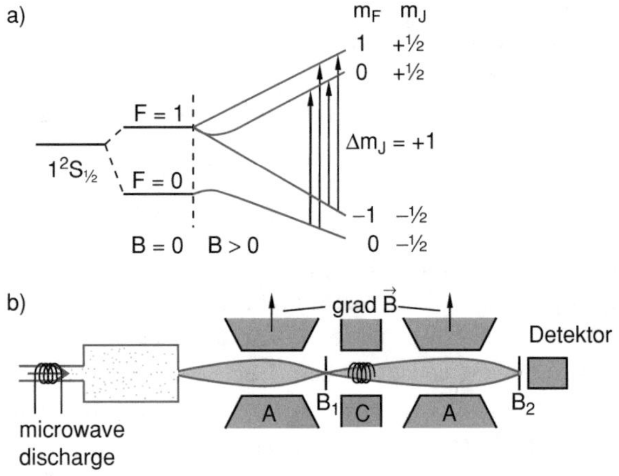

Fig. 2.18 Illustration of the measurement of the proton magnetic moment using the Rabi-method in a H-atomic beam. **a** Level scheme. **b** Experimental arrangement

μ_{el} is very precisely known with a very small uncertainty. Due to their magnetic moment, which is predominantly caused by the electron spin, the H-atoms are deflected in the inhomogeneous magnetic field A_1 (Fig. 2.18b).In the homogeneous field C the atoms are irradiated by radiation with the proper frequency ν. This causes a spin flip of the electron spin and the atoms are refocused in a second inhomogeneous field A_2. They can pass through the aperture B_2 and reach the detector. Since the nuclear spin is ½ħ, there are 4 possible transitions with $\Delta m_J = +1$ (Fig. 2.18a). The difference frequency of these transitions yields the contribution of the magnetic moment of the proton to the Zeeman splitting.

Now the frequencies $\nu(B)$ are measured at different magnetic fields B. By extrapolation to $B \to 0$ the field-free hyperfine splittings are obtained. These measurements revealed the magnetic moment of the proton as

$$\mu_p = +2.79278\,\mu_N.$$

where μ_N is the nuclear magneton. The level scheme of Fig. 2.18a shows that the spin quantum number of the proton must be $I = ½$, i.e. the proton spin has the components $I_z = \pm ½.\hbar$, which gives the amount $|I| = \sqrt{I}\cdot (I+1) = \sqrt{½}\cdot(3/2)\,\hbar = \sqrt{3}/4\cdot\hbar$.

The positive sign of μ_P implies that the proton spin and the magnetic moment point into the same direction, they are parallel.

The Landé-factor of the proton is then

$$g_p = \frac{\mu_p/\mu_N}{I_p/\hbar} = \frac{2.79278}{1/2} = 5.58556. \qquad (2.28b)$$

The proton has an anomalous large magnetic moment [12]. If it were an elementary particle with no internal structure, like the electron but with opposite charge one would expect a Landé-factor $g \approx 2$. The larger experimental value indicates already that the proton must be composed of several charged particles.

This hint became even more obvious after it was discovered that also the neutron possesses a magnetic moment

$$\mu_n = -1.91304\mu_N \Rightarrow g_n = -3.82608$$

whereas one would not expect a magnetic moment of a neutral particle. This indicates that also the neutron must be composed of charged particles, where the charges of these sub-particles just add up to zero (see Sect. 7.4.5).

More detailled experiments prove, whether the neutron has also an electric dipole moment [16]

Measurements of the magnetic moment of free neutrons is more difficult than for the proton, because it is experimentally tricky to realize an intense collimated neutron beam. The principle of such measurements is, however, similar to that of measurements of the proton magnetic moment. It is based on the fact, that the cross section for the scattering of neutrons at target nuclei depends on the relative orientation of neutron spin and target nuclear spin (see Sect. 5.2). For parallel spins the cross section is about twice as large as for anti-parallel spins. This results of a larger loss of neutrons with parallel spin which are more likely scattered out of the neutron beam (Fig. 2.19). The neutron beam transmitted through an iron sample in an external magnetic field is therefore partly polarized and has a preferential spin direction parallel to the direction of the magnetic field $\boldsymbol{B}_1$. (Fig. 2.20).The iron sheets in the external magnetic field act as neutron polarizers.

The polarization of the transmitted neutron beam is

$$P = \frac{N_+ - N_-}{N_+ + N_-} \tag{2.30}$$

which shows that $-1 \leq P \leq +1$. Because $N_- > N_+$ P is negative. For the example of Fig. 2.19a is $P = (6 - 8)/(6 + 8) = -1/7$.

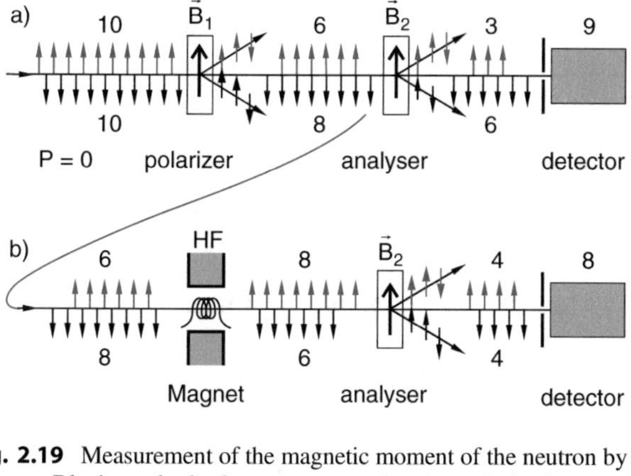

Fig. 2.19 Measurement of the magnetic moment of the neutron by the Alvarez-Bloch method of neutron scattering by polarized targets. **a)** principle of measurement with the schematically indicated dependence of the scattering cross section on the relative spin orientation. **b)** Scheme of the experiment, where in the HF-magnetic field a spin flip occurs (2.17)

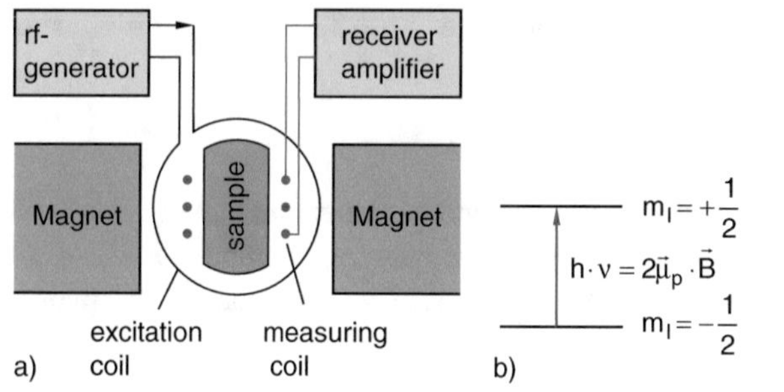

Fig. 2.20 Schematic drawing of magnetic nuclear resonance. **a** Experimental design. **b** Level scheme

When the partly polarized neutron beam passes through a second iron target in a magnetic field $\boldsymbol{B}_2$ the transmitted intensity is larger for $\boldsymbol{B}_2 \uparrow\uparrow \boldsymbol{B}_1$ but smaller for $\boldsymbol{B}_2 \uparrow\downarrow \boldsymbol{B}_1$. For our example in Fig. 2.19 out of 20 incident neutrons reach 9 neutrons the detector. If now the neutrons are irradiated in a homogeneous magnetic field between B_1 and B_2 by rf-transitions which induce transitions between the two Zeeman levels with energies.

$$E_{1,2} = \pm\boldsymbol{\mu}_n \cdot \boldsymbol{B} \tag{2.31}$$

the neutron spins flip and the scattered rate in the second target in B_2 changes if the rf-frequency equals the resonance frequency $\nu = \Delta E_{12}/h = 2\mu_n \cdot B_2/h$,

Now for our example only 8 neutrons reach the detector.

The magnetic moment of nuclei is measured nowadays generally with the method of nuclear magnetic resonance (NMR). Its basic principle is illustrated in Fig. 2.20. The nuclei of molecules in a liquid phase are placed in a static homogeneous magnetic field $\boldsymbol{B}_0$, which is super-imposed by a magnetic ac-field $\boldsymbol{B}_{rf} = B_1(1 + \cos\omega t)$ with $\boldsymbol{B}_1 \perp \boldsymbol{B}_0$. If the frequency $\omega = 2\pi \cdot \nu$ is tuned, the spins of the neutrons and with them also their magnetic moments precess about the direction of $\boldsymbol{B}_0$. For the resonance frequency $\omega = 2\pi \cdot \nu = 2\mu_n \cdot B/\hbar$ the magnetic moments flip into the plane $\perp \boldsymbol{B}_0$ which means into the plane of the ac-field $\boldsymbol{B}_1$. This causes a change of the total magnetic moment of the sample which causes an induction voltage in the detector coil surrounding the sample. This voltage is amplified and fed to the detector, e. g an oscilloscope or a computer (**nuclear induction method**). The detector coil should not receive any signal from the ac-magnetic field $\boldsymbol{B}_1$. Therefore its plane is aligned perpendicular to $\boldsymbol{B}_1$.

The nuclear spin I and the magnetic moment μ_I of a nucleus ${}^A_Z\mathrm{X}$ with Z protons and (A-Z) neutrons are related according to

$$\boldsymbol{\mu}_I = \sum_i \left(g_p \cdot \boldsymbol{I}_{pi} + g_n \boldsymbol{I}_{ni} + a \cdot \boldsymbol{L}_i\right) \tag{2.32}$$

Both quantities depend on the vector sum of all nucleon-angular momenta (spin + orbital angular momentum). In Table 2.3 the nuclear spins and the magnetic moments

Table. 2.3 Characteristic properties of proton, neutron and electron

Quantity	p	n	e^-
Mass (kg)	1.672623×10^{-27}	$1.6749286 \times 10^{-27}$	$9.1093898 \times 10^{-31}$
Spin quantum number	1/2	1/2	1/2
magnetic moment	1.4106×10^{-26} J/Tesla $= +2.79278\mu_K$	-9.6629×10^{-27} J/Tesla $= -1.91315\mu_K$	9.2847×10^{-24} J/Tesla $= 1.001\mu_B$
Gyro-magnetic ratio $[s^{-1}\,T^{-1}]$	2.67522×10^8	1.83247×10^8	$1.7.608597 \times 10^{11}$
lifetime	$>5 \times 10^{32}$a	881.5s	∞

of some light and heavy nuclei are listed. The comparison of the last two columns illustrates that the experimental values of the nuclear spins in column 3 often differ from those expected form the vector sum of the spins of free protons and neutrons (second column), and also the experimental values of the magnetic moments in column 5 differ from the expected values in column 4. The magnetic moment must be influenced by the interaction between the nucleons in the nucleus.

If the orbital angular momentum is $\boldsymbol{L} \neq \boldsymbol{0}$ the measured nuclear spins differ considerably from the sum of proton-and neutrons spins. Also for the magnetic moments the term $a \cdot \boldsymbol{L}_i$ may change the expected values. Examples are the nuclei $^7_3$Li, $^9_4$Be, $^{85}_{37}$Rb or $^{115}_{49}$In in Table 2.4, which have been marked with exclamation points.

It catches one's eye that all nuclei with even numbers of protons and neutrons (g-g-nuclei from the German gerade = even) have the nuclear spin $I = 0$ and therefore have no magnetic moment. This indicates that protons as well as neutrons arrange themselves in pairs with antiparallel spins. This hint will be proved later on by further arguments.

2.5.2 Electric Quadrupole Moments

If the electric charge distribution of the protons in the nucleus is not spherical symmetric the nucleus has an electric quadrupole moment (see Vol. 2, Sect. 1.4). For different isotopes often the charge distribution differs even if the total charge is equal. This causes that the different isotopes of the same chemical elements have different electric quadrupole moments. Only for nuclei with $I \neq 0$ a preferential direction is defined about which the nucleus can rotate. Nuclei with $I = 0$ have no preferential direction and therefore a time-averaged spherical charge distribution which implies that they have no observable electric quadrupole moment, even if the momentary charge distribution is not spherical

Table. 2.4 Nuclear spin quantum numbers and magnetic moments of some nuclei (given in units of the nuclear moment). The values [x] given in parenthesis deviate from the vector sum $\Sigma I_n + \Sigma l_n$

Nuclear	Spin quantum number		Magnetic moment	
	expected from $\Sigma I_n + \Sigma l_n$	Experimental	expected from $\Sigma\mu_p + \Sigma\mu_n$	Experimental
$^2_1$H	$\frac{1}{2}+\frac{1}{2}$	1	0.880	0.857
$^3_1$H	$\frac{1}{2}+0$	$\frac{1}{2}$	2.793	2.978
$^3_2$He	$0+\frac{1}{2}$	$\frac{1}{2}$	−1.913	−2.127
$^4_2$He	$0+0$	0	0	0
$^6_3$Li	$\frac{1}{2}+\frac{1}{2}$	1	0.880	0.822
$^7_3$Li	$\frac{1}{2}+0$	$\frac{3}{2}$(!)	(2.793)	3.256
$^9_4$Be	$0+\frac{1}{2}$	$\frac{3}{2}$(!)	(−1.91)	−1.177
$^{12}_6$C	$0+0$	0	0	0
$^{85}_{37}$Rb	$\frac{1}{2}+0$	$\frac{5}{2}$(!)	(+2.793)	1.353
$^{115}_{49}$In	$\frac{1}{2}+0$	$\frac{9}{2}$(!)	(2.793)	5.523

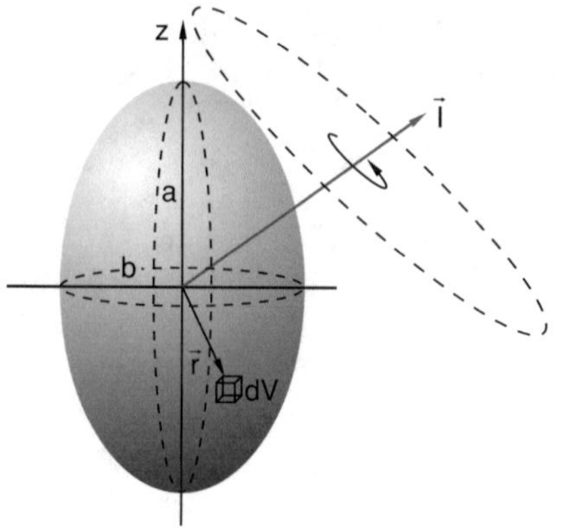

Fig. 2.21 Illustration of the quadrupole moment of a prolate charge distribution and the precession of the nuclear symmetry axis around the space fixed nuclear spin if no magnetic field is present

symmetric. One has to distinguish between the "intrinsic quadrupole moment QM", which is a measure of the deviation of the momentary charge distribution from a spherical distribution, and the observable time-averaged moment $\overline{QM}$ (see Sect. 5.6) When the symmetry axis of the rotationally symmetric charge distribution $\rho_e(r)$ is the z-direction, the quadrupole moment is

$$QM \propto \int \varrho_e(\boldsymbol{r})\left[3z^2 - r^2\right]d^3\boldsymbol{r} \qquad (2.33)$$

where $\boldsymbol{r}$ is the vector pointing from the charge centrum (= zero point of our coordinate system) to the charge element $dq = \rho_e(r)\, dV$ (Fig. 2.21).

Remark In the literature the quadrupole moment is often defined as the product $e \cdot Q$ of elementary charge e and a factor Q which has the dimension m^2 and is a measure for the deviation of the charge density from a spherical distribution (see below).

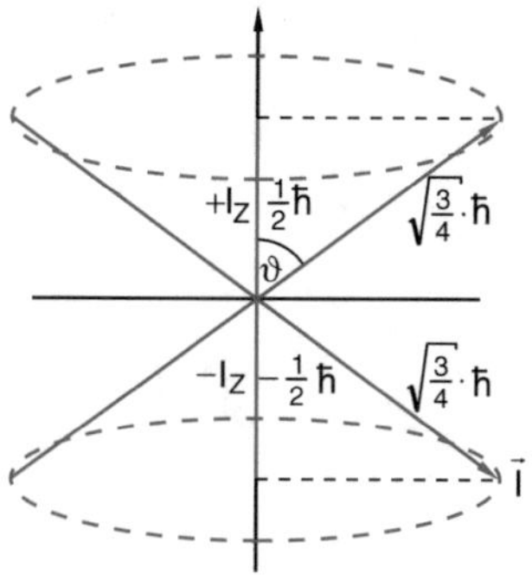

Fig. 2.22 For nuclei with $I \leq$ ½ the averaged charge distribution is rotationally sysmmetric and therefore the observable electric quadrupole moment is zero

If the nucleus has the spin $\boldsymbol{I}$ it will rotate without external magnetic field around the space fixed axis of $\boldsymbol{I}$ (Fig. 2.22).

The amount of $\boldsymbol{I}$ is $|\boldsymbol{I}| = \sqrt{I \cdot (I+1)} \cdot \hbar$ The projection of $\boldsymbol{I}$ onto the z-axis is $M \cdot \hbar$ with $-I \leq M \leq +I$. The observable quadrupole moment is maximum, if I points into the z-direction.

Nuclei with the spin quantum number I = ½ have the two different projections onto the quantization axis (z-axis) $I_z = \pm\frac{1}{2}\hbar$. For the angle ϑ of the spin directions against the z-axis we get $\cos\vartheta = \pm 1/\sqrt{3}$ (Fig. 2.22). According to (2.33) we obtain with $z = r \cdot \cos\vartheta$ for the quadrupole moment: QM = 0.

Only nuclei with a nuclear spin quantum number I> = 1 have a nonvanishing electric quadrupole moment.

In Fig. 2.23 measured quadrupole moments of nuclei with an odd number of protons and neutrons are plotted. Mainly the unpaired nucleons contribute to the nuclear spin (see Table 2.5) since for even numbers of protons and neutrons the spins of proton-or neutron-pairs arrange themselves antiparallel. Therefore the spin of g-g-nuclei is generally zero, which means that they have no electric quadrupole moment (Q = 0).

When the charge distribution $\rho_e(r)$ is approximated by a rotational ellipsoid with the semi-axes a and b and the difference $\Delta R = a - b$, a mean radius of the charge distribution

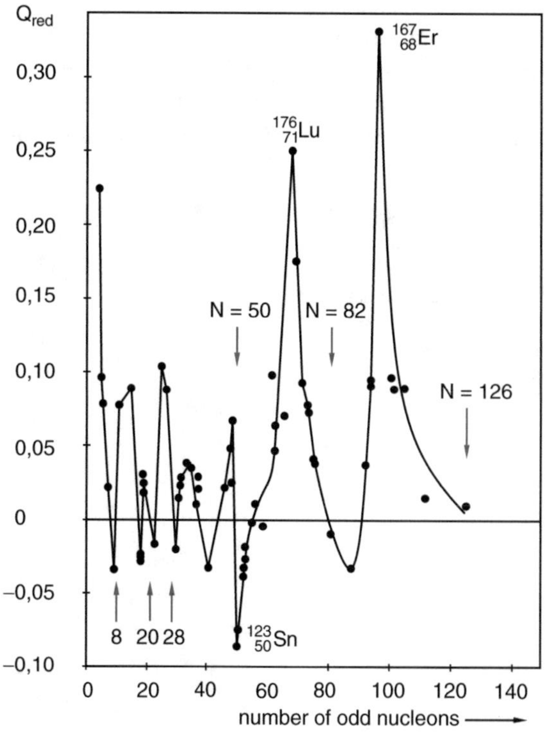

Fig. 2.23 Reduced quadrupole moments of g-u nuclei as a function of the odd neutron-resp. proton number. The arrows indicate the minima of the quadrupole moments for g-g-nuclei with the neutron number N (compare with Fig. 5.36)

Table. 2.5 Quadrupole Moments QM/e of some nuclei, (given in units of 1 barn = 10^{-28}m^2)

Nucleus	QM/e/10^{-28} m^2	Nucleus	QM/e/10^{-28}m^2
$^{2}_{1}D$	+0.0028	$^{35}_{17}Cl$	−0.8
$^{6}_{3}Li$	−0.002	$^{37}_{17}Cl$	−0.06
$^{7}_{3}Li$	−0.1	$^{79}_{35}Br$	+0.33
$^{23}_{11}Na$	+0.1	$^{113}_{49}In$	+1.14
$^{27}_{13}Al$	+0.15	$^{235}_{92}U$	+4.0

$$\langle R\rangle = \left(a \cdot b^2\right)^{1/3} \tag{2.34}$$

can be defined. The deviation from a spherical distribution is characterized by the **deformation parameter**

$$\delta = \frac{\Delta R}{\langle R\rangle} \tag{2.35}$$

ΔR = a-b. For most nuclei the values of δ range between 0.01 and 0.02. For heavy nuclei with the number A of nuclei in the range $150 < A < 192$ strongly deformed nuclei are found with $\delta \approx 0.1$. Figure 2.23 illustrates that more prolate ($a > b =>$ QM > 0) than oblate ($a<b \Rightarrow \boldsymbol{QM}<0$) deformed nuclei are found. The absolute value of the quadrupole moment depends on the total charge of the nucleus and on its volume. In order to have a good comparison between the deformation of different nuclei the reduced quadrupole moment

$$Q_{\text{red}} = \frac{\text{QM}}{Z \cdot e \cdot \langle R\rangle^2} \tag{2.36}$$

is introduced, which is chosen as ordinate in Fig. 2.23.

The values of the nuclear quadrupole moments are obtained from measurements of the hyperfine structure in the corresponding atomic spectra of different isotopes [11]. The nuclear quadrupole moment causes a small shift of the atomic energy terms.. This can be seen quantitatively in the multipole expansion of the potential energy of an electron in the electric field of the nuclear charge distribution. For a mixture of different isotopes different shifts of the energy terms appear for the isotopes with different quadrupole moments. This appears in the spectrum as hyperfine splitting (Fig. 2.24), but has a different underlying cause than the magnetic hyperfine splitting.

The absolute values of the QM's are listed in Table 2.5 for some nuclei.

In recent years detailed investigations have been performed in order to clear whether the neutron has an electric dipole moment, because of its composition of charged quarks. The experiment so far proved that a possible electric dipole moment must be smaller than $2.9 \times 10^{-28} \cdot e \cdot m \approx 5.5 \times 10^{-47}$ C m [13–16].

2.6 Binding Energy of Nuclei

Because of the attractive strong force between the nucleons one has to spend energy in order to decompose the nucleus into the separated nucleons. This energy is called the total binding energy E_B of the nucleus. Dividing E_B by the number A of all nucleons one obtains the mean binding energy per nucleon $\langle E_b\rangle = E_B/A$. If the state of the completely separated nucleons is chosen as the zero point of energy, the potential energy of the nucleus becomes negative. This is quite analog to the situation in atomic physics, where the energy of bound states is negative.

According to Einstein's relation $E = mc^2$ the negative binding energy corresponds to a mass defect $\Delta M = E_B/c^2$ of the nucleus against the sum of the masses of all separated nucleons. The mass of the nucleus

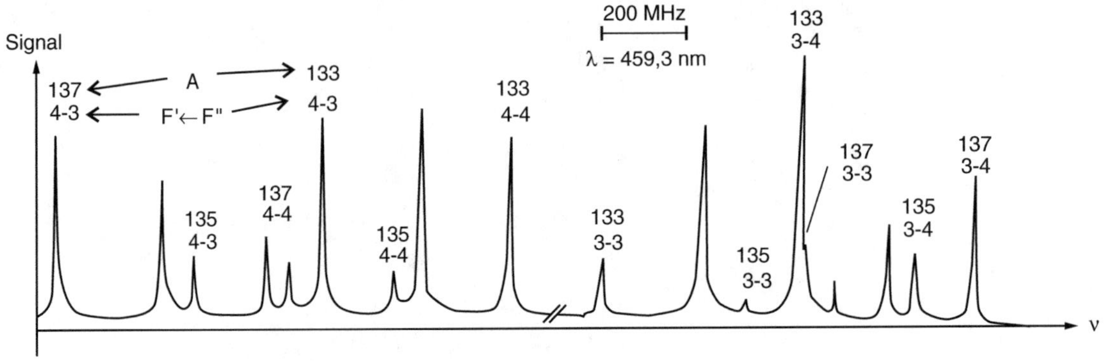

Fig. 2.24 Superimposed hyperfine-spectra of the isotopes $^{133}_{55}Cs$ (I = 7/2, QM = −e · 0.003 barn) and $^{125}_{55}Cs$ (I = 7/2, QM = e · 0.0441 barn) of the transition $F' = J' + 1 \leftarrow F'' = J'' + 1$. This spectrum represents a superposition of the magnetic hyperfine structure and the line shifts due to the electric quadrupole moments. It has been measured with the Doppler-free saturation spectroscopy (*From* H. Gerhardt, E. Matthias, F. Schneider, A. Timmermann, Z. Physik A288, 327 1978)

$$M_N = \sum m_p + \sum m_n - \Delta M \qquad (2.37)$$

is therefore smaller by ΔM compared to the total mass of all separated nucleons.

Remark In the literature often the sign of E_B is defined as positive. The binding energy is then $-E_B$.

The ratio $\Delta M/A$ gives the mass defect per nucleon.

2.6.1 Experimental Results

The mass defect ΔM can be obtained by accurate measurements of the different masses M_N, m_p, m_n in Eq. (2.37). The nuclear mass M_N is determined from the ion mass $M^+ = M_N + (Z-1)\cdot m_e - E_B^{el}/c^2$, which can be measured very accurately with a mass spectrometer. Also the proton mass m_p which represents the H^+-ion, is known very precisely from measurements with a mass spectrometer.

More difficult is the measurement of the neutron mass m_n. It can be indirectly determined by recoil experiments, where a neutron collides with a proton and transfers part of its energy and momentum onto the proton. This part can be deduced from the energy- and momentum conservation. The proton produces in a cloud chamber, placed in an external magnetic field, a visible spur which gives information about the momentum of the proton after the collision.

Another method is based on the mass difference of isotopes which differ by one neutron. However, here the different binding energies of the isotopes have to be taken into account which have to be determined by other methods.

The most accurate measurement is based on the photo-induced fission of the deuterium nucleus, which consists just of one proton and one neutron (Sect. 5.1).

$$1^2\mathrm{D} + \gamma \rightarrow \mathrm{p}^+ + \mathrm{n} + e\text{-} + \mathrm{E}_{kin}$$

measuring the kinetic energy of the proton (that of the neutron must be the same because of momentum conservation) and the energy $h\bullet\nu$ of the γ-quant allows the determination of the neutron mass, which is given by

$$m_n = M_d - m_p + E_B/c^2 \qquad (2.38)$$

In Table 2.6 the experimental values of the nuclear masses of some nuclei are compiled. A more recent method of the determination of the neutron mass is based on the deflection of a collimated neutron beam in an inhomogeneous magnetic field. Due to its magnetic moment μ_n the force on the neutron in the field with a gradient grad $\boldsymbol{B}$ is $\boldsymbol{F} = \boldsymbol{\mu}_n \cdot \mathbf{grad}\ \boldsymbol{B}$.

The magnetic moment can be obtained with the Rabi-method (see Sect. 2.5.1) from the Zeeman splitting of the two Zeeman components in a homogeneous magnetic field.

In Fig. 2.25 the measured total binding energies of some nuclei are plotted as a function of the total number A of nucleons, and in Fig. 2.26 the mean binding energy per nucleon is shown. This illustrates that the mean binding energy per nucleon has a maximum around nuclear masses $M_N = 62$.

The curves $E_B(A)$ show, that energy can be won by two processes:

Table. 2.6 Nuclear Masses (in atomic mass units AMU), total binding-energy E_B, and average binding energy per nucleon E_B/A for some nuclei

Nucleus	Masse/AMU	E_B (MeV)	E_B/A (MeV)
$^2_1$H	2.0141018	2.225	1.112
$^3_1$H	3.0160493	8.482	2.827266
$^4_2$He	4.0026032	28.295	7.073915
$^7_3$Li	7.016004	39.2445	5.606291
$^{12}_6$C	12 (definiert)	92.161	7.680144
$^{16}_8$O	15.994915	127.617	7.976206
$^{27}_{13}$Al	26.981538	224.951	8.33155
$^{35}_{17}$Cl	34.968853	298.2098	8.52028
$^{56}_{26}$Fe	55.934942	492.2539	8.79025
$^{58}_{26}$Fe	57.933280	509.9444	8.79214
$^{62}_{28}$Ni	61.92835	545.2588	8.79450
$^{238}_{92}$U	238.05078	1801.6947	7.57014

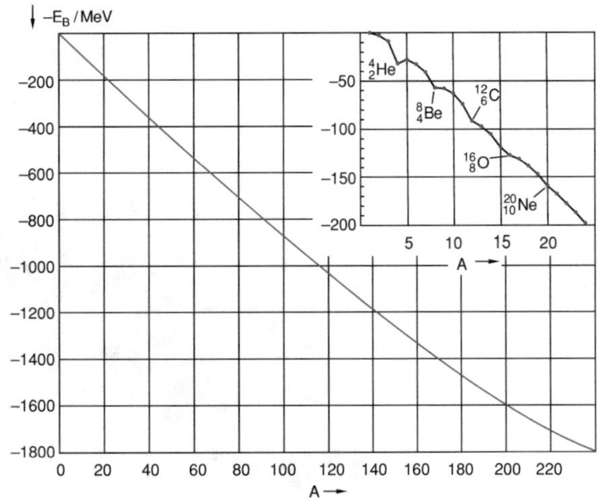

Fig. 2.25 Total binding energy of a nucleus as a function of the mass number A(sum of proton and neutron number)

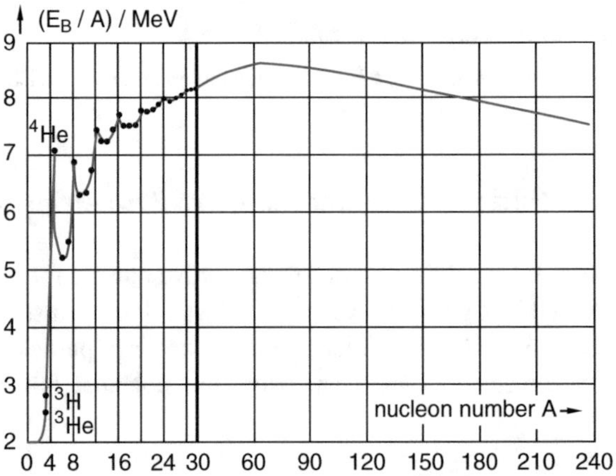

Fig. 2.26 Mean binding energy per nucleon as a function of the nucleon number A. Note the expansion of the abscissa scale for A < 30. The maximum of the curve is at the nickel isotope $^{62}_{28}$Ni

(1) The fusion of light nuclei
(2) The fission of heavy nuclei with $A > 60$.

The lower section of the curve $E_B(A)$ in Fig. 2.25 is shown in the upper right insert with a spread abscissa in order to illustrate the large variations of $E_B(A)$ for light nuclei. This shows that nuclei with even numbers of protons and neutrons, such as the helium nucleus $^4_2$He, the beryllium nucleus $^8_4$Be or the carbon nucleus $^{12}_6$C have relative maxima of their binding energy. The reason for this variation of E_B will be discussed in Sect. 2.6.3.

The most strongly bound nucleus is $^{62}_{28}$Ni with 28 protons and 34 neutrons. In Vol. 5 it is shown that all elements up to $Z = 28$ (except hydrogen, Helium and Lithium which had already been produced shortly after the big bang 14 Billion years ago) are synthesized by nuclear fusion in the central part of stars. For the upper limit of these fusion processes the iron isotope $^{56}_{26}$Fe plays the essential role because it is much more abundant than the Nickel isotope.

In Fig. 2.27 the binding energies are plotted for nuclei in the vicinity of the maximum binding energy of ^{62}Ni.

All known stable nuclei are plotted in Fig. 2.28 as black squares and in Fig. 2.29 as points with the neutron numbers

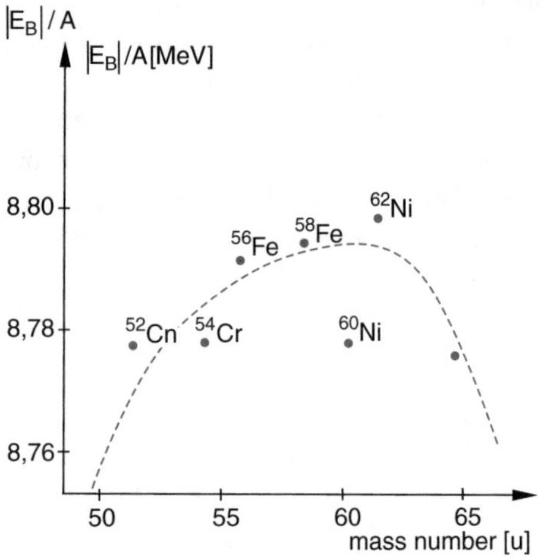

Fig. 2.27 Bindings energy around the nuclei which are most strongly bound

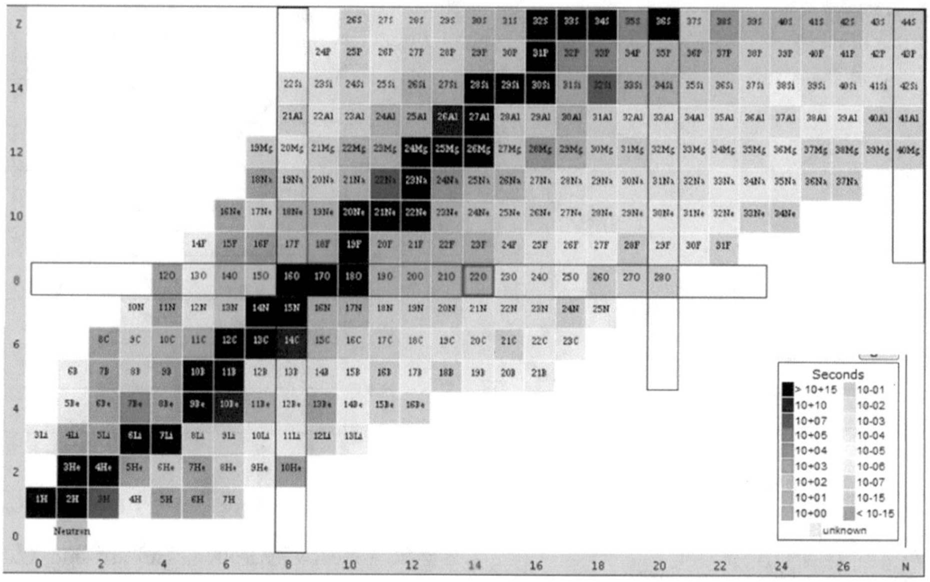

Fig. 2.28 Nuclide table. The ordinate gives the proton number, the abzsissa the neutron number.The black squares represent stable nuclei, the colored ones unstable nuclei (https://www.bing.com/images/search?view)

N as abscissa and the proton number as ordinate. The figure illustrates that with increasing number $A = Z + N$ of nucleons in a nucleus the neutron number N becomes larger than the proton number Z. For comparison the straight line $N = Z$ is shown in Fig. 2.29. Only some light nuclei lie on this line. The reason is that with increasing Z the repulsive long range Coulomb forces between the protons increases so strongly that the nucleus becomes unstable. The Pauli-principle which governs both protons and neutrons as Fermi-particles with spin $I = ½\hbar$ explains why for light nuclei the proton- and neutron numbers can be equal.

In Fig. 2.28 a section of the nuclide chart with stable and unstable nuclei is shown. The black squares depict the stable nuclei, the colored ones unstable nuclei. The numbers give the lifetimes of the corresponding nucleus in seconds. They have to be multiplied by a factor given in the insert.

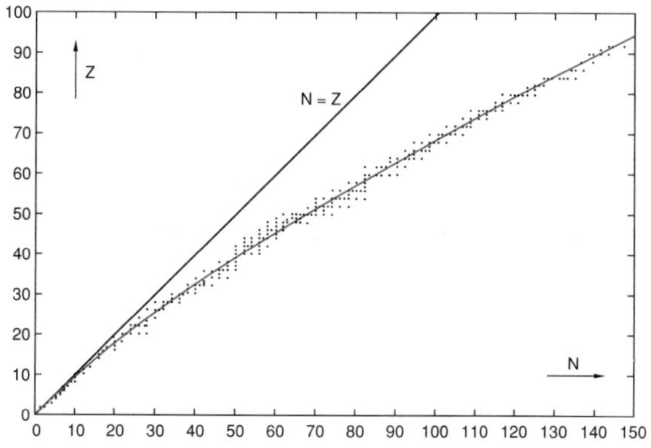

Fig. 2.29 Neutron number N and proton number Z of all stable nuclei

2.6.2 Nucleon Configuration and Pauli-Principle

In a simple model we describe the attractive nuclear force by a potential box with width a (Fig. 2.30). As has been shown in Vol. 3, Sect, 4.3 the stationary energy levels of particles in a three-dimensional spherical-symmetric potential box are given by

$$E_n = \frac{\hbar^2}{2m} k_n^2 \quad \text{with} \quad k_n = n \cdot \pi / a, \tag{2.39}$$

where $\hbar \cdot k = h/\lambda$ is the amount of the momentum of nucleons with deBroglie wavelength $\lambda_{dB} = 2a/n$ ($n = 1, 2, 3\ldots$). Protons and neutrons both have the spin $I = ½\hbar$. They are Fermions and obey the Pauli principle. This implies that each energy level can be occupied by at most two protons and two neutrons both with anti-parallel spins. Since between the protons the repulsive Coulomb force acts, their energy levels must be higher by the Coulomb repulsive energy E_C than those of the neutrons (Fig. 2.31).

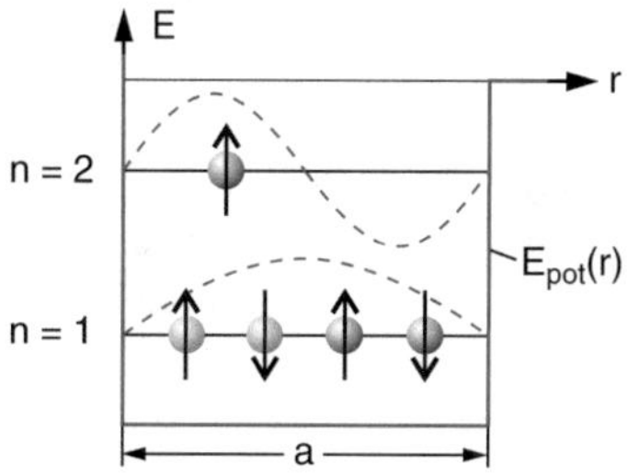

Fig. 2.30 Potential box with energy levels $n = 1$ and $n = 2$ occupied by protons and neutrons

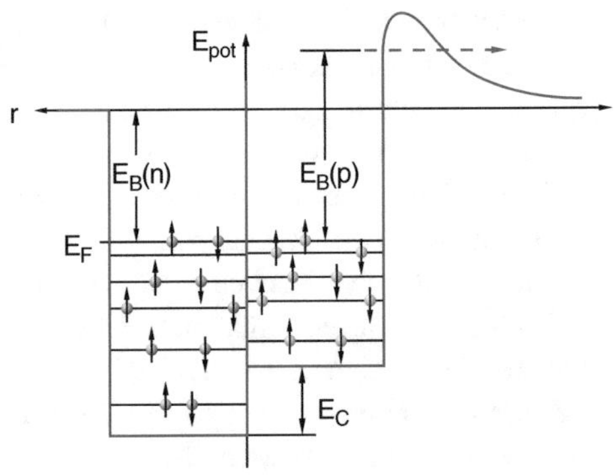

Fig. 2.31 Comparison of the potentials for protons and neutrons with occupied energy levels. Fermi-energy E_F, Coulomb repulsion energy E_c, and bindings-energy E_B for protons and neutrons

The energy levels can be occupied with nucleons up to the Fermi-energy E_F. In order to remove a neutron from the nucleus one has to apply the binding energy $E_B(n)$. For protons the potential barrier has to be overcome which originates from the superposition of attractive nuclear force and repulsive Coulomb force. Because of the quantum–mechanical tunnel-effect (see Vol. 3, Sect. 4.2.3) they can, however, penetrate through the barrier. Their effective binding energy is therefore slightly smaller as indicated in Fig. 2.31 by the dashed horizontal line below the maximum of the potential barrier.

Note The effective binding energy is slightly smaller than the depth of the potential box, because the lowest energy level is not at $E = 0$ but takes into account the zero-point energy of the nucleons. According to the uncertainty principle, the nucleons, which are restricted to the volume $(4/3)\pi(a/2)^3$ of the spherical potential box with radius a/2 have the zero-point energy $E_0 = \pi^2\hbar^2/(2 \cdot m \cdot a^2)$. The effective binding energy of nucleons in the energy state $E_n = n^2\pi^2\hbar^2/(2ma^2)$ is then equal to the potential depth minus E_n.

We can now, completely analog to the build-up of the electron shell, proceed with the build-up of the atomic nucleus; filling the different energy levels for increasing nucleon number A with nucleons, but taking into account the Pauli principle. The lowest energy level E_1 of our potential box contains for the hydrogen nucleus just 1 proton, for the deuteron nucleus 1 proton and 1 neutron. As will be shown in Sect. 5.1 the binding energy is larger, if proton and neutron have parallel spins. The ground state of the deuteron nucleus therefore has the spin $I = 1 \cdot \hbar$ (Fig. 2.32). In the tritium nucleus ${}_1^3$H (1 proton and 2 neutrons) the two neutrons must have antiparallel spins. The nuclear spin is therefore $I = ½\hbar$. The same is true for the helium isotope ${}_2^3$He (2 protons and 1 neutron) where the two protons have antiparallel spins.

In the ${}_2^4$He nucleus the two protons as well as the two neutrons have antiparallel spins and the total nuclear spin is then $I = 0$ in accordance with the experimental result (Fig. 2.32d). The attractive nuclear force between the 4 nucleons is larger than that between the two nucleons of the deuteron nucleus. Therefore the binding energy per nucleon is larger than for the deuteron nucleus or the ${}_2^3$He nucleus (Fig. 2.26).

When a 5th nucleon is added, it has to be placed in the higher energy level E_2. Detailed calculations have shown that its kinetic energy (zero point energy due to its enclosure in the nucleus) is larger than its potential energy, i.e. the nucleus is not stable. If this additional nucleon is a neutron, the resulting nucleus is the isotope ${}_2^5$He, which decays within 2×10^{-21} s into the stable isotope ${}_2^4$He. If the 5th nucleon is a proton the unstable isotope ${}_3^5$Li is formed which decays also into ${}_2^4$He:

$${}_2^5\text{He} \xrightarrow[\text{n}]{7.6\times10^{-22}\text{s}} {}_2^4\text{He}, \quad {}_3^5\text{Li} \xrightarrow[\text{p}]{3.7\times10^{-24}\text{s}} {}_2^4\text{He}.$$

However, adding to the He nucleus a proton and a neutron both additional nucleons can be placed in the same energy level E_2 with parallel spins. The binding energy per nucleon is now larger than for the case where only one nucleon is placed in level E_2 and the stable isotope ${}_3^6$Li is formed (Fig. 2.32e). Adding another neutron increases the binding energy and results in the stable isotope ${}_3^7$Li (Fig. 2.32f). The nuclear spin $I = 3/2\hbar$ is the sum of the nuclear spin $I = ½\hbar$ and its orbital angular momentum $l = 1 \cdot \hbar$ of the additional neutron (see Sect. 2.5.1). Adding another proton, which still fits into the same energy level, and another neutron, which has to occupy the next higher

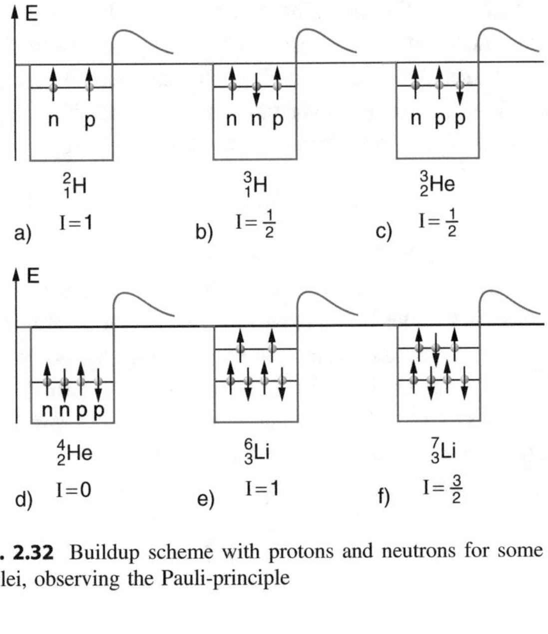

Fig. 2.32 Buildup scheme with protons and neutrons for some light nuclei, observing the Pauli-principle

energy level E_3 results in the beryllium isotope ${}^9_4$Be, which is less stable because of the neutron in the higher level.

An exception from the rule that g-g-nuclei are always stable is the ${}^8_4$Be nucleus. It decays with a half lifetime $\tau = 2 \times 10^{-16}$ s into two α-particles. The reason for its instability is the large binding energy of the α-particles. The total binding energy of the two α-particles is larger than that of the ${}^8_4$Be nucleus, i.e. the mass of the two α-particles is smaller than that of the ${}^8_4$Be-nucleus. The decay is therefore energetically favorable.

2.6.3 Liquid Drop Model of Nuclei; Bethe-Weizsäcker Formula

The homogeneous mass density within the nucleus suggests the comparison of the nucleus with a liquid drop, where the main part of the interaction takes only part between the nearest neighbors.

In order to explain the observed dependence of the binding energy $E_B(A)$ on the mass number A and its different amount for various isotopes we will discuss the different contributions to the binding energy. The nuclear model for nuclei with medium to large masses first proposed by *Friedrich von Weizsäcker* (1912–2007) treats the nucleus like a liquid drop and assumes that five different parts contribute to the total binding energy (this is the energy that has to be spent in order to separate the nucleus ${}^A_Z$X into Z free protons and $N = (A - Z)$ free neutrons).

1. Since the attractive nuclear forces are short range forces, it is assumed that the main part of the binding energy of a nucleon is due to its interaction only with its nearest neighbors. Therefore the binding energy per nucleon should be, at least for nuclei with medium to large mass numbers A, independent of A and the total binding energy should be therefore proportional to A. This also explains the constant nucleon density (see Sect. 2.10) which implies that the nuclear volume $V \sim A$ is proportional to the total number of nucleons $A = Z + N$. This contribution to the binding energy

$$E_{B_1} = +a_V \cdot A \qquad (2.40a)$$

is therefore called the volume part.
2. For nucleons at the surface of the nucleus the binding energy is smaller because part of the nearest neighbors is missing. Since the surface is proportional to the square R^2 of the radius R of the nucleus which in turn is proportional to $A^{2/3}$ (due to the constant nucleon density), we define this surface part, which diminishes the total binding energy, as

$$E_{B_2} = -a_S \cdot A^{2/3}. \qquad (2.40b)$$

3. The third contribution to the binding energy is related to the Pauli- Principle for the nucleons as Fermi-particles with half-integer spin. As illustrated in Fig. 2.32 the occupation of the different energy levels with nucleons has to obey the Pauli principle, which states that each energy level can be occupied at most with two identical particles with opposite spins. This implies that the Fermi energy (this is the energy of the highest occupied level) is minimum for equal proton- and neutron numbers, i.e. for $Z = N$. The additional neutrons for $N > Z$ increase the Fermi-energy and with it the mean kinetic energy of the nucleons which is

$$E_K = (3/5)E_F(N_n + N_p). \qquad (2.40c)$$

The Fermi energy is proportional to the number of protons N_P resp. Neutrons N_N, where we assume, that it has the same mean value $E_F \sim (N/V)^{2/3}$ for protons and for neutrons. It is then

$$E_F = (\hbar^2/2m)\cdot\left(3\pi^2/V\right)^{2/3}\left(N_p^{2/3} + N_n^{2/3}\right), \qquad (2.40d)$$

Inserting this into (2.40c) we obtain

$$E_K = \left(3\hbar^2/10m\right)\left(3\pi^2\right)^{2/3} V^{-2/3}\left(N_n^{5/3} + N_p^{5/3}\right). \qquad (2.40e)$$

With $\Delta N = N_n - N_p$ and $A = N_n + N_p$ and expressing N_p and N_n as

$$N_n = (A/2)(1 + \Delta N/A);$$
$$N_p = (A/2)(1 - \Delta N/A),$$

the brackets $(\ldots)^{5/3}$ can be expanded into a Taylor series

$$(1+x)^{5/3} = \sum a_n x^n = 1 + (5/3)x + (1/3)x^2 + \ldots$$
$$\text{with } x = \Delta N/A \ll 1$$

This yields

$$E_K = C_1 \cdot A(A/V)^{2/3}\left[2 + (2/3)(\Delta N/A)^2 + \ldots\right], \qquad (2.40f)$$

where $C_1 = (3\hbar^2/10\,m)\cdot(3\pi^2)^{2/3}$ combines the different constants in (2.40e). We then finally get

$$E_K = C\cdot\left[A + (1/3)\Delta N^2/A + \ldots\right]. \qquad (2.40g)$$

The first term $\sim A$ has the same dependence on A as the volume contribution and can be therefore included in the constant a_V. The second term which increases the kinetic energy of the nucleons and therefore lowers the binding energy, can be written as

$$E_{B_3} = -a_F(Z - N)^2/A, \qquad (2.40h)$$

where $a_F = -C/3$. It is called the *asymmetry energy* because it depends on the difference between proton- and neutron numbers. It diminishes the binding energy and becomes zero for $Z = N$. For most nuclei is $N > Z$.

4. The Coulomb repulsion between the protons also reduces the binding energy. Analog to the derivation of the gravitational energy of a spherical mass (see Vol. 1, Sect. 2.9.5) the electrostatic energy of a homogeneously charged sphere with the total charge $Q = Z \cdot e$ and the radius R can be obtained as follows: The potential energy of a test charge q in the electric field of the sphere with charge Q with spherical symmetric charge density $\rho_e(r)$ is

$$E_{pot}(r) = \frac{q}{4\pi \cdot \varepsilon_0} \cdot \frac{Q(r)}{r} \tag{2.41a}$$

with

$$Q(r) = 4\pi \int_0^r \varrho_e(r) r^2 dr.$$

For the charge within a sphere with radius $r < R$ (Fig. 2.33) we get

$$Q(r) = \frac{r^3}{R^3} \cdot Z \cdot e \quad \text{and} \quad \varrho_e = \frac{Z \cdot e}{\frac{4}{3}\pi \cdot R^3}. \tag{2.41b}$$

The potential energy of a spherical shell with the charge $dQ = 4\pi r^2 \cdot \rho_e(r) \cdot dr$ is according to (2.41a)

$$E_{pot} = \frac{\varrho_e}{\varepsilon_0} \cdot Q(r) \cdot r dr.$$

Inserting of (2.41b) and integration over r yields the total Coulomb repulsion energy

$$E_C = \frac{3Z^2 e^2}{4\pi \cdot \varepsilon_0 R^6} \cdot \int_0^R r^4 dr = \frac{3Z^2 \cdot e^2}{5 \cdot 4\pi \cdot \varepsilon_0 \cdot R}. \tag{2.42}$$

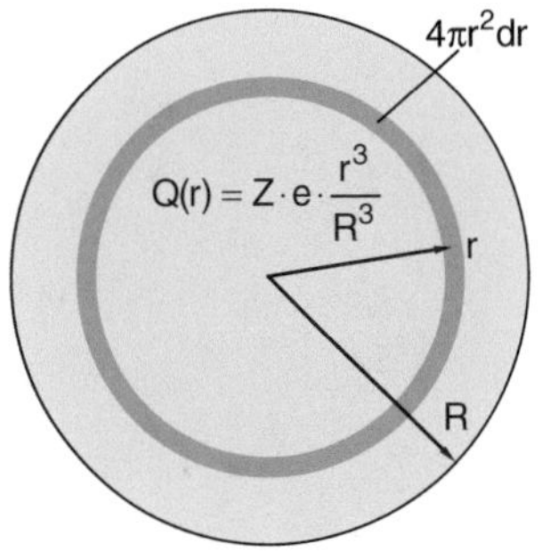

Fig. 2.33 Illustration for the derivation of the Coulomb repulsion energy

With $R = r_0 \cdot A^{1/3}$ we then obtain the share of the Coulomb energy which reduces the total binding energy as

$$E_{B_4} = -a_C \cdot \frac{Z^2}{A^{1/3}} \tag{2.43}$$

with

$$a_C = \frac{3}{5} \cdot \frac{e^2}{4\pi \cdot \varepsilon_0 \cdot r_0}.$$

5. It had been shown above that g-g-nuclei with even numbers of protons and neutrons are stable because always two protons resp. two neutrons form a pair with opposite spins of the partners which occupy the same energy level. On the other hand u-u-nuclei are less stable because there exists always an unpaired nucleon which has to occupy the next higher energy level. We write this "pair-contribution" which can increase or decrease the binding energy as

$$E_{B_5} = a_P \cdot A^{-1/2} \cdot \delta \tag{2.44}$$

with

$$\delta = \begin{cases} +1 & \text{for g-g-Nuclei} \\ 0 & \text{for g-u-, u-g-Nuclei}, \\ -1 & \text{for u-u-Nuclei} \end{cases}$$

where the dependence $A^{-1/2}$ has been introduced purely empirically in order to fit the binding energies of g-g- and u-u-nuclei to the experimental values of the difference

$$E_B^{gg}(A) - E_B^{uu}(A).$$

Adding all these 5 terms we obtain the **Bethe-Weizsäcker formula** for the total binding energy of a nucleus with mas m_N, proton number Z, Neutron number N and total number of nucleons A:

$$\begin{aligned} E_B &= \Delta M \cdot c^2 = (Z \cdot m_p + N \cdot m_n - M_K)c^2 \\ &= a_V A - a_S A^{2/3} - a_F (N-Z)^2 \cdot A^{-1} \\ &\quad - a_C Z^2 \cdot A^{-1/3} + \delta \cdot a_p A^{-1/2}. \end{aligned} \tag{2.45}$$

The mean binding energy per nucleon is then E_B/A. It is plotted in Fig. 2.34 for the lightest 30 nuclei. This figure illustrates that the g-g-nuclei (^4_2He, ^8_4Be, $^{12}_6C$, $^{16}_8O$) have maxima of the binding energy. It furthermore shows that the binding energy per nucleon of ^8_4Be is slightly smaller than that of two α-particles and therefore it decays into two α-particles.

In Fig. 2.35 the different contributions of the Bethe-Weizsäcker formula are plotted as a function of A.

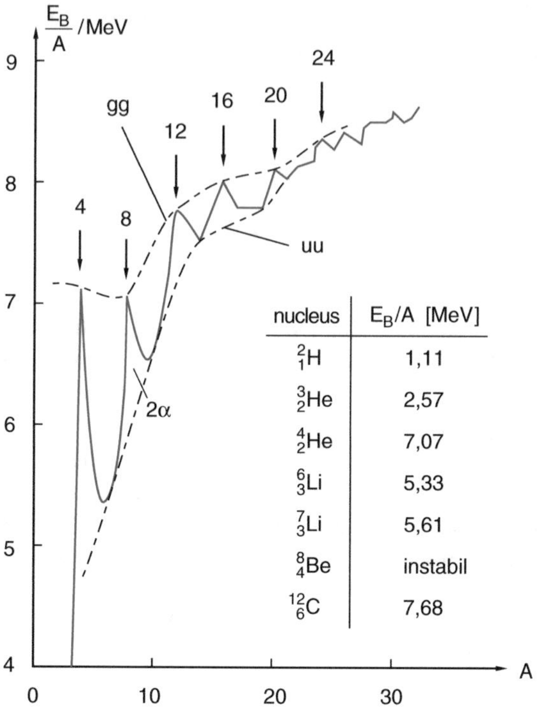

Fig. 2.34 Mean binding-energy per nucleon for the lightest nuclei

It shows that the Coulomb share increases with A and explains why there is an upper limit of A for stable nuclei. The constants in (2.45) had been originally regarded as pure fit constants which could be obtained from a fit of the formula (2.45) to the experimental values. These empirical values depend slightly on the fit procedure. Their mean values are

$$a_V = 15.84\ \text{MeV}, \quad (2.46a)$$

$$a_S = 18.33\ \text{MeV}, \quad (2.46b)$$

$$a_F = 23.2\ \text{MeV}, \quad (2.46c)$$

$$a_C = 0.714\ \text{MeV}, \quad (2.46d)$$

$$a_p = 12\ \text{MeV}. \quad (2.46e)$$

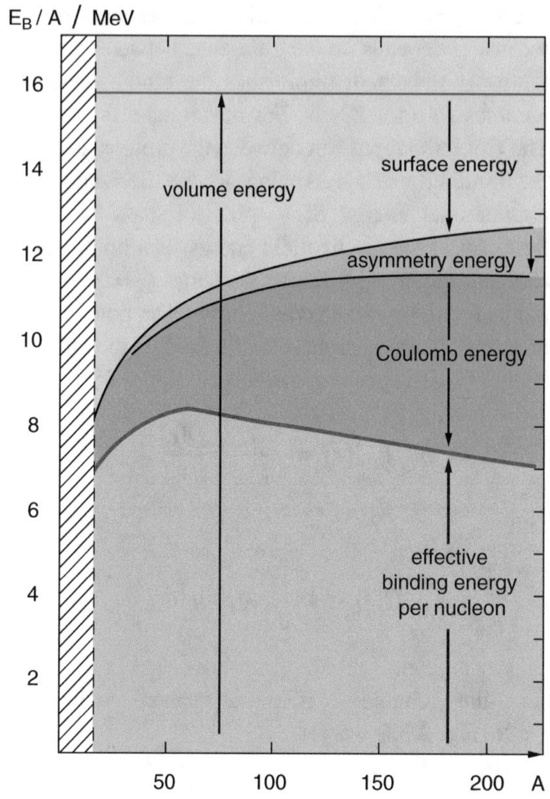

Fig. 2.35 The different contributions to the total binding energy of nuclei for A > 20

Only after a more detailed understanding of the nuclear structure had been achieved (see Chap. 5) the values of these constants can now be calculated based on more thorough theoretical models of the nucleus. The total mass of a nucleus is then

$$M_K(A,Z) = Z \cdot m_p + (A-Z)m_n - E_B/c^2 \quad (2.47a)$$

whereas the total mass of the corresponding atom is

$$M_a(A,Z) = M_K(A,Z) + Z \cdot m_e - E_B^{el}/c^2, \quad (2.47b)$$

More information about nuclear Physics can be found e.g. in [17–20]

Summary

- The nuclear charge can be obtained from the wavelengths of the characteristic X-ray radiation. The wavenumbers $\overline{\nu} = 1/\lambda$ of atomic transitions between energy levels with principal quantum numbers n_1 and n_2 are given by

$$\overline{\nu} = Ry \cdot (Z - S)^2 \left(\frac{1}{n_1^2} - \frac{1}{n_2^2} \right),$$

  where Ry is the Rydberg constant and S the screening constant. For $n_1 = 1$ is $S \approx 1$.
- The radius R of nuclei can be determined from scattering experiments. For nuclei with mass number A one obtains $R = r_0 \cdot A^{1/3}$-. The constant r_0 depends slightly on the assumed mass distribution in the nucleus. For the model of a hard sphere is $r_0 = 1.3 \times 10^{-15}$ m = 1.3 fm.
- The mass density of nuclear matter is about $\rho_m = 10^{17}$ kg/m^3', which is larger by 13–14 orders of magnitude than that of normal solids.
- A nucleus ${}^A_Z X$ consists of $A = Z + N$ nucleons: Z protons and $N = A - Z$ neutrons.
- Protons and neutrons possess a nuclear spin $\boldsymbol{I}$ with the amount $\boldsymbol{I} = \sqrt{I(I+1)}\hbar$ with $I = 1/2$ is the spin quantum number, and a magnetic moment $\mu_p = +2.79\mu_K, \quad \mu_n = -1.91\mu_K,$ where $\mu_N = 5.05 \times 10^{-27}$ J/T is the nuclear magneton.
- The total nuclear spin is the vector sum of the spins of protons and neutrons plus a possible orbital angular momentum of the nucleons.
- Nuclei with $\boldsymbol{I} \neq \boldsymbol{0}$ have a magnetic dipole moment, those with $\boldsymbol{I} \geq 1$ also an electric quadrupole moment.
- The build-up principle of nuclei with nucleons occupying the different energy levels follows the Pauli-Principle which leads to the shell model of nuclei, quite similar to the build-up principle of the atomic electron shell.
- The binding energy of nucleons in a nucleus is the sum of negative potential energy and positive kinetic energy, where the latter is determined by the uncertainty relation due to the enclosure within the limited volume of the nucleus.
- The binding energy of stable nuclei is caused by the attractive nuclear force. It is diminished by the repulsive Coulomb force between the protons. Therefore stable nuclei with large values of the mass number A have more neutrons than protons.
- The mean binding energy per nucleon has a maximum at the element Nickel ($A = 62$) and decreases for $A < 62$ to lower values as well as for $A > 62$. One gains energy for the fission of heavy nuclei or for fusion of light nuclei.

Problems

2.1. What is the nearest distance of an α-particle with the Kinetic Energy $E_{kin} = 50$ MeV to the center of a gold nucleus
  (a) For a central collision?
  (b) For a deflection angle $\vartheta = 60°$ if a Coulomb potential is assumed?
  (c) At which deflection angle ϑ does the scattering distribution deviate from that for a pure Coulomb-potential for a nuclear radius $R = 6.5 \times 10^{-15}$ m and a kinetic energy of the α-particle $E_{kin} = 50$ MeV?

2.2. α-particles with the kinetic energy $E_{kin} = 5$ MeV are shot through a gold foil with thickness $d = 10^{-7}$ m ($\rho_{gold} = 19.32$ g/cm^3). Which fraction of the incident α-particles is scattered by the angle $\vartheta \geq 90°$?

2.3. Calculate the mean square of the nuclear radius $\langle r^2 \rangle$ in (2.12) for the spherical density distribution $\rho_m(r)$ with the total mass $M = 4\pi\rho \int r^2 dr$.
  (a) $\varrho_m(r) = \varrho_0 \cdot (1 - ar^2)$, $0 \leq r \leq 1/\sqrt{a} = R_0$
  (b) $\varrho_m(r) = \varrho_0 \cdot e^{-r/a}$ $0 \leq r \leq \infty$

2.4. Neutrons with the kinetic energy $E_{kin} = 15$ MeV are scattered by lead nuclei. At which scattering angle ϑ appears the first diffraction minimum?

2.5.
  (a) What is the minimum energy of $(A - Z)$ electrons enclosed within a spherical volume with radius R?
  (b) What is their potential energy if Z protons are enclosed in the same volume?
  Numerical values
  (1) $R = 1.8 \times 10^{-15}$ m; $A = 4$ $Z = 2$
  (2) $R = 6.5 \times 10^{-15}$ m; $A = 200$, $Z = 80$.

2.6. A Deuterium Nucleus ${}^2_1H$ with the Binding Energy $E_B = -2.2$ MeV is Irradiated with γ-quanta of Energy $h \cdot \nu = 2.5$ MeV.
  (a) What are energy and velocity of the proton released at the photon-induced fission of deuterium?
  (b) What is the magnetic field strength B of a 60°-sector field, if the proton should be guided on a circle with radius $R = 0.1$ m? At which distance from the deuteron source (which is assumed to be point-like) are the protons refocused?

2.7. The magnetic moment of the proton is $\mu_p = 2.79$ μ_N, that of the neutron $\mu_n = -1.91$ μ_N. The nuclear spin of the deuteron is I = $1 \cdot \hbar$. What is the ratio of the hyperfine-splitting in the $1^2S_{1/2}$ state of the two isotopes ${}^1_1H$ and ${}^2_1H$?

2.8. What is the Frequency Shift Between the Lyman-α Lines in the two Isotopes

(a) due to the mass defect ?

(b) due to the quadrupole moment $QM(^2_1H) = 4.5 \times 10^{-50}$ cm^2, if the shift of the $2P$-term is neglected?

2.9. Prove, that the internal quadrupole moment of a homogeneous charge distribution in a rotationally symmetric ellipsoid with the axes a and b is given by $QM = (2/5)Ze\cdot(a^2 - b^2)$ or by $QM = (4/5)\ Z\cdot e\cdot \overline{R}^{\,2}\cdot\delta$ with $a = \overline{R} + ½\Delta R$, $b = \overline{R} - ½\ \Delta R$ and $\delta = \Delta R/\overline{R}$.

2.10. What is the positive Coulomb energy of a nucleus with homogeneous charge distribution and radius R? Compare for $Z = 80$ and $R = 7$ fm the energy levels of protons and neutrons in such a nucleus. What is the energy difference?

References

1. William R. Leo: Techniques for Nuclear and Particle Physics (Springer. Berlin, Heidelberg 1994)
2. L.C.L. Yuan, CH. Sh. Wu (eds). Methods of Experimental Physics Vol. 5 (Academic Press 1963)
3. E. Rutherford: The Structure of the Atom: Phil. Mag. **22**, 488u (1914)
4. H.E. Wegener, R.M. Eisberg, G. Igo: Elastic Scattering of 40 MeV α-particles from Heavy Elements. Phys. Rev. **99**, 825 (1955)
5. G.W. Farwell, H.E. Wegener: Elastic Scattering of Intermediate-Energy α-Particles by Gold. Phys. Rev. **93**, 356 (1954)
6. R. Hofstadter: Nuclear and Nucleon Scattering of High Energy Electrons. Ann. Rev. Nucl. Sci. **7**, 303 (1957)
7. R. Hofstadter: Electron Scattering and Nuclear Structure. Rev. Modern Physics **28**, 214 (1957)
8. R. Hofstadter (ed): Electron Scattering and Nuclear and Nucleon Structure (Benjamin, New York 1963)
9. R. Pohl et al. Nature **466**, 213 (2010) A. Antognini et.al. Science **339**, 417 (2013)
10. J. Chadwick: Existence of a Neutron, Proc. of Royal Soc. **A136,** 692, (1932)
11. H. Kopfermann: Nuclear Moments. (Academic Press 1958)
12. N. Ramsey: Nuclear Moments (Wiley and Sons, 1953)
13. L. Alvarez, F. Bloch: A quantitative determination of the Neutron Moment in absolute Nuclear Moments. Phys. Rev. 51, 111 (1940)

13b. J.M.B.Kellog, I.I.Rabi, J.R.Zacharias: Gyromagnetic properties of the hydrogens. Phys. Rev. 50, 472 (1936).

14. B.R. Martin, G. Shaw: Nuclear and Particle Physics (Wiley 2019)
15. H. Sommer, H.A. Thomas, J.A. Hhippie: The Measurement of e/m by Cyclotron Resonance. Phys. Rev.82, 697 (1951)
16. C.A. Baker et al.: Improved Experimental Limit on the Electric Dipole-Moment of the Neutron. Phys. Rev. Lett. **97**, (13) 13180 (2006)
17 G. Musiol: Kern-und Elementarteilchen-Physik. 2:Auflage Verlag Harry Deutsch, 1995
18. S.B.Crane: Nuclear Physics (Wiley 1991)
19. Hari M- Agrawal: Nuclear Physics_ Problem-based Approach (Asoke K, Gosh 2016) D.C. Tayal: Nuclear Physics (Skylab Textbook Comanian 2023)
20. Leonel Wore Nuclear Physics (Larsen and Keller Education 2019)

Unstable Nuclei, Radioactivity

3

Since generally several isotopes exist for every number Z of protons, there are altogether more than thousand different nuclei. We distinguish between **stable nuclei**, which do not convert spontaneously into other nuclei, and **unstable nuclei,** which, after a finite lifetime transform themselves into other nuclei by emitting α-particles, electrons or positrons. They can also undergo fission-processes where they generate two or more lighter nuclei as fission fragments. Examples of unstable nuclei are the radioactive elements *Uranium, Radium, Thorium* occurring in nature, or the artificially produced Trans-Uranium elements as well as many more unstable elements created by bombardment with particles of high energy. Figure 3.1 shows a section of the *Karlsruhe-nuclide map* for the 8 smallest nuclei, which lists all stable (red) and unstable particles with their decay times, their main decay paths and the energies of the emitted particles [1].

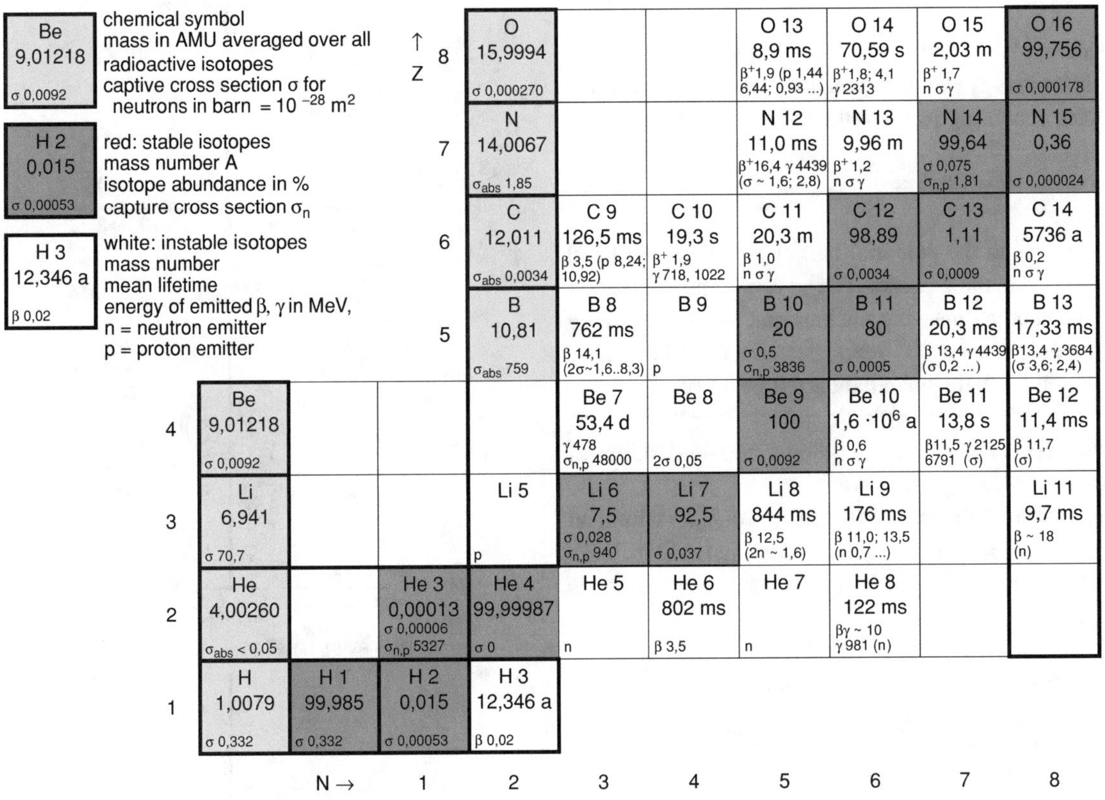

Fig. 3.1 Section of the Karlsruhe nuclide chart. The stable isotopes are marked in red. Above these stable isotopes are unstable ß⁺-emitting nuclei, below are ß⁻emitting nuclei. Quoted are furthermore half-lifetime NS Maximum energy of the ß⁺- and ß⁻emitting nuclei (KIT Karlsruhe Germany)

W. Demtröder, *Nuclear and Particle Physics*, Undergraduate Lecture Notes in Physics,
https://doi.org/10.1007/978-3-030-58313-2_3

3.1 Stability Criteria; Stable and Unstable Nuclei

A nucleus can decay by emission of particles or by fission only if its mass M is larger than the sum of the masses of the decay products, i.e. if

$$M({}^{A}_{Z}X) \geq M\left({}^{A'}_{Z'}Y\right) + M_2, \tag{3.1}$$

where M_2 is the mass of the emitted particle or in case of a fission process it is the sum of the fragment masses plus that of emitted neutrons. For the case of emission of a γ-quantum with energy $h \cdot \nu$ is $M_2 = h \cdot \nu/c^2$.

The spontaneous reaction ${}^{A}_{Z}X \rightarrow {}^{A'}_{Z'}Y_1 + {}^{A-A'}_{Z-Z'}Y_2$ has to be exothermic.

Equation (3.1) is a necessary but not sufficient condition. Even if (3.1) is fulfilled, the reaction can be hindered by a potential barrier (Fig. 3.2) or by symmetry conditions (selection rules).

Example A nucleus can in principle emit an α-particle, if the condition

$$\Delta M = m(Z,A) - m_1(Z-2, A-4) - m_\alpha > 0. \tag{3.1a}$$

is fullfilled.

The kinetic energy of the fragments is shared between the two particles according to the ratio of their masses, i.e.

$$E_{\text{kin}}(m_\alpha) = \frac{m_1}{m_\alpha} E_{\text{kin}}(m_1). \tag{3.1b}$$

However, due to the potential barrier (see Figs. 2.29 and 3.2) the emission of the α-particle can be very improbable even for $\Delta M > 0$. Such an unstable nucleus can have a very long lifetime (for example $\tau > 10^6$ years) because it can decay only by tunneling through the potential barrier (see Sect. 3.3).

The Bethe-Weizsäcker formula allows the derivation of some important stability criteria.

We regard at first the mass dependence $M(A, Z)$ within an isobar-row with A = const. for even A and odd A (Fig. 3.3). From (2.45) it follows that the binding energy per nucleon is dependent on the square of the proton number Z. For odd numbers A (gu- or ug nuclei) there is only one parabola for the plot $M(A, Z)$ against Z with a minimum at $Z = Z_0$ (Fig. 3.3a), while for even numbers A there are two parabolas, one for uu-nuclei and one for gg-nuclei (Fig. 3.3b).

From (2.47) we obtain the condition for the nuclear charge number Z_0 for which the mass $M(Z, A$ = const.) becomes minimum:

$$\left(\frac{\partial M_a(A,Z)}{\partial Z}\right)_{A=\text{const.}} = 0, \tag{3.2a}$$

where we regard Z as continuous variable. This yields the (not necessarily integer) value

$$Z_{\min} = \frac{A}{2} \frac{\left(m_n - m_p - m_e\right) c^2 + a_F}{a_F + a_C \cdot A^{2/3}}. \tag{3.2b}$$

Inserting the numerical values we obtain

$$Z_{\min} = \frac{A}{1{,}972 + 0{,}015 \cdot A^{2/3}}. \tag{3.2c}$$

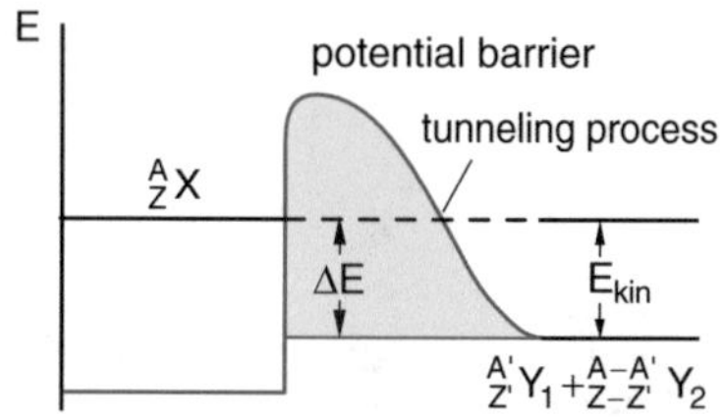

Fig. 3.2 Suppression of energetically possible decay by the potential barrier (Coulomb Barrier)

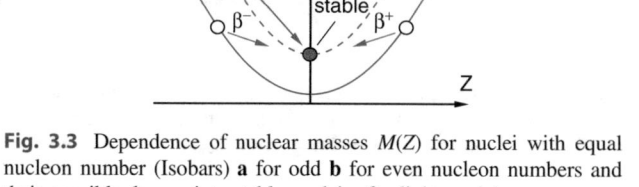

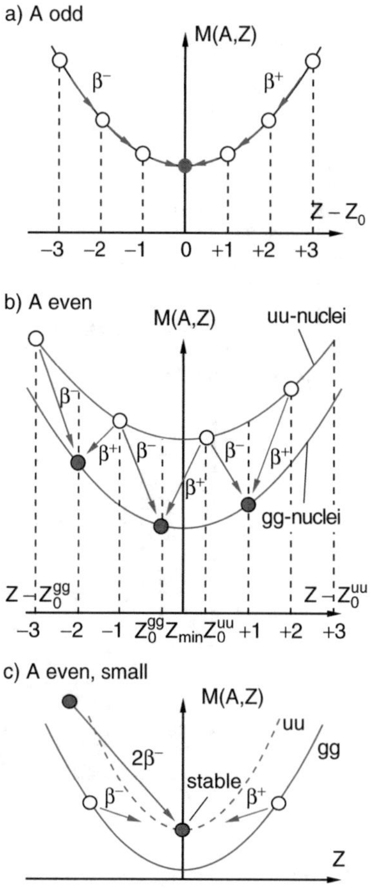

Fig. 3.3 Dependence of nuclear masses $M(Z)$ for nuclei with equal nucleon number (Isobars) **a** for odd **b** for even nucleon numbers and their possible decays into stable nuclei **c** for light nuclei

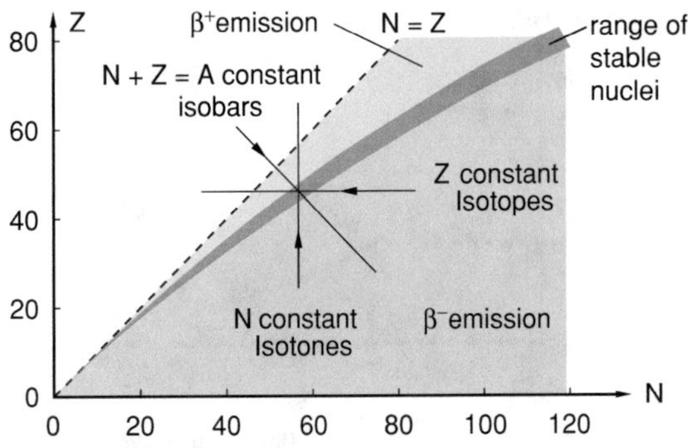

Fig. 3.4 Range of stable nuclei in a Z, N-diagram (from T.Mayer Kuckuk: Kernphsik Teubner, Stuttgart 1995)

That nucleus of the isobar row A = const. which comes closest to Z_{min} has the minimum mass and should be therefore stable.

Plotting these values of Z_0 in a diagram $Z(A - Z)$ one obtains the curve of highest stability in Fig. 3.4, which matches the experimentally found chart of all stable nuclei in Fig. 3.1. Above this stability curve the nuclei decay by $ß^+$-emission, below the curve by $ß^-$-emission.

The experimental search for stable nuclei brought the following result:

Above $Z = 7$ there are no stable uu-nuclei. The 1. Mattauch's isobar rule states that all uu-nuclei with $Z > 7$ decay always either by $ß^+$ or $ß^-$-emission into gg-nuclei. This implies that the only stable uu-nuclei are ${}^2_1D, {}^6_3Li, {}^{10}_5B$ and ${}^{14}_7N$.

For Z $\leq$ 7 the mass dependence M(Z) is so strong (i.e. the curvature of M(Z) is so large) that the mass of a gg-nucleus is larger than that of the adjacent uu-nucleus (Fig. 3.3c). In this case an uu-nucleus can only decay by a double ß-emission into a lighter uu-nucleus. This process, however, a very small probability.

This can be also understood as follows:

All uu-nuclei have an unpaired proton as well as an unpaired neutron. A neutron can decay by $ß^-$ emission into a proton. If this proton can be placed in an energy level lower than that of the original neutron without violating the Pauli-principle, the newly formed gg-nucleus is more stable than the original uu-nucleus (Fig. 3.5).

The question now arises how many stable isotopes of an element with a given atomic number Z are possible:

The experimental answer is:

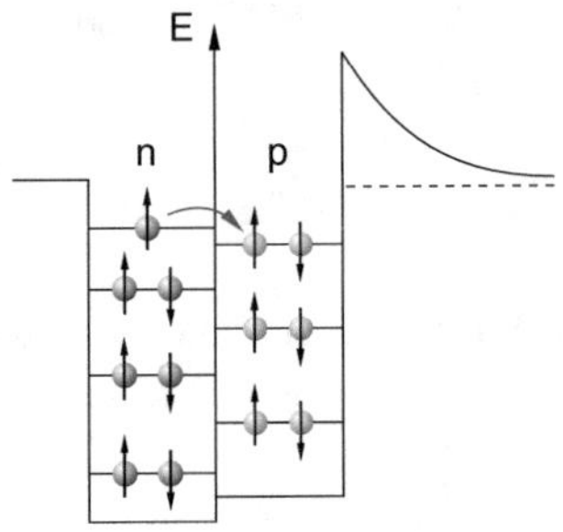

Fig. 3.5 Explanation of the $ß^-$decay of u-u nuclei

For all elements with odd Z there are at most two stable isotopes which differ by $\Delta A \geq 2$ for $Z > 7$.

The reason for this rule is that there are no stable uu-nuclei for $Z > 7$. For odd proton numbers $Z > 7$ the number of neutrons must be therefore even.

Example Stable isotopes with ug-nuclei are for example:

$${}^{39}_{19}K \text{ and } {}^{41}_{19}K;\ {}^{69}_{31}Ga \text{ and } {}^{71}_{31}Ga;\ {}^{35}_{17}Cl \text{ and } {}^{37}_{17}Cl$$

For elements with even number of protons often more than two isotopes exist.

Example

1. ${}^A_{30}Zn$ has 5 stable isotopes with A = 64; 66–68; 70
2. ${}^A_{50}Sn$ has 10 stable isotopes with A = 112; 114–120, 122 and 1124.

The main reason for these empirical rules is the large negative pair binding energy between equal nucleons (see 2.44) which results in a larger stability of gg-nuclei.

These rules can be summarized as 2. Mattauch's isobar rule:

For elements with odd nucleon numbers $A = 2n + 1$ there exists generally only one isobar. For elements with even numbers $A = 2n$ and $Z > 7$ there are no stable uu-nuclei but at least two stable gg-isobars.

Similar to the isobar rules there exist also isotope rules:

- Elements with even Z have at least 2 isotopes, those with odd Z at most 2 isotopes.
- The same is true for neutron numbers $N = A - Z$: There are at least 2 stable isotopes for elements with even N and at most 2 for odd N.

Exception: For $N = 2$ and $N = 4$ only 1 stable isotope exists: ${}_2^4He$ and ${}_3^7Li$

- For some odd neutron numbers $N = 19, 21, 35, 39, 45, 61, 89, 115$ and 123 no stable isotope exists.

The question, under which conditions neutrons in stable nuclei do not decay can now be answered as follows (Fig. 3.5):

For the reaction

$$n \rightarrow p + \beta^- + \bar{\nu}$$

the highest energy level E_a in the nucleus ${}_Z^AX$ is converted into the lowest energy level E_f of the nucleus ${}_{Z+1}^{A}Y$ which can be occupied by the proton without violating the Pauli principle. The energy difference

$$\begin{aligned} \Delta E &= E_a - E_e \\ &= (M_X - M_Y + m_e)c^2 + E_{kin}(e, \bar{\nu}) \end{aligned} \quad (3.3)$$

is partly transferred into kinetic energy of the emitted particles β^- and $\bar{\nu}$ (see Sect. 3.4).

If $\Delta E < (M_X - M_Y + m_e)c^2$ the ß-decay cannot occur because the energy would not be sufficient. In this case the neutrons exist forever as stable particles within the nucleus although they would decay as free neutrons with a lifetime of 887 s.

3.2 Unstable Nuclei and Radioactivity

Atomic nuclei are named unstable, if they decay after a finite lifetime spontaneously into other nuclei, either by emission of particles (α; n; β^-; β^+) or by fission into fragments. Excited nuclei can also pass over into lower energy states by emission of γ-radiation. All these processes are possible only if the energy of the initial state is higher than that of the final state (see Sect. 3.1).

Unstable nuclei can be found in nature (*natural radioactivity*) but they can be also produced by nuclear reactions induced by collisions or by fission of heavy nuclei (*artificial radioactivity*).

Investigations of such unstable nuclei and their decay channels has essentially contributed to our understanding of the structure of atomic nuclei. Historically such investigations started 1896 with the discovery of radio-activity of uranium minerals by Henry Bequerel, (Fig. 1.1) before any idea existed about atomic nuclei [2]. Meanwhile radio-active nuclei are largely understood and have found various applications for solving technical problems and in Biology and Medicine (see Chap. 8).

We will now at first discuss general decay laws of radioactive nuclei, before we deal with specific decay mechanisms.

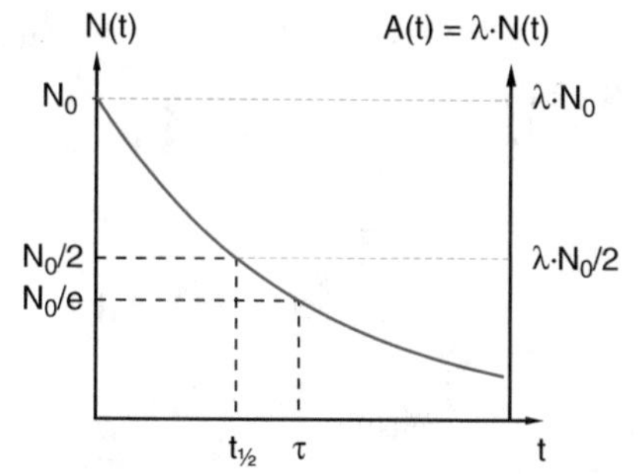

Fig. 3.6 Exponential decay curve of the number $N(t)$ of unstable nuclei and its relation with the activity $A(t)$

3.2.1 Decay Laws

We regard an ensemble of N unstable particles: The probability $\lambda = dP/dt$ that one particle decays is for spontaneous decays equal for all particles of the same kind. The total number of decays per sec is therefore Fig. 3.6.

$$\frac{dN}{dt} = -\lambda \cdot N = -A(t). \quad (3.4)$$

The time-dependent quantity A is the activity of the sample. Its unit is 1 Bequerel (1 Bq). A radioactive sample has the activity $A = n$ Bq if n nuclei decay per sec.

Integration of (3.1) yields the decay function

$$N(t) = N_0 \cdot e^{-\lambda t} \quad (3.5)$$

which gives the number N(t) of present unstable particles at time t, if at $t = 0$ N_0 particles were existent in the sample (Fig. 3.6).

The activity $A = \lambda \cdot N$ at time t then becomes

$$A(t) = A_0 e^{-\lambda t} \quad (3.6)$$

with

$$A_0 = A(t = 0) = \lambda \cdot N_0.$$

Plotting the logarithm $\ln(A(t))$ of the measured activity against the time t the slope of the straight line

$$\ln(A(t)) = \ln A_0 - \lambda \cdot t \quad (3.7)$$

gives the decay constant λ (Fig. 3.7).

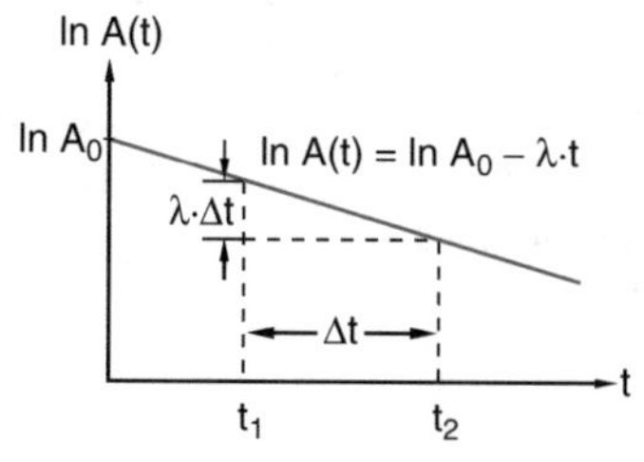

Fig. 3.7 Determination of the decay constant λ

The **mean lifetime** τ of the unstable particles is the time-average of the statistically varying decay times t of the single particles, integrated from $t = 0$ up to $t = \infty$, weighted with the number $\lambda \cdot N(t)$ of particles decaying within the time interval dt:

$$\tau = \bar{t} = \frac{1}{N_0} \cdot \int_0^\infty t \cdot \lambda \cdot N(t)\mathrm{d}t = \int_0^\infty t \cdot \lambda \cdot \mathrm{e}^{-\lambda t}\,\mathrm{d}t = \frac{1}{\lambda} \tag{3.8}$$

This shows that $\tau = 1/\lambda$. The mean lifetime τ is equal to the inverse decay constant λ. At $t = \tau$ the original number N_0 of unstable particles has decreased to N_0/e (Fig. 3.6).

Often the half- life-period $t_{1/2}$ is used instead of the mean lifetime, where $N(t_{1/2}) = N_0/2$. The relation between the to times is

$$t_{1/2} = \tau \cdot \ln 2. \tag{3.9}$$

From (3.6) we get the relation

$$A(t) = \lambda N(t) = N(t)/\tau \tag{3.10}$$

between activity $A(t)$ and the number $N(t)$ of radioactive particles at time t.

It is therefore possible to deduce the number of emitting particles from the measured activity A(t) and the known mean lifetime τ, if the sample contains only particles of the same kind.

In Table 3.1 the half-life period in years of some radioactive nuclei are compiled. They vary over many orders of magnitude.

Note: The product nucleus $|2\rangle$, into which the original nucleus $|1\rangle$ decays, may be also unstable when it decays with the decay constant λ_2 into a product $|3\rangle$ (Fig. 3.8). The decay products $|2\rangle$ and $|3\rangle$ can be different energy levels of the nucleus $|1\rangle$ or different "daughter nuclei" X and Y. For the different populations we obtain.

$$\frac{dN_1}{dt} = -\lambda_1 N_1. \tag{3.11a}$$

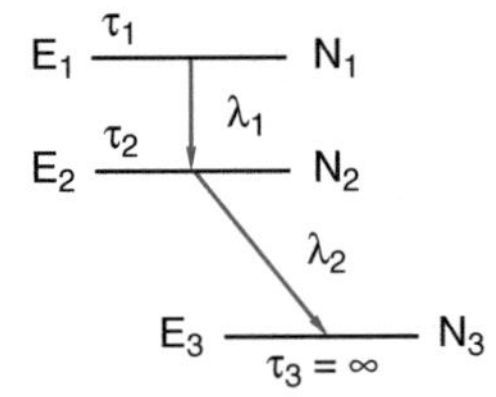

Fig. 3.8 Term-diagram for a cascade decay

Table 3.1 Most abundant radio-active nuclei in nature

Element	Symbol	kind of radiation, energy/MeV	$t_{1/2}$/a
Tritium	$^3_1$H	β^- 0,0286	12,3
Potassium	$^{40}_{19}$K	β^- 1,35	$1{,}5 \cdot 10^9$
Rubidium	$^{87}_{37}$Rb	β^- 0,275	$5 \cdot 10^{10}$
Iodine	$^{129}_{53}$I	β^- 0,15	$1{,}7 \cdot 10^7$
Cesium	$^{135}_{55}$Cs	β^- 0,21	$3{,}0 \cdot 10^6$
Lead	$^{205}_{82}$Pb	α 2,6	$\approx 1{,}4 \cdot 10^{16}$
Polonium	$^{209}_{84}$Po	α 4,87	103
Radium	$^{226}_{88}$Ra	α 4,77	1620
Thorium	$^{230}_{90}$Th	α 4,5–4,7	$8 \cdot 10^4$
Uranium	$^{234}_{92}$U	α 4,6–4,8	$2{,}5 \cdot 10^5$
	$^{235}_{92}$U	α 4,3–4.6	$7.1 \cdot 10^8$
	$^{238}_{92}$U	α 4,2	$4{,}5 \cdot 10^9$

$$\frac{\mathrm{d}N_2}{\mathrm{d}t} = +\lambda_1 N_1 - \lambda_2 N_2, \tag{3.11b}$$

$$\frac{\mathrm{d}N_3}{\mathrm{d}t} = +\lambda_2 N_2, \tag{3.11c}$$

where we have assumed that the nucleus $|3\rangle$ is stable and $N_2(0) = N_3(0) = 0$, which implies that the nuclei 2 and 3 are produced solely by decay of the nucleus 1.

Since the total number of nuclei must be constant (there are no particles lost), we get the condition

$$N_1(t) + N_2(t) + N_3(t) = N = \text{ const.}$$

Multiplication of (3.11b) with $\mathrm{e}^{\lambda_2 t}$ and rearrangement of the different terms gives

$$\frac{dN_2}{dt} \cdot e^{\lambda_2 t} + \lambda_2 N_2 e^{\lambda_2 t} = \lambda_1 N_{10} e^{(\lambda_2 - \lambda_1)t}. \tag{3.12a}$$

The left side of this equation equals the time derivative $\frac{\mathrm{d}}{\mathrm{d}t}\left(N_2 \cdot \mathrm{e}^{\lambda_2 t}\right)$. We can therefore write (3.12a) in the form

$$\frac{\mathrm{d}}{\mathrm{d}t}\left(N_2 e^{\lambda_2 t}\right) = \lambda_1 N_{10} e^{(\lambda_2-\lambda_1)t}. \quad (3.12b)$$

Integration yields

$$N_2 e^{\lambda_2 t} = \frac{\lambda_1}{\lambda_2 - \lambda_1} N_{10} e^{(\lambda_2-\lambda_1)t} + C. \quad (3.12c)$$

With $N_2(0) = 0$ the integration constant C becomes $C = -\frac{\lambda_1}{\lambda_2-\lambda_1} N_{10}$ and we get the time dependent population $N_2(t)$ after multiplication with $e^{\lambda_2 t}$

$$N_2(t) = \frac{\lambda_1 \cdot N_{10}}{\lambda_2 - \lambda_1}\left(e^{-\lambda_1 t} - e^{-\lambda_2 t}\right). \quad (3.13a)$$

Inserting (3.13a) into (3.11c) we get with $N_3(0) = 0$ after integration the population

$$N_3(t) = N_{10}\left[1 - \frac{1}{\lambda_2 - \lambda_1}\left(\lambda_2 e^{-\lambda_1 t} - \lambda_1 e^{-\lambda_2 t}\right)\right]. \quad (3.13b)$$

The time progression of the population $N_2(t)$ depends on the ratio λ_1/λ_2 of the decay constants. It is illustrated in Fig. 3.9 for different ratios λ_1/λ_2. It reaches a maximum at the time

$$t = \ln(\lambda_2/\lambda_1) \,/\, (\lambda_2 - \lambda_1).$$

Artificial radioactivity can be realized by collisions of nuclei with other particles, for example by neutrons from a nuclear reactor. As example we assume that P nuclei per sec are activated, which decay with the decay constant λ. The time dependent number $N(t)$ of activated nuclei can be obtained from the equation

$$\frac{\mathrm{d}N}{\mathrm{d}t} = P - \lambda N \quad (3.14)$$

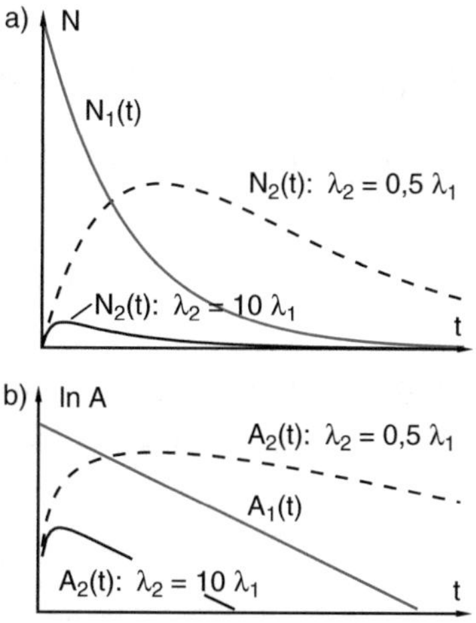

Fig. 3.9 Population number (**a**) and activity (**b**) in case of a cascade decay for two different ratios λ_2/λ_1 of the decay constants

Integration yields with the initial condition $N(0) = 0$

$$N(t) = \frac{P}{\lambda}\left(1 - e^{-\lambda t}\right). \quad (3.15)$$

After the time $t = 3/\lambda$ $N(t)$ has already nearly reached its final value $N(\infty) = P/\lambda$ (Fig. 3.10).

3.2.2 Natural Radioactivity

Natural radioactivity means the spontaneous transformation of unstable nuclei into other nuclei with emission of radioactive radiation. The first observation of spontaneous radioactive radiation was reported 1896 by H. Bequerel. He observed that uranium salts spontaneously emit a radiation which can penetrate layers of material that are opaque for visible radiation and which blackens a photo plate.

This radiation does not depend on the visible light emitted by some uranium ores and it has other characteristics as the X-rays discovered at the same time by W. Röntgen (Fig. 3.10X).

Using the ionization chamber (see Sect. 4.3) Bequerel and later Marie and Pierre Curie could quantitatively measure the intensity of this radiation for different radioactive

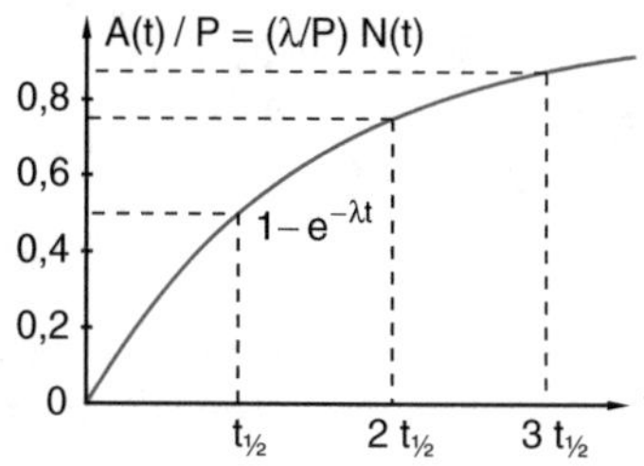

Fig. 3.10 Activity A(t) of a sample for a constant activation rate P

Fig. 3.10X Wilhelm Conrad Röntgen (1845–1923) First Nobel-Price in Pysics 1901

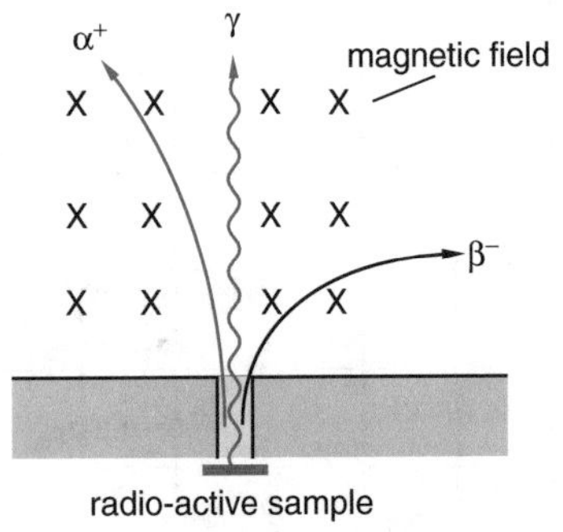

Fig. 3.11 Simple experimental distinction between α, ß, and γ-radiation based on their different deflection in a magnetic field

substances. Rutherford discovered 1899 that the uranium radiation consists of two components which show different penetration depths. He named them α- and ß-radiation. In the same year several other scientists found that these two components are deflected in a magnetic field into opposite directions (Fig. 3.11). They concluded that they must carry charges with opposite signs.

One year later *P*. Villard found a third kind of radiation which he called γ-radiation. It is not affect by magnetic fields. In the following ten years detailed measurements with mass spectrometers (see Vol. 3 Sect. 2.7) allowed the determination of the charge-mass ratio e/m for α- and ß-particles. This brought the insight that α-particles are helium nuclei He^{++} and ß-particles are electrons which are emitted by unstable nuclei.

Max von Laue (Fig. 3.11X) and coworkers could show 1914 with diffraction experiments on crystals that X-rays as well as γ-radiation are both electromagnetic radiation, where γ-radiation has a much shorter wavelength and a higher penetration depth than X-rays. This was confirmed by William Bragg (Fig. 3.11X), who measured together with his son the distances between atomic planes in crystals by X-ray-diffraction, proving that X-rays were electro-magnetic waves .

This radiation emitted by elements found in nature, is called **natural radioactivity** contrary to the **artificial radioactivity** which is induced by nuclear reactions (for example by bombardment of nuclei with protons, neutrons α-particles or by fission).

Fig. 3.11X Max von Laue (1879–1960) Nobel-Price 1914, William Bragg (1862–1942), Nobelprize 1915

We will at first present the decay chains of natural radioactive elements and subsequently discuss the different decay types in more detail. This will give us a deeper insight into the structure of unstable nuclei and the reasons why they are unstable.

3.2.3 Decay Chains

As has been discussed above, an unstable nucleus can change over into a new nucleus by emission of α-, ß-, or γ-radiation. In these processes the following reactions take place:

- α-decay: Emission of a He-nucleus ${}^4_2He$ by the nucleus ${}^A_ZX$ which is transferred into the nucleus ${}^{A-4}_{Z-2}Y$

$$ {}^A_ZX \xrightarrow{\alpha} {}^{A-4}_{Z-2}Y. \tag{3.16}$$

- ß-decay: Emission of an electron by the nucleus

$$ {}^A_ZX \xrightarrow{e^-} {}^{A}_{Z+1}Y, \tag{3.17}$$

  By this process a neutron in the nucleus is converted into a proton. This implies that the nuclear charge number Z is increased to $Z + 1$ but the atomic mass number A remains constant (see Sect. 3.4).
- γ-decay: Emission of a γ-quant with energy $h \cdot \nu$ by an excited nucleus, which undergoes a transition into a lower energy level.

$$ {}^A_ZX^* \xrightarrow{\gamma} {}^{A}_{Z}X. \tag{3.18}$$

The intense investigation of all radioactive substances revealed that all natural radioactive elements can be arranged in *decay chains* (often called *radioactive families*). Each chain starts with the heaviest in nature existing unstable element and it ends with a stable lead or bismuth isotope. In Fig. 3.12a–d the four existing decay channels are shown. The isotopes ${}^{242}Pu$; ${}^{239}U$; ${}^{239}Np$ and ${}^{241}Am$ have such a short lifetime that they are already nearly completely degraded since the time of their generation. The naturally existing long lived isotopes ${}^{238}U$; ${}^{235}U$; ${}^{232}Th$; and ${}^{237}Np$ are the actual progenitors of the different decay chains (in Fig. 3.12 they are marked by red points).

By artificially generated heavy radioactive elements e.g. Plutonium or Americium, the chains can be extended to heavier nuclei.

3.3 Alpha-Decay

After Rutherford and Geiger had shown, that α-particles are identical with helium nuclei which are emitted by radioactive substances, the following experimental results essentially contributed to the explanation of the α-decay:

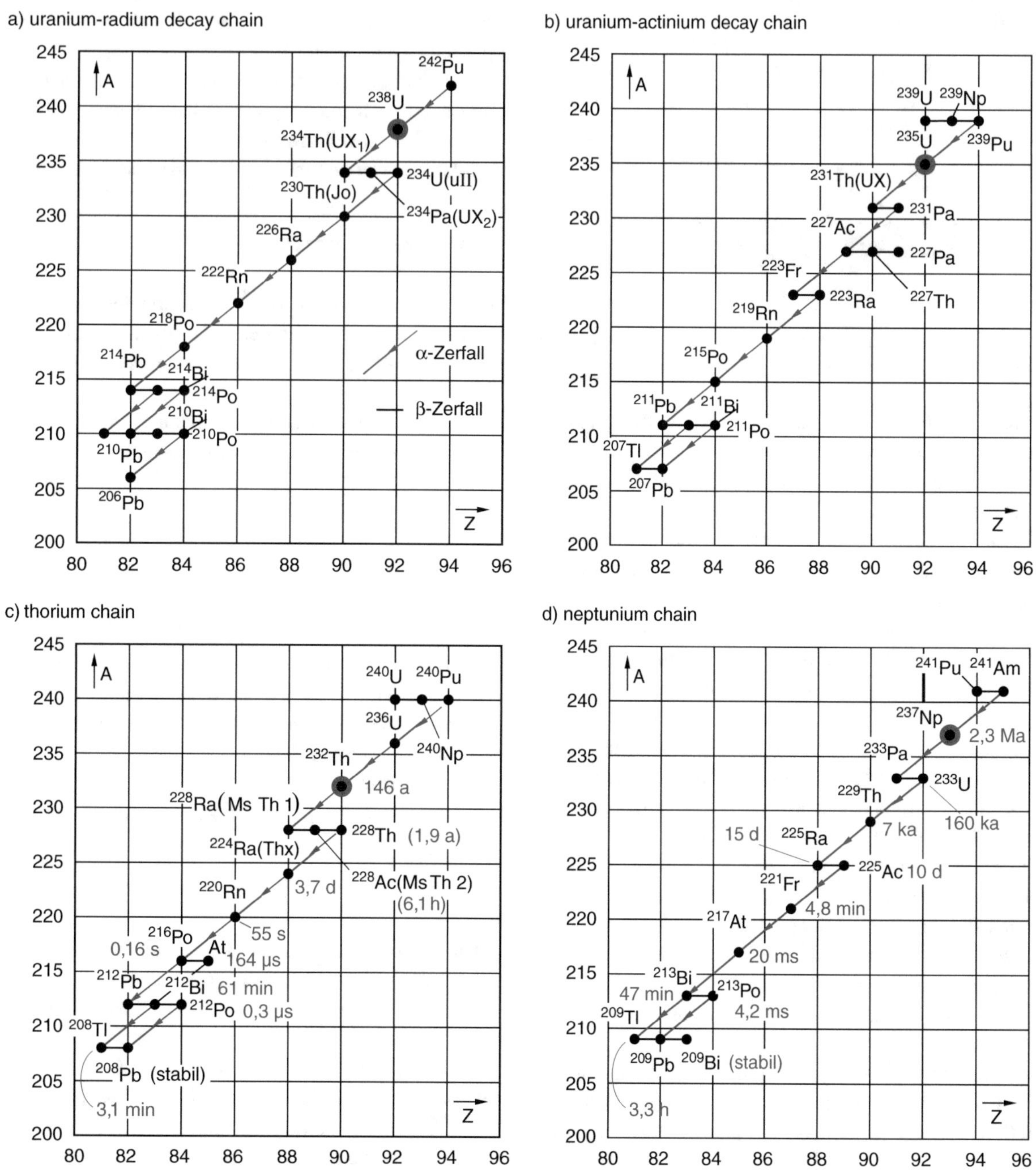

Fig. 3.12 The 4 natural radio-active decay chains, starting from **a)** ^{238}U, **b)** ^{235}U, **c)** ^{232}Th and **d)** ^{237}Np

- Measuring the penetration range of α-particles emitted in air (for instance in a cloud chamber, see Sect. 4.3.4) one finds that in most cases all α-particles emitted by a specific substance have the same penetration range (Fig. 3.13), which means that they have the same kinetic energy. For some radioactive samples there are groups of α-particles with different, but discrete energies. The analysis of the energy with an energy-selective detector (e.g. an ionization chamber, a scintillation counter or a semiconductor detector) reveals a discrete energy spectrum (Fig. 3.14). The kinetic energy of the α-particles

$$E_{\text{kin}} = E_1 - E_2$$

is determined by the energy difference $E_1(M_1) - E_2(M_2)$ between two energy levels of initial and final state of the process

$$E_1(M_1) \rightarrow E_2(M_2) + \alpha$$

Fig. 3.13 Cloud chamber photograph of equally long tracks of α-particles which are emitted by $^{212}_{84}Po$ which is transferred into $^{208}_{82}Pb$. There appears a single longer track which is caused by an α-particle with higher energy emitted by an excited level of the $^{212}_{84}Po$ nucleus. (from W. Finkelnburg: Einführung in die Atomphysik (Springer, Bertlin Heidelberg 1954))

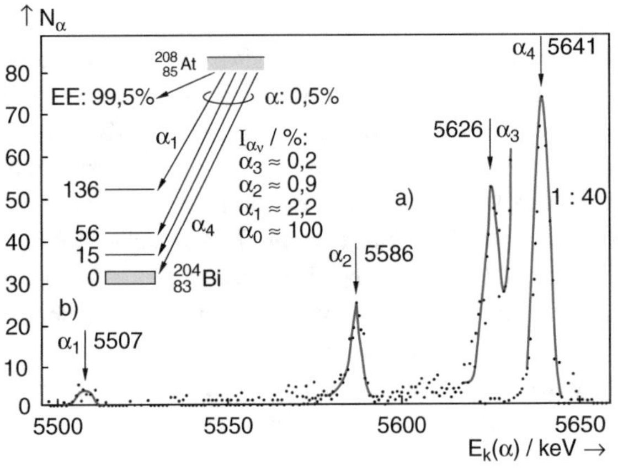

Fig. 3.14 Level scheme and energy spectrum of α-particles emitted by the Astat isotope $^{208}_{85}At$ which decays by electron capture with a probability of 99.5% and by α-emission only with 0.5%. (from G. Musiol,J. Ranft, R. Reif and D. Seliger: Kern-und Elementarteilchen-Physik, Deutscher Verlag der Wissenschaften Berlin 1988)

and depends on the energy levels of initial and final nuclei, which can be the ground state or an excited state. If the emitting state is an excited state, the α-particles have a higher kinetic energy; if the daughter nucleus remains in an excited state, the excitation energy is missing in the kinetic energy of the α-particles.

Fig. 3.14X Hans Geiger (1882–1945)

- The kinematics of the observed α-decays and the spectroscopy of parent and daughter nucleus proved, that all conservation laws, such as energy conservation, momentum and angular momentum conservation are valid under the consumption that only one particle (the α.particle) is emitted contrary to the ß-decay), where the emitting nucleus (energy E_1, momentum p_1 angular momentum ℓ_1) is transferred into the state (E_2,p_2, ℓ_2) of the daughter nucleus.

- Geiger (Fig. 3.14X) and Nuttal found 1911 a relation between the decay constant λ of the initial nucleus and the penetration range R_α of the α-particles

$$\log \lambda = A + B \cdot \log R_\alpha, \qquad (3.19a)$$

where the constants A and B are identical for all members of the same decay chain, although the lifetimes of the different nuclei in this chain can range from 10^{-7} s up to 10^{15} years. The constant B has for all 4 decay chains the

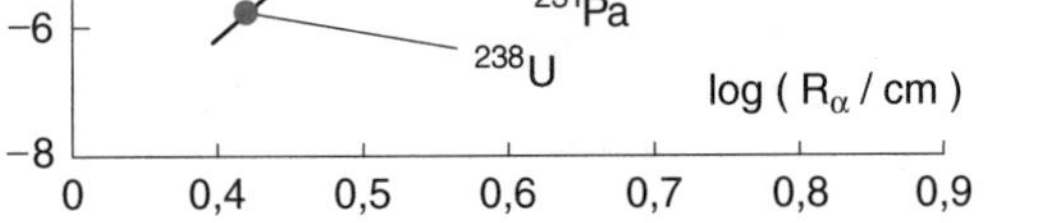

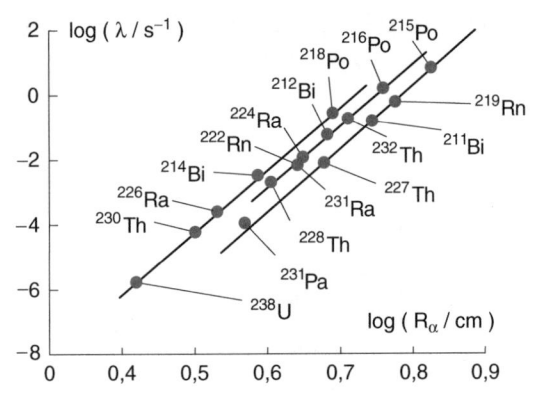

Fig. 3.15 Geiger-Nuttal plot with experimental results (red points) of the range of α-particles from three different decay chains (from Musiol et al. Kern-und Elementarteilchen-Physik, Deutscher Verlag der Wissenschaften, Berlin 1988)

same value, while A varies at most up to 5% (Fig. 3.15). The penetration range R_α is proportional to the power $E_\alpha^{3/2}$ of the kinetic energy of the α-particles, (see Sect. 4.2). Equation (3.19a) can be therefore also written as

$$\log \lambda = C + D \cdot \log E_\alpha \qquad (3.19b)$$

$$\Rightarrow \log E_\alpha = a + b \log \lambda \qquad (3.19c)$$

with $a = -C/D$ and $b = 1/D$.

Fig. 3.16X George Gamow, 1904–1968

The quantitative explanation of the α-decay was given 1928 by Gamow (Fig. 3.16X), Condon and Henry.

A nucleus ${}_Z^A X$ with energy levels occupied with protons and neutrons, according to the Pauli-principle up to the energy level E_{max} forms an α-particle with the probability W_1. Since the binding energy of α-particles is very large the formation of the α-particle releases energy that can be used to excite the α-particle into a higher energy level E_α (Fig. 3.16), which lies, however, still below the potential maximum, which is due to the superposition of the positive Coulomb energy and the negative nuclear bindings energy of the α-particle in the nucleus.

Example The binding energy of the nucleons in the upper occupied energy level is about 5.5–6 MeV. The binding energy of the α-particle (Fig. 2.25) is 28.3 MeV and therefore larger than the sum 22–24-MeV of the binding energies of 2 protons and 2 neutrons. The α-particle in the nucleus has therefore a positive energy of 4.3–6.3 MeV above E = 0.

The spatial range of the nuclear force is according to (2.7) $r = r_0 \cdot A^{1/3}$. The maximum of the potential barrier is then at the distance $r_1 = r_0\left(A_1^{1/3} + A_2^{1/3}\right)$ where $A_1 = 4$ is the mass number of the α-particle and $A_2 = A-4$ that of the daughter nucleus. The potential energy at this distance is

$$E_{pot} = \frac{Z_1 Z_2 e^2}{4\pi\varepsilon_0 r_1} = \frac{Z_1 Z_2 e^2}{4\pi\varepsilon_0 r_0\left(A_1^{1/3} + A_2^{1/3}\right)}. \qquad (3.20)$$

In a classical model the α-particle could leave the nucleus only if its energy exceeds this maximum. The kinetic energy should then be that of the potential maximum minus the recoil energy of the daughter nucleus. Inserting into (3.20) the numerical values of Z and A for an α-emitter, one obtains values of the kinetic energy, which are way above the experimental results.

A solution of this discrepancy is given by the quantum–mechanical explanation, where the α-particles can tunnel through the potential wall (see Vol. 3. Sect. 4.2.3). The particle is described by its de-Broglie wavelength λ_{DB}. The probability W_2 that it tunnels through the potential barrier depends on its wavelength $\lambda_{DB} = p/h$, its kinetic energy E, the heights $H = E_{pot} - E$ and the width $d = r_2 - r_1$ of the barrier at the energy E (Fig. 3.17). After tunneling through the barrier the potential energy of the α-particle is converted into kinetic energy which is shared by the α-particle and the daughter nucleus (recoil). Conservation of energy and momentum yields

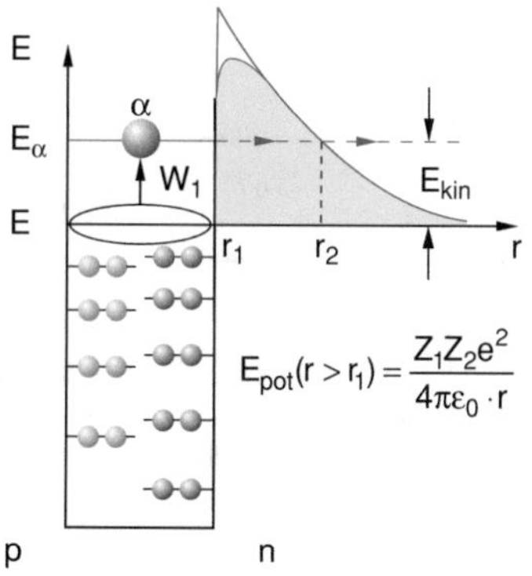

Fig. 3.16 Gamow's explanation of the α-decay

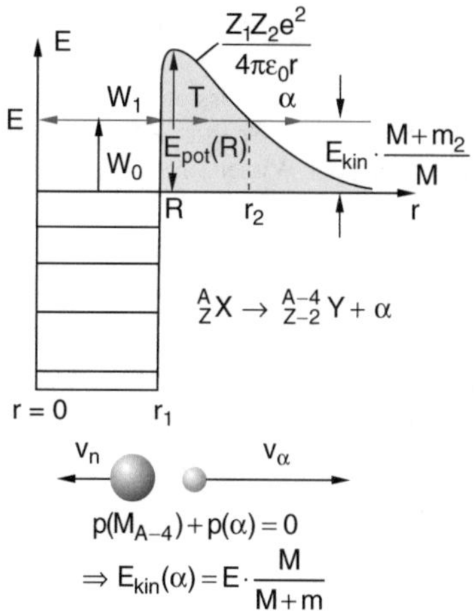

Fig. 3.17 Energy-and momentum balance at the α-decay

$$E_{kin}(\alpha) = \frac{M}{M+m} \cdot E; \quad E_{kin}\left({}^{A-4}_{Z-2}Y\right) = \frac{m}{M+m} \cdot E. \quad (3.21)$$

The total kinetic energy can be expressed by the mass difference

$$E = \left[M\left({}^{A}_{Z}X\right) - M\left({}^{A-4}_{Z-2}Y\right) - m(\alpha)\right] \cdot c^2. \quad (3.22)$$

In Vol. 3, Eq. (4.22b) it was shown, that for $\lambda_{dB} << d$ the probability for tunneling becomes very small. In this case the approximation

$$T = T_0 \cdot e^{-2/\hbar \int_{r_1}^{r_2} \sqrt{2m(E_{pot}(r)-E)}dr} = T_0 \cdot e^{-G}. \quad (3.23)$$

is valid. The exponent

$$G = \frac{2 \cdot \sqrt{2m}}{\hbar} \int_{r_1}^{r_2} \sqrt{\left(\frac{Z_1 Z_2 e^2}{4\pi\varepsilon_0 r} - E\right)} dr \quad (3.24)$$

is called **Gamow factor**. The integral can be solved analytically and gives

$$G = \frac{2}{\hbar}\sqrt{\frac{2m}{E}\frac{Z_1 Z_2 e^2}{4\pi\varepsilon_0}} \cdot \left[\arccos\left(\sqrt{\frac{r_1}{r_2}}\right) - \sqrt{\frac{r_1}{r_2}\left(1 - \frac{r_1}{r_2}\right)}\right] \quad (3.25)$$

with $r_1/r_2 = E/E^{max}_{pot}(r_1)$. If $Z \cdot e$ is the nuclear charge of the parent nucleus, that of the daughter nucleus is ($Z_1 = Z - 2$) and that of the α-particle is ($Z_2 = 2$). The Gamow factor then becomes

$$G \propto (Z-2)/\sqrt{E}. \quad (3.25a)$$

The probability W for the emission of an α-particle per sec, which determines the lifetime $\tau = 1/W$ of the parent nucleus, is given as the product of the three factors.

$$W = W_0 \cdot W_1 \cdot T \quad (3.26)$$

where W_0 gives the probability that an α-particle with the energy E is formed in the nucleus, W_1 is the rate with which the α-particles bumps against the inner Wall of the potential barrier, and T gives the probability that it tunnels per bump through the barrier.

In Table 3.2 some examples are compiled for the experimentally determined half-lifetimes $t_{1/2} = \tau/\ln 2$ of some radioactive elements, the emission energies of the α-particles and the transmission coefficients T. With $\tau \propto 1/T \propto e^{-G}$ and $G \propto 1/\sqrt{E}$ it follows from (3.23)

$$\ln t_{1/2} \propto -G \propto -E_\alpha^{-1/2}. \quad (3.27)$$

Table 3.2 Characteristic data (energy E_α half-lifetime $\tau_{1/2}$, tunnel-probability T_α) of some α-emitter

Isotop	E_α/MeV	$t_{1/2}$	T_α
${}^{212}_{84}Po$	8,78	0,3 μs	$1,3 \cdot 10^{-13}$
${}^{224}_{88}Ra$	5,7	3,64 d	$5,9 \cdot 10^{-26}$
${}^{228}_{90}Th$	5,42	1,91a	$\sim 3 \cdot 10^{-28}$
${}^{238}_{94}Pu$	5,5	$8,8 \cdot 10^1$ a	$\sim 10^{-29}$
${}^{230}_{90}Th$	4,68	$7,5 \cdot 10^4$ a	$\sim 10^{-32}$
${}^{235}_{92}U$	4,6	$7,1 \cdot 10^8$ a	$\sim 10^{-36}$

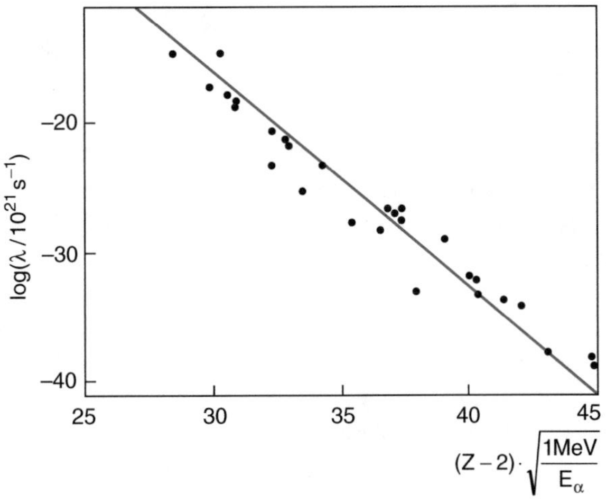

Fig. 3.18 Experimental verification of the Gamow model for the α-decay (from C.I. Gallagher, J O. Rasmussen: J. Inorg.Nucl, Chemistry 3, 333 (1957)

Plotting In λ against $1/\sqrt{E}$ (Fig. 3.18) all α-emitting nuclei lie on the straight line $\log\lambda = (Z-2) \cdot \sqrt{E_\alpha}$ because it follows from (3.25a) that the Gamow factor is $G \propto (Z-2)/\sqrt{E}$.

In order to match the calculated decay probabilities with the measured values the potential box in Fig. 3.16 must be altered into more realistic potentials (as for instance that shown in Fig. 5.35).

For the relation between half-lifetime $t_{1/2}$ and the energy E_α the Eqs. (3.19b) and (3.27) give nearly identical results.

3.4 Beta-Decay

Investigations of the ß-decay have played a very important role in the development of our understanding of possible interactions in nature. It represents not only a prominent example for the weak interaction, but has also given new insight into symmetry principles (parity violation) and has essentially contributed to our knowledge about a new particle, the neutrino, which was postulated by Wolfgang Pauli (Fig. 1.9), already 26 years before it was actually discovered experimentally.

3.4.1 Experimental Results

Many radioactive substances emit, besides α-radiation also electrons e^- or positrons e^+. Examples are:

$$^{225}_{88}\text{Ra} \xrightarrow{\beta^-} {}^{225}_{89}\text{Ac} \quad {}^{208}_{81}\text{Tl} \xrightarrow{\beta^-} {}^{208}_{82}\text{Pb} \quad {}^{15}_{8}\text{O} \xrightarrow{\beta^+} {}^{15}_{7}\text{N}.$$

Altogether there are about twice as many ß–emitter than α-emitter. Measurements of the energy of the electrons or positrons always show a continuous energy distribution dN/dt(E) (Fig. 3.19), contrary to the α-emission, where sharp lines are observed.

The maximum energy depends on the ß-active nucleus. It ranges from some keV up to several MeV. The explanation of the experimental results meets with the following difficulties:

- Besides the ß-particles no other emitted particles were observed. This leads to the assumption, that the ß-decay should be a two-body decay like the α-decay. However, this results in the following discrepancies:
  For a two body decay (e.g. the α-decay) of a nucleus at rest the momentum of the emitted particle is $p_1 = -p_R$ where p_R is the recoil momentum of the daughter nucleus. Cloud chamber pictures of ß-emitting nuclei at rest show, however, that in some cases the momenta of ß and of the daughter nucleus point into the same half space (Fig. 3.20), i.e their momentum sum is not zero.
- With $|p_1| = |p_R|$ the kinetic energies of the emitted ß-particle and the recoil nucleus are

$$\begin{aligned} E_{\text{kin}_1} &= \frac{p_1^2}{2m_1}, \\ E_{\text{kin}_2} &= \frac{p_R^2}{2m_2} = \frac{p_1^2}{2m_2} \Rightarrow \frac{E_{\text{kin}_1}}{E_{\text{kin}_2}} = \frac{m_2}{m_1}. \end{aligned} \tag{3.28}$$

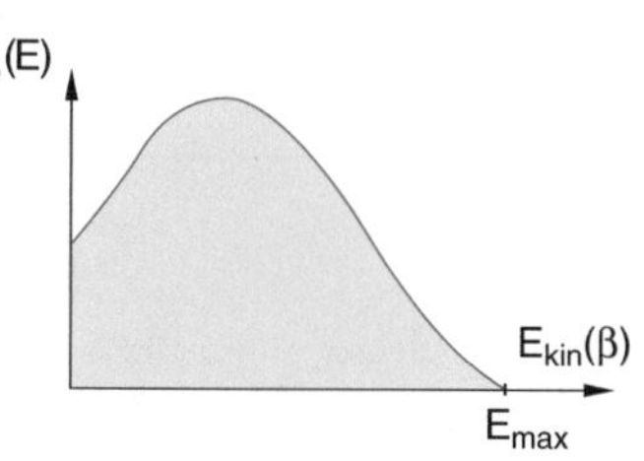

Fig. 3.19 Continuous energy distribution of the electrons emitted at the ß-decay

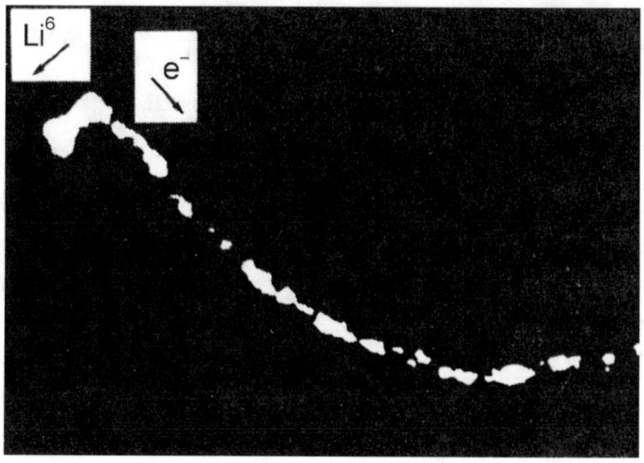

Fig. 3.20 Cloud Chamber photograph of the $ß^-$decay of a helium isotope at rest in a magnetic field for the reaction $^6_2\text{He} \rightarrow {}^6_3\text{Li} + e + \bar{\nu}_e$. The $^6_3$Li-nuclei as well as the electron are pushed into the same half space. Without the neutrino the momentum conservation would be violated. (from Simonyi: Kulturgeschichte der Physik Harry Deutsch Frankfurt 1995)

Energy conservation demands

$$E_{\text{kin}_1} + E_{\text{kin}_2} = (M - m_1 - m_2)c^2 = E_0, \tag{3.29}$$

For a two-body decay this determines the energy of the emitted ß-particle to the fixed value

$$E_{\text{kin}_1} = \frac{m_2}{m_1 + m_2} E_0. \tag{3.30}$$

The observed continuous energy distribution **is not compatible** with the sharp energy of a two-body decay, unless one is willing to sacrifice the validity of energy and momentum conservation.

- The validity of energy conservation was experimentally proved by the measurements of double-decay paths where the initial nucleus decays on two different ways into the same final nucleus. One example is the reaction

$$\begin{array}{ccc} ^{212}_{83}\text{Bi} & \xrightarrow{\beta} & ^{212}_{84}\text{Po} \\ \downarrow\alpha & & \downarrow\alpha \\ ^{208}_{81}\text{Tl} & \xrightarrow{\beta} & ^{208}_{82}\text{Pb} \end{array}$$

- For a two-body ß-decay of a nucleus with odd nucleon number, which has a half integer spin, an u-g-nucleus is transferred into a g-u nucleus (because a proton is transferred into a neutron). Since the emitted ß-particle has the spin ½ħ the final nucleus should have an integer spin. However, all g-u nuclei, observed so far have a half-integer spin. This means, if the ß-decay were a two-body decay also the angular momentum conservation would be violated.

3.4.2 Neutrino-Hypothesis

in order to explain the experimental facts without violating the proven conservation laws for momentum, energy and angular momentum one has to assume that besides the electron another up to now invisible particle is emitted which carries the missing energy E_{max} – E(ß), and the missing momentum and angular momentum.

Pauli postulated therefore 1930 in a letter to his "*radioctive collegues*" at the physics conference in Tübingen, that in the ß-decay an up to now not observed neutral particle with spin ½$\hbar$ is emitted which he called tentatively "neutron".

After Chadwick discovered the neutron as a neutral component of the nucleus with a mass nearly equal to that of the proton ($m_n \approx m_p$) it soon became obvious, that the postulated third particle in the ´ß-decay cannot be the neutron but must have a much smaller mass which must be even smaller than that of the electron, otherwise the maximum energy E(ß) = $E_{max} \approx$ observed in the ß-spectrum could not be the same as expected from the mass difference between (parent nucleus + electron mass) and the mass of the daughter nucleus. Therefore the hypothetical particle was named "**neutrino** = small neutron".

Because of symmetry arguments there should be also, as for all elementary particles the corresponding anti-particle the anti-neutrino $\overline{\nu}$ (see Sect. 7.3).

The experimental search for neutrinos ν was unsuccessful for many years, until 25 years after Pauli had postulated this curious particle *E. Reines* (Nobel-Price 1995) and *C. L. Cowan* (Fig. 3.21X) succeeded 1956 in using the strong flux of neutrinos from the ß$^-$-decay of the fission products in nuclear reactors [3]. For the detection of the neutrinos the reaction

$$\overline{\nu}+p\rightarrow n+e^{+} \qquad (3.31a)$$

where an anti-neutrino is captured by a proton in a H_2O + $CdCl_2$-Solution, which converts into a neutron that emits a positron e^+.

The positron collides with an electron in the atomic shell and is annihilated with the emission of two γ-quanta h · ν.

$$\mathrm{e}^{+}+\mathrm{e}^{-}\rightarrow\gamma+\gamma \quad \left(h\nu_{\gamma}=0,5\mathrm{MeV}\right). \qquad (3.31b)$$

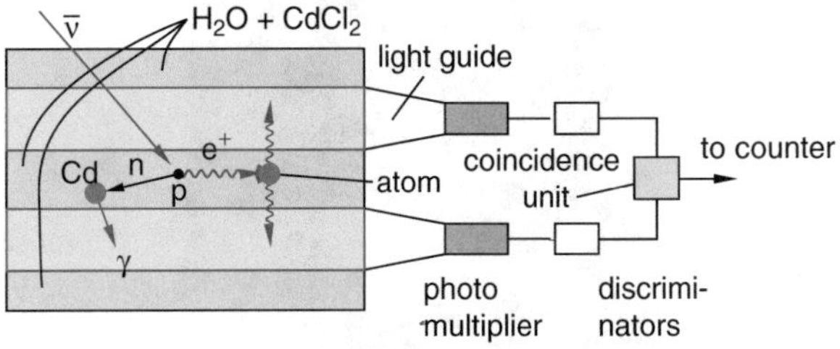

Fig. 3.21 Schematic representation of the experimental arrangement for the neutrino detection by Reines and Cowan. The device has rotational symmetry around the horizontal symmetry axis

Fig. 3.21X Frederic Reines (Nobelprize 1995), Clyde Gowan (1920–1974)

The experimental design is shown in Fig. 3.21. In two water tanks (200 ℓ) are about 2·10^{28} protons. The anti-neutrino- flux from the reactor passes the water tanks and produces neutrons and positrons according to Eq. (3.31a). With scintillation detectors (see Sect. 4.3.2) arranged around the water tanks the annihilation radiation is monitored. The two γ-quants are emitted into opposite directions and can be detected by a coincidence detector. The neutrons are decelerated in the water. They can be detected by adding CdCl to the water which captures the neutrons and induces the reaction

$$^{113}\mathrm{Cd}+\mathrm{n}\rightarrow{}^{114}\mathrm{Cd}^{*}\rightarrow{}^{114}\mathrm{Cd}+\gamma \qquad (3.31c)$$

This neutron capture is temporarly delayed against the annihilation reaction $e^+ + e^- \rightarrow 2\gamma$. The γ-quantum of reaction (3.31c) with E = 9.1 MeV is detected with delayed coincidence against the 2γ-signal with E(2γ) = 1.02 MeV. The two signals can be separately measured because of the different energies and their time-delay.

3.4.3 Model of Beta-Decay

According to the neutrino model the ß$^-$-decay can be understood as the reaction

$$^{A}_{Z}\mathrm{X}\overset{\beta^{-}}{\rightarrow}{}^{A}_{Z+1}\mathrm{Y}+\mathrm{e}^{-}+\nu \qquad (3.32a)$$

and the ß$^+$-decay (Positron emission) as

$$^{A}_{Z}\mathrm{X}\overset{\beta^{+}}{\rightarrow}{}^{A}_{Z-1}\mathrm{Y}+\mathrm{e}^{+}+\nu. \qquad (3.32b)$$

Remark

As will be shown in Chap. 7, for all reactions between elementary particles the baryon number B (number of heavy particles, such as p or n) as well as the number of leptons

L (light particles such as e^-, e^+ ν, $\overline{\nu}$, μ) must be preserved. A lepton gets the lepton number $L = 1$ and its anti-particle $L = -1$. On both sides of Eq. (3.32a) or (3.32b) the baryon number is B = A while the lepton number is $L = 0$.

In Sect. 2.4 we have discussed, that, because of the uncertainty relation, electrons cannot stay in the nucleus. According to the model of ß$^-$-decay the electron is just generated by the conversion of a neutron into a proton

$$n \rightarrow p + e + \bar{\nu}. \tag{3.33}$$

It leaves together with the anti-neutrino the nucleus immediately after its creation (Fig. 3.22).

A free neutron has a slightly larger mass then the free proton and decays spontaneously according to (3.33) after a mean lifetime of $\tau = 887$ s while a free proton is stable ($\tau > 10^{32}$ years).

Inside the nucleus the reaction (3.33) is only possible, if the energy of the parent nucleus $^A_Z X$ is higher than that of the daughter nucleus $^A_{Z+1}Y$. This is always the case if the neutron at the Fermi-level can occupy after conversion into a proton a free place in the proton levels allowed by the Pauli-principle (Fig. 3.22). The energy balance is

$$\begin{aligned} \Delta E &= \left[M\left(^A_Z X\right) - M\left(^A_{Z+1}Y\right)\right] \cdot c^2 \\ &> (m_e + m_{\bar{\nu}}) \cdot c^2 \\ E^{max}_{kin} &= E_{kin}(e) + E_{kin}(\bar{\nu}) \\ &= \Delta E - (m_e + m_{\bar{\nu}}) \cdot c^2. \end{aligned} \tag{3.34}$$

In spite of its smaller mass also a proton inside the nucleus can be converted into a neutron

$$p \rightarrow n + e^+ + \nu \tag{3.35}$$

emitting a positron and a neutrino. This process becomes possible if the neutron to be formed can occupy a lower

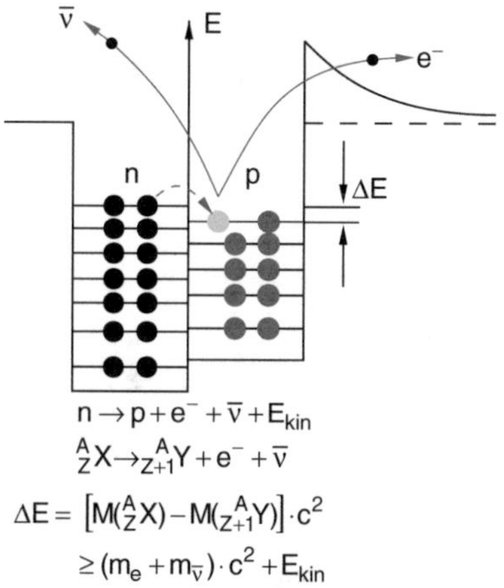

Fig. 3.22 Beta-decay of an unstable nucleus by conversion of a neutron into a proton

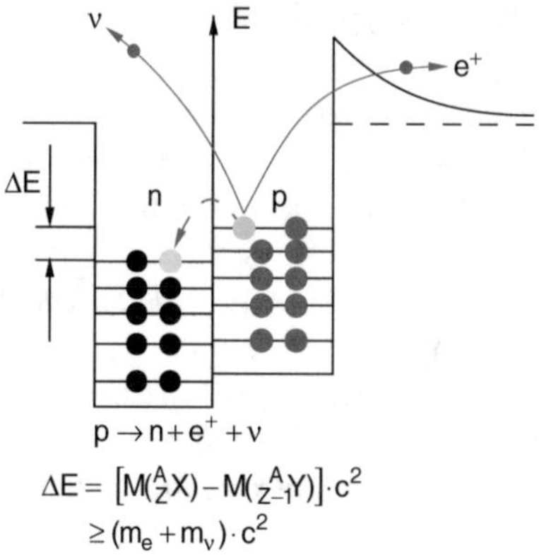

Fig. 3.23 Nuclear model of positron emission at the conversion of a proton in a neutron in ß$^+$-emitting unstable nuclei

energy level as the original proton (Fig. 3.23). By this process the binding energy of the daughter nucleus becomes larger than that of the original nucleus. The excess energy is converted into kinetic energy of positron and neutrino.

$$\begin{aligned} \Delta E &= \left[M\left(^A_Z X\right) - M\left(^A_{Z-1}Y\right) - m_e - m_\nu\right] \cdot c^2 \\ &= E_{kin}(e^+) + E_{kin}(\nu) = E_{max}(e^+) \end{aligned} \tag{3.36}$$

Such unstable nuclei with positron emission are located in Fig. 3.4 on the left side of the stable nuclei while the ß$^-$ emitters lie on the right side.

The detailed form of the continuous energy distribution observed for the emitted ß-particles can be explained by a model developed by Fermi (Fig. 3.24). It is based on the following considerations [4].

The probability $W(E)$, that a nucleus converts from its initial state $|i\rangle$ to its final state $|f\rangle$ and emits an electron with the kinetic energy E_e and a neutrino with energy E_ν $E_{max} = E_0 = E_e + E_\nu$ can be written as product of three factors:

Fig. 3.24 Enrico Fermi (1901–1954) (Wikipedia)

$$W(E) \cdot \mathrm{d}E = g_e \cdot g_\nu \cdot W_{if} \mathrm{d}E. \tag{3.37}$$

The factors $g_e(E_e)$ and $g_\nu(E_\nu)$ are the statistical weights for electron and neutrino, i.e. the number of possible momentum states $\boldsymbol{p}_e$ and $\boldsymbol{p}_\nu$ which can be realized for statistically distributed directions of the particle emission at fixed energies E_e and E_ν. The factor W_{if} gives the probability for the nuclear transformation ${}^A_Z X \to {}^A_{Z+1} Y$ which approximately does not depend on the energies E_e and E_ν for a given total energy E_0.

All momentum vectors $\boldsymbol{p}$ of a free particle which lie in momentum space within the spherical shell $4\pi p^2 \mathrm{d}p$ lead to the same energy $E = p^2/2\,m$ within the interval from E to $E + \mathrm{d}E$. Since the momenta of electron and neutrino are statistically independent (the recoil of the daughter nucleus can always compensate the two momenta to make the total momentum $\mathbf{p} = \mathbf{p}_e + \mathbf{p}_\nu + \mathbf{p}_r = \mathbf{0}$) the statistical weights of the two emitted particles are

$$g_e = 4\pi p_e^2\, \mathrm{d}p_e \quad \text{and} \quad g_\nu = 4\pi p_\nu^2\, \mathrm{d}p_\nu. \tag{3.38}$$

Since the rest mass of the neutrino is very small ($m_\nu < < m_e$) it is

$$E_\nu = c \cdot p_\nu.$$

With $E_\nu = E_0 - E_e$ we get

$$p_\nu^2 = \frac{1}{c^2}(E_0 - E_e)^2. \tag{3.39}$$

With (3.37) and (3.38) we finally obtain for the probability that the electron in the ß-decay has the kinetic energy $E_e\, ss\, p_e = \sqrt{(2\,\mathrm{m} \cdot \mathrm{E})}$

$$\begin{aligned} &W(p_e)\mathrm{d}p_e \\ &= a_1 \cdot W_{if} \cdot p_e^2 \cdot (E_0 - E_e)^2\, \mathrm{d}p_e, \end{aligned} \tag{3.40a}$$

where a_1 conflates the constant prefactors.

Taking into account the relativistic relation (see Vol. 3. Sect. 3.1)

$$E = \sqrt{p_e^2 c^2 + (m_e c^2)^2}$$

between total energy $E = E_e + m_e c^2$ and momentum p_e of the electron we can write (3.40a) for the energy interval $\mathrm{d}E = c^2 p_e \mathrm{d}p_e / E$ as

$$\begin{aligned} &W(E)\mathrm{d}E \\ &= a_2 \cdot W_{if} \cdot E \cdot \sqrt{E^2 - m_e^2 c^4}. \end{aligned} \tag{3.40b}$$

Replacing E by $E_e + m_e c^2$ we finally arrive at the energy distribution

$$\begin{aligned} N(E_e)\mathrm{d}E_e &= a_3 W_{if}\left(E_e + m_e c^2\right) \\ &\cdot \sqrt{E_e^2 + 2 m_e c^2 E_e}(E_0 - E_e)^2\, \mathrm{d}E_e \end{aligned} \tag{3.40c}$$

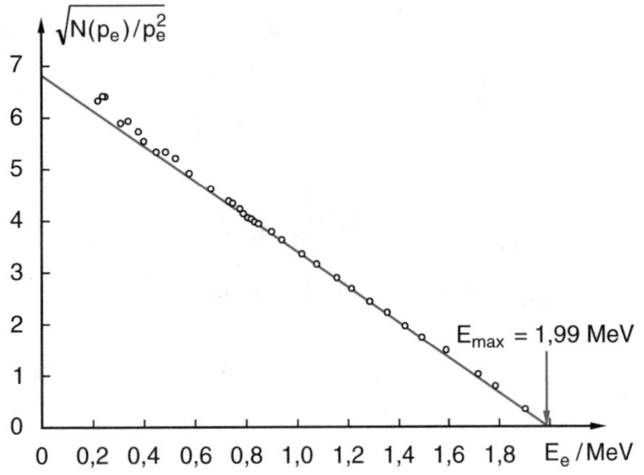

Fig. 3.25 Fermi-Kurie Diagram of the ß-decay of ${}^{114}_{49}$In. (From J.L Lawson, J-M.Cork: Phys. Rev. 57, 982 1940)

of the electrons as a function of their kinetic energy where the factors a_i are constants.

The number of detected electrons within the energy interval $\mathrm{d}E_e$ is proportional to the probability $W(E_e)\mathrm{d}E_e$.

For a more detailed consideration we have to take into account that the nucleus has a finite size and that the emitted electron on its way from the nucleus through the electron cloud of the atom to the detector suffers a variable attraction by the nucleus and a repulsion by the electron cloud. This can be taken into consideration by a correction factor $F(E, Z)$ called **Fermi-factor**.

Plotting $[N(E_e)/F]^{1/2}$ against the energy E_e of the electrons (Fig. 3.25) one obtains a straight line with a slope that gives the probability W_{if} for the ß-decay for the nuclear transition ${}^A_Z X \to {}^A_{Z+1} Y$ (**Fermi-Kurie Plot**, named after *E. Fermi* and *F.N.D. Kurie* [4]

The ß-decay becomes possible due to the weak interaction. Its study therefore gives information about the strength and the range of the weak interaction (see Sect. 7.6).

3.4.4 Experimental Methods for the Investigation of the ß-Decay

The energy distribution of the electrons, emitted at the ß-decay can be measured with a magnetic spectrometer (Fig. 3.26). Part of the electrons, emitted by the ß-source pass through the entrance slit of the spectrometer and enter into the homogeneous 180° magnetic field. The transmitted electrons are focused onto the detector (see Vol. 3. Sect. 2.7.4 and Vol. 2, Sect. 3.3.2). The detector output signal gives the number $N(B)$ of the transmitted electrons as a function of the magnetic field strength B. The electrons traverse the magnetic field on circles with radius R.

The condition: Centripetal force = Lorentz-force gives the equation

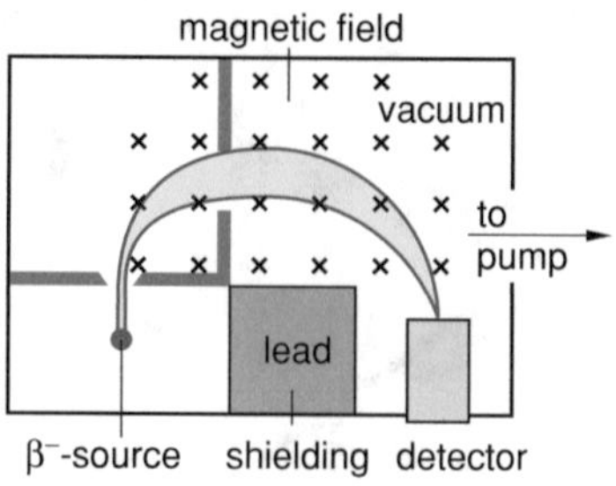

Fig. 3.26 Measurement of the ß-spectrum with a magnetic spectrometer

$$\frac{m \cdot v^2}{R} = e \cdot v \cdot B \Rightarrow m \cdot v = R \cdot e \cdot B.$$

We then get for the kinetic energy of the electrons

$$E_{\text{kin}} = \frac{m}{2}v^2 = \frac{1}{2m}R^2e^2B^2. \tag{3.41}$$

In order to shield the detector against γ-quanta and to prevent electrons to reach the detector on a direct path from source to detector without passing through the magnetic field, the source is shielded by lead walls.

Another method for measuring the energy distribution uses energy selective detectors, such as scintillation counters or semiconductor detectors (see Chap. 4).

The exact form of the Fermi-Curie plot $N(E)$ close to the maximum energy allows the determination of an upper limit for the neutrino mass. This is illustrated by the example of the ß-decay of tritium in Fig. 3.27

$${}^3_1\mathrm{T} \rightarrow {}^3_2\mathrm{He} + \mathrm{e} + \bar{\nu}$$

For $m_\nu = 0$ the plot would be a straight line up to the maximum energy. Assuming a rest mass of 250 eV/c^2 the curve would noticeably deviate from a straight line close to the maximum energy.

A new experiment in Karlsruhe Germany [7] has increased the accuracy to limit the neutrino mass to $m_\nu <$

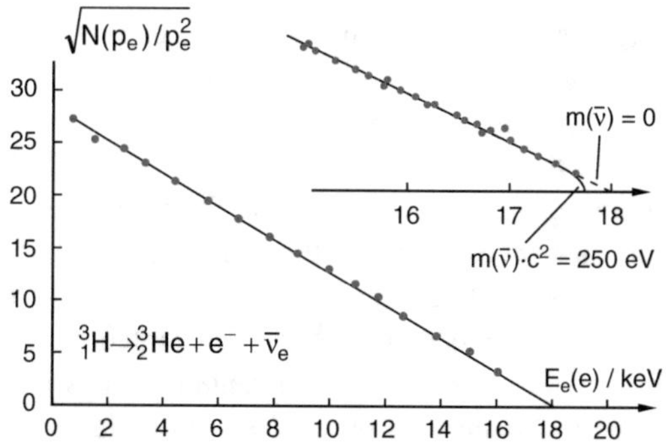

Fig. 3.27 Fermi-Kurie diagram for the ß⁻ decay of tritium ${}^3_1$H [7]

Fig. 3.28 Transport of the huge mass spectrometer for presice measurements of the upper limit of the neutrino mass on its way to Karlsruhe (KATRIN experiment KIT)

0.2 eV/c². The transport of the huge magnetic mass spectrometer has met some difficulties (Fig. 3.28).

In Heidelberg the double ß-decay of the isotope ^{76}Ge was observed, where two electrons are emitted but no neutrino [8]. One would expect that also two neutrinos $\nu + \bar{\nu}$ should be emitted. However, if these two neutrinos are identical particles, they annihilate and transfer their mass energy to the two electrons. Such experiments are very difficult because they are seldom and their signals have to be filtered out of a large background signal. Because of the great importance for fundamental physics these experiments are now repeated in the underground lab in the *Grand Sasso* in the Italian Apennine mountains with an improved equipment and a higher accuracy because the underground can be much better suppressed by several hundred meters rock above the lab [9].

3.4.5 Electron Capture

The probability density $|\psi(r)|^2$ to find an atomic electron in the 1s-shell at the distance r from the nucleus has a maximum for $r = 0$ (see Vol. 3 Sect, 5.1). During its stay in the nucleus the electron can be "captured" by a proton (K-capture, which is then transformed into a neutron:

$$\mathrm{e} + \mathrm{p} \rightarrow \mathrm{n} + \nu_\mathrm{e}. \tag{3.42}$$

This creates a hole in the K-shell (energy E_1), which can be filled by the transition of an electron from higher shells with energy E_2. This process causes the emission of radiation with the frequency $\nu = (E_2 - E_1)/h$. For the atomic nucleus this electron capture implies the reaction

$${}^A_Z\mathrm{X} + \mathrm{e}^- \rightarrow {}^A_{Z-1}\mathrm{Y} + \nu_\mathrm{e}. \tag{3.43}$$

Electron capture is possible if the energy balance is positive, i.e. when the mass $\left(M({}^{A}_{Z}X) + Z\, m_e\right)$ of the neutral atom $\left({}^{A}_{Z}X + Z \cdot e\right)$ is larger than the mass of the new atom $\left({}^{A}_{Z-1}Y + (Z-1)e\right)$ + emitted photon $h \cdot \nu/c^2$:

$$\Delta E = \Delta M \cdot c^2 > E_B(1s) \geq h \cdot \nu_K, \quad (3.44)$$

This means that the energy difference $\Delta E = \Delta M \cdot c^2$ between parent and daughter atom must be larger than the binding energy of the electron in the 1s- state. This energy difference is converted into kinetic energy of daughter nucleus and neutrino.

$$\Delta E = E_{kin}\left({}^{A}_{Z-1}Y\right) + E_{kin}(\nu), \quad (3.45)$$

Because of the small neutrino mass $(m_\nu \ll m_e)$ we conclude (Fig. 3.29)

$$E_{kin}(\nu) \gg E_{kin}\left({}^{A}_{Z-1}Y\right).$$

Note For the K-capture a neutral atom with nuclear charge Z is transformed into a neutral atom with nuclear charge $(Z-1)$ (Fig. 3.29). For the ß-emission of a neutral Atom with nuclear charge number Z a positively charged ion is created with the nuclear charge number $(Z+1)$ but with Z electrons. For the $ß^+$-decay of a neutral atom with the nuclear charge number Z a negatively charged ion with the nuclear charge number $(Z-1)$ and Z electrons is created.

3.4.6 Energy Balance and Decay Types

The energy scale in Fig. 3.30 illustrates, under which conditions the different decay types can occur. We choose as the zero of the energy scale the mass energy $E_0 = M\left({}^{A}_{Z+1}X^+\right) \cdot c^2$ of the positively charged ion with Z electrons and the nuclear charge number $Z+1$. There are three possibilities for the mass $M\left({}^{A}_{Z}Y\right)$ of the created neutral atom:

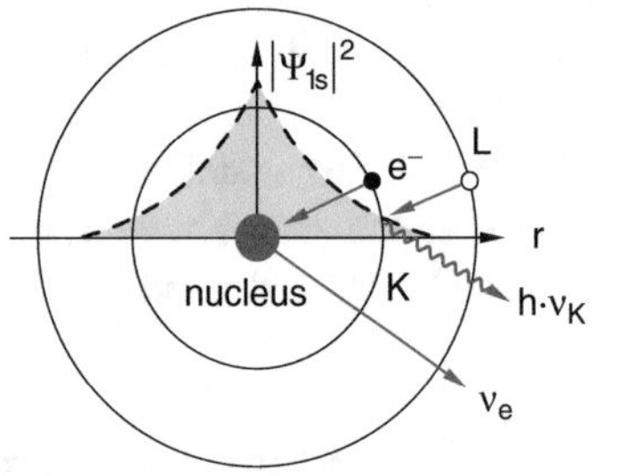

Fig. 3.29 K-Capture

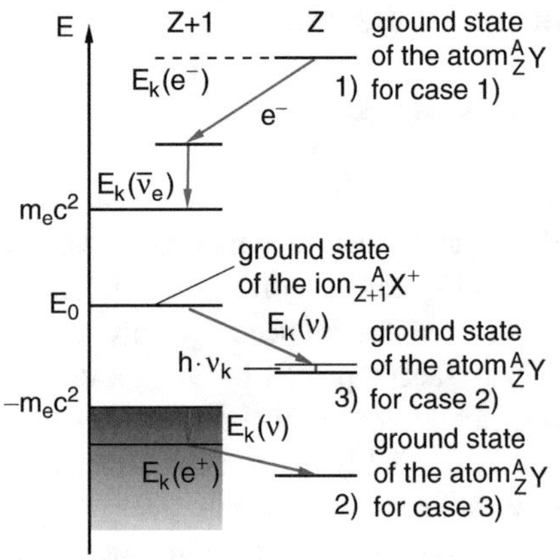

Fig. 3.30 Energy relations and level scheme for the $ß^-$, the $ß^+$-decay and the *K*-capture

1. $\Delta M = M\left({}^{A}_{Z}Y\right) - M\left({}^{A}_{Z+1}X^+\right) > m_e$

   where m_e is the rest mass of the electron. In this case the neutral atom ${}^{A}_{Z}Y$ can decay by $ß^-$-decay into the positively charged ion ${}^{A}_{Z+1}X^+$. The energy balance is

$$\left[M\left({}^{A}_{Z}Y\right) - M\left({}^{A}_{Z+1}X^+\right)\right] \cdot c^2 = m_e c^2 + E_{kin}(e) + E_{kin}(\bar{\nu}_e). \quad (3.46a)$$

2. $\Delta M = M\left({}^{A}_{Z}Y\right) - M\left({}^{A}_{Z+1}X^+\right) < 0; |\Delta M| > m_e$

   Here the positron decay of the ion ${}^{A}_{Z+1}X^+$ is energetically possible and a neutral atom ${}^{A}_{Z}Y$ is formed.

$${}^{A}_{Z+1}X^+ \xrightarrow{\beta^+} {}^{A}_{Z}Y + e^+ + \nu_e.$$

The total kinetic energy of the emitted particles is

$$E_{kin}(e^+) + E_{kin}(\nu_e) = \left(M\left({}^{A}_{Z+1}X^+\right) - M\left({}^{A}_{Z}Y\right) - m_e\right)c^2. \quad (3.46b)$$

3. $\Delta M < 0$ and $|\Delta M| < m_e$

   In this case the positron decay is no longer possible because the energy difference is not sufficient for producing the positron. Now the ion ${}^{A}_{Z+1}X^+$ with Z electrons can pass by K-capture into the positive ion ${}^{A}_{Z}Y$ with $Z-1$ electrons. The energy balance is now:

$$\left[M\left({}^{A}_{Z+1}Y^+\right) - M\left({}^{A}_{Z}X\right)\right] \cdot c^2 = E_{kin}(\nu_e) + h\nu_K, \quad (3.47)$$

where $h \cdot \nu_K$ is the energy of the X-ray quantum which is emitted when the hole in the *K*-shell is refilled by electrons from higher energy states.

3.5 Gamma Radiation

3.5.1 Observations

Among the naturally existing radioactive substances one finds besides α- and ß-emitters also nuclei which emit high energetic photons called γ-radiation. The spontaneous γ-radiation, that is not induced by any external energy supply, is always accompanied by α- or ß-emission.

Measuring the energy differences between the different α-components (Fig. 3.14) one finds a correlation between these energy differences and the energy $h \cdot \nu_k$ of the γ-quanta (Fig. 3.31).When collecting many experimental results it became clear that γ-radiation is produced when a nucleus passes from a higher energy state E_k into a lower state E_i. The energy $h \cdot \nu_k = E_k - E_i$ is equal to the energy difference of the two nuclear states. This process of emitting γ-radiation is completely similar to the emission of photons from the atomic electron shell when the atomic electron passes from an excited state to a lower state (see Vol. 3 Sect. 7), the only difference is the energy of the emitted photons. While the photon energy for transitions in the atomic shell ranges from 1–10 eV that of the γ-quanta ($E = 10^4 - 10^7$ eV) is larger by several orders of magnitude. As example Fig. 3.32 shows the γ-spectrum of the nucleus ${}^{22}_{10}Ne^*$ which is generated by $ß^+$ -decay of ${}^{22}_{11}Na$ into an excited as well as into the ground state of Ne.

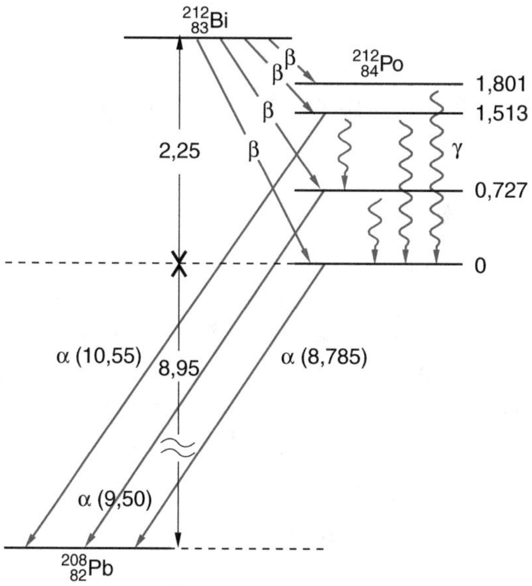

Fig. 3.31 Gamma radiation of the ${}^{212}_{84}Po$ isotopes. The excited states of the nucleus are populated by $ß^-$-decay of the ${}^{212}_{83}Bi$ isotope and they decay by γ-radiation and by α-emission. The number in brackets give the energy in MeV. Note that the energy of the α-particles is slightly lower than that of the γ-radiation on the same transition, because of the recoil

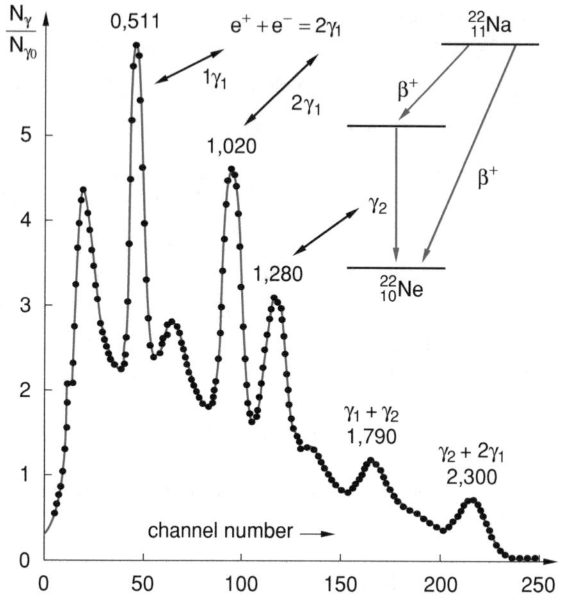

Fig. 3.32 γ-spectrum of the ${}^{22}_{10}Ne$ isotope which is generated by $ß^+$-decay of ${}^{22}_{11}Na$. The measured maxima correspond to the energies $h \cdot \nu_2$ of the γ-quanta emitted by the excited Ne*, the annihilation radiation $e^- + e^+ \to \gamma_1$ at the positronium capture and the combinations $2\gamma_1$, $\gamma_1 + \gamma_2$, $2\gamma_1 + \gamma_2$) from G. Musiol et al. Kern-und Elementarteilchen-Physik Deutscher Verlag der Wissenschaften berlin 1988)

Actually one would expect only a single line for the $ß^+$-emission into the ground state of Na. However, the emitted $ß^+$ positrons annihilate when colliding with electrons and the corresponding energy $E(\gamma_1)$ of the emitted γ-radiation is

$$e^+ + e^- \to 2\gamma_1 \quad \text{with} \quad h\nu_1 = E(\gamma_1) = m_e c^2,$$

where the two γ-quanta are emitted into opposite directions. A detector enclosing the sample will measure the energies $h\nu_1$; $2h\nu_1$; $h\nu_2$; $h\nu_1 + h\nu_2$; and $h\nu_2 + 2h\nu_1$.

While the excitation energy in the atomic shell is due to the potential and kinetic energy of a single electron (for doubly excited states also that of two electrons) the excitation energy of atomic nuclei consists of rotational and vibrational energy, where the nucleus has a larger angular momentum or the nucleons in the nuclei vibrate against each other.

These excited nuclear levels are generated either as the result of the α- or ß-decay of unstable radioactive nuclei (Figs. 3.14 and 3.31) (natural γ-radiation) or they are created by bombardment of stable nuclei with energetic particles (neutrons, protons etc.) (artificial γ-radiation).

3.5.2 Multipole Transitions and Transition Probabilities

As has been shown in Vol. 2 Sect. 1.4 the potential of an arbitrary charge distribution can be expressed by a Taylor expansion with the different terms representing the different multipoles. This multipole expansion is constituted by spherical harmonic functions or Legendre polynoms $Y_l^m(\vartheta, \varphi)$. If the charge distribution suffers a periodic change electro-magnetic waves are emitted, which can be regarded as a superposition of multipole modes corresponding to the multipole expansion of the oscillating charge distribution. The same applies to the time varying electric current distributions which give rise to magnetic multipoles.

Transitions between different rotational levels ⟨1| and ⟨2| of nuclei with nuclear spins I_1 and I_2 cause a change of the nuclear angular momentum $\boldsymbol{I}$ by $\Delta \mathbf{I} = \mathbf{I_1} - \mathbf{I_2}$. Since the total angular momentum must be preserved the emitted γ-quant must carry an angular momentum $\boldsymbol{L} = \Delta \boldsymbol{I}$, such that the total system preserves its angular momentum. The spatial orientation of the vector L is quantized with the number of spatial projections L_z obeying the condition

$$|I_a - I_e| \le I_z \le I_a + I_e, \tag{3.48}$$

These modes of the electro-magnetic field of the emitted γ-quanta classified according to the angular quantum number L are the multipole modes with the multipolarity 2^L (with the quantum number $L = 0,1,2,\ldots$). The emitted γ-quant has the angular momentum $\boldsymbol{L}$ (with reference to the nucleus) with the amount $|\boldsymbol{L}| = \sqrt{L(L+1)}/\hbar$.

There are no transitions with $L = 0$ (monopole transitions). Quanta with $L = 1$ constitute the dipole radiation, those with $L = 2$ the quadrupole radiation, with $L = 3$ the octupole radiation etc. The transition probability A_{ik} for a transition between the nuclear levels E_i and E_k is according to a model developed by Victor Weisskopf [6]

$$A_{ik} \propto \left(\frac{R}{\lambda\!\!\!^{-}}\right)^{2L}, \tag{3.49}$$

proportional to the power $2L$ of the ratio of radius R of the charge distribution (in this case of the nucleus) and the reduced wavelength $\lambda\!\!\!^{-} = \lambda/2\pi$ of the emitted radiation.

Examples

1. For the emission of light at transitions between levels of the atomic electron shell is $R = 0.1$ nm, $\lambda\!\!\!^{-} = 100$ nm $\rightarrow R/\lambda\!\!\!^{-} \approx 10^{-3}$. The quadrupole radiation with $L = 2$ is then less probable by the factor 10^{-6} than the dipole radiation with $L = 1$.
2. For the γ-radiation emitted by nuclei with $R = 5 \cdot 10^{-15}$ m and $h \cdot \nu = 1$ MeV ($\lambda\!\!\!^{-} = 1.5 \cdot 10^{-13}$ m) is the probability of quadrupole radiation smaller only by a factor of 10^{-3} than that of dipole radiation.

The lifetime τ_i of an excited nuclear state which can decay only by γ-radiation with the transition probability $A_i = \Sigma A_{ik}$ is given by the relation

$$\frac{1}{\tau_i} = \sum_k A_{ik}$$

where the index k runs over all lower states accessible to the γ-radiation from level ⟨i|. The lifetime can be expressed by [7]

$$\tau_i = \frac{1}{\alpha} \cdot \frac{\hbar}{E_\gamma} \cdot \frac{1}{S(L)} \cdot \left(\frac{\lambda\!\!\!^{-}}{R}\right)^{2L}, \tag{3.50}$$

where $\alpha = 1/137$ is the fine structure constant (see Vol. 3 Sect. 3.5) and $E_\gamma = E_i - E_k$ is the energy of the γ-quant. The factor S(L) depends on the angular momentum L and the statistical weights of the levels ⟨i| and ⟨|k with nuclear spin orientations I_{zi} and I_{zk}).The more lengthy calculation gives

$$S(L) = \left(\frac{3}{L+3}\right)^2 \cdot \frac{2(L+1)}{L \cdot \left[\prod_{n=1}^{L} (2n+1)\right]^2} \tag{3.51}$$

Nuclei with $I = 0$ have, besides the monopole $Z \cdot e$ no higher multipoles because their charge distribution is spherical symmetric. There are therefore no γ-transitions between nuclear levels with $I = 0$. For $I \neq 0$ the transition probability increases according to (3.50) (the lifetime τ_i becomes smaller) with decreasing values of L (see (3.51) and therefore also with ΔI (3.48), as long as $\lambda\!\!\!^{-} > R$.

The transition probability depends not only on the ratio $R/\lambda\!\!\!^{-}$ but also on symmetry selection rules. In particular the parity Π of the levels involved in the transition plays an important role.

For your remembrance: The parity of a level is $\Pi = +1$. if the wave function of this level is preserved under reflection at the origin ($\mathbf{r} \rightarrow -\mathbf{r}$); it is $\Pi = -1$ if it undergoes a change of its sign i.e. $\psi \rightarrow -\psi$ for $\boldsymbol{r} \rightarrow -\boldsymbol{r}$.

For an electric multipole transition the parity changes by $\Delta\Pi = (-1)^L$, for a magnetic multipole transition it changes by $\Delta\Pi = -(-1)^L = (-1)^{L+1}$ (see Vol. 3, Sect. 7.2).

The multipole modes M1; E2; M3 and E4.... have even parity (i.e. the parity Π does not change at the transition) while the modes E1; M2; E3....) have odd parity (Table 3.3).

In Fig. 3.33 some examples for multipole-γ-transitions are listed. The multipolarity can be determined from the angular distribution of the emitted γ-quanta (see Sect. 4.5).

Inserting the numerical values of the natural constants in (3.50) and the value $R = r_0 \cdot A^{1/3}$ one obtains the lifetimes of nuclear levels which can decay by γ-radiation into lower levels. They are compiled in Table 3.4. These values illustrate that the lifetimes of excited nuclear levels can vary by many orders of magnitude.

Table 3.3 Multipole-transitions

L	2^L	ΔL	Denotation	Symbol	Parity conservation
0	1	0	Monopol	–	0 → 0 forbidden
1	2	1	Dipol	E1	no (–1)
2	4	2	Quadrupol	M1	yes (+ 1)
				E2	yes (+1)
				M2	no (–1)
3	8	3	Oktupol	E3	no (–1)
4	16	4	Hexadekapol	M3	yes (+1)
				E4	yes
				M4	no

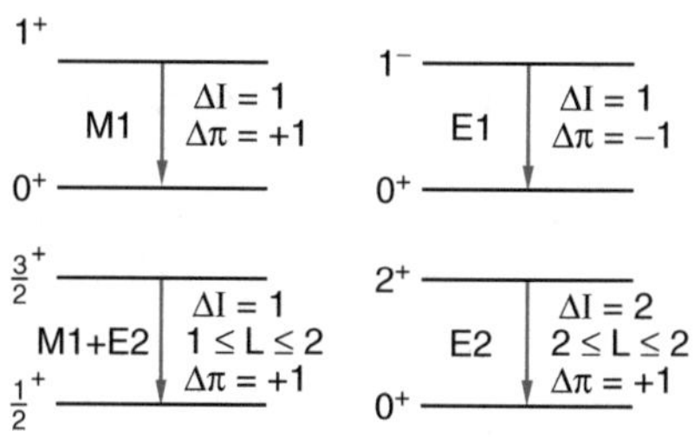

Fig. 3.33 Different multipole-transitions of the γ-radiation

Table 3.4 Reciprocal transition probabilities (in seconds) for different multipole-transitions at different γ-energies)after K.Bethge kernphysik (Springer Heidelberg 1996)

E_γ/Mev	E1	E2	E3	E4
0,1	10^{-13}	10^{-6}	10^{+2}	10^{+9}
1	10^{-15}	10^{-10}	10^{-5}	1
10	10^{-18}	10^{-15}	10^{-12}	10^{-9}

In a similar way the transition probabilities for magnetic multipole transitions can be obtained. The result of such calculation is

$$\frac{1}{\tau_M} \approx E_\gamma^{2L+1} \cdot A^{2(L-1)/3}, \tag{3.52a}$$

where A is the number of nucleons in the nucleus (= atomic mass number).

For electric multipole transitions one gets

$$\frac{1}{\tau_E} \approx E_\gamma^{2L+1} \cdot A^{2L/3}. \tag{3.52b}$$

This gives the ratio τ_M/τ_E for equal γ-energies and the same nucleus (taking into account the statistical weights of the levels)

$$\frac{A_{ik}^{(E)}}{A_{ik}^{(M)}} = \frac{\tau_M}{\tau_E} \approx 4,5 \cdot A^{2/3}. \tag{3.52c}$$

Example For nuclei with $A = 125$ the transition probability for a magnetic transition ML is about 100 times smaller than for the corresponding transition EL.

Note For heavy nuclei (large A) the probability of magnetic transitions becomes nearly negligible compared to electric transitions.

3.5.3 Conversion Processes

Besides bγ γ-radiation an excited nucleus can transfer its excitation energy also by direct energy transfer to an electron in the atomic shell (generally in the *K*-shell). If the acquired energy is large enough, the electron will leave the atom and a positively charged ion is left (**internal conversion**).

$${}^A_Z X^* \rightarrow {}^A_Z X^+ + e, \tag{3.53}$$

This process can be regarded as an internal photo-effect by a virtual γ-quant which is emitted by the nucleus but immediately absorbed by the electron. The electron then has the kinetic energy

$$E_{kin}(e) = E\left({}^A_Z X^*\right) - E({}^A_Z X) - E_B(e), \tag{3.54}$$

where the last term represents the binding energy of the electron in the atom ${}^A_Z X$ (Fig. 3.34).

Since the excited nucleus ${}^A_Z X^*$ is often produced by $ß^-$-decay of the nucleus ${}^A_{Z-1}Y$ one observes in these cases a $ß^-$-spectrum with a continuous energy distribution, overlapped by sharp lines at the energies (3.54), with a separation that equals the energy differences of the binding energies of the shell electron in the *K*-resp. the *L*-shell.

The probability of internal conversion depends on the overlap of the electronic wave function with the nucleus (Fig. 3.34). This overlap is maximum for the 1s-function and therefore the internal conversion occurs mainly for electrons in the 1s-state. The subsequent emission of the electron produces a hole in the 1s-state. Electrons in higher energy states of the electron shell can undergo transition into this free place in the 1s-shell, producing X-ray emission with the wavelength.

$$\lambda = c/\nu = v/(E_i - E_k)/h$$

The ratio

$$\eta = \frac{N_e}{N_\gamma} \tag{3.55}$$

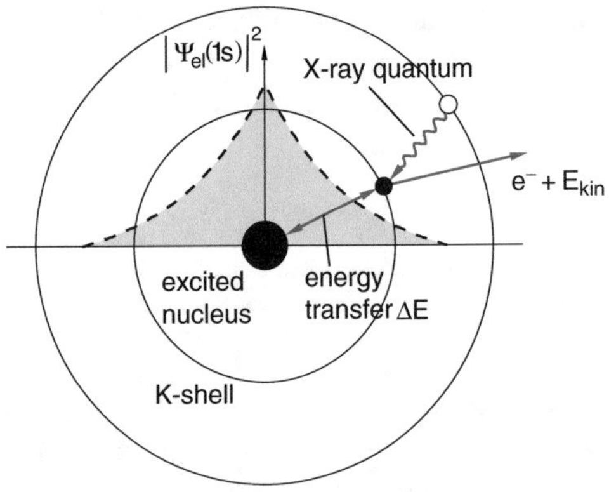

Fig. 3.34 Transfer of the excitation energy of the nucleus onto an electron of the atomic shell (conversion process)

of the emission rates of electrons to γ-quanta is called the **conversion efficiency**.

This ratio becomes particular large if the γ-transition is forbidden and the nucleus has no chance to release its energy other than by internal conversion. One example is the first excited state of $^{72}_{32}$Ge at E = 691 keV with a spin parity 0^+. Since the ground state has also the same parity 0^+ the γ-transition is forbidden by parity selection rules.

Another possibility of energy conversion is the **internal pair formation**. For nuclear excitation energies above E = 1.02 MeV this energy can be converted in the strong Coulomb field of the positive nucleus into the formation of an electron–positron pair. An example of this pair formation is the excited state (I = 0; Π = + 1) of the $^{16}_{8}$O -nucleus with E = 6.06 MeV which cannot decay by γ-emission into the 0^+-ground state with $I = 0$. Because of angular momentum conservation the transition with ΔI = 0 is forbidden since the γ-quant has the angular momentum 1·ħ. However, the excitation energy can be converted into the production of an e^-e^+ pair according to

$$^{16}_{8}\mathrm{O}^* \rightarrow {}^{16}_{8}\mathrm{O} + \mathrm{e} + \mathrm{e}^+ + E_{\mathrm{kin}} \qquad (3.56)$$

where the kinetic energy E_{kin} = (6.06 − 1.02) MeV is equally shared by electron and positron. One finds in the $ß^-$-spectrum a sharp line at E = 2.52 MeV.

In principle the excited 0^+-state of $^{16}_{8}$O* could decay by simultaneous emission of two γ-quanta into the 0^+-ground state, because this would not violate angular momentum- and parity selection rules. The probability of this process is, however, very small. The branching ratio of two-photon emission and internal conversion is only $2.5 \cdot 10^{-4}$.

3.5.4 Nuclear Isomers

Long living excited nuclear states are called **Isomers** of the nuclei in the ground state, because they have the same number of protons and neutrons. They differ from the nuclei

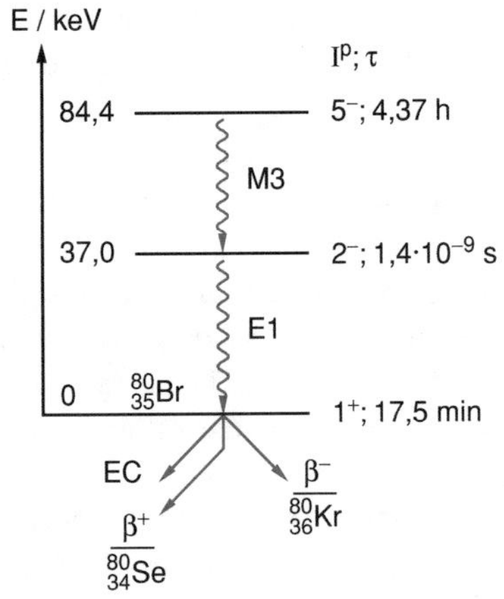

Fig. 3.35 Isomeric state $^{80}_{35}$Br* of the unstable $^{80}_{35}$Br ground state (EC = electron capture)

in the ground state only regarding energy, and eventually angular momentum and parity. The long lifetimes of such isomers can be due to forbidden dipole and quadrupole transitions into the ground state. This is, for instance, true for excited states with high spin quantum numbers I, where γ-transition into the ground state would demand large changes ΔI. For illustration Fig. 3.35 shows the level scheme of the isomer of the $^{80}_{35}$Br -nucleus. The excited state with I = 5 and Π = −1 has a lifetime of τ = 4.37 h, because it can only decay by a magnetic octupole transition into the lower state with I = 2 and Π = −1 (see Table 3.3).

Summary

- Nuclei can spontaneously pass into other nuclei, if the mass of the parent nucleus is larger than the sum of the masses of the decay products

$$M(^{A}_{Z}\mathrm{X}) \geq M_1\left(^{A'}_{Z'}\mathrm{Y}\right) + M_2$$

  and if the decay is not hindered by a potential barrier or forbidden by symmetry selection rules.
- The different possible decay channels of an unstable nucleus are α-, $ß^-$- or $ß^+$-, γ- emission, K-capture, internal conversion and nuclear fission.
- Stable nuclei with a nucleon number A have a narrow stability range for the ratio N/Z of neutron number N to proton number Z. If N becomes too large the nucleus becomes deformed and unstable and decays by $ß^-$ emission, if N becomes too small, $ß^+$- emission occurs.
- Nuclei with even numbers N and Z (gg-nuclei) are especially stable, nuclei with odd numbers N and Z (uu-nuclei) are generally unstable. There are only 4 exceptions with $Z \leq 7$. All uu-nuclei with $Z > 7$ are unstable.

- The number N of unstable nuclei in a sample decreases exponentially with time according to $N(t) = N(0) \cdot e^{-\lambda t}$. After the mean lifetime $\tau = 1/\lambda$ the number of nuclei has decreased to $N(\tau) = N(0)/e$.
- The activity $A(t) = \lambda \cdot N(t)$ (number of decays per sec) of a sample of unstable nuclei) has decreased after the half-lifetime $t_{1/2} = \tau \cdot \ln 2 = \ln 2/\lambda$ to $A(0)/2$ of its initial value. The unit of the activity is the Bequerel (1 Bequerel = 1 decay per sec) The decay constant λ and the half-lifetimes $t_{1/2}$ vary for the different unstable nuclei over many orders of magnitude.
- The natural occurring radioactive elements can be arranged into 4 decay series where in all series the parent element (after which the series is named) decays successively by α- or ß -decay into other elements The final element of each series is a stable lead isotope ${}^{A}_{Z}Pb$.
- Natural α-emitters exist only for heavy elements with $A > 205$. During the α-decay at first 2 protons and 2 neutrons inside the nucleus form a stable α-particle. Because this α-particle has a much higher binding energy the energy released at its formation is converted into kinetic energy. This enables the α-particle to tunnel through the potential barrier and leave the nucleus.
- Since the probability for tunneling increases strongly with increasing energy short-lived α- radio-active elements emit α-particles with a higher energy than long-lived elements.
- The energy levels in the nucleus have discrete values. Therefore the emitted α-particles show a discrete energy spectrum.
- The ß-decay produces a continuous energy spectrum of the emitted electrons. Conservation of energy, momentum and angular momentum demand a three-body decay:

$$ {}^{A}_{Z}X \xrightarrow{\beta^-} {}^{A}_{Z+1}Y + e + \overline{\nu}, $$

$$ {}^{A'}_{Z'}X' \xrightarrow{\beta^+} {}^{A'}_{Z'-1}Y' + e^+ + \nu $$

and therefore the existence of a new particle, the **neutrino,** postulated by W.Pauli, which had not been discovered at that time. Neutrinos are leptons with a rest mass $m_\nu < 4 \cdot 10^{-6}\, m_e = 0.5\ eV/c^2$. They have a very small interaction with other particles [8].

- Gamma radiation is a short wavelength electromagnetic radiation with photon energies $h \cdot \nu = 10$ keV – 10 MeV. It is produced by transitions between different energy levels of a nucleus (rotational or vibrational states).
- The multipole modes of order 2^L of the γ-radiation depend on the change $\Delta I = L$ of the nuclear spin and on the parity of the wave-functions of the nuclear states involved in the ransition.

Problems

3.1 (a) Assume that at time $t = 0$ there are $N_A(0)$ radioactive nuclei, which decay with the half-lifetime $t_{1/2} = 10$ d into nuclei B with a half- lifetime $t_{1/2} = 5$ d. How large are $N_A(t)$ and $N_B(t)$ after a time $t = 1$ d, 10 d and $t = 100$ d if $N_B(0) = 0$?

(b) After which time is only 1 g tritium ($t_{1/2} = 12.3$ y) left, if $M(0) = 1$ kg?

3.2 (a) What are the decay constant λ and the half-lifetime $t_{1/2}$ of a radio-active α-emitter, if E_1 is the energy of the α-particles within the nucleus, and the Coulomb-wall is approximated by a rectangular barrier with width a and heights E_0 and the nuclear potential is described by a potential box with depth $-E_2$ and width b. *Numerical values*: $E_1 = 6$ MeV; $E_2 = +15$ MeV; $E_0 = +11$ MeV; $a = 4 \cdot 10^{-14}$ m, $b = 6 \cdot 10^{-15}$ m.

(b) Which decay constant λ is obtained, if for $r < b$ the attractive potential box of a) is chosen and for $r \geq b$ the repulsive Coulomb potential for $Z_1 = 90$, $Z_2 = 2$?

3.3 Calculate the probability P for the case that an α-particle with kinetic energy $E_{kin} = 8.78$ MeV for a central collision with the nucleus ${}^{208}_{82}Pb$ penetrates through the Coulomb barrier.

3.4 The nucleus ${}^{62}_{30}Zn$ can decay by e^+-emission as well as by electron capture. Calculate the maximum neutrino energy for both kinds of decay if the maximum positron energy $E(e^+)$ is 0.66 MeV. How much differs the result, if recoil and the bindings-energy of K-electrons are neglected resp. taken into account?

3.5 Assume that the mass difference between parent nucleus ${}^{A}_{Z}X$ and ${}^{A}_{Z+1}X_2$ is $\Delta M = 3\ MeV/c^2$. What is the maximum energy of the electron of the $ß^-$-decay with and without neglecting the recoil energy of the daughter nucleus with mass M_2? Numerical example: $M_2 = 70$ AMU. What is the maximum energy of the neutrino?

3.6 Assume that the binding-energy of an electron in the K-shell is 50 keV. What is the maximum mass difference ΔM for the nucleus and for the atom at the K-capture $M_1 + e^- \rightarrow M_2 + \nu_e + (h \cdot \nu)_K$?

3.7 An excited nucleus with mass M transfers its excitation energy onto an electron in the *K*-shell which is emitted. In an external magnetic field *B* it follows a circular path with radius R. What is the recoil energy of the nucleus and the excitation energy of the nucleus?
Numerical example: $R = 10$ cm, $B = 0.05$ T, $M\left({}^{137}_{55}\mathrm{Cs}\right) = 137$ AMU.

3.8 The tritium nucleus ${}^{3}_{1}\mathrm{H}$ has a binding energy $E_B =$ 8.4819 MeV, which is larger than the binding energy $E_B = -7.7180$ of ${}^{3}_{2}\mathrm{He}$. Why can tritium still decay into ${}^{3}_{2}\mathrm{He}$ by β^--emission? What is the maximum energy of the electron and the recoil energy of helium if the neutrino mass is assumed as $m_\nu = 0$?

3.9 The nucleus ${}^{12}_{5}\mathrm{B}$ decays by β^--emission into the nucleus ${}^{12}_{6}\mathrm{C}$. The nucleus ${}^{12}_{7}\mathrm{N}$ decays by β^+-emission also into the same nucleus ${}^{12}_{6}\mathrm{C}$. Both decays occur with high probability into the ground state of ${}^{12}_{6}\mathrm{C}$. The maximum energies are E(β^-) 13.3695 MeV, E (β^+) = 16.3161 MeV What are the energies of the ground states of ${}^{12}_{5}\mathrm{B}$ and ${}^{12}_{7}\mathrm{N}$ relative to the ground state of ${}^{12}_{6}\mathrm{C}$? Draw a sketch true to scale!

3.10 At time $t = 0$ 10 g of the isotope ${}^{226}\mathrm{Ra}$ (density $\rho = 5.5$ g/cm^3) are filled into a glass tube with a volume of 5 cm^3.The glas tube is then sealed and stored at a temperature of T = 20 °C. the isotope ${}^{226}\mathrm{Ra}$ decays with a half lifetime $t_{1/2} = 1600$ a into the radio-active gas ${}^{222}\mathrm{Rn}$ which further decays with a half-lifetime $t_{1/2} = 3.825$ d resulting in the final stable product ${}^{206}\mathrm{Pb}$.
1. Calculate the Number of Rn Nuclei as a Function of Time, Assuming that N(Rn) is 0 at $t = 0$.
2. At which time is the partial pressure of radon maximum? How large is it then? What is the maximum pressure increase in % of the initial pressure?.

References

1. The latest edition of the Nuclide-chart of Karlsruhe KIT can be obtained from marktdienste@haberbeck.de. Further informatiobn on "Karlsruhe Nuclid chart. Wikipedia"
2. H. Becquerel: Compte Rendue 122, 420 and 689 (1896)
3. F. Reines, C. L. Cowan: Free Anti-Neutrino Cross section. Phys. Rev. 113, 273 (1959)
4. F. Fermi: Versuch einer Theorie der ß-Strahlen. Z. Physik 88, 161 (1934)
5. Katrin Wikipedia
6. J. M. Blatt, V. F. Weißkopf: Theoretical Nuclear Physics (Dover, New York 1991)
7. C. Weinheimer, et. al: Improved Limit on the Electron Antineutrino Rest mass from Tritium Beta Decay. Phys. Lett. **B300**, 210 (1993), and Annal Phys. **525**, 565 (2013), F. W. Otten et.al. Int. J. Mass Spectrom. **251,** 173 (2006)
8. H. V. Klapdor. Kleingrothaus: Is the Neutrino a Majorana Particle? Physik in uns, Zeit 33 (4) 155 (2002).
9. W. M-. Yao: Particle data group J. Phys. G33, 1 (2006)
10. J. C. Bernauer, R. Pohl: The proton radius Puzzle. Scientific American (Febr. 2014)

General Literature

11. J. Magil; Radioactivity, Radionuclides Radiation. Springer, Berlin, Heidelberg 2005
12. M. Malley: Radioactivity (Oxford Univ. Press 2011)
13. K. S. Krane: Introductory Nuclear Physics (Wiley and Sons 1992)
14. E. Browne, R. B. Firestone: Table of Radioactive Isotopes (Wiley 1986)
15. R. F. Mould: A Century of X-Rays and Radioactivity in Medicine. (Taylor and Francis 1993)
16. M. L'Annunziata: Radioactivity (Elsevier 2016)
17. Klaus Bethge, Gertrud Walter, B. Wiedemann: Kernphysik (Springer, Heidelberg 2008)

4 Experimental Techniques and Equipment in Nuclear- and High Energy Physics

After having introduced in the previous chapters the basic perceptions about stable and unstable nuclei we will now discuss the instrumentation and the experimental techniques which have brought about this knowledge of structure and composition of nuclei and further information on the characteristics of elementary particles. We will start with particle accelerators and detectors which belong to the most important devices of nuclear and high energy physics.

In order to study the structure of atomic nuclei in more detail the spatial resolution power of the used method must be sufficiently large. For scattering experiments this means that from measurements of the differential scattering cross section information about the structure of the mass- and charge distribution within the nucleus can be only obtained if the de-Broglie wavelength $\lambda_{\mathrm{dB}} = h/p$ of the incident particle with momentum p is small compared to the diameter $D_K = 2r_0 \cdot \mathrm{A}^{1/3}$ of the target nucleus. Therefore the kinetic energy E_{kin} of the projectile with mass m has to be correspondingly high. For $\lambda_{\mathrm{dB}} > D_K$ diffraction effects wash out all existing structures of the nucleus, which then cannot any longer be resolved. This is completely similar to the spatial resolution in optics where the resolution limit of the microscope is given by the wavelength of the illuminating light. (See Vol. 2. Sect. 11.3).

Also for the excitation of nuclei by collisions with projectile particles or for the production of new particles in reactive collisions the kinetic energy of the collision partners in the center of mass system must be above a certain minimum value which lies in the energy range of some keV to many GeV, depending on the specific reaction. As has been shown in Vol. 1, Sect. 4.3, only the kinetic energy in the center of mass system can be converted into excitation energy of the collision partners.

One therefore has to accelerate particles up to the wanted energy before they undergo central collisions.

4.1 Accelerators

In the first part of this chapter we will introduce the most important types of particle accelerators. For more detailed information the reader is referred to the cited literature [1–9].

4.1.1 Velocity, Momentum and Acceleration of Particles at Relativistic Energies

Since most accelerators bring the particles up to kinetic energies that are larger than their rest energy one has to use relativistic formulas for the determination of the relations between m_0c^2, velocity, momentum, kinetic energy and relativistic mass of the particles. Also for the treatment of collision processes at relativistic energies the Newton laws have to be extended (see Vol. 1, Sect. 4.4).

For instance the relativistic relation between total energy E and momentum p is

$$E = E_{\mathrm{kin}} + E_0 = \sqrt{(c \cdot p)^2 + (m_0c^2)^2}, \qquad (4.1)$$

where the "**rest mass energy** $\boldsymbol{E_0}$" of the resting particles with rest mass m_0 is

$$E_0 = m_0c^2 = \sqrt{E^2 - (c \cdot p)^2} \qquad (4.2)$$

For very high energies $(E \gg m_0c^2)$ the energy E becomes $E \approx c \cdot p$.

The kinetic energy is then

$$\begin{aligned} E_{\mathrm{kin}} &= E - m_0c^2 \\ &= \sqrt{(c \cdot p)^2 + (m_0c^2)^2} - m_0c^2. \end{aligned} \qquad (4.3)$$

W. Demtröder, *Nuclear and Particle Physics*, Undergraduate Lecture Notes in Physics,
https://doi.org/10.1007/978-3-030-58313-2_4

Often the ratio

$$\alpha = \frac{E_{kin}}{m_0 c^2} \tag{4.4}$$

of kinetic energy to the rest mass energy is used as parameter of characteristic quantities at relativistic collision processes. This is advantageous because one can express several relevant quantities of collisions as a function of α, independent of the specific particle.

At relativistic energies the increase of the particle mass with the particle velocity becomes essential. It is

$$\frac{m(v)}{m_0} = \frac{1}{\sqrt{1 - v^2/c^2}} = \gamma = 1 + \alpha, \tag{4.5}$$

With $E = E_{kin} + m_0 c^2$ we get

$$1 + \alpha = \frac{E}{m_0 c^2} = \frac{mc^2}{m_0 c^2} = \frac{m(v)}{m_0}. \tag{4.6}$$

The velocity v of a particle with kinetic energy E_{kin} is derived from (4.5) as

$$v = \frac{c}{1+\alpha}\sqrt{\alpha^2 + 2\alpha}. \tag{4.7}$$

From (4.7) and Fig. 4.1a one can see that for $\alpha > 2$ the velocity v nearly reaches the veolocity c of light $(v(\alpha = 2) = 0{,}943c)$.

For the momentum

$$\begin{aligned} \mathbf{p} &= m(\mathbf{v}) \cdot v = m_0 \mathbf{v}(1+\alpha) \\ |\mathbf{p}| &= m_0 c \cdot \sqrt{\alpha^2 + 2\alpha} \end{aligned} \tag{4.8}$$

the de Broglie wavelength λ_{dB} can be expressed by

$$\lambda_{dB} = \frac{h}{p} = \frac{h}{m_0 c \cdot \sqrt{\alpha^2 + 2\alpha}} \tag{4.9}$$

In Fig. 4.2 the reduced momentum $p(\alpha)/(m_0 c^2) = (v/c) \cdot (m/m_0)$ and the ratio v/c are plotted over a large range of α.

This illustrates that for $\alpha > 2$ i.e. $E_{kin} > 2m_0c^2$ the increase of the momentum with increasing α is mainly due to the mass increase since the velocity remains nearly constant.

In Fig. 4.3 the de Broglie wavelength is plotted for different masses as a function of the parameter $\alpha = E_{kin}/m_0c^2$. This illustrates that for electrons the de Broglie wavelength decreases below 1 fm only for $\alpha > 2000 \to E_{kin} > 1000$ MeV. In order to resolve nuclear structures by electron scattering, the electrons have to be accelerated at least above E_{kin} = 1 GeV.

Examples:

1. For electrons with E_{kin} = 500 MeV is $\alpha = 10^3$, this implies $m = 10^3\ m_0$ and λ_{dB} = 2.5 fm.
2. For protons with E_{kin} = 500 MeV is $\alpha \approx 0.5$, m = 1.5 m_0 and λ_{dB} = 1.0 fm

4.1.2 Basic Physics of Accelerators

All charged particles, such as electrons, protons and their anti-particles positrons and anti-protons, as well as all q-fold charged positive ions A_ZX^{q+} or negative ions A_ZX^{q-} can be accelerated by an electric potential difference $U = \phi_{el}(r_1) - \phi_{el}(r_2)$. As long as the acceleration energy $q \cdot U$ is small compared with the rest mass m_0 one can use the nonrelativistic expressions

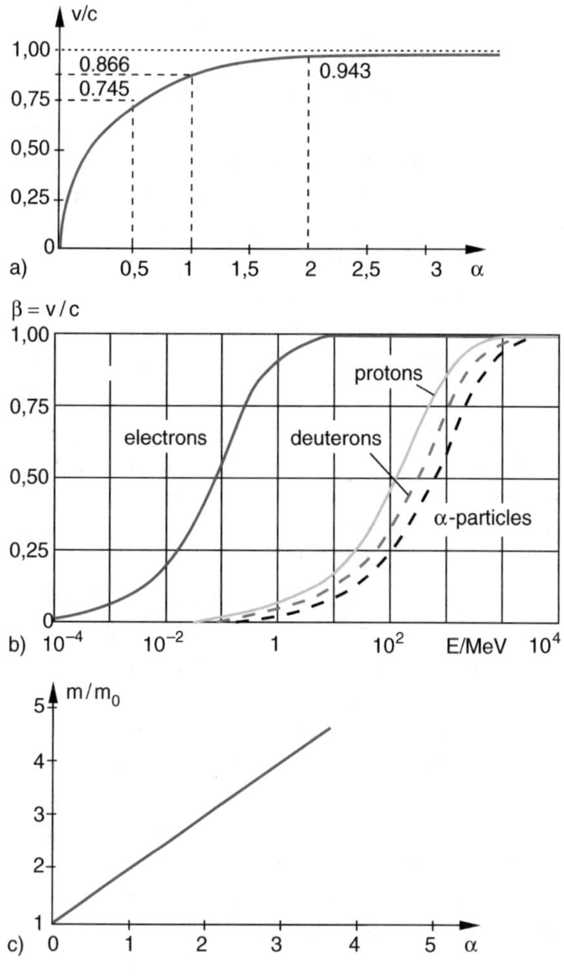

Fig. 4.1 The ratio $\beta = v/c$ **a** as a function of α, independent of the mass m, **b** as a function of the energy E for different masses, **c** Mass ratio $m(v)/m_0$ as a function of α a) b) c)

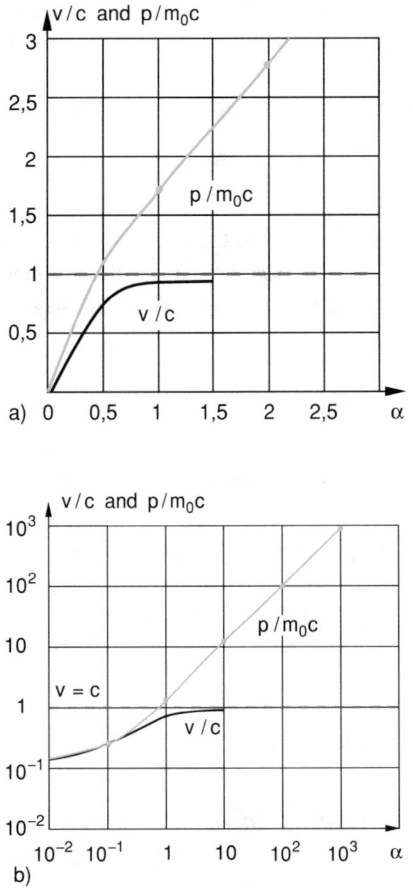

Fig. 4.2 Reduced momentum p/m_0c and ratio v/c as a function of $\alpha = E_{kin}/m_0c^2$. **a** plotted on a linear scale, **b** on a log–log diagram

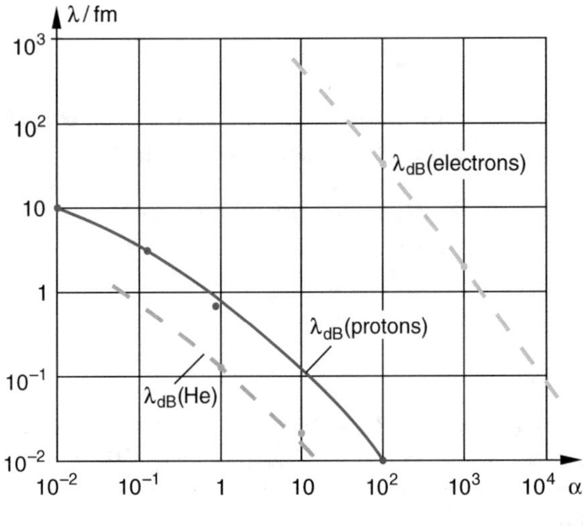

Fig. 4.3 De-Broglie Wavelength λ_{DB} as a function of α for electrons, protons and α-particles

$$E_{kin} = \frac{m}{2}v^2 = q \cdot U \Rightarrow v = \sqrt{\frac{2q \cdot U}{m}} \qquad (4.10)$$
$$p = m \cdot v$$

for the determination of kinetic energy, velocity v and momentum p of the accelerated particles. In this nonrelativistic range is $m \approx m_0$ and $p \propto v$.

For $\alpha > 0.1$ m becomes noticeable larger than m_0. With $m(\alpha = 0.1) = 1.1m_0)$ the moving mass is already 10% larger than the rest mass. In Fig. 4.2 the curves $v(\alpha)/c$ and $p(\alpha)/m_0c$ start to separate for $\alpha = E_{kin}/m_0c^2 > 0.1$.

At higher energies the velocity v approaches the velocity of light c (see Fig. 4.1). An increasing fraction of the acceleration energy

$$q \cdot U = E_{kin} = (m - m_0)c^2 = \alpha \cdot m_0c^2 \qquad (4.11)$$

is used for the relativistic increase of the mass $m = \gamma \cdot m_0 = m_0/\sqrt{(1 - v^2/c^2)}$.

In order to accelerate charged particles they can be either pass once through a region with high potential difference U **(electrostatic accelerators)** or they can be guided successively through N acceleration lines with voltage U where their final energy is then $E_{kin} = N \cdot q \cdot U$ (**high frequency accelerators**).

One distinguishes between **linear accelerators** (linac) where the particles during their acceleration run on a straight line, and **circular accelerators** where the particles are forced by an external magnetic field to run many times on a circular path. The choice of the optimum accelerator depends on the type of the accelerated particles, the wanted final energy, the desired particle current density and last but not least on the building and running costs of the accelerator.

The accelerated particles are used as projectiles for the investigation of collision processes. With the particles flux $\phi_A = \dot{N}_A$, describing the number of incident particles A per sec and unit area onto the n_B target particles per unit volume δV in the reaction volume (Fig. 4.4) we get the differential scattering cross section $d\sigma/d\Omega$ which gives the number of particles per sec

$$\frac{dN(\vartheta)}{dt} = \phi_A \cdot n_B \cdot \Delta V \cdot \frac{d\sigma}{d\Omega}\Delta\Omega \qquad (4.12)$$

scattered into the solid angle $\Delta\Omega$ and accepted by the detector. The reaction volume ΔV should be as small as possible, in order to approximate it by a point source. This allows the accurate determination of the scattering angle ϑ.

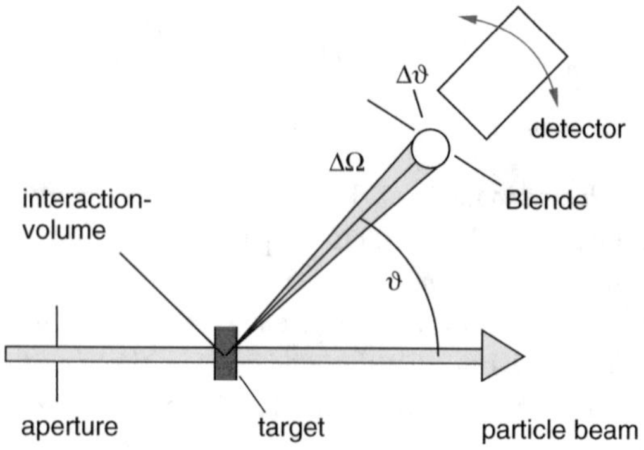

Fig. 4.4 Experimental scheme for measuring differential cross sections. $d\sigma/d\Omega$

For the measurement of small differential cross sections $d\omega/d\Omega$ the incident particle flux ϕ_A should be sufficiently large.

At the collision of the incident particles A (energy E_A, momentum $\boldsymbol{p}_A$) with target particles B at rest ($E_B = m_{0B} \cdot c^2$, $E_{kin} = 0$, $\boldsymbol{p}_B = 0$) the total momentum of the collision pair $\boldsymbol{p} = \boldsymbol{p}_A$ must be conserved. The maximum energy transfer ΔE for inelastic or reactive collisions equals the energy of the collision pair measured in the center of mass system (see Vol. 1, Sect. 4.2).

Within the nonrelativistic energy range ($v << c$) this maximum energy transfer for inelastic collisions which can be converted into internal energy of the collision partners is

$$\Delta E_1 = \frac{1}{2}\mu \cdot v_A^2 \qquad (4.13a)$$

with the reduced mass $\mu = m_A \cdot m_B/(m_A + m_B)$. The part

$$\Delta E_2 = \frac{1}{2}(m_A + m_B)v_S^2 \qquad (4.13b)$$

remains $v_S = m_A \cdot v_A/(m_A + m_B)$ as kinetic energy of the center of mass motion. The conservation of the total energy demands $\Delta E_1 + \Delta E_2 = E_A = 1/2 m_A \cdot v_A^2$.

Example

$m_A = 10\, m_B \rightarrow \mu = 0{,}091\, m_A \rightarrow \Delta E_1 = (0.91/10) \cdot E_A^{kin}$.
This means that only 9.1% of the kinetic energy of A can be transferred into internal energy of B.

In order to calculate the relations for relativistic energies we start with the energy–momentum vector

$$\mathcal{P} = \{\boldsymbol{p}, iE/c\}$$

(see Vol. 1, Sect. 4.4.5). The square of the energy–momentum vector $\mathcal{P}^2 = E^2/c^2 - p^2$ is invariant and does not change under the transformation from one inertial system to another, for example from the laboratory system to the center of mass system. The relativistic relation between energy and momentum of a particle with rest mass m_0 is

$$E = \sqrt{m_0^2 c^4 + c^2 p^2}. \qquad (4.13c)$$

For the special case that the particle A with the rest mass m_A moves in the laboratory system with the velocity v_A and the particle B with rest mass m_B is at rest we get the relation in the lab-system

$$E_{AL} = m_{A0}c^2 + E_{kin}; \quad E_{BL} = m_{B0}c^2,\ p_{AL} \neq 0;$$
$$p_{BL} = 0.$$

This replaces (4.13c) for the energy in the lab system by

$$E_L = \sqrt{m_A^2(v_A)c^4 + m_{B0}^2 c^4 + p_A^2 c^2} \qquad (4.13d)$$

In the center of mass system is $\boldsymbol{p}_{AS} + \boldsymbol{p}_{BS} = 0$. We therefore obtain for the square of the energy–momentum vector

$$\begin{aligned} c^2 \cdot \mathcal{P}_S^2 &= (E_A + E_B)^2 = E_S^2 \\ &= c^2 \cdot \mathcal{P}_L^2 = \left(E_{AL} + \left(m_B c^2\right)^2 - (c p_{AL})^2\right. \end{aligned} \qquad (4.13e)$$

With $E_{AL}^2 - c^2 p_{AL}^2 = m_{A0}^2 c^4$ the maximum energy available for transfer into excitation energy of B becomes

$$\begin{aligned} E_S &= \sqrt{m_{0A}^2 c^4 + m_{0B}^2 c^4 + 2 m_{0B} c^2 \cdot E_A} \\ &= \sqrt{E_{0A}^2 + E_{0B}^2 + 2E_{0B} \cdot E_A} \end{aligned} \qquad (4.13f)$$

(see Problem 4.7). For $m_{0A} = m_{0B} = m_0$ this reduces to

$$\begin{aligned} E_S &= m_0 c^2 \sqrt{2 + (2E_A/m_0 c^2)} \\ &= m_0 c^2 \sqrt{2 + 2\alpha}. \end{aligned} \qquad (4.13g)$$

The fraction of the total energy which can be converted into energy for collision-induced reactions is then

$$\begin{aligned} \eta &= E_S/m_A(v_A)c^2 = \sqrt{(2 + 2\alpha)/(1 + \alpha)} \\ &= \sqrt{2/(1 + \alpha)}. \end{aligned} \qquad (4.13h)$$

Example

Protons with energy E_A = 50 GeV collide with protons at rest. Here is α = 53 and the fraction η is η = 0.19.

For high energies is $E_A >> m_0 c^2 \rightarrow \alpha = >> 1$ and we can approximate

$$E_S \approx \sqrt{2E_A \cdot m_0 c^2} \qquad (4.13i)$$

$$E_S/E_A \approx \sqrt{2m_0c^2/E_A}. \qquad (4.13j)$$

The energy available for reactions increases for high energies only as the square root of the total energy of the accelerated incident particles.

Example

Protons (m_0c^2 = 940 MeV) are accelerated to E = 100 GeV and collide with protons at rest of a hydrogen target. The energy available for the production of new particles (e.g. generation of mesons) is according to (4.13b).

$$E_S \approx \sqrt{2 \cdot 100 \cdot 0{,}94}\,\text{GeV} \approx 13{,}7\,\text{GeV},$$

this means that in case of target nuclei at rest only 13.7% of the projectile energy can be converted in the reaction $p+p \rightarrow x_1 + x_2 + \ldots$ into reaction energy for the production of new particles. One reason for this small efficiency is the much larger mass of the accelerated projectile which is about 100 times larger than the rest mass of the target.

In order to achieve higher reaction energies the collision partners both have to be accelerated to high energies and collide anti-collinear with opposite directions but equal amounts of their momenta. In this case the total energy of the colliding particles with equal rest mass can be completely converted into reaction energy. This can be realized in colliders or large storage rings (see Sect. 4.1.8) where central collisions between particles from opposite directions occur. Here the total kinetic energy of the collision partners can be transferred into excitation energy or the production of new particles (see Sects. 4.1.8–4.1.10).

4.1.3 Electrostatic Accelerators

The principal design of an electrostatic accelerator is shown in Fig. 4.5. The acceleration of electrons, which are emitted from a hot cathode at the negative potential $V = -U$ occurs on their way through an evacuated tube of isolating material (Fig. 4.5a) to the anode at the potential $U = 0$. A voltage divider chain with feeds through to conducting ring apertures ensures a constant voltage gradient along the whole tube.

For the acceleration of positively charged ions the thermionic cathode is replaced by an ion source (see Vol. 3, Sect. 2.5.4) which is now at a positive voltage. By a correct ion optics the ions emitted by the ion source can be focused and form a nearly parallel ion beam, which passes through a hole in the anode which is kept at ground potential. Using magnetic or electric fields the ions can be deflected and guided through different beam lines to the wanted experimental apparatus.

A relatively simple technique for the generation of high voltages is realized in the **van-de-Graaff Generator** (Fig. 4.6, see also Vol. 2, Fig. 1.31), which can generate high voltages up to several million volts. An isolating conveyor belt is positively charged at the point A where a sharp tip sprays charges from a 20 kV source onto the belt. These charges are transported by the moving belt into the interior of a conducting sphere where the charges are transferred to a tip at the point B which is connected to the inner part of the conducting sphere.

Due to influence the charges are pushed to the outer part of the sphere, which can be charged up to very high voltages until a discharge starts between the sphere and its surroundings. The sphere is connected by a resistance chain with the common ground of the ion source.

The disadvantage of the van-de-Graaff Generator is the small charge dQ/dt that can be transported per sec to the conducting sphere. When the current I of the accelerated particles exceeds the charge transport $I > dQ/dt$ the high voltage breaks down. One can reach ion currents of 0,1–1 mA. For sufficiently small currents the constancy of the high voltage is very good, because the charge transport by the conveyor belt, moving with constant velocity, is very stable.

With a cascade circuit (see Vol. 2, Sect. 5.7.4) fed by a transformer and rectifier delivering the dc-voltage U_0 and a cascade of N rectifiers and capacitors (Fig. 4.7) a voltage of $N \cdot U_0$ can be obtained. At the points 1; 3, 5 in Fig. 4.7 the voltage changes at each period of the ac voltage of the transformer between 0 and $2U_0$, while the voltage at the points 2, 4 and 6

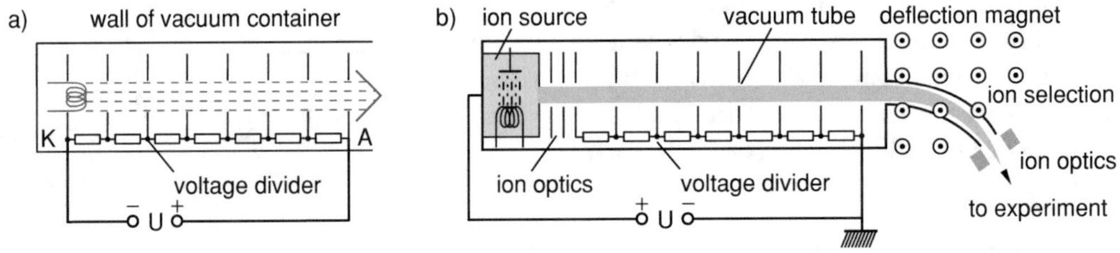

Fig. 4.5 Schematic design of an electrostatic Accelerator. **a** for electrons. **b** for ions

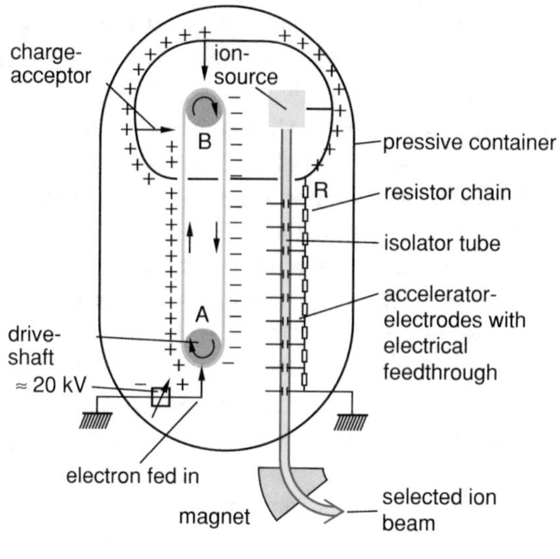

Fig. 4.6 Van de Graaff accelerator

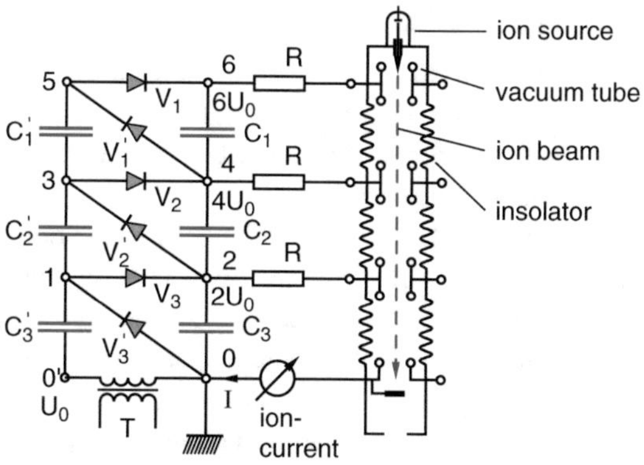

Fig. 4.7 Cascade high voltage generator

has the constant values $2U_0$, $4U_0$ $6U_0$. This cascade circuit is used in the **Cockroft-Walton accelerator** which delivers voltages of several million volts at much higher currents than the van-der-Graaff-generator. However, at larger currents the output voltage shows a ripple ΔU with $\Delta U/U \approx 0.1$–1% typical to all rectifiers (see Vol. 2, Sect. 5.7).

A widely used version of electrostatic accelerators is the Tandem Accelerator (Fig. 4.8) which uses the acceleration voltage U twice.

In an ion source at ground potential negative ions with charge $-q_1$ are produced which are accelerated by the positive voltage U. In a charge exchange chamber the ions suffer streaking collisions with neutral atoms or molecules which wrench from the negative ions some electrons, converting them to positive ions with the charge $+q_2$. These ions are now accelerated to the cathode at ground potential. Their total kinetic energy is now

$$E_{\max} = (+q_1 + q_2)U.$$

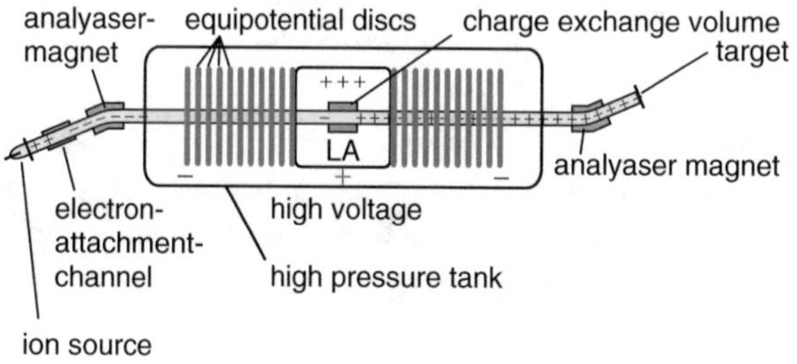

Fig. 4.8 Tandem accelerator

4.1.4 High-Frequency Accelerators

In high frequency accelerators the charge particles pass through several acceleration sections which are connected to a high frequency source. The phase of the high frequency is chosen such, that the particles always experience an accelerating voltage. In the **drift tube accelerator**, designed by *Wideroe*, the particles fly through a series of metal tubes with increasing length L (Fig. 4.9). Between the tubes a HF-voltage with the correct phase is applied. Inside the tubes the potential is constant and the particles experience there no accelerating force. The acceleration occurs only between the tubes. The optimum acceleration is reached if the half-period $T/2 = 1/(2f)$ of the high frequency f is equal to the time of flight $\Delta t = L/v$. Because L and v both increase, the frequency f of the accelerating HF can be kept constant if the length increase ΔL from tube to tube is chosen correctly.

Such accelerators are nowadays predominantly used as circular rings for the acceleration of heavy Z-fold ionized particles up to energies $E_{\text{kin}} << m_0c^2$. They then have velocities $v << c$. With

$$v = \sqrt{(2E_{kin}/m)} = \sqrt{(2n \cdot U \cdot Z \cdot e/m)}$$

the frequency of the accelerating HF-field must be

$$f = \frac{v}{2L} = \frac{1}{2L}\left(\frac{2n \cdot U \cdot Z \cdot e}{m}\right)^{1/2}, \tag{4.14}$$

where $n = 1, 2, 3, \ldots..N$ is the number of the nth tube and U is the voltage between subsequent tubes. In order to keep

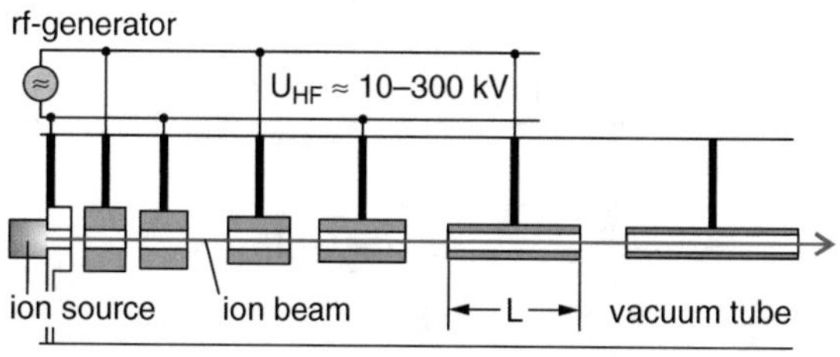

Fig. 4.9 Drift-tube accelerator

the frequency f constant (this has technical advantages) because of the increasing velocity v the length L of the tubes must increase such that $L/v \propto L/\sqrt{n}$ remains constant.

Example

Threefold charged carbon ion C^{+++} should be accelerated in a drift tube accelerator with $N = 10$ and $U = 100$ kV. With $Z = 3$, $m = 12 \cdot 1.66 \cdot 10^{-27}$ kg $\rightarrow f = 4.9$ MHz. The length L of the tubes increases from $L_1 = 0.44$ m to $L_{10} = 1.4\ m$. The final energy of the ions is $E_{kin} = N \cdot e \cdot U = 10^6 \cdot 1,6 \cdot 10^{-19}\ J = 1$ MeV.

For the acceleration of leptons (electrons and positrons) to high energies ring accelerators are not the best choice because the accelerated charged particles emit radiation. The intensity of this radiation is proportional to the square of the acceleration a. When the particles move with the velocity v on a circle with radius R the radiated power is proportional to $a^2 = v^4/R$ (see Vol. 2, Sect. 6.5.4) and becomes at high energies ($v \approx c$) intolerably large. A solution are linear accelerators.

One alternative to the drift tube accelerator is the **Travelling wave Accelerator** (also named Scanning Field Accelerator), which is used for fast particles that had been already accelerated to velocities close to the velocity of light c. It is used predominantly for the acceleration of electrons or positrons. Here one takes advantage of the fact that an electromagnetic wave travelling through a waveguide is no longer a pure transverse wave but has a longitudinal component E_z of the electric vector E (Fig. 4.10). Its phase velocity in a cylindrical waveguide

$$v_{Ph} = \frac{c}{\sqrt{1 - (\lambda/2D)^2}} \tag{4.15}$$

depends on the wavelength λ and the diameter D of the waveguide and is larger than the velocity of light c in free space (see Vol. 2, Sect. 7.9).

The phase velocity can be reduced by placing apertures into the waveguide. It depends on the ratio $2d/D$ of aperture diameter $2d$ and tube diameter D and can be adapted to the particle velocity. Furthermore the correct spatial arrangement can enhance a wanted mode and suppress unwanted ones.

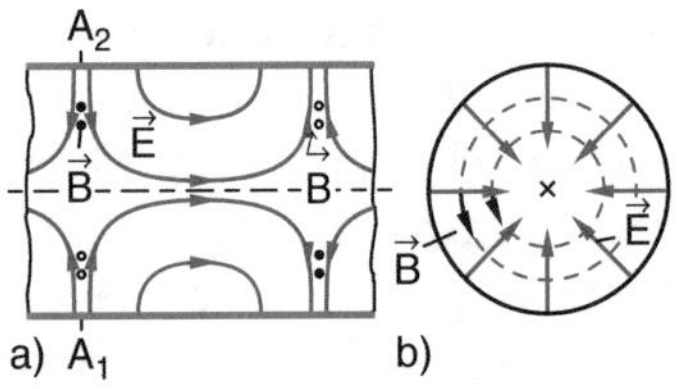

Fig. 4.10 Amplitude distribution of an electro-magnetic wave in a circular waveguide. **a** longitudinal section. **b** transverse section

Such a travelling wave accelerator was designed by *L. Alvarez*. It is called "*Runzelröhren-Accelerator*".

In Fig. 4.11 a snapshot of the momentary state of the accelerating wave is shown in an arrangement with 18 apertures, where the wavelength λ equals 12 times the distance between the apertures. This figure also shows that the component E_z of the accelerating field and therefore the accelerating force is maximum on the axis of the device.

When the electrons are taken along by the travelling wave they accumulate in the hatched area in Fig. 4.12 i.e. in the phase range from $\varphi = 0$ to $\varphi = \pi/2$ mod. 2π. This can be seen as follows:

Particles which are slightly faster velocity than the phase velocity of the wave reach a range with smaller electric field strength, which means that they experience a smaller acceleration, for $\varphi < \pi/2$ while in the phase range $\varphi > \pi/2$ they are even decelerated. They are therefore pushed back into the range $0 < \varphi < \pi/2$. Particles which are slower than v_{ph} reach a range with higher field strength and are therefore more accelerated. These stable ranges are repeated at phase differences $2n \cdot \pi$. The accelerated particles therefore do not arrive as continuous flux but in packets with a time separation that equals a multiple $n (n = 1, 2, \ldots)$ of the HF-period T.

The worldwide largest linear travelling wave accelerator is the SLAC (Stanford Linear Accelerator) of the Stanford University, California (Fig. 4.13). It is 3 km long and can accelerate electrons and positrons up to energies of 50 GeV. With a special design (see Sect. 4.1.8) electrons and positrons can be both accelerated and guided into a circular ring, where they run into opposite directions and collide anti-collinearly in special intersection points. This extends

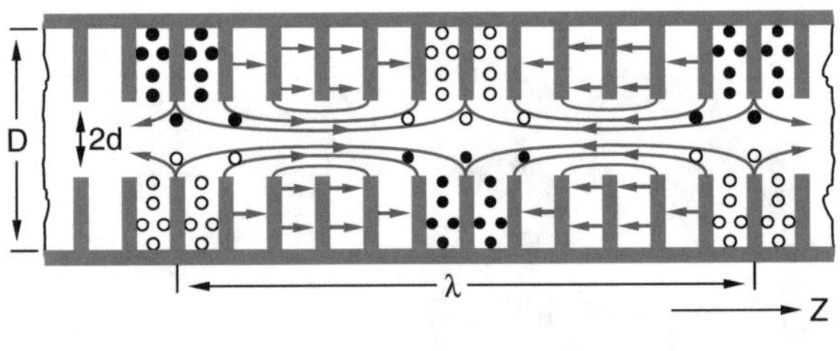

Fig. 4.11 Momentary picture of electric and magnetic field distribution in an Alvarez travelling wave accelerator

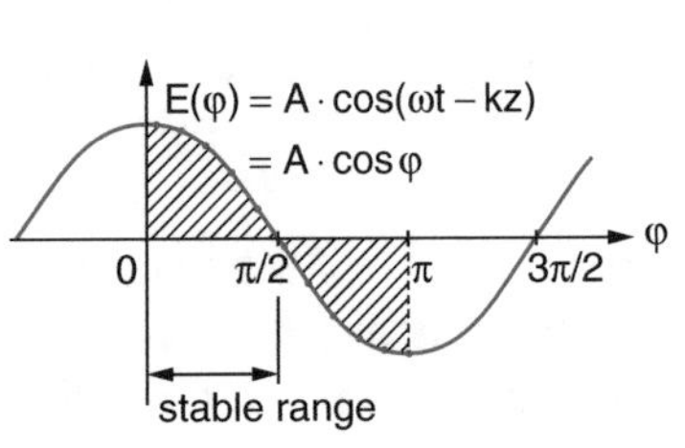

Fig. 4.12 Principle of particle acceleration by a travelling wave

Fig. 4.13 The Stanford linear accelerator SLAC (Courtesy of Stanford University)

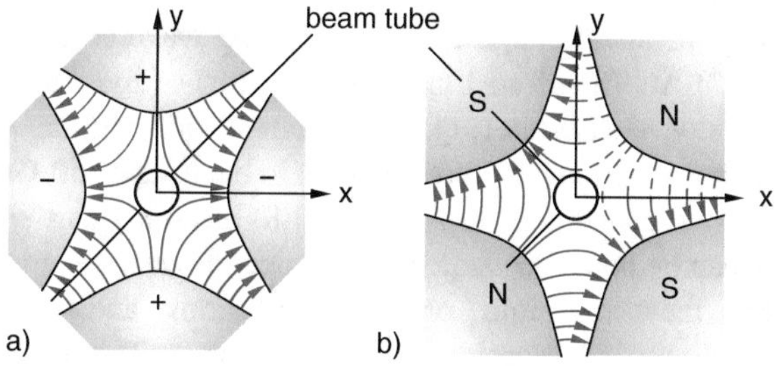

Fig. 4.14 **a** Electric, **b** magnetic quadrupole particle lenses

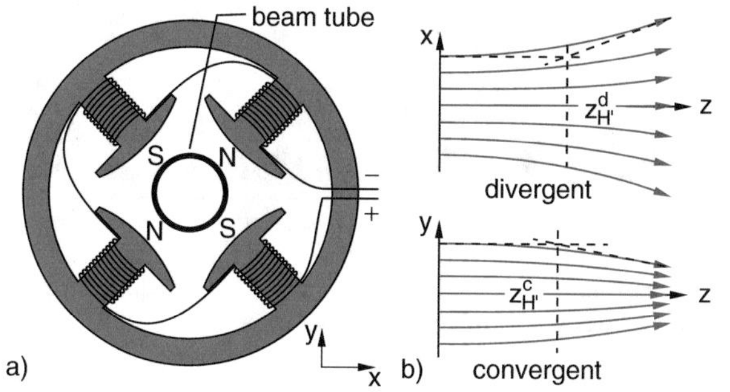

Fig. 4.15 **a** Experimental realization of a magnetic quadrupole. **b** Particles tracks in the xz—and the yz plane of a quadrupole lens

the linear accelerator to a collider and opens the possibility for collisions of leptons at very high energies.

In order to prevent the particles to deviate on their long way in the accelerator too much from the axis where they could collide with the walls of the vacuum tube, they have to be focused. This is realized by electric or magnetic lenses (Fig. 4.14), which generate fields symmetric to the planes $x = 0$ and $y = 0$. Each lens acts convergent (focusing) in one direction and divergent (defocusing) in the other direction′ (Fig. 4.15). Arranging two quadrupole lenses which are twisted around the z-axis by 90° against each other, the combined system acts as focusing lens (see Vol. 3. Sect. 2.6).

4.1.5 Acceleration by Lasers

The development of high power lasers (see Vol. 3, Chap. 8) allows the generation of extremely high electric fields up to 10^{11}V/m [10–13]. This is higher by many orders of magnitude than what can be realized with classical methods for the acceleration of charged particles. Therefore several proposals for laser accelerators have been made, where the length of the accelerator can be reduced to a few meters. This would reduce the costs considerably.

There are several methods which can be utilized for acceleration by lasers. One example is based on the irradiation of a thin metal foil by the focused radiation from a high power laser, which generates a hot plasma (Fig. 4.16). The electrons emitted from the plasma are focused by the strong transverse magnetic field of the laser radiation and are guided along the laser beam direction. The resulting charge

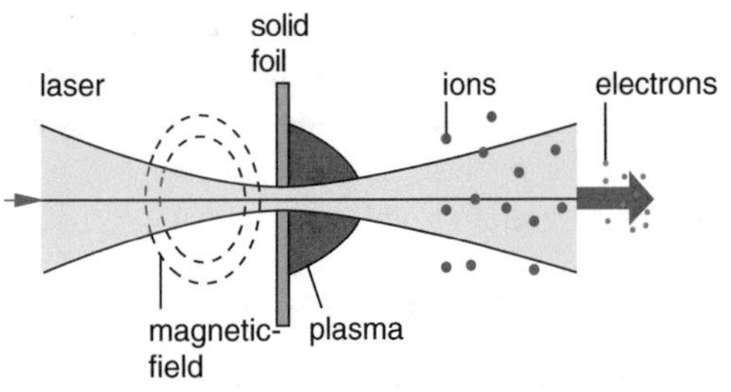

Fig. 4.16 Particle acceleration by a laser

separation in the plasma creates a strong electric field, which accelerates the positive ions (e.g. protons).

At very high laser intensities ($I > 10^{20}$ W/cm^2) the magnetic field strength B becomes comparable with the electric field strength E. This means that the Lorentz force $\boldsymbol{F_L} = q(\boldsymbol{v} \times \boldsymbol{B})$ nearly equals the electric force $\boldsymbol{F}_\mathrm{E} = q{\cdot}\boldsymbol{E}$ and the velocity v of the electrons becomes nearly equal the velocity of light c. In Fig. 4.17 the electric and magnetic fields of the laser pulse are illustrated as well as the path of the electrons traversing with the laser pulse as seen by an observer at rest (left oscillating curve) and for an observer moving with the mean velocity of the electrons (the loop-like black curve). Due to the Lorentz-force the electrons get an acceleration in the forward direction [13].

Another proposal uses the direct acceleration of charged particles by the laser field. Its basic principle is illustrated in Fig. 4.18 [14]. The laser pulse passes through a cylindrical lens and is, after reflection by an optical grating, focused into the middle plane of a periodic wave guide structure. The phase of the laser wave experiences at the periodic structure alternatively a shift of π resp. π/2. If the transit time of the fast electrons over one segment of the periodic structure equals the propagation time of the laser pulse the electrons are always accelerated by the longitudinal electric field of the laser wave. They can reach final energies up to several GeV. A critical point is the phase stability of the laser, which can be, however, handled by phase stabilization techniques [13x].

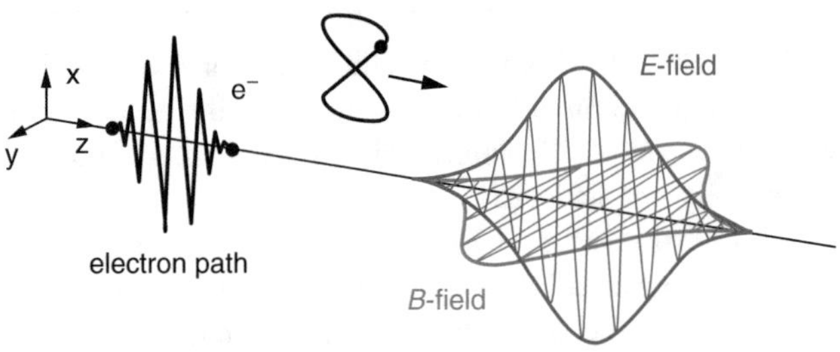

Fig. 4.17 Illustration of the impact of an intense electro-magnetic field on an electron. E, B and the propagation direction z form a right-handed coordinate system. The acceleration of a negatively charged electron into the negative direction of E causes the acceleration into the + z-direction, due to the Lorentz-force. The figure shows a corresponding electron path (black curve) for an observer at rest (left) and for one moving with the mean velocity to the right [13]

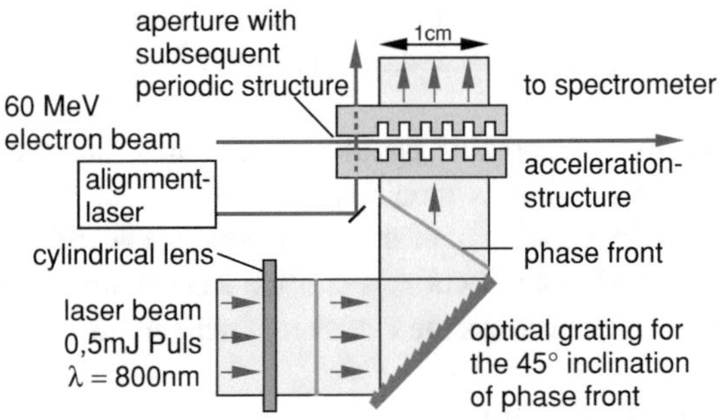

Fig. 4.18 Acceleration of electrons by a laser wave with inclined phase front in a periodic slit structure [12]

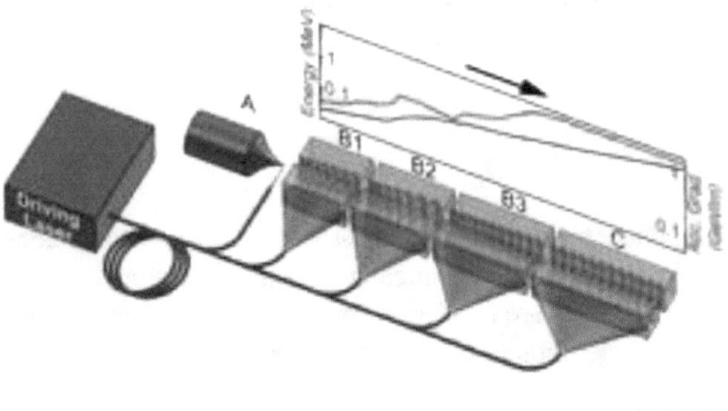

Fig. 4.19 Laser.-particle acceleration by an evanescent wave [MPI for Quantum Optics, Garching]

Another version of laser acceleration uses the evanescent wave, generated by a laser wave that is diffracted by an optical grating and runs parallel to the grating surface. First experiments showed an energy gain of 25 MeV per meter [12]. Figure 4.19 shows the basic principle: The laser beam is coupled into different optical fibers and guided to optical gratings. The light diffracted by the gratings into directions parallel to the grating surface is used to accelerate the electrons. In the upper diagram the blue curve gives the local energy of the electrons at the position of the grating and the red curve gives the energy gain in GeV/m.

4.1.6 Circular Accelerators

In circular accelerators the charged particles are forced by magnetic fields on a circular path in a plane perpendicular to the magnetic field.

The particles perform many circulations and are accelerated by n acceleration devices per circulation (n = 1, 2, 3, ...).

The most important circular accelerators are the **zyclotron** in its various modifications and the **synchrotron**. For the acceleleration of electrons to moderate energies up to about 100 **MeV** the **betatron** and the **microtron** are used, while for high energies the **electron-synchrotron** is chosen.

4.1.6.1 Cyclotron

The cyclotron consists of a flat cylindrical vacuum chamber in the x–y-plane between the poles of an electromagnet which generates a magnetic field in z-direction (Fig. 4.20). The cylindrical chamber is divided into two D-shaped parts. A high frequency voltage

$$U = U_0 \cdot \cos \omega t$$

is applied between the two D-sections. The positive ions produced in the ion source in the gap between the two D-shaped parts are accelerated by the voltage U and perform a half cycle through the D-part with radius r determined by the condition (Fig. 4.20x)

$$\frac{mv^2}{r} = q \cdot v \cdot B \Rightarrow r = \frac{mv}{qB} \tag{4.16}$$

which means that: Lorentz force q·v·B = centripetal force m·v^2/r. When reaching the intersection space between the two D's they are accelerated again and their velocity increases and therefore, according to (4.16) also their radius r.

The transit time through one D

$$t = \pi \cdot r/v = \pi \cdot m/(qB) \tag{4.17}$$

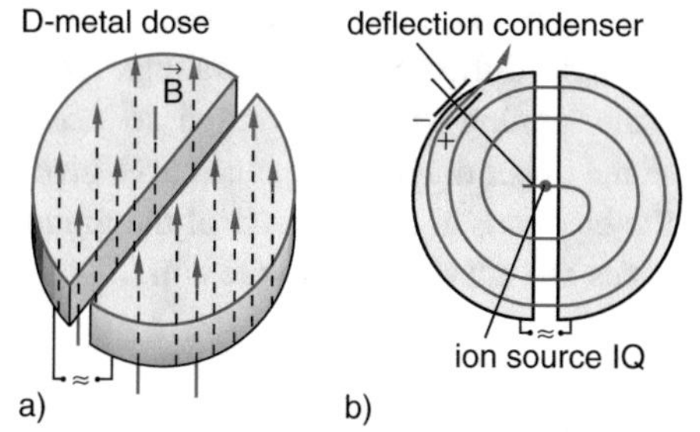

Fig. 4.20 **a** Principle of the cyclotron. **b** Half-circles of the particle paths in an magnetic field with successively increasing radii after each acceleration

is independent of the radius r. This means that the time $T_{½}$ for a half cycle is independent of the kinetic energy of the accelerated particles, as long as the mass remains approximately constant. This has the great advantage that the frequency of the accelerating voltage

$$2\pi f_{HF} = \omega_{HF} = (q/m)B \tag{4.18}$$

can be kept constant over the whole acceleration time. After each half roundtrip time the polarity of the HF changes sign and therefore the ions are always accelerated. Their energy increases after each half-circle by $\Delta E = q \cdot U$. The ions therefore traverse approximately a spiral path consisting of half cycles with increasing radii, until they have reached with $r = R$ the edge of the chamber, where they are extracted by an electric field out of the cyclotron. Their maximum energy

$$E_{kin} = \frac{mv^2}{2} = \frac{q^2}{2m} \cdot (R \cdot B)^2 \tag{4.19}$$

depends on the radius R of the chamber, on the magnetic field strength B and on the ratio (q^2/2 m).

Example

U = 50 kV, B = 2 T, R = 1 m.

1. For protons the necessary HF is $f_{HF} = (e/m) \cdot (B/2\pi) = 30\,\text{MHz}$. They can reach a kinetic energy of 200 MeV. The protons traverse about 4000 circulations for which they need a total time of 130 μs.
2. For α-particles with $Z = 2$ and $m \approx 4m_p$ the factor q^2/m is the same as for protons and therefore the same kinetic energy is reached for equal values of R and B. The necessary HF is, however, only f_{HF} = 15 MHz.
3. For deuterons is q/m = ½(e/m_p). Therefore only E_{kin} = 100 MeV is reached, which is half of the energy for Protons. The HF is f_{HF} = 15 MHz.

These numerical values are valid, if the particles pass the acceleration section at the maximum U_0 of the acceleration voltage. Due to possible phase shifts this optimum condition is generally not achieved.

For higher energies the relativistic mass increase can no longer be neglected. The circulation time of the particles increases with increasing energy and the frequency of the accelerating HF voltage must be therefore adapted. This can be done synchronously with the increasing circulation period (**Synchro-Cyclotron**). One drawback is the fact that the ions cannot be extracted at each HF-period as in the standard cyclotron, but only as pulses with time intervals ΔT which

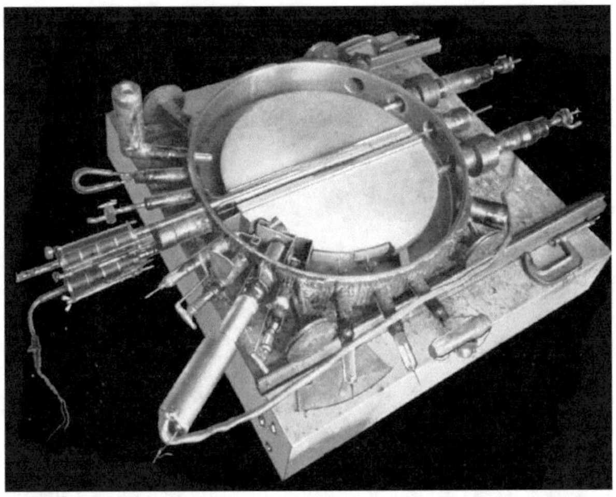

Fig. 4.20x Open vacuum chamber of the 27 inch Cyclotron built 1935 by E Lawrence in Berkeley. At the right side are the two connections for the acceleration voltage. At the lower left side is the connection for the deflection voltage. [John B. Livingood/Public domain]

equal the acceleration period of an ion packet. For the example above this would be $\Delta T = 130\ \mu s$.

4.1.6.2 Betatron

The Betatron is used for the acceleration of electrons to energies up to 10^7 eV. Contrary to the situation in the cyclotron, here the radius of the electron circular path remains constant over the whole acceleration period (Fig. 4.21). Therefore the magnetic field has to increase with increasing velocity of the electrons. This increasing magnetic field induces, according to the Maxwell equation

$$\mathbf{rot}E = -\frac{d\mathbf{B}}{dt} \tag{4.20}$$

the electric field E (see Vol. 2, Sect. 4.6). The tangential component of the electric field, which is directed along the tangent to the electron circular path with radius r_0 induces the acceleration voltage

$$\begin{aligned} U_{\text{ind}} &= \oint_s E \cdot \mathrm{d}s = \int_F \mathbf{rot}E \cdot \mathrm{d}F \\ &= -\frac{\mathrm{d}}{\mathrm{d}t}\Phi = -\frac{\mathrm{d}}{\mathrm{d}t}\int_F B \cdot \mathrm{d}F = \pi \cdot r_0^2 \cdot \frac{\mathrm{d}\langle B\rangle}{\mathrm{d}t} \end{aligned} \tag{4.21}$$

which is determined (according to Faraday's Law) by the time derivative of the magnetic flux through the area inside the circular electron path. The mean magnetic field in the area is

$$\langle B\rangle = \left(\int B \cdot \mathrm{d}F\right) / \left(\pi \cdot r_0^2\right)$$

The acceleration voltage is then obtained from (4.20) as

$$U = E \cdot 2\pi \cdot r_0 = \pi \cdot r_0^2 \cdot \frac{\mathrm{d}}{\mathrm{d}t}(\langle B\rangle) \tag{4.22}$$

The electrons are accelerated solely by the voltage induced by the increasing magnetic field, without any external voltage. They accumulate in the time interval from the time t_1 when the magnetic field starts its rise until the time t_2 when the final energy is reached. Their momentum is

$$\begin{aligned} p &= \int F\mathrm{d}t = e\int E \cdot \mathrm{d}t = \frac{e \cdot r_0}{2}\int \frac{\mathrm{d}\langle B\rangle}{\mathrm{d}t}\mathrm{d}t \\ &= e \cdot r_0 \cdot \langle B\rangle/2 \end{aligned} \tag{4.23}$$

The particles are kept on the nominal orbit determined by the Lorentz force only if the condition

$$\begin{aligned} &\tfrac{mv^2}{r_0} = e \cdot v \cdot B(r_0) \\ &\Rightarrow p = m \cdot v = e \cdot r_0 \cdot B(r_0). \end{aligned} \tag{4.24}$$

is fulfilled at any time during the acceleration. The comparison of (4.23) with (4.24) gives the *Wideroe-Condition*

$$B(r_0) = \frac{1}{2} \cdot \langle B\rangle. \tag{4.25}$$

The magnetic field $B(r_0)$ at the electron path must be always equal to one half of the average magnetic field inside the circular radius r_0 of the electron trajectory (Fig. 4.20y).

This can be only fulfilled, if $\mathrm{d}B/\mathrm{d}r < 0$ which can be realized by a suitable form of the magnetic pole pieces (Figs. 4.21a and 4.22).

The Betatron can be regarded as a transformer with a single secondary winding, which is the circular vacuum tube in which the electrons are accelerated. If the primary winding is supplied by a 50 Hz ac-current the time-interval for the acceleration of the electrons is restricted to the red area in Fig. 4.23 with a time interval of 5 ms where the accelerating voltage increases. At the time t_1 the electrons from an external source are injected with about 40 keV tangentially into the nominal circle with radius r_0 inside the Betatron. At the later timer t_2 shortly before the maximum of the voltage is reached, they are extracted out of the betatron by an additional electric field and guided to the experiment.

During these 5 ms the electrons perform about 10^6 circulations and travel on a circular path with radius $r_0 = 0.5$ m over a total distance of 3000 km. They gain per circulation the energy of about 50 eV where the energy gain decreases with each circulation because of the increasing radiation losses.

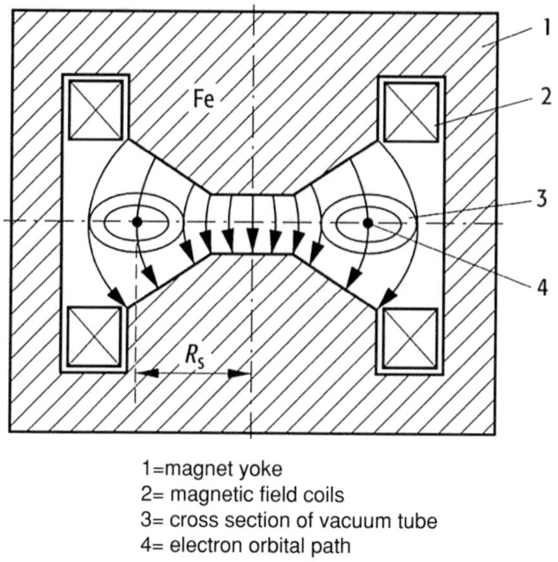

Fig. 4.20y Cross section through a Betatron

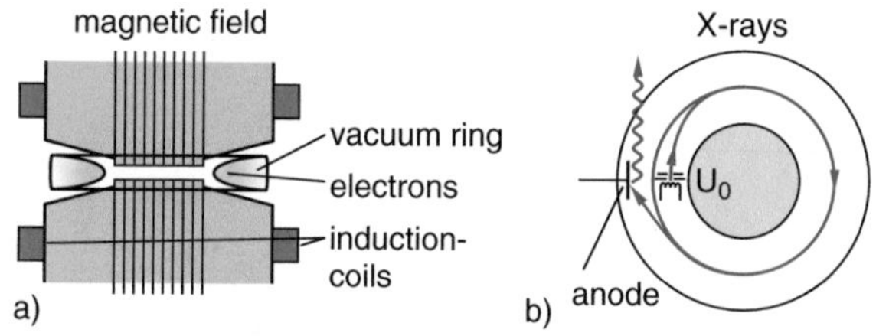

Fig. 4.21 Schematic drawing of a Betatron. **a** Side view. **b** top view

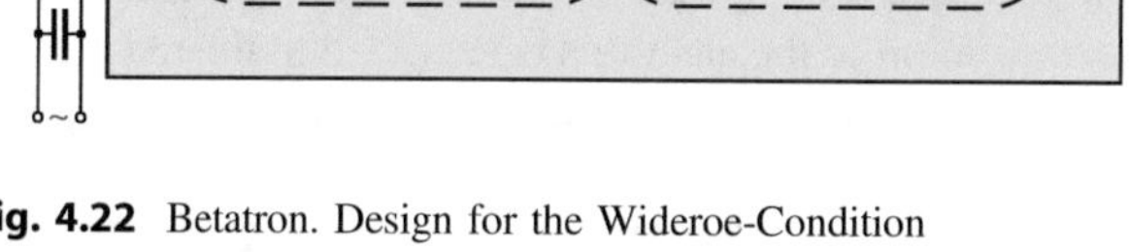

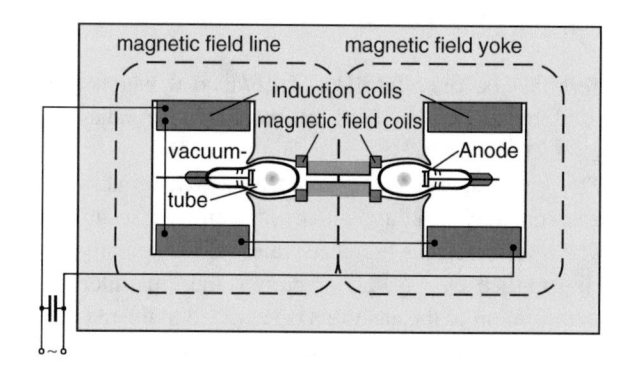

Fig. 4.22 Betatron. Design for the Wideroe-Condition

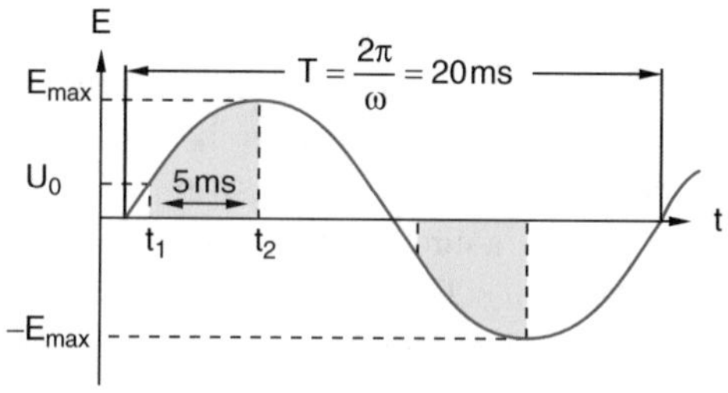

Fig. 4.23 Time windows for the acceleration phase

The final achievable energy is limited by the maximum magnetic flux Φ_m through the area enclosed by the electron path. For the relativistic energy range we obtain with (4.24, 4.25) and $\Phi(t) = \pi r_0^2 B(t)$ the relations

$$E = \sqrt{(pc)^2 + (m_0c^2)^2} = \left[\left(\frac{e \cdot c}{2\pi \cdot r_0}\Phi(t)\right)^2 + \left(m_0c^2\right)^2\right]^{1/2} \tag{4.26}$$

Typical numerical values are $E = 10 - 50$ MeV. The largest Betatrons reach even $E = 320$ MeV.

4.1.6.3 Synchrotron

In order to accelerate electrons, protons or heavy ions up to energies of 10^8–10^{12} eV, Betatrons or cyclotrons are not appropriate because of the following reason: The radius r of the circular path for relativistic particles with the momentum $p = m \cdot \mathrm{v}$ in the magnetic field B can be derived from (4.6) when the formulas for the relativistic mass (4.5) and the velocity (4.7) are used. With $\alpha = E_{kin}/m_0 c^2$ we obtain for the radius

$$r = \frac{m\mathrm{v}}{qB} = \frac{m_0 \cdot c}{q \cdot B} \cdot \sqrt{\alpha^2 + 2\alpha} = \frac{E_{\mathrm{kin}}}{q \cdot c \cdot B} \cdot \sqrt{1 + 2m_0c^2/E_{\mathrm{kin}}}. \tag{4.27}$$

This shows that for $\alpha >> 1$ i.e. $E_{kin} >> m_0c^2$ the radii of circular accelerators increases proportional to the ratio E_{kin}/B but is independent of the rest mass m_0. For large values of E_{kin} the radius r becomes for technical realizable magnetic fields very large, which implies that the extension of the magnetic field must be accordingly large. This is very costly and demands huge and heavy magnets. In other words: For a given radius $r = r_0$ and a maximum feasible magnetic field B_{max} the maximum possible kinetic energy becomes

$$E_{\mathrm{kin}}^{\mathrm{max}} = q \cdot c \cdot B_{\mathrm{max}} \cdot r_0. \tag{4.28}$$

Example

In order to accelerate electrons up to the energy $E = 30$ GeV the radius in a magnetic field of $B = 1$ T becomes $r = 100$ m, for protons is $r = 103$ m which is only slightly larger than for electrons. The reason is that the relativistic mass is nearly the same for both particles. For $E_{kin} = 300$ GeV the radius becomes $r = 1$ km for electrons as well as for protons.

The material consumption for a cyclotron, where the whole area must be covered by the magnetic field, which would allow such high kinetic energies would be unjustifiably high. Therefore nowadays mainly synchrotrons are used for the acceleration of particles to high relativistic energies. Here the path of the particles is a circle with fixed radius in a vacuum tube with narrow cross section between the pole-pieces of many electro-magnets arranged on a circle with radius r. These magnets produce a spatially confined magnetic field.

In Fig. 4.23x the new synchrotron in the basic principle is illustrated in Fig. 4.24. The particles are pre-accelerated in a linear accelerator up to velocities $v \approx (0.8 - 0.9) \cdot c$ and are injected tangential into the synchrotron, which consists of circular arcs where the magnets are arranged. Between these arcs are linear sections for cavity resonators, which accelerate the particles and for ion-lenses which focus the particles in order to keep them on their nominal path (Fig. 4.24). During the acceleration period from $t = t_1$ to $t = t_2$ the magnetic field must continuously increase in order to keep the particles always on their nominal circle (Fig. 4.25). After having reached their maximum energy at $t = t_2$ the magnetic field is kept constant during the experimental time interval t_2–t_3 where the particle energy is constant and the particles are colliding with the target. At $t = t_3$ the magnetic field is ramped down in order to start a new acceleration cycle.

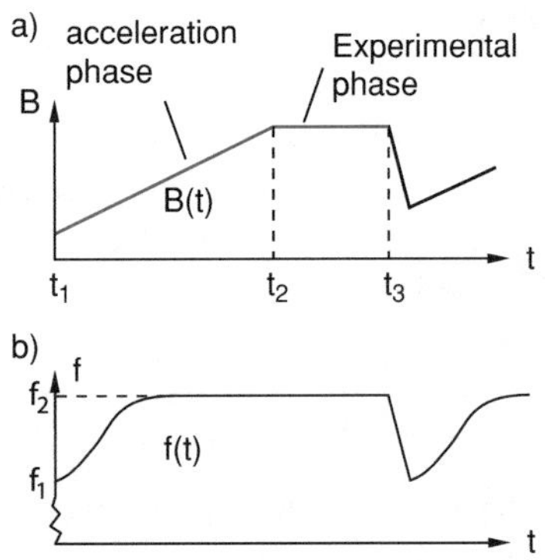

Fig. 4.24x Time variation **a** of the magnetic field **b** of the frequency of the acceleration voltage

Although the particles are injected into the ring with the energy 50–500 MeV depending on the pre-accelerator, their velocity still changes during their acceleration by $\Delta v(t) = c - v(t_1)$. For electrons at $E_{\text{kin}} = 500$ MeV is $\Delta v = 5 \cdot 10^{-4} \cdot c$ and therefore very small while for protons at 500 MeV Δv amounts to $0.2 \cdot c$. The frequency of the HF-accelerating cavities must be tuned from an initial value

$$f_1 = f(t_1) = k \cdot v_1 / (2\pi \cdot r_0) \tag{4.29a}$$

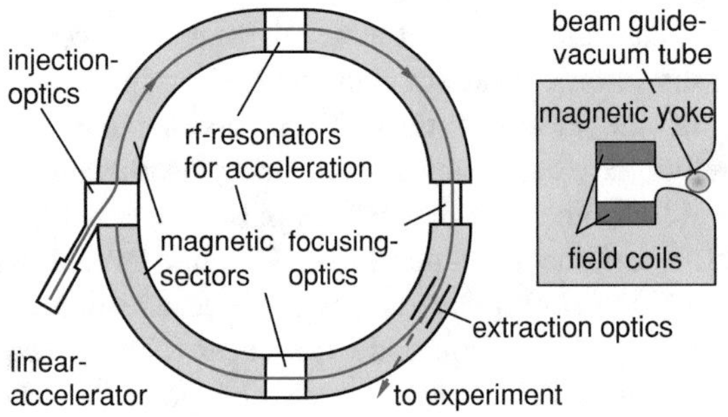

Fig. 4.23X Basic design of a synchrotron

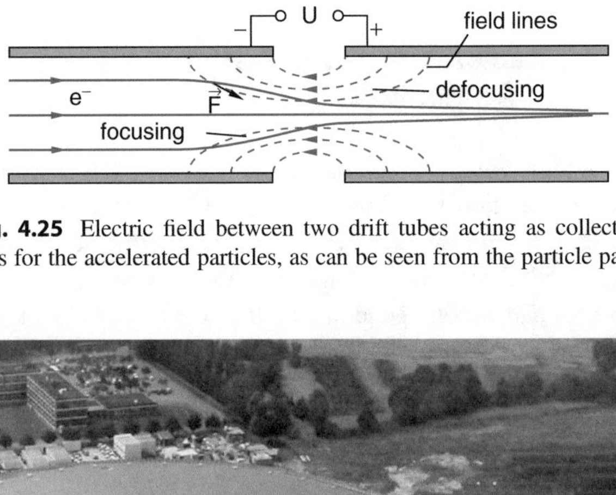

Fig. 4.25 Electric field between two drift tubes acting as collecting lens for the accelerated particles, as can be seen from the particle paths

Fig. 4.24 Synchrotron, in Soleil France https://commons.wikimedia.org/w/index.php?curid=376902

to the final value

$$f_2 = f(t_2) = k \cdot c/(2\pi \cdot r_0) \quad \text{for} \quad v \to c \tag{4.29b}$$

where the integer k gives the number of acceleration cavities along the circumference of the ring. Because also the necessary magnetic field

$$B(v) = \frac{mv}{qr_0} = \frac{v}{qr_0} \frac{m_0}{\sqrt{1 - v^2/c^2}} \tag{4.30}$$

which keeps the particles on their nominal circular path with radius r_0 depends on v, the acceleration frequency and the magnetic field are not independent. From (4.29a, b) and (4.30) one obtains

$$f = \frac{k \cdot q \cdot B}{2\pi m_0} \sqrt{1 - v^2/c^2}. \tag{4.31}$$

Elimination of v from (4.30) gives

$$f = \frac{k}{2\pi \cdot \sqrt{\left(\frac{r_0}{c}\right)^2 + \left(\frac{m_0}{q \cdot B}\right)^2}} \tag{4.32}$$

The acceleration frequency $f(v)$ has to be synchronized with the increasing magnetic field $B(v)$.

4.1.7 Stabilization of the Particle Path in Accelerators

In all high energy accelerators the particles travel during their acceleration very long distances. One has to take care that they do not deviate from their stable orbit due to collisions with the residual gas atoms in the vacuum chamber or due to small inhomogeneities of the magnetic field or the accelerating electric fields. Even small deviations would accumulate in large aberrations during their long travel distance resulting in collisions of the particles with the walls of the vacuum tube.

Examples

1. In the Stanford linear accelerator the travel distance is 5 km.
2. In a synchro-cyclotron the particles can circulate 10^3-times on a radius $r = 2$ m which gives an acceleration distance of $s = 1.2 \cdot 10^4$ m.
3. In a synchrotron the particles travel during the acceleration period $\Delta t \approx 1\ s$ with $v \approx c$ a distance of $3 \cdot 10^5$ km.
4. In a storage ring (see Sect. 4.1.8) the particles can be kept on their stable orbit for about $\Delta t > 10$ h, During this time they travel about $4 \cdot 10^9$ km. This corresponds to 4 times the path of the earth around the sun

These examples demonstrate that measures have to be taken to stabilize the particles paths. There are several possibilities:

- A suitable shaping of the electric acceleration fields to make them act as focusing lenses
- Installation of additional electrical or magnetic lenses
- Special shaping of the magnets in circular accelerators.

We will illustrate these possibilities by some examples.

In the drift-tube accelerators (see Sect. 4.1.4) the accelerating electric fields between the tubes act as lenses for the particle trajectories. Since the accelerating force $F = q \cdot E$ is always tangential to the electric field lines, the force is focusing in the first half but defocusing in the second half of the acceleration distance. Since the particle velocity is smaller in the first half than in the second half the radial deflection by the focusing force is larger in the first half and the total lens is convergent.

The situation is different in a high frequency accelerator. Here the electric field strength changes during the transit time of the electrons through the accelerating section. Therefore the electrons must pass during the rise time of the electric field otherwise the electron pulse would not be hold together and the time profile of the pulse would widen (Fig. 4.23). A slower electron passes the acceleration section at a higher field strength and experiences a higher acceleration, bringing it back to the center of the pulse, while a faster electron reaches the acceleration section earlier at a lower field strength and is therefore less accelerated.

Since the electric field rises during the transit time of the electrons the defocusing action in the second half of the lens becomes stronger than the focusing effect in the first half and the total lens becomes divergent. Therefore one has to invent other methods to focus the electron beam.

Often a homogeneous magnetic field is used between the acceleration sections, which points into the forward direction and acts as focusing lens (see Vol. 3, Sect. 2.6.4). In the drift-tube accelerator, for example, this field is generated by a solenoid around the drift tubes. Also magnetic quadrupole lenses are often used (Figs. 4.14 and 4.15).

In circular accelerators in particular in the synchrotron with narrow beam guiding tubes, the particles have to be stabilized in the radial as well as in the vertical direction, in order to keep them at their nominal circle in the x–y plane with radius r_0.

4.1.7.1 Vertical Stabilization

Vertical stabilization to the plane $z = 0$ can be achieved with an inhomogeneous magnetic field

$$B_z(r) = B_0\left(1 - n\cdot\frac{r - r_0}{r_0}\right) \quad (4.33)$$

which decreases with increasing r (Fig. 4.26), where $B_0 = B(r_0)$. The quantity n is the field index. With $dB/dr = -n \cdot B_0/r_0$ the field index is

$$n = -\frac{dB/B_0}{dr/r_0} \quad (4.34)$$

This shows that n gives the relative field decrease dB/B_0 per relative change dr/r_0 of the radius r.

Figure 4.26 illustrates that, because of the curvature of the field lines there exists for $z \neq 0$, besides the z-component also a radial component of the magnetic field. This component causes a Lorentz force F q · $(v \times B)$. With $B = \{B_r, 0. B_z\}$ we obtain

$$F_z = q\cdot v\cdot B_r \quad (4.35)$$

For a static field is $\mathbf{rot}B = 0$ which yields

$$(\mathbf{rot}B)_\varphi = \frac{\partial B_r}{\partial z} - \frac{\partial B_z}{\partial r} = 0$$

This gives

$$\frac{\partial B_r}{\partial z} = \frac{\partial B_z}{\partial r} = -n\cdot\frac{B_0}{r_0} \quad (4.36)$$

For $z << r_0$ the field B is approximately independent of z. The integration then gives

$$B_r(z) = -n\cdot\frac{B_0}{r_0}\cdot z \quad (4.37)$$

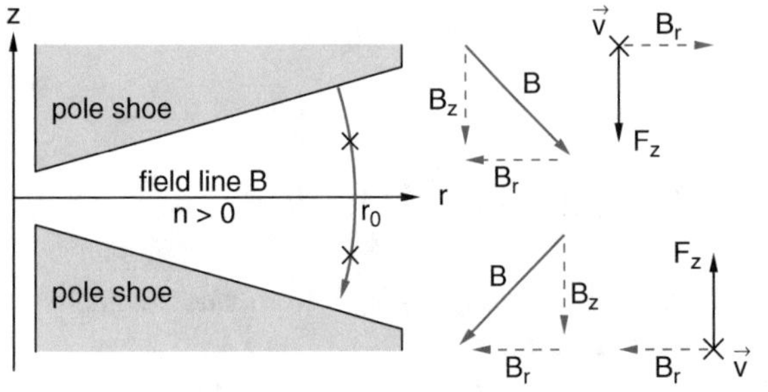

Fig. 4.26 Vertical stabilization of ions in z-direction in an inhomogeneous radially decreasing magnetic field

For magnetic fields with $n > 0$ the field exerts a restoring force on particles at $z \neq 0$ with the z-component

$$F_z = -\frac{q\cdot v\cdot B_0}{r_0}\cdot n\cdot z \quad (4.38)$$

which induces vertical oscillations of the particles through the plane $z = 0$ with the frequency

$$f_z = \frac{1}{2\pi}\sqrt{\frac{q}{m}\cdot\frac{v\cdot B_0}{r_0}\cdot n}. \quad (4.39a)$$

With the equilibrium condition $m \cdot v^2/r_0 = q \cdot vB_0$ we obtain from (4.39a)

$$f_z = \frac{1}{2\pi}\frac{v}{r_0}\sqrt{n}. \quad (4.39b)$$

Example

$v \approx c$, $r_0 = 100\ m$, $n = 1$ => $f_z = 5 \cdot 10^5\ s^{-1}$.

During their circulation on a circle with radius r_0 the particles perform vertical oscillations with the spatial period

$$L = \frac{v}{f_z} = 2\pi r_0/\sqrt{n}. \quad (4.40)$$

For $n < 1$ the period is larger than the circumference of the accelerator ring. The amplitude z_{max} of the vertical oscillation must be smaller than half of the heights of the vacuum tube. Otherwise the particle would collide with the wall.

4.1.7.2 Radial Stabilization

How is the situation with the radial stability, i.e. what happens with a particle that is pushed out of its nominal circle with radius $r = r_0$ but still remains in the plane $z = 0$? For the nominal circle is the equilibrium condition Lorentz force = centripetal force

$$\text{Lorentz force} = \text{centrifugal force},$$
$$q\cdot v\cdot B_0 = \tfrac{m\cdot v^2}{r_0},$$

For $r = r_0$ the two forces just cancel. This is no longer true for $r \neq r_0$. Here a difference ΔF between the two forces appears which is with (4.33)

$$\begin{aligned}\Delta \boldsymbol{F} &= \left[\frac{mv^2}{r} - q\cdot v\cdot B_0\left(1 - n\frac{r-r_0}{r_0}\right)\right]\hat{e}_r \\ &= \frac{mv^2}{r_0}\left[\frac{1}{1+\frac{r-r_0}{r_0}} - \left(1 - n\frac{r-r_0}{r_0}\right)\right]\hat{e}_r.\end{aligned} \quad (4.41a)$$

For $\Delta r/r_0 << 1$ we can approximate (4.41a) using $1/(1 + x) \approx 1 - x$ and obtain

$$\Delta \boldsymbol{F} = -\frac{mv^2}{r_0^2}(1-n)(r-r_0)\hat{e}_r. \qquad (4.41b)$$

For $n < 1$ this force is restoring, i.e. for $r > r_0$ its direction points to the center of the accelerator ring, for $r < r_0$ it points to larger values of r. The particles perform radial oscillations with the frequency

$$f_r = \frac{1}{2\pi}\sqrt{\frac{mv^2(1-n)}{r_0^2 \cdot m}} = \frac{v}{2\pi r_0}\sqrt{1-n}. \qquad (4.42)$$

As long as the amplitude of these oscillations does not exceed half of the width of the vacuum tube the particles remain on their stable path.

For $n = 0$ (homogeneous magnetic field without radial gradient) the oscillation period $L = v/f_r$ equals the circumference of the accelerator ring, but there is no longer a radial as well as a vertical stabilization.

For $n < 0$ the period length L and the oscillation ampltudes become small.

4.1.7.3 Alternating Field Gradient

Synchrotrons for very high energies use nowadays mostly the principle of alternating field gradients. Here the magnets along the particle path are divided in segments which have alternately large positive and negative field indices n (with $|n| >> 1$) (Fig. 4.27). In the field segments with positive n values the particles are stabilized in the vertical direction but defocused ′in the radial direction, while in the segments with $n < 0$ ($dB/dr > 0$) they are stabilized in the radial direction but destabilized in the vertical direction (Fig. 4.28). In order to illustrate that, in spite of the alternating destabilization altogether a stable particle path is realized, an example in optics might be helpful (Fig. 4.29): A system of a collecting lens with the focal length $+f$ and a diverging lens with focal length –f has the total focal length $f_t = f^2/d$, where d is the distance between the two lenses (Vol. 2. Sect. 9.5). For $d > 0$ the system is focusing, while for $d \to 0$, $f_t \to \infty$.

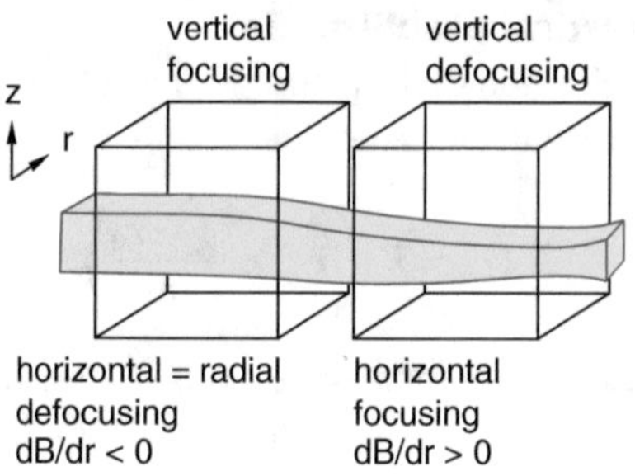

Fig. 4.28 Principle of strong focusing. Vertical focusing for $dB/dr < 0$ and radial focusing for $dB/dr > 0$

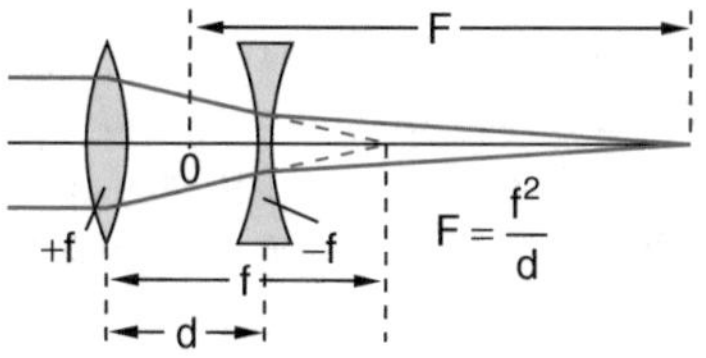

Fig. 4.29 Resulting focusing properties of a system of converging lens with focal length f_1 and diverging lens with $f_2 = -f_1$ as optical analogue to the strong focusing in a synchrotron

The quantitative treatment of the focusing properties in the alternating fields demands the solution of the equation of motion of the particles in the alternating gradient field **E**. We use cylindrical coordinates $\{r,\varphi,z\}$ (Fig. 4.30a). The velocity of a particle is then

$$\mathrm{v} = \{\underline{\mathrm{r}}, \mathrm{r} \cdot \underline{\varphi}, \underline{\mathrm{z}}\},$$

where $r \cdot \dot{\varphi} \approx (q/m) r_0 \cdot B_0$ is very large compared to $\dot{r}$ and $\dot{z}$. With the Lorentz force $F = q(v \times B) = m \cdot \ddot{\mathrm{v}}$ and the magnetic field $B = \{B_r, 0, B_z\}$ the equation of motion for the two relevant components becomes

$$q \cdot r \cdot \dot{\varphi} \cdot B_z = m \cdot \ddot{r}, \qquad (4.43a)$$

$$-q \cdot r \cdot \dot{\varphi} \cdot B_r = m \cdot \ddot{z} \qquad (4.43b)$$

With

$$B_z = B_0\left(1 - n\frac{r-r_0}{r_0}\right); \quad B_r = -n \cdot \frac{B_0}{r_0} \cdot z,$$

The index n changes its sign from one magnetic field sector to the next one.

We will here discuss only the vertical stabilization in $\pm$ z direction. For the radial stabilization a similar derivation is valid.

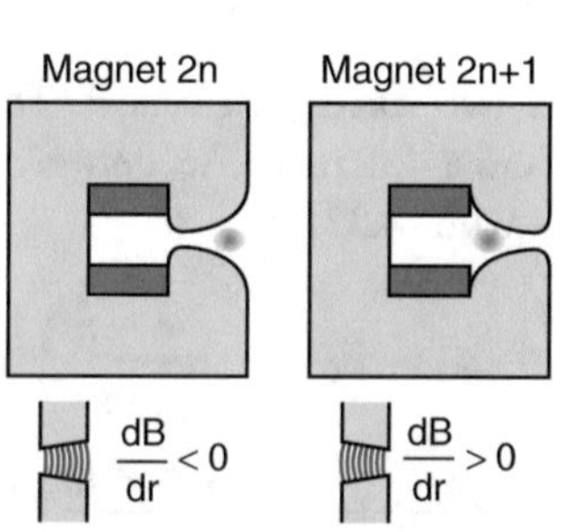

Fig. 4.27 Principle of alternating field gradients in a synchrotron

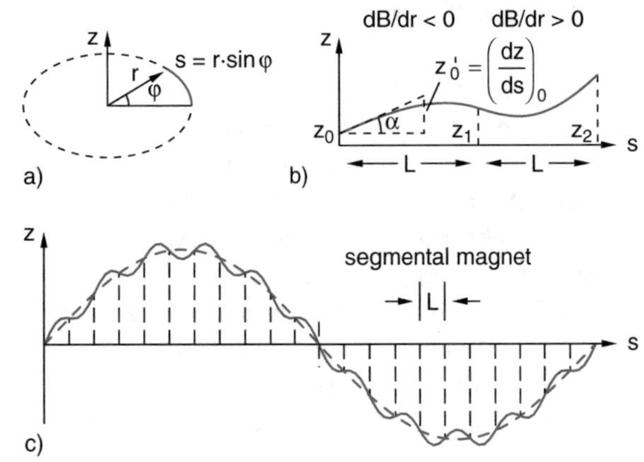

Fig. 4.30 Illustration of Betatron oscillations

Replacing according to (4.37) B_r by B_0, Eq. (4.43b) can be written with $v = r \cdot \dot{\varphi}$ and $q \cdot B_0 = m \cdot v / r_0$ as

$$\ddot{z} = -\frac{q}{m} \cdot v \cdot \frac{B_0}{r_0} \cdot n \cdot z = -\frac{v^2}{r_0^2} \cdot n \cdot z. \tag{4.43c}$$

The solutions are vertical harmonic oscillations

$$z = A \cdot \sin(2\pi f_z \cdot t) + B \cdot \cos(2\pi f_z \cdot t) \tag{4.44a}$$

with the frequency

$$f_z = \frac{1}{2\pi} \frac{v}{r_0} \sqrt{n},$$

which the particles perform during their way through a magnetic sector with the length $L = r_0 \cdot \varphi$ (Fig. 4.30c). With (4.39b) and $s = v \cdot t$ (4.44a) converts to

$$z = A \cdot \sin\left(\sqrt{n} \cdot s/r_0\right) + B \cdot \cos\left(\sqrt{n} \cdot s/r_0\right). \tag{4.44b}$$

These oscillations are called Betatron-oscillations. The vertical slope of the oscillation plane against the circular path of the particles is given by the derivation

$$\begin{aligned} z' = \frac{dz}{ds} &= A \cdot \frac{\sqrt{n}}{r_0} \cos\left(\sqrt{n} \cdot s/r_0\right) \\ &- B \cdot \frac{\sqrt{n}}{r_0} \sin\left(\sqrt{n} \cdot s/r_0\right) \end{aligned} \tag{4.44c}$$

If the initial values z and dz/ds at the beginning of the segment ($s = 0$) are known, the constants A and B and therefore the vertical component of the particle curve can be determined.

In the subsequent sector with $n < 0$ the magnetic field acts defocusing in z-direction and the solution of (4.43c) is the exponential function

$$z = C \cdot e^{\sqrt{n} \cdot s/r_0} + D \cdot e^{-\sqrt{n} \cdot s/r_0}, \tag{4.45a}$$

$$\frac{dz}{ds} = \frac{\sqrt{n}}{r_0}\left(C \cdot e^{\sqrt{n} \cdot s/r_0} - D \cdot e^{-\sqrt{n} \cdot s/r_0}\right). \tag{4.45b}$$

With the length L of the path through one sector the solutions of (4.44b, 4.44c) for $s = L$ must be identical to the solutions of (4.45a, 4.45b) for $s = 0$. It is therefore possible to calculate the particle path through the different segments similar to the path of a light ray through an optical system of different optical elements in geometrical optics using the matrix method (see Vol. 2, Sect. 9.6). With the values z_1 and $(dz/ds)_1$ at the entrance into a segment and z_2,$(dz/ds)_2$ at the exit of the segment we get the matrix equation

$$\begin{pmatrix} z_2 \\ r_2' \end{pmatrix} = \begin{pmatrix} a & b \\ c & d \end{pmatrix} \cdot \begin{pmatrix} z_1 \\ r_1' \end{pmatrix} \tag{4.46}$$

The matrix (abcd) gives the characteristic properties of the segment, which can be calculated with (4.44a–4.44c) and (4.45a, 4.45b). The quantities z_n and $(dz/ds)_n$ after n segments can be obtained by multiplication of the corresponding matrices.

The quite similar treatment of the radial motion in the plane $z = 0$ gives in the radial focusing segments (n < 0) the radial oscillation

$$r = r_0[1 + a \cdot \sin(2\pi f_r \cdot t + \varphi_r)] \tag{4.47}$$

around the nominal radius r_0. In the radially defocusing segments ($n > 0$) we get again the exponentially increasing radial deviations from the nominal radius $r = r_0$.

The maximum deviation in radial and vertical direction are much smaller in the alternating gradient field devices than in the conventional non-alternating accelerators.

Taking all these considerations into account one can describe the particle path as superposition **of** vertical and radial oscillations around a point M on the nominal circle ($r = r_0$, $z = 0$), which moves with the velocity $v_M = r_0 \cdot \dot{\varphi}$. In a coordinate system moving with M as zero point the particle path would be an ellipse.

4.1.7.4 Synchrotron Oscillations

The acceleration of the particles by the HF-voltage $U = U_0 \cdot \cos(2\pi f t + \varphi_0)$ of the electric field in the cavity resonators depends on the phase $\varphi(t_n) = 2\pi f t_n + \varphi_0$ of the field during the flight time t_n of the particles through the accelerating cavity. The energy gain per circulation is with k cavities around the circle

$$\Delta E = k \cdot q \cdot U_0 \cdot \cos\varphi(t_n). \tag{4.48}$$

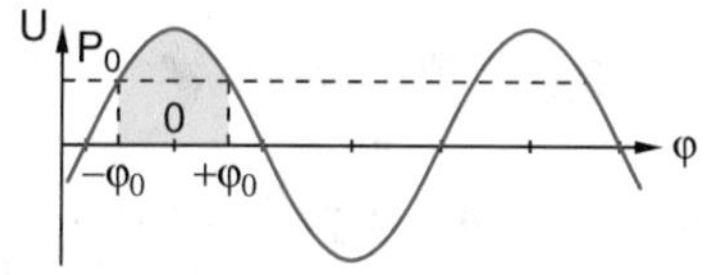

Fig. 4.31 Stable phase range of the HF-voltage in the acceleration sections for the passage of the particle package

So that the particles are always kept on their nominal path, the increase of the magnetic field must be synchronized with the energy increase. With $m \cdot v^2/r_0 = q \cdot v \cdot B_r$ follows:

$$m \cdot v = p = q \cdot r_0 \cdot B. \tag{4.49}$$

Because of $E^2 = m^2c^4 = m_0^2c^4 + p^2c^2$ one obtains the increase in energy per cycle:

$$\Delta E = (\mathrm{d}E/\mathrm{d}t) \cdot T \quad \text{with} \quad T = 2\pi r_0/v \tag{4.50}$$

The necessary increase of the magnetic field per unit time can be obtained from (4.49) with

$$E \cdot \mathrm{d}E/\mathrm{d}t = c^2 p \cdot \mathrm{d}p/\mathrm{d}t$$

$$\frac{\mathrm{d}B}{\mathrm{d}t} = \frac{1}{qr_0} \cdot \frac{\mathrm{d}p}{\mathrm{d}t} = \frac{E}{qr_0 \cdot c^2 p} \frac{\mathrm{d}E}{\mathrm{d}t}. \tag{4.51}$$

Inserting $\mathrm{d}E/\mathrm{d}t = \frac{\Delta E}{T} = \frac{\Delta E \cdot v}{2\pi r_0}$ and taking ΔE from (4.48) this gives

$$\frac{\mathrm{d}B}{\mathrm{d}t} = \frac{E}{qr_0 \cdot c^2 p} \frac{\Delta E \cdot v}{2\pi r_0} = \frac{E(t) \cdot k \cdot U_0 \cdot \cos\varphi(t_n)}{2\pi r_0^2 \cdot c^2 \cdot m}.$$

If a particle deviates from its nominal circle it will pass through the accelerating cavity at another phase φ of the accelerating field and will therefore experience another energy gain ΔE. In order to stay on a stable path particles which arrive later must experience a larger accelerating voltage, those which arrive earlier a lower voltage. The point P_0 in Fig. 4.31 represents a stable phase φ_0 for an ideal particle which moves on the nominal circle. The real particles perform phase oscillations (synchrotron oscillations) around such a stable phase φ_0. The amplitude φ_{max} of these phase oscillations is restricted to the red colored range in Fig. 4.31. If $|\varphi_{max}| > |\varphi_0|$ the particle becomes unstable and leaves its particle packet.

Remark Note the difference of the stable phase range of the acceleration voltage fixed in space in Fig. 4.31 against the phase range of the travelling wave in Fig. 4.12.

4.1.8 Storage Rings

We have discussed in the previous sections that in synchrotrons only one particle packet can be accelerated during one acceleration cycle which takes the time $\Delta t = t_3 - t_2$ (Fig. 4.24x). After having reached their final energy the particles can be kept for a short time interval at the maximum energy where they are available for planned experiments. In order to increase the time available for experiments and in particular the flux of the high energy particles, storage rings have been developed. The accelerated particles are injected into the storage ring where they circulate for long times at constant energy while continuously particle packets are injected from the synchrotron into the storage ring thus increasing the particle flux in the storage ring (Fig. 4.32). This procedure increases the current of the charged particles considerably and one can reach currents of more than 10 A. Because of radiation losses the particles have to be accelerated in order to compensate the losses and keep the particles at a constant energy. Here also the particles run as packets around the storage ring, because the ring is fed by packets from the synchrotron and all particles have to pass through the accelerators in the storage ring during the right time interval where the phase of the HF- field has the correct value (Figs. 4.31 and 4.33). The spatial extension of the packets depends on the velocity spread of the particles and on the focusing properties of the storage ring.

Even if the energy of the particles remains constant, the accelerator sections are necessary in order to compensate for the energy losses per circulation

$$\Delta E = (Z \cdot e)^2 \cdot E^4 / \left(3\varepsilon_0 \cdot r_0 \left(mc^2\right)^4\right)$$

caused by the radiation emitted by charged particles, travelling with constant amount of their velocity $v \approx c$ in a ring with radius r_0.(see Vol. 2, Sect. 6.5.4). In Fig. 4.32 such a storage ring is shown schematically.

Storage rings offer another essential advantage: They allow the simultaneous storage of negatively and positively charged particles which run into opposite directions through

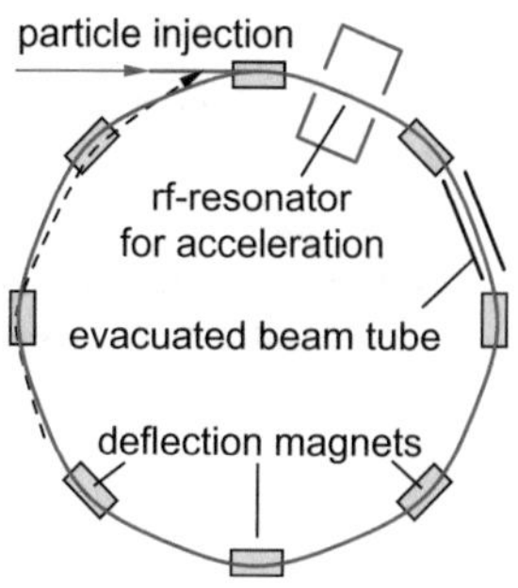

Fig. 4.32 Schematic drawing of a storage ring

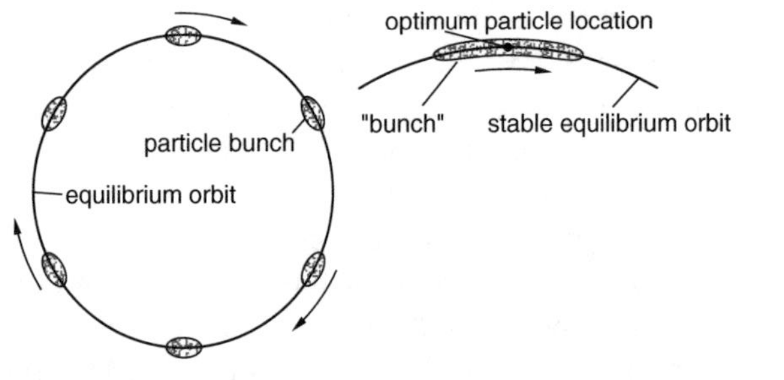

Fig. 4.33 Particle bunches in a storage ring

the same vacuum tube. At special places in the storage ring both particle bunches are collimated to a very narrow focus where they have a larger probability to collide with each other. Therefore such devices are called "**colliders**". Examples for such colliders are DORIS (Double Orbiting Ring System) in Hamburg, Germany or LEP (large electron–positron collider) at CERN close to Geneva. Both colliders investigate electron–positron collisions at high energies. DORIS has a circumference of 289 m and a collision energy of 4.5 GeV, while LEP is with 27 km the worldwide largest electron–positron collider with energies above 200 GeV. High energy collisions between protons p and antiprotons $\overline{p}$ can be measured for example, in the colliders TEVATRON in Batavia, Illinois, or in the large hadron collider LHC at CERN.

At the crossing points huge detectors are installed (see Sect. 4.3.6) which measure the reaction products resulting from the electron–positron collisions resp. the hadron collisions.

Figure 4.34 shows as example the DORIS storage ring where alternately electrons and positrons are accelerated in the synchrotron and injected into opposite directions into the storage ring.

For central collision the total energy 2E of the collision partners can be converted into reaction energy (for instance for the production of new particles), while for collisions of a particle A with high kinetic energy with another particle B at rest with the same rest mass as A according to (4.13h) only a small fraction

$$\eta = E_s/\left(m(v_A)\cdot c^2\right) = \sqrt{2/(1+\alpha)}$$

with $\alpha = (m/m_0) - 1$ and $E_s = m_0 \cdot c^2 \times \sqrt{(2+2\alpha)}$ can be converted.

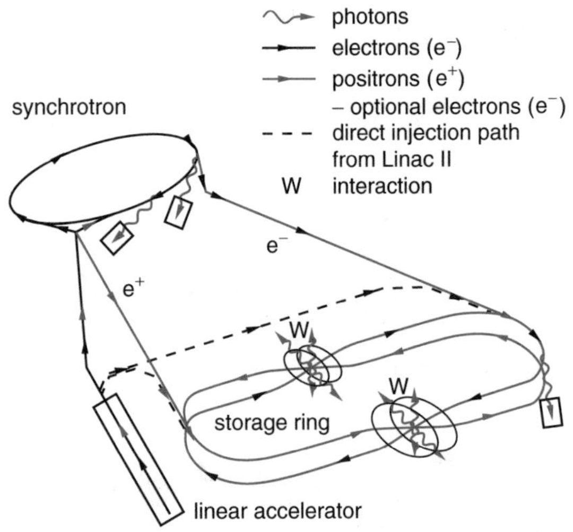

Fig. 4.34 Storage ring DORIS (double orbiting ring system) in Hamburg with synchrotron and collision zones into which the two anti-collinear particle beams are focused (from E.E.Koch, C,Kunz: Synchrotron Strahlung bei DESY, Hamburg 1974)

Example

1. A proton with E_{kin} = 100 GeV ($v \approx c$) collides with a proton at rest. Here is m = 101 GeV/c^2, m_0 = 1 GeV/c^2, $\Rightarrow$ α = 100 $\Rightarrow$ $\eta = \sqrt{(}2/101$ = 1/7. Only 14% of the kinetic energy of A $\Rightarrow \Delta E$ 14 GeV can be converted.This should be compared with ΔE = 200 GeV for an anti-collinear central collision of two protons.
2. Positrons with E_A = 1 GeV collide with electrons at rest (m_0c^2 = 0.5 meV). $\Rightarrow$ α = 2000 $\Rightarrow$ = $\sqrt{(}0.001$ = 0.03 $\Rightarrow$ only 3.3% of the positron energy is available for excitation.

$$\Rightarrow \frac{E_S}{E_A} = 0{,}137.$$

3. Electrons with E_A = 30 GeV collide with protons at rest (m_0c^2 = 936 MeV). In this case the energy efficiency becomes 0.25 i.e E_s/E_A = 0.25. Only 25% of the electron energy can be used for exciting the protons.

$$\Rightarrow \frac{E_S}{E_A} = 0{,}25.$$

With the colliders one pays for the much larger available excitation energy with a much smaller collision rate, because for a target at rest one can use a solid or liquid target (for example liquid hydrogen) with a density, that is several orders of magnitude higher than achievable in the crossing point of two counter-propagating beams.

In order to realize higher particle densities in the colliding beams, both beams ore tightly focused by special ion optical elements increasing the particle flux densities considerably. Around these crossing point the detectors are arranged, measuring the reacting particles.

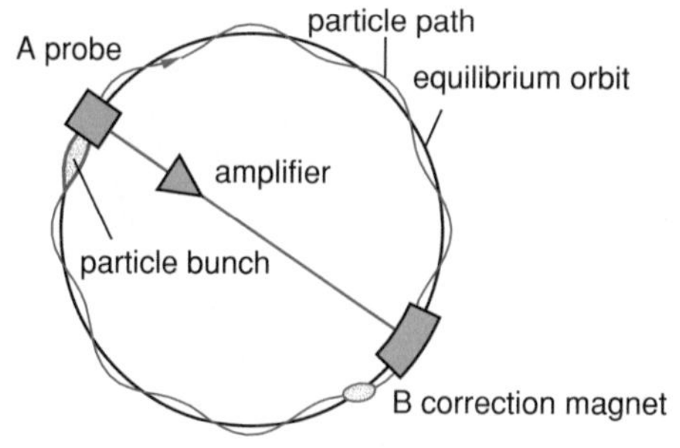

Fig. 4.35 Principle of Stochastic Cooling

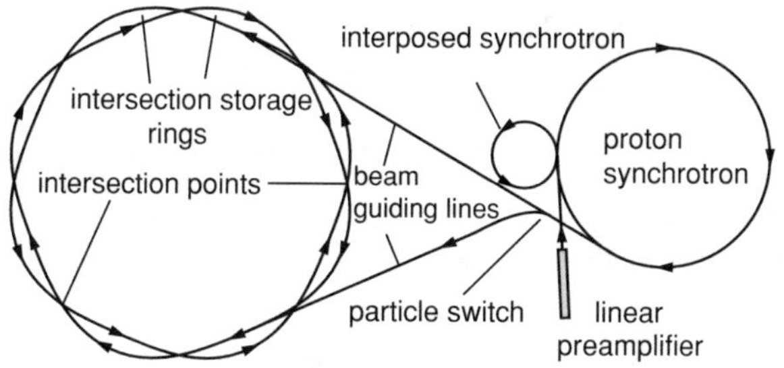

Fig. 4.36 Synchrotrons and storage ring at CERN, Geneva, Switzerland

The reaction rate $\dot{N}$ for the investigated reaction with the reaction cross section σ and a beam cross section A is

$$\dot{N} = (N_1 \cdot N_2/A) \cdot f \cdot \sigma = L \cdot \sigma. \qquad (4.52)$$

where N_1 and N_2 are the number of particles in one of the particle bunches, and f is the number of colliding bunches per sec. The factor

$$L = (N_1 \cdot N_2/A) \cdot f \qquad (4.53)$$

is the **luminosity** of the storage-ring.

We can infer from (4.52) that the cross section A of the two beams in the crossing points should be as small as possible (this demands a very careful alignment of the two counter-propagating beams). Furthermore the time-synchronization of the particle pulses in the two beams must be very accurate in order to bring the collisions exactly to the focal points where the detectors are located.

In order to determine the location of a collision, which limits the accuracy of the scattering angle, as accurate as possible, the length of the particle bunches should be short. This length is given by the velocity distribution of the particles in the bunch. Here the new technique of **stochastic cooling** [14] has been developed by *S. van der Meer* (Nobel prize 1984), which was first applied to the proton-antiproton storage ring at CERN (**C**onceil **E**uropeen pour las **R**echerché **N**ucleaire) close to Geneva Switzerland. The time distribution of the particles is measured at the point A (Fig. 4.35). The detector signal is sent on a diameter of the accelerator ring to the point B where it controls a correction magnet which changes the path of the particles in such a way that the slow particles traverse a shorter path and the fast particles a longer path, thus constricting the time distribution of the particles. This occurs at every circulation until the maximum possible constriction is reached.

A new design for proton-proton collisions at high energies, which is realized at the double storage ring collider at CERN, is based on the following construction: Two storage rings with opposite directions of the magnetic fields, where the protons run into opposite directions, are closely placed on top of each other (Figs. 4.36 and 4.36x). At specific points the two particle beams intersect each other. This allows anti-collinear collisions between particles of the same kind. The synchrotron, which feeds the protons into the

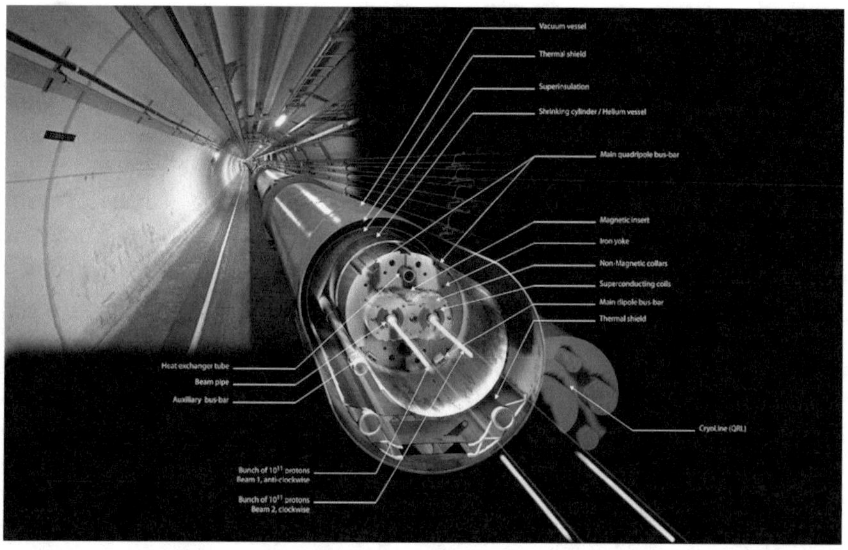

Fig. 4.36x LHC, double storage ring: The two beam lines for the anticollinear proton beams (yellow lines) (CERN)

storage rings must inject the protons alternately into opposite directions into the two rings. This can be achieved with properly designed magnets and beam guiding lines. The protons are pre-accelerated in a linear accelerator, further accelerated in a small synchrotron up to 800 MeV and brought to their final energy of 30 GeV in the large proton synchrotron. A deflecting magnet extracts the protons out of the synchrotron and guides them to the injection device of the storage rings where they are alternately injected into the two storage rings into opposite directions.

In another method the high energy particle bunches are accumulated in a storage ring and then extracted and by special guiding lines alternately in opposite directions injected into the double storage ring. One has to take care that the oppositely running bunches meet exactly at the crossing points.

4.1.9 Synchrotron Radiation

The electron storage rings offer, besides their use in high energy physics an intense source of synchrotron radiation, which is emitted by the electrons on their way through the circular segments (see Vol. 2, Sect. 6.5.4).The radiation power, emitted by a charge e with energy E accelerated on a circle with radius r_0 is [15, 16]

$$P(E, r_0) = \frac{e^2 \cdot c}{6\pi\varepsilon_0 r_0^2}\left(\frac{E}{m_0c^2}\right)^4 = \frac{e^2 c\gamma^4}{6\pi\varepsilon_0 r_0^2} \tag{4.54a}$$

It is proportional to the 4th power of the ratio $\gamma = (E/m_0c^2)$. The energy loss ΔE of the charge e per circulation is

$$\Delta E = \frac{P(E, r_0)\cdot 2\pi r_0}{c} = \frac{e^2\gamma^4}{3\varepsilon_0 r_0}. \tag{4.54b}$$

Inserting typical numerical values one gets

$$\Delta E = 9{,}64 \cdot 10^{-28}\left(\gamma^4/r_0[m]\right)[\text{Joule}]. \tag{4.54c}$$

With E measured in units of MeV and r_0 in meters, this gives

$$\Delta E/\text{MeV} = 6\cdot 10^{-15}\frac{E^4/\text{GeV}}{(m_0c^2)^4/\text{GeV}^4\cdot r_0/m}. \tag{4.54d}$$

With the electron mass energy m_0c^2 = 0.5 MeV we get

$$\Delta E/\text{MeV} = 0{,}0885\cdot\frac{E^4/\text{GeV}^4}{r_0/\text{m}}. \tag{4.54e}$$

Example

With E = 6 GeV and r_0 = 100 m we get for one electron $(\gamma = mc^2/m_0c^2 = 1.2\cdot 10^4)$.

$$\Delta E = \frac{e^2\cdot\gamma^4}{3\varepsilon_0 r_0} = 1,25\,\text{MeV}$$

per circulation. This energy loss has to be compensated by the acceleration sections, in order to keep the energy of the electrons constant.

For protons (m_0c^2 = 938 MeV) under the same conditions E and r_0 ΔE becomes smaller by the factor 1839^4 which gives.

$$\Delta E = 10^{-7} eV = 10^{-13}\,\text{MeV}.$$

This is completely negligible.

This example illustrates that the synchrotron radiation is significant only for electrons while for protons it is negligible. Even for very high proton energies the energy ΔE of the synchrotron radiation is very small. For example with E = 1000 GeV and r_0 = 5000 m →

$$\Delta E = 1.5\text{eV.per circulation.}$$

Since the total power of the synchrotron radiation is proportional to the electron flux, i.e. the number of electrons passing through a cross section of the accelerator, storage rings with their much higher electron flux (up to 10 A) are superior to synchrotrons. Furthermore this current stays constant over many hours while in the synchrotron the electron bunches are available only for a few seconds per acceleration cycle.

In order to keep the electron energy in a storage ring constant the energy loss has to be replaced by accelerating sections along the storage ring.

> The synchrotron radiation power increases with the 4th power of the ratio mc^2/m_0c^2. It also increases with decreasing radius r_0 of the ring radius (according to (4.54a) proportional to r_0^{-2}), while the energy loss per circulation increases as $1/r_0$.

The spectral distribution of the synchrotron radiation was first calculated by J Schwinger. The result of his calculations is for the electron current I in the storage ring

$$\frac{dP}{d\omega} = \frac{e\cdot\gamma^4\cdot I}{3\varepsilon_0\cdot r_0\cdot\omega_c}\cdot S_S(\omega/\omega_c). \tag{4.54f}$$

where

$$\omega_c = 3c\gamma^3/(2r_0) \qquad (4.54g)$$

is the critical radiation frequency.

The spectral function S_S can be written as

$$S_S(x) = \frac{9 \cdot \sqrt{3}}{8\pi} \cdot x \cdot \int_x^\infty K_{5/3}(x)\mathrm{d}x,$$

With $\int_0^\infty S_S(x)\mathrm{d}x = 1$ and $K_{5/3}$ is the modified Bessel-function. It is $\int_0^1 S_S(x)\mathrm{d}x = 1/2$. The critical frequency ω_c divides the whole spectral range into two parts with equal radiation power.

The spectral distribution of the synchrotron radiation is shown in Fig. 4.37 for the example of the DORIS storage ring in Hamburg for different electron energies between 1 and 6 GeV. It illustrates that the radiation wavelength ranges from 100 nm in the near vacuum UV to 10^{-2} nm in the X-Ray region. Here the radiation power emitted within the wavelength interval $\mathrm{d}\lambda = 1$ nm into the solid angle $\mathrm{d}\Omega = 1$ Sterad is plotted as a function of the wavelength λ.

The angular range $\Delta\vartheta$ of the emitted radiation is given by

$$\tan\vartheta = 1/\gamma = m_0c^2/E \qquad (4.55)$$

it depends on the energy E of the electrons and decreases with increasing energy E.

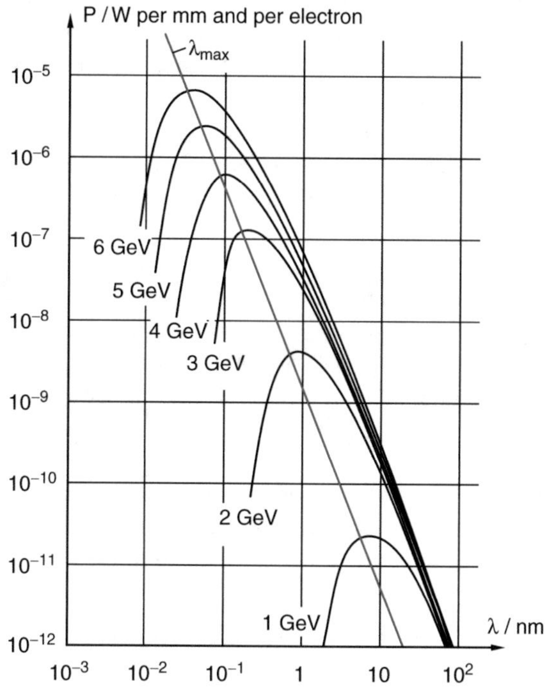

Fig. 4.37 Spectral intensity distribution of the synchrotron radiation of the electron storage ring DORIS for different electron energies (from E. E.Koch, C.Kunz: Synchrotron Strahlung bei DESY, Hamburg 1974)

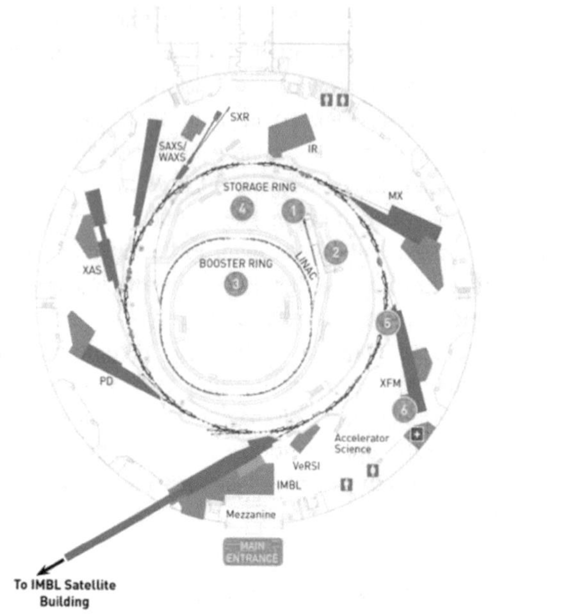

Fig. 4.37x Outlets in the storage ring to guide the synchrotron radiation to different experiments

The synchrotron radiation, which represents an inevitable energy loss can be used for atomic and molecular spectroscopy while, the high energy experiments can be performed parallel. It is used for the excitation of molecules to high lying energy levels, to many experiments in Molecular- and Solid State Physics as well as in Biophysics [18].

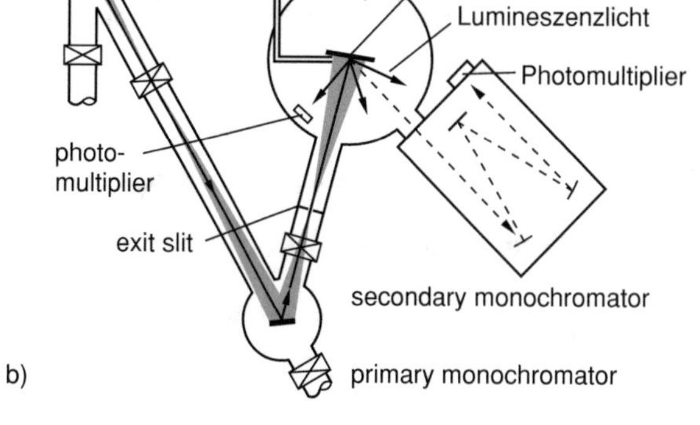

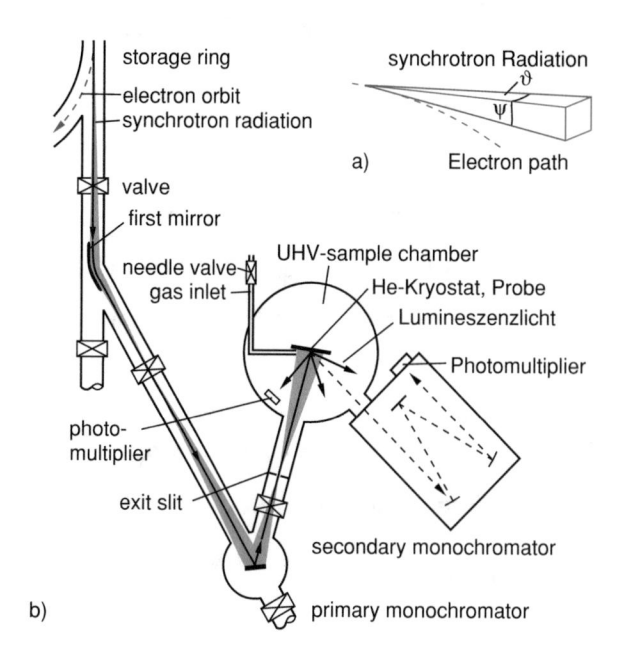

Fig. 4.38 **a** Emission of synchrotron radiation into a narrow angular range around the tangent to the circular electron path. **b** focusing of the radiation by a toroidal mirror onto the entrance slit of a VUV spectrograph

In order to optimize this "byproduct using" many beam lines are attached to the circular parts of the storage ring to collect and guide the synchrotron radiation to several experiments (Fig. 4.37x).

These radiation outlets guide the radiation to a mirror which bends the radiation into the wanted directions where it can be focused by a concave mirror and spectrally selected by a primary spectrographs (Fig. 4.38), before the sample is irradiated by selected wavelengths. The fluorescence emitted by the excited molecules can be again spectrally resolved by a secondary monochromator.

4.1.10 The Large Colliders

At the end of this section we will give a survey on existing and planned large accelerators and storage rings in order to illustrate the considerations above by specific real numerical values of size and achievable energies. All experiments where new particles are created by collisions between electrons or hadrons in the energy range above 20 GeV are based on colliders where the colliding particles run into opposite directions in large storage rings. This concept allows the conversion of the total kinetic energy of the colliding particles into excitation energy or the production of new particles.

The concept of the new large colliders was designed in such a way that the already existing accelerators could be used as pre-accelerators. This reduces the costs of the big colliders considerably. The disadvantage of this concept is that often large distances between the different accelerators and storage rings have to be accepted. This demands not only a very good vacuum to avoid collisions of the particles with the residual gas on their many thousand kilometer long travel paths, but also a well-designed ion optics to focus the particle beam in order to keep the particles on their correct path and to prevent collisions with the walls of the beam guiding tubes.

The total complex therefore represents a master stroke of modern engineering and surveying. This becomes evident when the demanded accuracy of positioning of the particle beams is regarded. For example the LEP storage ring at CERN with a circumference of 28 km does not tolerate a deviation of more than 1 mm from the stable orbit.

Figure 4.39 shows an aerial view of the storage rings PETRA (positron–electron tandem ring accelerator), circumference 2.3 km, and HERA (hadron-electron ring accelerator) with 6.3 km circumference at DESY (Deutsches Elektronen-Synchrotron in Hamburg Germany). The electrons and positrons are injected into PETRA from the electron synchrotron DESY, and PETRA serves as pre-accelerator for the larger storage ring HERA.

Fig. 4.39 Aerial view of DESY in Hamburg, indicated are the Positron-Electrons ring accelerator PETRA and the large hadron-electron ring accelerator HERA (With kind permission of DESY)

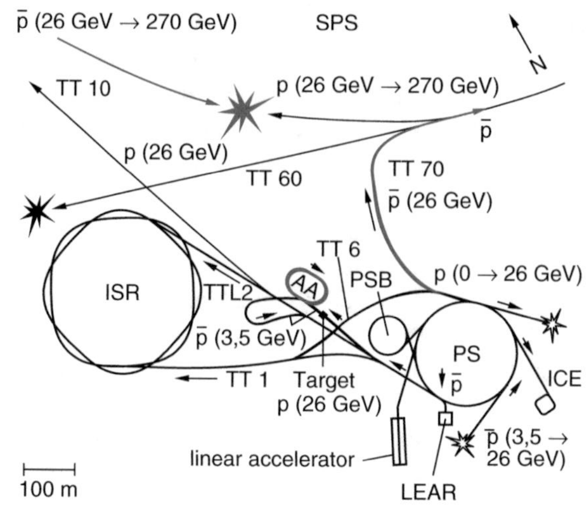

Fig. 4.40 26 GeV proton synchrotrons PS, 270 GeV super-proton synchrotron SPS, intersecting storage ring ISR, anti-proton accumulator AA, proton storage booster PSB, where the protons are concentrated and accelerated (see text). Note the numerous beam guide tubes which connect the different machines (CERN)

An even more complex facility is the proton-anti-proton collider at CERN, which is illustrated in Fig. 4.40. The protons are pre-accelerated in a linear accelerator, injected into the proton synchrotron PS where they reach a maximum energy of 26 GeV. These high energy protons are focused onto a tungsten disc where they produce many secondary particles, with a fraction of about 1% anti-protons. These anti-protons are focused into a collimated beam which is accelerated in the PS and stored in the storage ring AA (*anti-proton accumulator*). With a large cross section. For each PS bunch about 10^7 anti-protons can be accumulated. The anti-protons have initially a broad energy distribution, caused by their production mechanism, which is narrowed in the AA-ring by stochastic cooling down to $\Delta E/E \approx 6\%$. After 24 h about 10^{12} anti-protons are accumulated and cooled in the AA-ring. Now the anti-protons are transported back into the PS and there accelerated to 26 GeV. After having reached this energy they are guided into the super-proton synchrotron SPS which runs at a fixed energy and serves as storage ring. Here they meet the anti-collinearly running protons which had been accumulated in the SPS during the storage time of the anti-protons in the AA. Now the protons and anti-protons are accelerated in the SPS up to the final energy of 270 GeV. Both particle beams are focused into the observation points where they collide and produce new particles. These reactions are monitored by detectors which are arranged around the crossing points (Figs. 4.41 and 4.42).

Fig. 4.41 Aerial View of CERN. The three indicated rings show the proton synchrotron PS (small ring in the forefront, barely seen at the area of white houses). The super-proton synchrotoron SPS (medium ring) and the large hadron collider LHC8 (large ring). In the background is the lake Geneve). (with kind permission of CERN)

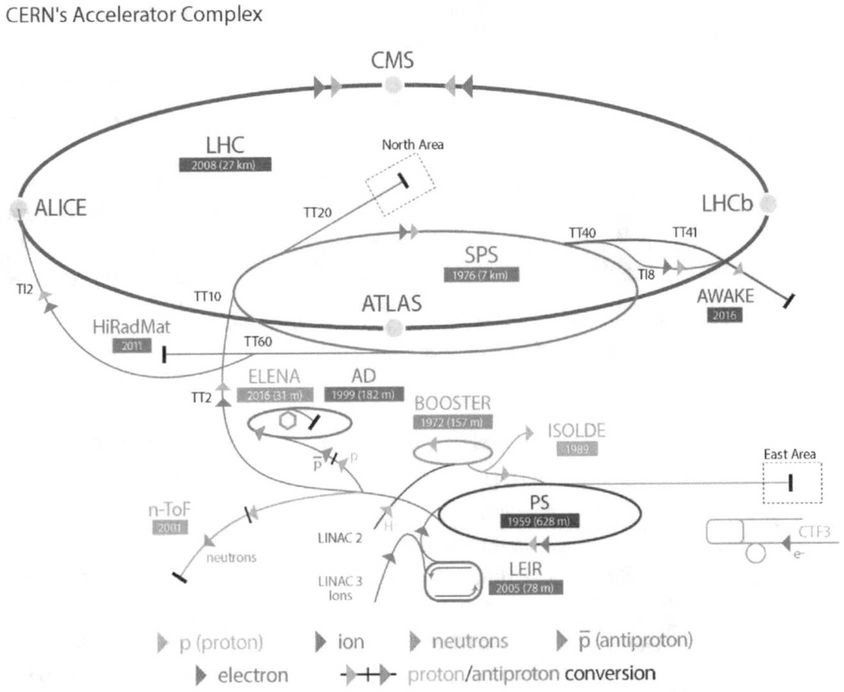

Fig. 4.42 Simplified drawing of the LHC with the 4 detectors at the crossing points and the various pre-accelerating systems. LA = linear accelerator, *B* = booster, PS = proton synchrotron SPS = super proton synchrotron (CERN)

4.1.11 The Large Hadron Collider LHC

The worldwide largest particle accelerator is the LHC at CERN near Geneva, Switzerland (Figs. 4.41 and 4.42), which accelerates protons up to 7000 GeV. The protons are guided into opposite directions in two tubes, which are arranged closely besides each other in opposite magnetic fields (Fig. 4.43 and 4.44). In four crossing points the protons are slightly deflected and focused into a common spot where they collide with each other (Figs. 4.43 and 4.46). A section of the large ring is shown in Fig. 4.45, where both beam lines with their opposite magnets are enclosed by large thermally isolated tubes allowing to cool down the whole interior to temperatures of a few Kelvin, in order to realize the advantages of superconducting magnets.

The energy available for the production of new particles in the collision area is.

$$\mathrm{E} = 2 \cdot 7000\,\mathrm{GeV} = 14.000\,\mathrm{GeV}.$$

Instead of the protons also other heavy ions such as C^{6+}, Au^{79+} or Pb^{82+} can be accelerated, giving a center-of-mass-energy of $2Z \cdot 7000$ GeV.

The protons are pre-accelerated in a linear accelerator up to 50 MeV and in a proton-synchrotron booster accumulated. From there they are transported into the already existing proton-synchrotron, accelerated and then injected into the updated super-proton-synchrotron SPS where they are further accelerated to 450 GeV (Fig. 4.40). Now they are alternately injected into opposite directions into one of the beam tubes of the LHC. Here they are further accelerated to their final energy of 7000 GeV.

The protons now traverse with a velocity $v = 0.9999999991 \cdot c = (1–9 \cdot 10^{-10})c$, are compressed into bunches with 16 μm diameter and 8 cm length, which run into opposite directions through the beam tubes which are evacuated to 10^{-10} mbar. Each bunch contains about 10^{11} protons and performs one circumference around the 28 km ring in a time of 90 μs. Within 1 s the protons run 11.245 times around the ring. They are kept on their circular path by 1232 superconducting magnets.

With a ring radius $r = 4.46$ km the centripetal force is

$$a_Z = \frac{v^2}{r} \sim \frac{c^2}{r} = 2.3 \cdot 10^{13}\mathrm{m/s^2}.$$

This corresponds to $2.4 \cdot 10^{12}$ times the gravitational acceleration on earth. Due to inhomogeneities of the magnetic and electric fields the particles deviate during their long path from their stable orbit. They are therefore refocused by 392 quadrupole magnets (see Fig. 4.14) along the ring.

At 4 crossing points both anti-collinear proton beams are focused and deflected in such a way that the counter-propagating particles can collide (Fig. 4.42). Around the collision zones huge detectors are arranged (Figs. 4.48, 4.49 and 4.50).

The luminosity L of the LHC, defined by (4.53) is with $N_1 = N_2 = N$ and a cross section $A = \mathrm{d}x \cdot \mathrm{d}y$

$$L = \left(N^2/A\right) \cdot f \quad \text{with} \quad A = \mathrm{d}x \cdot \mathrm{d}y \tag{4.53a}$$

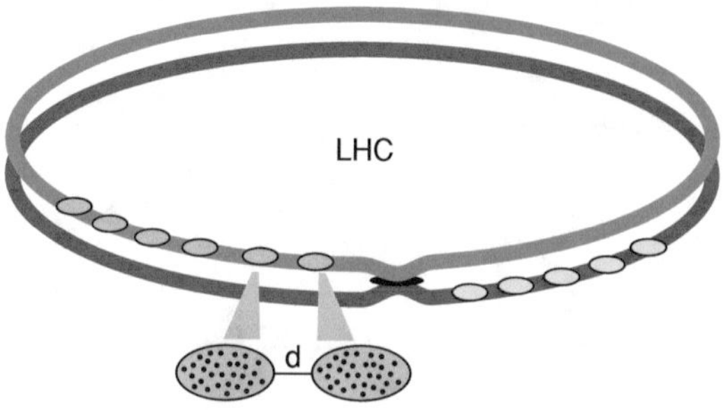

Fig. 4.43 Schematic drawing of the two channels with the anti-collinear proton bunches colliding in the crossing point

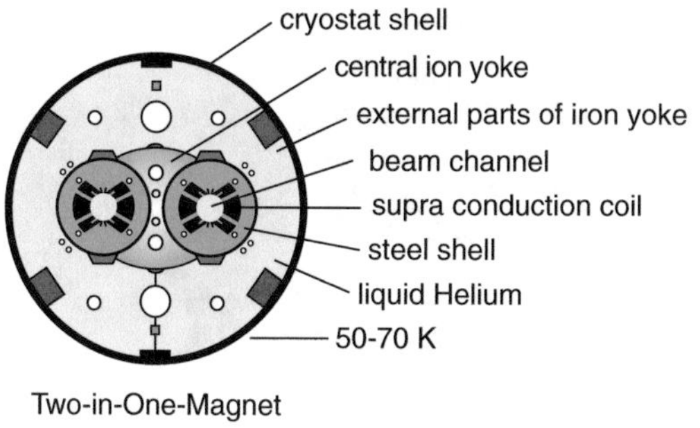

Fig. 4.44 The two beam lines with the opposite magnetic fields

for the anti-collinear particle beams running into ± z-directions, where f is the frequency of the particle bunches. Inserting the relevant data for the LHC one obtains a luminosity of $L = 10^{34}\ \mathrm{cm}^{-2}\ \mathrm{s}^{-1}$.

At each of the 4 crossing points indicated by yellow circle in Fig. 4.42 a huge detector is located which monitors with a sophisticated combination of different detection assemblies and electronic techniques energy and momentum of the produced particles and identify the kind of particles (see Sect. 4.3.6).

These detectors got names as abbreviations of their function:

ALICE = **A** **L**arge **I**on **C**ollider **E**xperiment.
ATLAS = **A** **T**orodial **L**HC **A**pparatu**s**.
CMS = **C**ompact **M**uon **S**olenoid.
LHCB = LHC beauty, which detects mesons containing the beauty quark.

More detailed information about the LHC and these detectors can be found in [18–19].

At the first test run in October 2008 unfortunately some of the supra-conducting magnets were destroyed, caused by

Fig. 4.45 Section of the tunnel with the Large Hadron Collider showing the cooling mantle containing the liquid helium to cool the magnets (CERN)

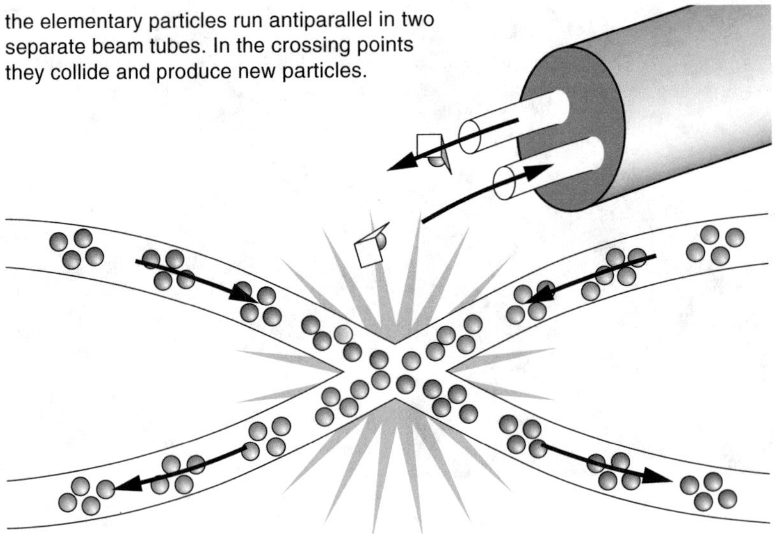

the elementary particles run antiparallel in two separate beam tubes. In the crossing points they collide and produce new particles.

Fig. 4.46 Illustration of the two beam lines where the protons travel into opposite direction. Both beams are focussed into the narrow collision zone

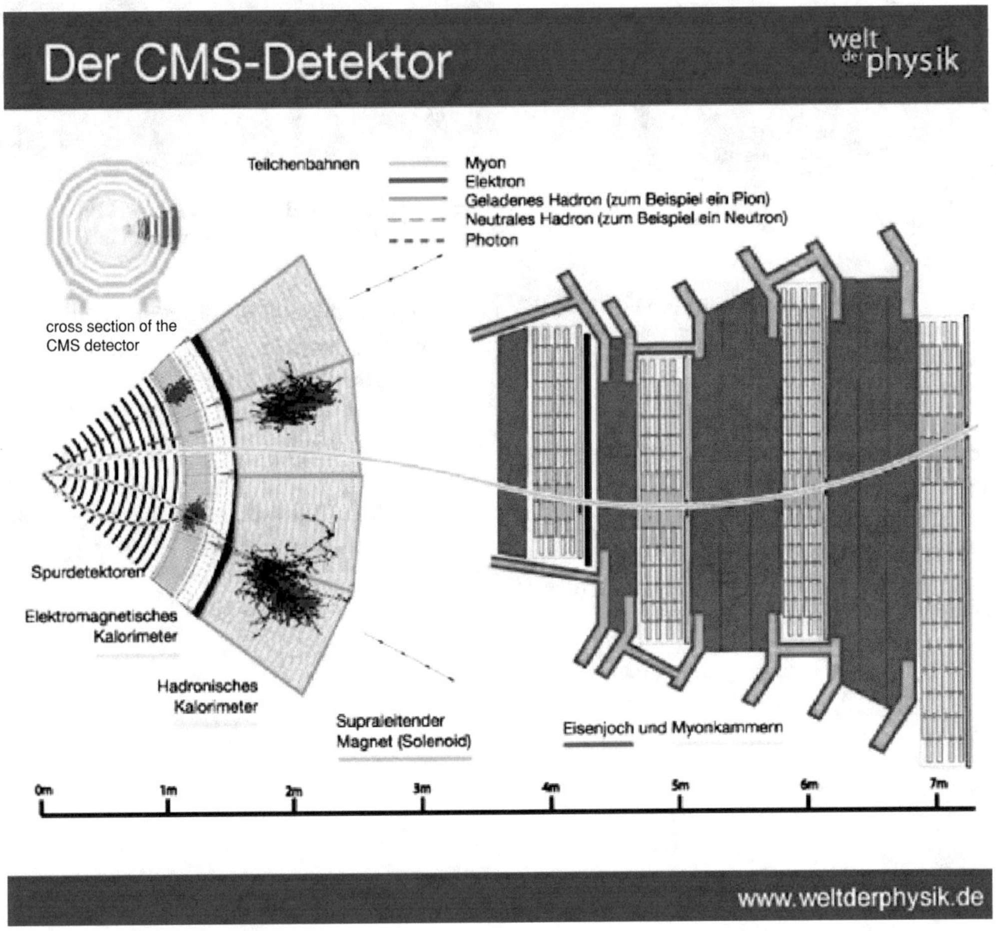

Fig. 4.47 The CMS-detector at the LHC (CERN)

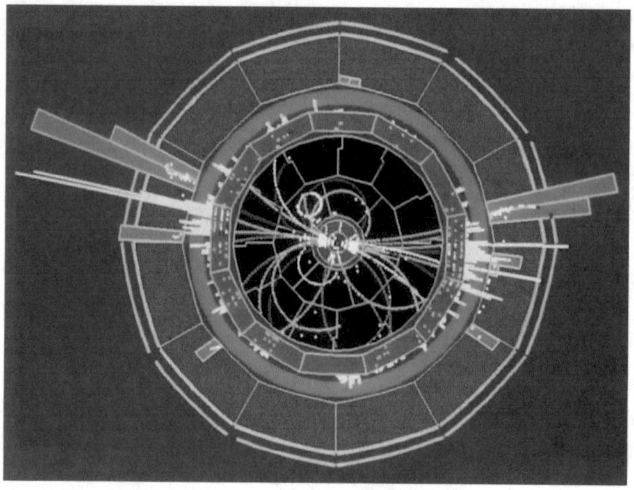

Fig. 4.48 Computer reproduction of the decay of a Z^0-vector boson via a quark-antiquark pair into a cascade of new particles in toe jets, flying into opposite directions. Monitored by the alpha-detector. (with kind permission of CERN)

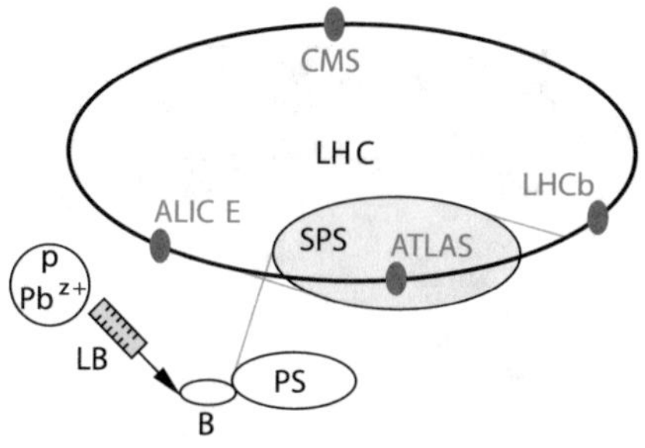

Fig. 4.49 The four large detectors at the crossing points of the LHC. LB = linear preacceleratgor, B = booster, PS = proton-syxnchrotron, SPS = super-proton synchrotron

a soldering defect. The regular operation of the LHC could be only started again in autumn 2009.

Due to its very high energy new particles have been discovered, for example the Higgs-boson (see Sect. 7) which represented the last missing particle in the standard model of particle physics [17].

Fig. 4.50 The ALEPH-detector at the LEP large electron–positron collider) at CERN (with kind permission of CERN)

The present investigations at the LHC concentrate on the discovery of possible particles which are responsible for the "**dark matter**". The hope is that possible results of such investigations might find out the limits of the Standard model and discover *New Physics* beyond the Standard Model.

In Table 4.1 the worldwide presently running or planned large colliders are compiled.

Table 4.1 The large Colliders

Collider	Location	Colliding particles	Maximum energy (GeV)	Center of Mass energy (GeV)	circumference (km)	Luminosity (cm^{-2} s^{-1})
PETRA	Hamburg	$e^- + e^+$	2×23.5	47	2.3	$1.6 \cdot 10^{31}$
HERA	Hamburg	$e^- + p$	30 + 820	314	6.336	$3.5 \cdot 10^{31}$
PEP	Stanford	$e^- + e^+$	2×22	44	Linear-Collider	$7 \cdot 10^{30}$
SLAC	Stanford	$e^- + e^+$	2×50	91	Linear-Collider	
LEP	CERN	$e^- + e^+$	2×100	200	26.6	$3.6 \cdot 10^{31}$
SPS	CERN	$p + \bar{p}$	2.270	540	6.9	
Tevatron	Batavia	$p + \bar{p}$	2.980	1960	6.3	$2 \cdot 10^{32}$
RHIC	Brookhaven	p + p Au + Au	2Z×8.86	Z.17.7	3.8	$2 \cdot 10^{26}$
UNK	Serphukov	$p + \bar{p}$	2×3000	6000	64	$3 \cdot 10^{30}$
LHC	CERN	p + p Pb + Pb	2×7000 $2 \times 1.5 \times 10^6$	14.000 3.10^6	26.6 26.6	$1.6 \cdot 10^{34}$ $1.6 \cdot 10^{34}$

4.2 Interaction of Particles with Matter

All detectors of micro-particles (electrons, protons, neutrons, mesons, neutrinos, photons etc.) use the interaction of these particles with the detector material. This interaction includes the following elementary processes:

- Elastic collisions with the electrons in the atomic shell
- Excitation or ionization of atomic electrons
- Deviation of charged particles in the Coulomb field of atomic nuclei, which leads to emission of X-rays.
- Elastic collisions with an atomic nucleus which suffers a recoil
- Inelastic collisions with an atomic nucleus, resulting in the excitation of the nucleus with subsequent emission of particles or γ-quanta.
- Emission of Cerenkov radiations when charged particles pass with a velocity $v > c/n$ larger than the velocity of light c/n through a transparent medium with refractive index n.

All these effects can be used, either individual or in combination, for the detection of particles, where the second last process has a very small cross section and plays only a role at high energies of the particles.

Particles with kinetic energies in the keV–MeV range loose at the ionization of atoms or molecules ($E_{ion} \approx 10$ eV) only a small fraction of their energy. They can therefore on their way through the detector material ionize many atoms, i.e. they produce many electron–ion-pairs. The specific energy loss dE/dx per length unit and therefore the number of electron–ion-pairs per length unit depend on the kind and density of the detector material and on the species of the ionizing particle (Fig. 4.51).

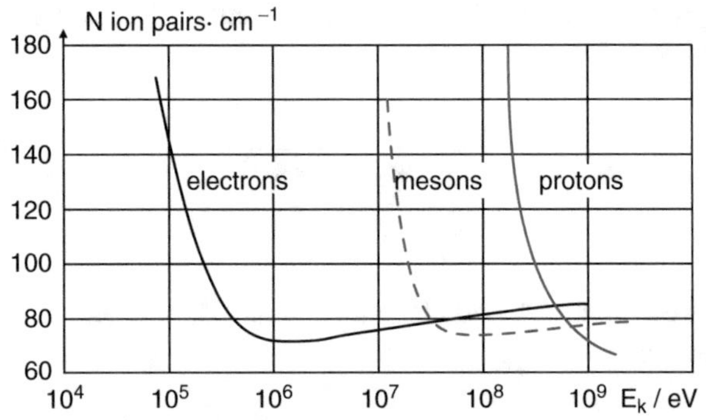

Fig. 4.51 Specific ionization (number of ion pairs generated per cm path length in air at $p = 1$ bar for electrons. π-mesons and protons as a function of the kinetic energy of the incident particles

4.2.1 Charged Heavy Particles

The order of magnitude of the specific energy loss and its dependence on the energy of heavy charged particles (protons, α-particles, mesons or fast ions) can be illustrated by a simple model.

We regard a particle with charge $Z_1 \cdot e$ which passes through the atomic electron shell (Fig. 4.52). If the energy E_{kin} of the particle is large compared to the binding energy of the atomic electrons the transfer of momentum $\Delta p/p$ at the collision of the heavy particle with an atomic electron is small. The path of the particle can be approximated by a straight line and the atomic electrons can be regarded as quasi-free because their binding energy is small compared to

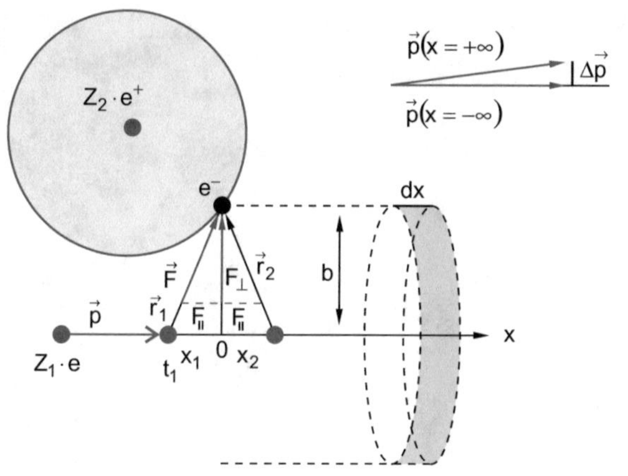

Fig. 4.52 Illustration of the derivation of the Bethe-formula for the energy loss of incident particles per cm path length

the kinetic energy of the particle. The momentum transferred from the particle to the atomic electron caused by the Coulomb force

$$F_C = \frac{Z_1 e^2}{4\pi\varepsilon_0 (x^2 + b^2)} r \tag{4.56}$$

between charged particle with charge $Z_1 \cdot e$ and a free electron is

$$\begin{aligned} \Delta p &= \int_{-\infty}^{+\infty} F \mathrm{d}t = \int_{-\infty}^{+\infty} F_\perp \mathrm{d}t \\ &= \frac{1}{v} \int F_\perp \mathrm{d}x = \frac{e}{v} \int E_\perp \mathrm{d}x \end{aligned} \tag{4.57}$$

Because the effect of the component $F_\parallel$ parallel to the particle path cancels. Integration over the surface A of the cylinder with radius b around the particle path as cylinder axis and the length dx yields, using the Gaussian rule (see Vol. 2. Sect. 1.2.2)

$$\int_A \mathbf{E} \cdot \mathrm{d}\mathbf{A} = 2\pi \cdot b \cdot \int E_\perp \mathrm{d}x = Q/\varepsilon_0 = Z_1 e/\varepsilon_0, \tag{4.58}$$

This gives the total momentum transfer to the electron

$$\Delta p = \frac{1}{2\pi \cdot \varepsilon_0} \cdot \frac{Z_1 e^2}{vb}. \tag{4.59}$$

Here the mass m of the atomic electron is regarded as negligibly small compared to the mass of the incident particle. The correct treatment has to replace the mass M of the incident particle by the reduced mass $\mu = (M \times m)/(M + m)$.

For the energy transfer onto the single arbitrarily selected electron we get

$$\Delta\epsilon = \frac{\Delta p^2}{2m_e} = \frac{1}{8\pi^2 \varepsilon_0^2 m_e} \left(\frac{Z_1 e^2}{vb}\right)^2 \tag{4.60}$$

The interaction with all electrons along the path of the particle is obtained by integration over all impact parameters b between the limits $b_{\min}$ and $b_{\max}$ which represent the validity limits of our simple model and which depend on the ratio of particle energy to binding- energy of the electron.

This gives for the energy loss of the particle along the path dx at the electron density n_e

$$\mathrm{d}E = -\left(\int_{b_{\min}}^{b_{\max}} \frac{\Delta p^2}{2m_e} n_e 2\pi \cdot b \cdot \mathrm{d}b\right) \mathrm{d}x, \tag{4.61}$$

We then obtain with (4.59) the specific energy loss

$$\frac{\mathrm{d}E}{\mathrm{d}x} = -\frac{Z_1^2 e^4 n_e}{4\pi \cdot \varepsilon_0^2 v^2 m_e} \cdot \ln \frac{b_{\max}}{b_{\min}} \tag{4.62}$$

This illustrates that the specific energy loss $\mathrm{d}E/\mathrm{d}x$ is proportional to the electron density n_e in the detector material. It increases with the square $(Z_1 e)^2$ of the particle charge but is inversely proportional to the square of the particle velocity.

The limits $b_{\min}$ and $b_{\max}$ of the impact parameter depend on the velocity v of the particle and on the binding energy E_b of the atomic electrons. A more quantitative quantum–mechanical calculation, performed by H. Bethe et. al. gives the more general formula for the specific energy loss, which is also valid for relativistic particles [26]

$$\begin{aligned} \frac{\mathrm{d}E}{\mathrm{d}x} &= -\frac{Z_1^2 e^4 n_e}{4\pi \cdot \varepsilon_0^2 v^2 m_e} \\ &\cdot \left[\ln \frac{2m_e v^2}{\langle E_b \rangle} - \ln\left(1 - \beta^2\right) - \beta^2\right] \end{aligned} \tag{4.63}$$

with $\beta = v/c$. For $\beta \ll 1$ (4.63) passes over to (4.62) if the ratio $b_{\max}/b_{\min}$ is replaced by the expression $(2m_e v^2/\langle E_b \rangle) = 4(E_t/m_t)/(E_b/m_e)$ where $\langle E_b \rangle$ is the mean binding energy of the electrons in the atomic shell and E_1, m_1 are kinetic energy and mass of the incident particle.

Equation (4.63) shows that the specific energy loss of charged heavy particles depends as $(1/E) \cdot \ln(E/E_b)$ on the kinetic energy of the incident particle. It decreases slightly with increasing energy E.

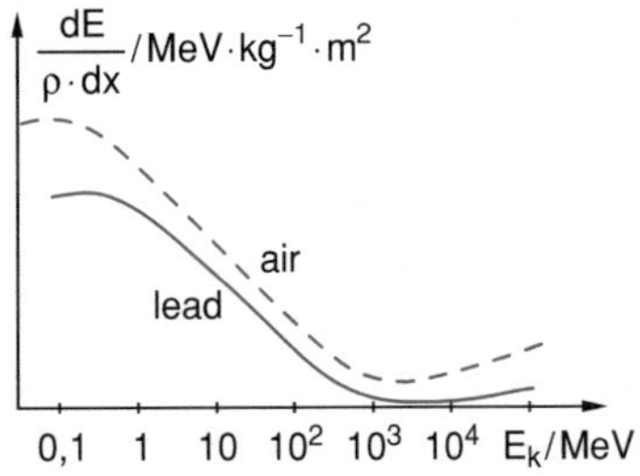

Fig. 4.53 Specific energy loss $(dE/dx)/\rho$ per mass layer density for protons in air and in lead

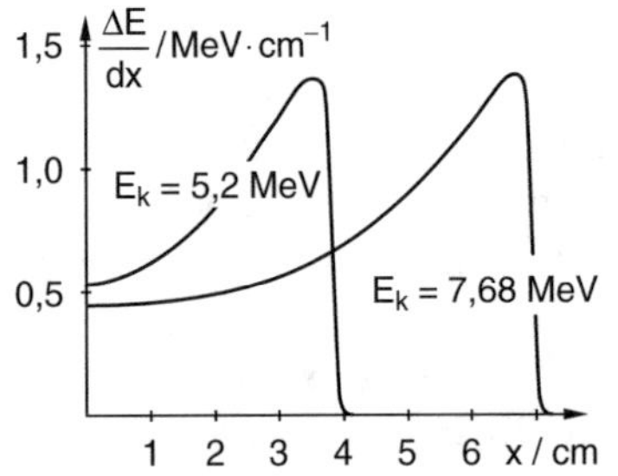

Fig. 4.54 Specific energy loss dE/dx in air at $p = 1$ bar for α-particles with two different energies (Bragg Curves)

If the electron density $n_e = Z \cdot n_a$ is expressed by the nuclear charge number Z, and the atomic density n_a, and the mass density $\rho = n_a \cdot M_a \approx n_a \cdot A \cdot m_p$ of the detector material is given by the atom density n_a and the atomic mass $M_a = A \cdot m_p$, we obtain for the electron density

$$n_e = Z \cdot n_a \approx \frac{Z}{A \cdot m_p} \cdot \varrho \approx (0{,}4 - 0{,}5) \cdot \frac{\varrho}{m_p}, \qquad (4.64)$$

One recognizes that the specific energy loss caused by excitation and ionization of the atomic electron shell

$$\frac{dE}{dx} \propto \varrho \cdot \left(\frac{Z_1 e}{v}\right)^2 \propto \varrho \frac{Z_1^2}{E_{kin}}, \qquad (4.65)$$

which a heavy particle suffers when passing through the detector material depends on the mass density of the absorbing material, on the charge $Z \cdot e$ and the velocity v of the particle. Therefore the stopping power $(dE/dx)/\rho$ of the material is often given in the units $1\ \mathrm{eV} \cdot \mathrm{kg}^{-1} \cdot m^2$.

From Fig. 4.53 one can see, that the stopping power $(dE/dx)/\rho$ depends only slightly on the detector material. Since the mean binding-energy $\langle E_B \rangle$ for the electrons of lead is higher than that of air (nitrogen + oxygen) $dE/dx/\rho$ is slightly smaller for lead than for air.

This means that for the same mass per unit area air stops heavy charged particles according to (4.63) more effective then lead.

According to (4.65) the specific energy loss dE/dx is inversely proportional to the kinetic energy of the particle.

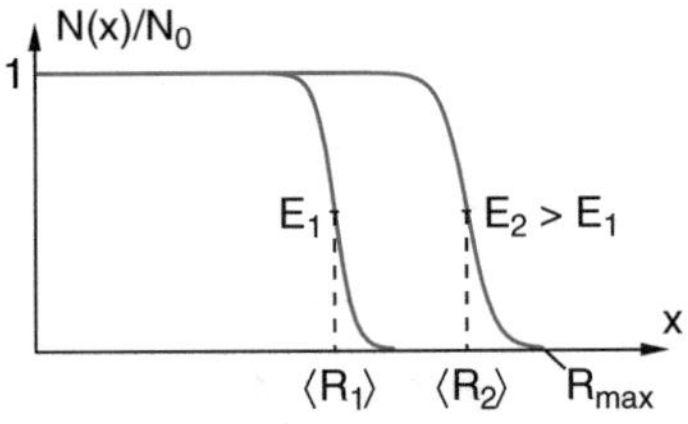

Fig. 4.55 Range of α-particles in air, shown as the number $N(x)/N_0$, surviving the distance x

This increase of dE/dx with $1/E$ is illustrated by the Bragg curves in Fig. 4.54, which show the energy loss of α-particles in air as a function of the passed distance. For larger penetration depth x the kinetic energy of the incident particles decreases and the energy loss increases. The penetration depth of α-particles is therefore sharply defined (Fig. 4.55). This can be clearly seen in the cloud chamber photograph in Fig. 3.13.

The mean range <R> of particles with the initial energy E_0 is given by

$$\langle R \rangle = -\int_{E_0}^{0} \frac{dE}{dE/dx}. \qquad (4.66)$$

For small kinetic energies ($E_{kin} << m_0 \cdot c^2$) of particles with mass m and velocity v one obtains

$$\langle R \rangle \approx f(v) \cdot \frac{E_{kin}^2}{m_1 Z_1^2} + R_\tau, \qquad (4.67a)$$

where the factor $f(v)$ takes care of the logarithmic term in (4.63). Its energy dependence can be approximated by $f(v) \propto E^{-1/2}$. This implies that the mean range can be written as

$$\langle R \rangle \propto \frac{E^{3/2}}{m_1 Z_1^2} + R_\tau \qquad (4.67b)$$

The residual range $\langle R_\tau \rangle$ depends on the detector material, on the species of the particle and its velocity v. One sees from (4.67b) that $\langle R \rangle$ increases with increasing energy as $E^{3/2}$.

For the same kinetic energy the range decreases with increasing mass m and charge Ze of the particle (Fig. 4.56).

The corresponding increase of the specific energy loss with the mass of the incident particle is in Fig. 4.58 indicated by the increasing track density of different particles on the photo-plate. In Table 4.2 the specific energy losses are compared for different particles in different media. In Fig. 4.57 the energy losses dE/dx are plotted for protons, pions and muons in different stopping media. If the abscissa is scaled as the ratio $p/(m \cdot c)$ of momentum p and $m \cdot c$ the same curves are valid for all particles.

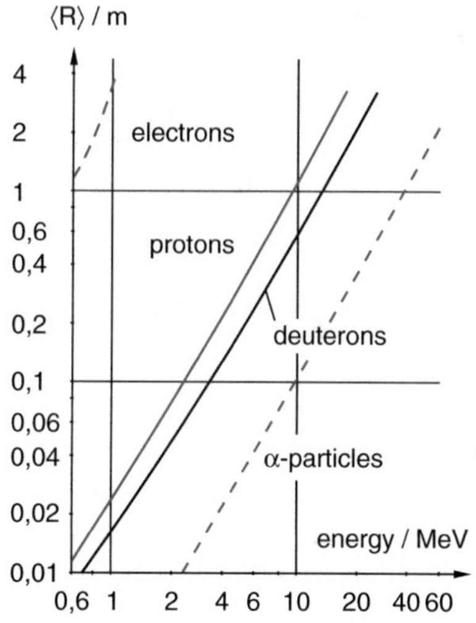

Fig. 4.56 Mean range in air for different particles in air as a function of the kinetic energy, plotted on a double-logarithmic scale

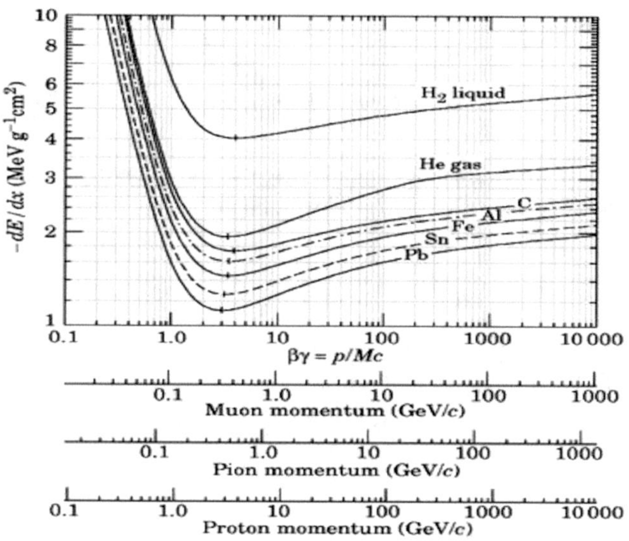

Fig. 4.57 Energy loss of protons, pions and muons as a function of their normalized momentum p/Mc) in different materials

4.2.2 Energy Loss of Electrons

For light particles (electrons, positrons) with velocities $v \ll c$ the deflection of the particles by collisions with atomic shell electrons can no longer be neglected as has been done for heavy particles. A collimated beam of incident light particles will be more affected by scattering and becomes diffuse. The specific energy loss has been calculated by H. Bethe as

$$\frac{\mathrm{d}E}{\mathrm{d}x} \approx \frac{Z_1^2 e^4 n_e}{4\pi \cdot \varepsilon_0^2 m_e v^2} \ln \frac{m_e v^2}{2\langle E_b \rangle}. \tag{4.68}$$

The comparison with (4.63) shows that for equal velocities v the energy loss $\mathrm{d}E/\mathrm{d}x$ is equal for heavy (mass m_a) and for light particles (mass m_e), because the mass of the particle does not enter into the two equations (m_e is here the mass of the atomic electrons of the detector material). For equal energies, however, $\mathrm{d}E/\mathrm{d}x$ is for electrons smaller by the factor m_e/m_a. For example $\mathrm{d}E/\mathrm{d}x$ is for electrons with $E_{kin} = 50$ keV about 10^3 times smaller than for protons with the same energy. Their track density in Fig. 4.58 is therefore smaller by the same factor.

The mean penetration depth $\langle R \rangle$ of electrons with kinetic energy E_{kin} is therefore very much larger in spite of their larger scattering cross section, than that of protons with the same energy (Fig. 4.56 and Table 4.2). The values of $\langle R \rangle$ have a larger distribution for electrons than for protons (because of the larger scattering), i.e. the number $N(x)$ does not suddenly decrease at the end of the penetration depth as it does for protons (Fig. 5.54), but has a flatter decrease (Fig. 4.59). For particles with relativistic energies, however, the differences between light and heavy particles diminish because in this case the mass is mainly determined by the kinetic energy and does not differ much for electrons and protons at the same energy...

Besides excitation and ionization of atoms of the detector material for electrons a further energy loss appears at higher

Table 4.2 Range of electrons, protons and α-particles in meter for different materials air, water, aluminum and lead

		Air	Water	Aluminum	Lead
Electrons	0.1 MeV	0.13	$1.4 \cdot 10^{-4}$	$7 \cdot 10^{-5}$	$2.7 \cdot 10^{-5}$
Protons	1.0 MeV 10.0 MeV 0.1 MeV	3.8 40 $1.3 \cdot 10^{-13}$	$4.3 \cdot 10^{-3}$ $4.8 \cdot 10^{-2}$ $1.2 \cdot 10^{-6}$	$2.1 \cdot 10^{-3}$ $2 \cdot 10^{-2}$ $7.7 \cdot 10^{-7}$	$6.7 \cdot 10^{-3}$ $5.3 \cdot 10^{-3}$
α -Particles	0.1 MeV 1.0 MeV 10. 0 MeV	 $5 \cdot 10^{-3}$ $1 \cdot 10^{-1}$	$3.5 \cdot 10^{-6}$ $9 \cdot 10^{-5}$	$3.3 \cdot 10^{-6}$ $6.6 \cdot 10^{-5}$	$8.8 \cdot 10^{-6}$ $3 \cdot 10^{-4}$ $2.4 \cdot 10^{-6}$ $3.5 \cdot 10^{-4}$

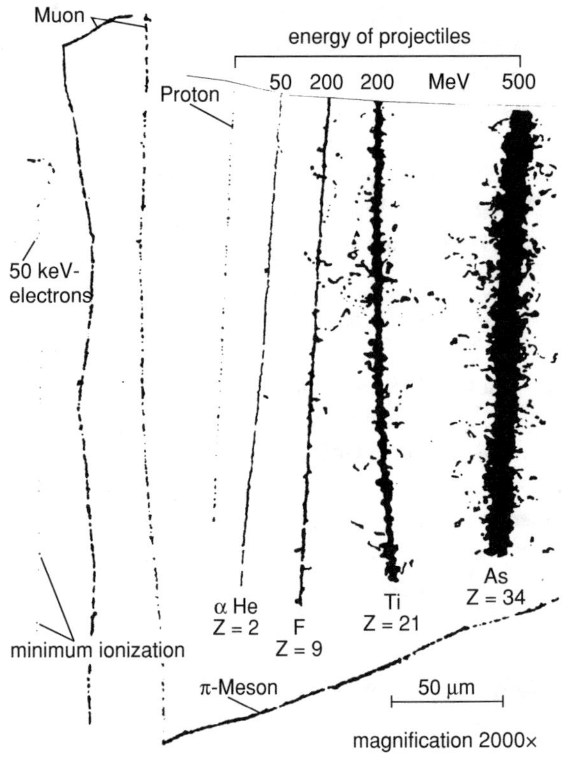

Fig. 4.58 Tracks of particles with different masses and energies recorded on a photo-plate. The blackness of the tracks is proportional to the specific energy loss d*E*/d*x*. (From Finkelburg: Einführung in die Atomphysik 12th Edition. Springer, Berlin, Heidelberg 1967)

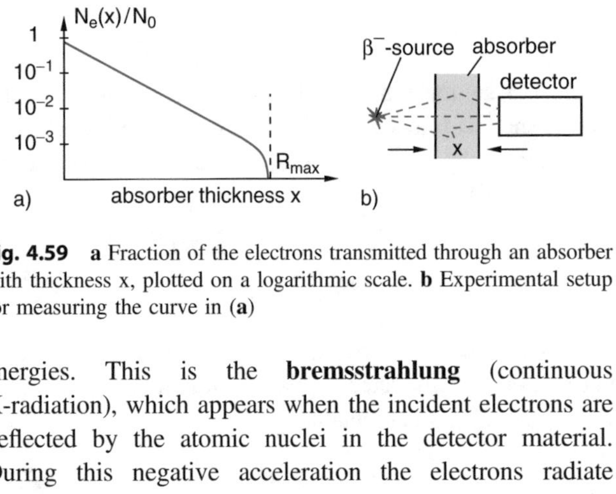

Fig. 4.59 **a** Fraction of the electrons transmitted through an absorber with thickness x, plotted on a logarithmic scale. **b** Experimental setup for measuring the curve in (**a**)

energies. This is the **bremsstrahlung** (continuous X-radiation), which appears when the incident electrons are deflected by the atomic nuclei in the detector material. During this negative acceleration the electrons radiate electro-magnetic waves with a power proportional to the square of the acceleration. The calculation gives for the energy loss d*E*/d*x* per unit path length of an electron with kinetic energy E_e in a medium with atomic density n_a and atomic charge $Z \cdot e$ [26]

$$\left(\frac{dE_e}{dx}\right)_{Str} = \frac{4n_a Z^2 \alpha^3 (\hbar c)^2 E_e}{m_e^2 c^4} \cdot \ln \frac{a(E)}{Z^{1/3}}, \qquad (4.69a)$$

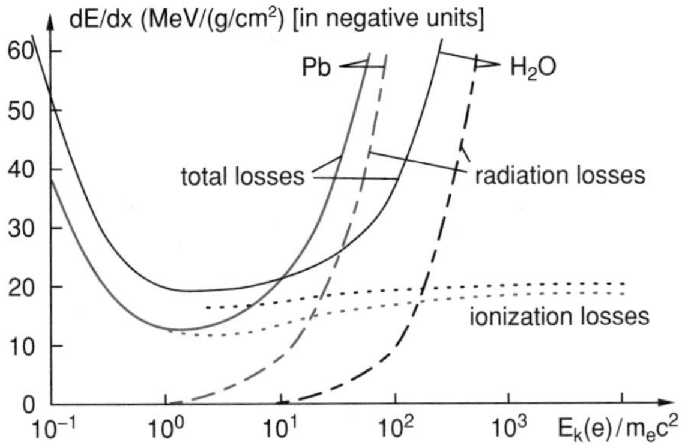

Fig. 4.60 Ionization losses, radiation losses and total losses d*E*/d*x* for electrons in lead (red curves) and in water (black curves) as a function of the kinetic energy of the incident electrons

where $\alpha = e^2/(4\pi\varepsilon_0 \hbar c)$ is the *fine-structure constant* and a(*E*) gives the maximum impact parameter where the deflection of the electron is still measurable. The radiation losses increase slightly more than linear with the kinetic energy of the electrons. At sufficiently high energies they predominate the ionization losses (Fig. 4.60).

Neglecting the weak energy dependence of the factor $a(E)$ in the logarithm we can integrate (4.69a) and obtain

$$E_e = E_e(0) \cdot e^{-C \cdot x}. \qquad (4.69b)$$

The distance $x = x_s$, where the energy of the electrons has dropped due to radiation losses to $1/e$ of its initial value, is called the radiation length

$$x_S = 1/C = \left[\frac{4n_a Z^2 \alpha^3 (\hbar c)^2}{m_e^2 c^4} \cdot \ln \frac{a(E)}{Z^{1/3}}\right]^{-1}. \qquad (4.69c)$$

Its numerical value can be calculated from (4.69c) for a material with nuclear charge number *Z* and atomic mass number *A*.

4.2.3 Interaction of Gamma Radiation with Matter

The detection of γ-radiation is based on the following interaction mechanisms:

- Elastic scattering (Rayleigh- and Thomson-scattering)
- inelastic scattering (Compton effect (Fig. 4.61a))
- Absorption in the atomic electron shell (Photo-effect, Fig. 4.61b)
- Absorption by atomic nuclei (nuclear photo-effect (Fig. 4.61c)
- Generation of particles through γ-quanta (pair formation), Fig. 4.61d)

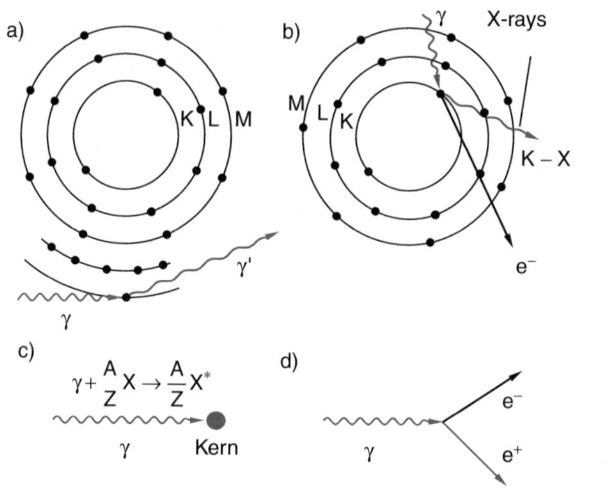

Fig. 4.61 **a** Compton Effect represented as inelastic scattering of γ-quanta at nearly free electrons. **b** Photo-effect explained as absorption of γ-quanta by electrons bound in atoms. **c** Absorption of γ-quanta by atomic nuclei. **d** pair formation

The Rayleigh scattering is dominant for small γ-energies ($h \cdot \nu < E_b$ = binding energy of the atomic electrons). Here the electromagnetic wave induces oscillations of the atomic electrons, which emit their energy in form of el.magn. waves of the same frequency as the inducing wave. The cross section of this process increases strongly with the frequency ν and is proportional to ν^4. As long as the wavelength $\lambda = c/\nu$ is large compared to the atomic diameter (this implies $h \cdot \nu \ll 3$ keV), the waves scattered by the different atomic electrons can superimpose coherently. The total scattered amplitude is then proportional to Z, their intensity proportional to Z^2.

For higher photon energies ($h \cdot \nu \gg E_B$) the inelastic scattering becomes important (Compton Effect, see Vol 3, Sect. 3.1). The cross section σ_c for Compton Scattering has been calculated by Oskar B. Klein and Yoshio Nishina [26–28].

For relativistic energies ($h \cdot \nu \gg m_0c^2$) the cross section can be expressed as

$$\sigma_c = \pi \cdot r_e^2 \cdot Z \cdot \frac{m_ec^2}{E_\gamma}\left[\ln\left(2E_\gamma/m_ec^2\right) + \frac{1}{2}\right] \propto Z/E_\gamma, \tag{4.70}$$

while for medium energies $\left(E_b \ll E_\gamma \ll m_ec^2\right)$ the Compton cross section can be described by the expansion

$$\sigma_c = \sigma_0 Z\left(1 - \frac{2E_\gamma}{m_ec^2} + \frac{26}{5}\left(\frac{E_\gamma}{m_ec^2}\right)^2 + \ldots\right) \tag{4.71}$$

where $\sigma_0 = (8/3)\pi r_e^2$ is the Thomson cross section for elastic scattering of photons with $h \cdot \nu < E_b$ at an electron with the classical electron radius $r_e = 1.4 \cdot 10^{-15}$ m.

Remark The classical electron radius r_e is defined by the assumption that the electrostatic energy of a sphere with radius r_e and charge $q = -e$ is equal to the rest mass energy $m_e \cdot c^2$ (see Vol. 2 Sect. 1.6).

This gives the equation

$$\frac{e^2}{8\pi\varepsilon_0 \cdot r_e} = m_ec^2 \Rightarrow r_e = \frac{e^2}{8\pi\varepsilon_0 m_e \cdot c^2}.$$

The **photo-effect** means the absorption of a photon with $h \cdot \nu > E_b$ by an electron in the atomic shell, The electron leaves the atom with a kinetic energy $E_{kin} = h \cdot \nu - E_b$. Since, different from the Compton effect, the photon is absorbed and therefore vanishes, energy- and momentum- conservation can be only fulfilled if the atom takes part of the momentum by its recoil.

Therefore the photo-effect on free electrons is not possible.

The importance of the binding energy E_B is illustrated by the fact that the predominant part of the total cross section

$$\sigma_{ph} = \sum_1^Z \left(\sigma_{ph}\right)_i$$

where the summation extends over all Z electrons in the atomic shell, is provided by the electrons in the K-shell. The calculation by Heitler [20] gives for γ-energies $\hbar \cdot \nu > E_b(K)$

$$\sigma_{ph} \sim \sigma_0 \cdot Z^5 \cdot \left(\frac{m_ec^2}{E_\gamma}\right)^{7/2}, \tag{4.72a}$$

This illustrates that σ_{ph} decreases strongly with increasing γ-energy. This decrease flattens for higher γ-energies. For $E_\gamma \gg E_b$(K) one obtains

$$\sigma_{ph} \sim Z^5/E_\gamma. \tag{4.72b}$$

Because of its dependence on Z^5 the photo-effect is for heavy elements the predominant absorption mechanism for γ-radiation with $E_b < E_\gamma < m_ec^2$.

When the energy of the γ-quants exceeds $2m_ec^2$ a new absorption channel opens, namely the **pair-production,** where a γ-quant generates in the Coulomb field of the atomic nucleus an electron–positron pair. Energy- and momentum-conservation can be only fulfilled, if the atomic nucleus takes part of the momentum of the γ-quantum.

According to a model proposed by *P. Dirac* [28] this process of the electron–positron pair production can be illustrated by a similar process in a solid crystal, when a photon is absorbed, exciting an electron from the valence band into the conduction band and leaving a positive hole in the valence band, thus creating an electron–hole-pair (Figs. 4.62a and 4.62X).

On the energy scale $E = m_ec^2 + E_{kin}$ there are, besides the positive energy values, also values at negative energies $E = -(m_ec^2 + E_{kin})$ which are occupied with electrons.

According to the Pauli-Principle transitions between these occupied energy states are in principle not observable, as long as the excitation energy $E_a < 2m_ec^2$. Absorption of a photon with $h \cdot \nu > 2m_ec^2$ causes, however, the excitation of an electron in an occupied negative energy state of the "non-observable anti-world" into a real state with positive energy. The result is an observable electron with positive energy in the "real world" and a hole in the "anti-world".

Note that this hole corresponds to a positive energy because the electron can recombine with the hole and the electron–hole-pair emits two γ-quanta with the total energy $2m_ec^2$ which fly into opposite directions (annihilation radiation, Fig. 4.62).

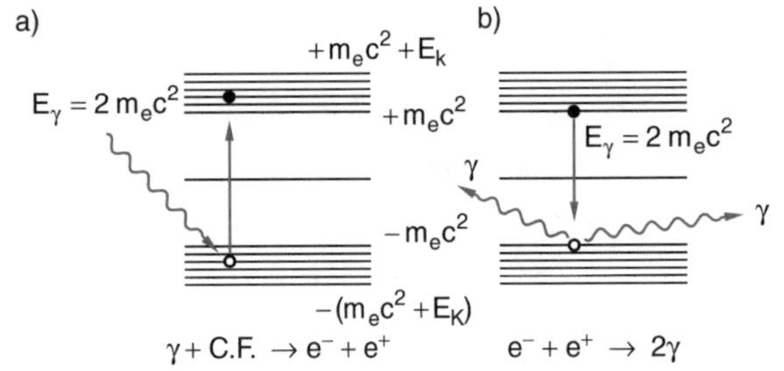

Fig. 4.62 **a** Schematic illustration of pair formation. **b** pair annihilation radiation according to the Dirac Model

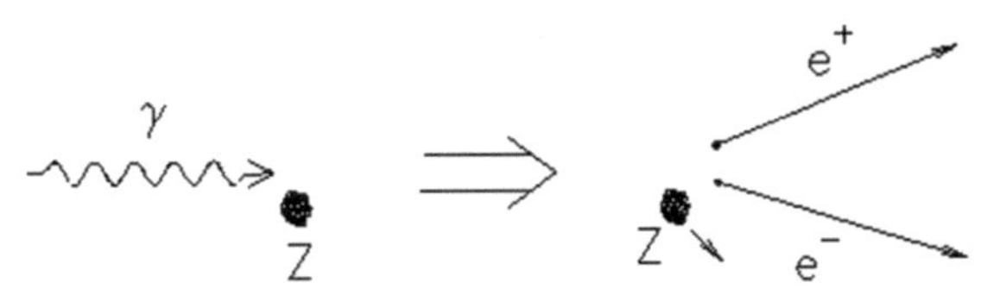

Fig. 4.62X Pair Production (By Jess H. Brewer—http://www.jick.net/~jess)

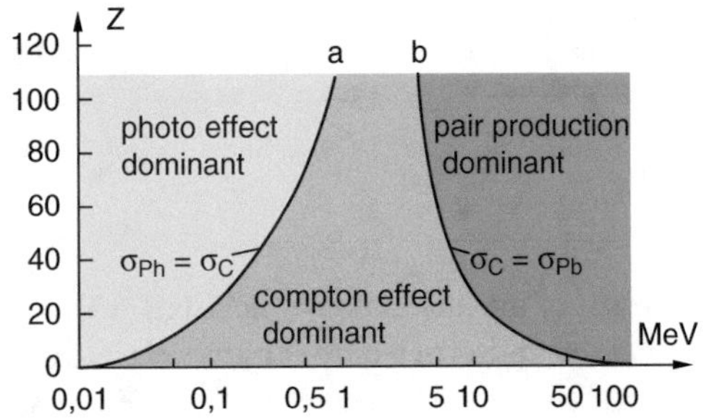

Fig. 4.63 The dominant ranges for Photo-effect, Compton effect and pair formation as a function of the energy of the γ-quantum and the charge number Z of the absorber

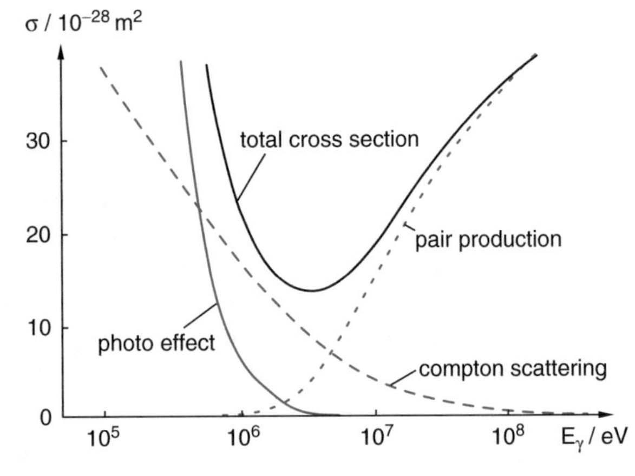

Fig. 4.64 Cross sections for Photo-effect, Compton effect and pair formation in lead $n(Z = 82)$ as a function of the γ-energy

$$\sigma_P \sim Z^2 \ln E_\gamma \tag{4.73}$$

rises initially with the logarithm of the photon energy E_γ and becomes at very high energies ($E_\gamma >> m_ec^2$) nearly independent on the photon energy $h \cdot \nu$.

The contribution of the different processes to the total absorption of γ-radiation is illustrated in Figs. 4.63 and 4.64. It depends on the γ-energy E_γ and on the charge $Z \cdot e$ of the absorber material. At the the curve a the cross sections for photo-effect and Compton-effect are equal, while for the curve *b* the cross sections for Compton-effect and pair-production are equal ($\sigma_C = \sigma_{PB}$). In Fig. 4.64 the total absorption cross section and its different contributions are plotted for the absorber material lead with $Z = 82$.

4.2.4 Interaction of Neutrons with Matter

Since neutrons carry no charges and therefore show no Coulomb interaction with the atomic electron shell they can only interacts with the atomic nuclei via magnetic forces due to their magnetic moments. This magnetic interaction is very weak compared to the electric interaction. Therefore the

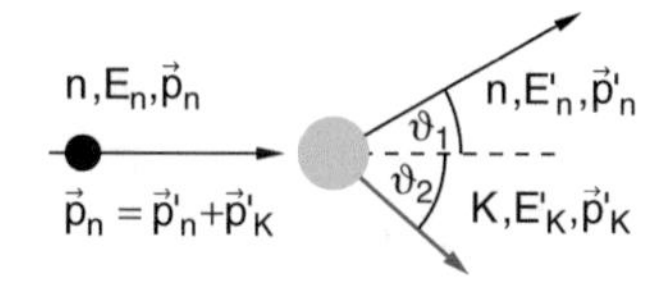

Fig. 4.65 Energy- and momentum balance for the elastic collision of a neutron against a nucleus K at rest

dominant processes of the interaction of neutrons with matter are scattering and absorption of neutrons by atomic nuclei, due to the strong interaction caused by nuclear forces. For charged particles on the other side, the electro-magnetic interaction with the atomic electron shells with a volume about 10^{15}-times larger than the nuclear volume, gives the major contribution.

For the elastic scattering of neutrons by nuclei the total kinetic energy is preserved. If a neutron with kinetic energy

$$E_n = m \cdot v^2/2 \quad \text{with}\, v << c$$

collides with a nucleus and is deflected by the angle ϑ_1, the nucleus suffers a recoil into the direction ϑ_2 against the incident direction of the neutron (Fig. 4.65). As has been shown in Vol. 1. Sect. 4.2 the energy of the neutron E_n before and E_n after the collision and the recoil energy E_R of the nucleus are related by the equations

$$E'_K = \frac{4m_K \cdot m_n}{(m_K + m_n)^2} E_n \cdot \cos^2 \vartheta_2, \tag{4.74a}$$

$$E'_n = \frac{m_n^2}{(m_K + m_n)^2} \cdot E_n \cdot \left(\cos \vartheta_1 + \sqrt{m_K^2/m_n^2 - \sin^2 \vartheta_1}\right)^2 \tag{4.74b}$$

For central collisions is $\vartheta_1 = \pi$, $\vartheta_2 = 0$. The transferred energy is then

$$E'_K = E_n - E'_n = E_n \frac{4m_K \cdot m_n}{(m_K + m_n)^2} \tag{4.74c}$$

For $m_n = m_K$ and $\vartheta_2 = 0$ (central collision of a neutron against a proton) the recoil energy E_R' becomes equal to the initial kinetic energy of the neutron ($E_K' = E_n$) and the total kinetic energy of the neutron is transferred to the proton.

At the recoil of the nucleus the electron shell of the atom losses electrons, thus a positive ion is produced. The interaction of this ion with the atoms of the absorber material generates new ions which produce a track visible in the cloud chamber and can be therefore used to monitor the initial neutron.

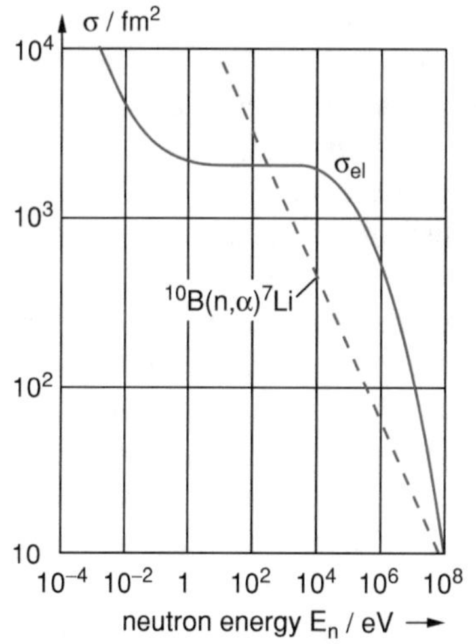

Fig. 4.66 Energy dependence of the cross section for elastic collisions between neutrons and protons, and capture cross section of neutrons in Bor (dashed curve)

The cross section of elastic neutron scattering depends on the mass of the absorber atoms and on the de-Broglie-wavelength of the neutron. i.e. on its kinetic energy. It decreases with increasing energy E_K (Fig. 4.66).

Neutrons can be also *absorbed* by atomic nuclei. The binding energy released when the neutron is attached to the nucleus (see Sect. 2.6) brings the nucleus into an excited state. The excited nucleus can either emit γ-radiation or particles when it returns into the ground state. In case of heavy fissionable nuclei the nucleus can fissure into lighter fragments (see Sect. 6.5.3).

Examples of such reactions are

$$\begin{aligned} & n + {}^{113}_{48}\text{Cd} \rightarrow {}^{114}_{48}\text{Cd}^* \rightarrow {}^{114}_{48}\text{Cd} + \gamma \\ & n + {}^{3}_{2}\text{He} \rightarrow {}^{3}_{1}\text{H} + p \\ & n + {}^{10}_{5}\text{B} \rightarrow {}^{7}_{3}\text{Li} + \alpha \\ & n + {}^{235}_{92}\text{U} \rightarrow {}^{136}_{53}\text{I}^* + {}^{98}_{39}\text{Y} + 2n. \end{aligned} \tag{4.75}$$

The absorption cross section is inversely proportional to the velocity of the incident neutron ($\sigma_a \propto 1/v$) and shows sharp maxima at certain energies, which are caused by resonances with energy levels of the nucleus (Fig. 4.67).

All reactions (4.75), where charged products or γ-quanta are generated, can be used for the detection of neutrons. Here in particular nuclei with large capture efficiency for neutrons as for example Bor or Cadmium are advantageous.

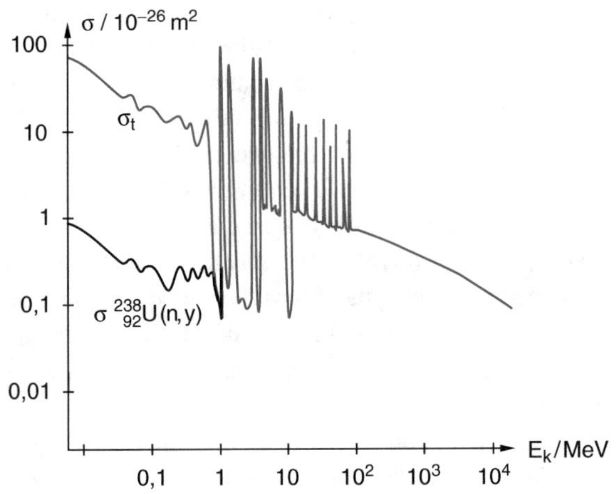

Fig. 4.67 Resonances of the total elastic scattering cross section $\sigma_{ela}(E_{kin})$ of neutrons at ${}^{238}_{92}U$ Uranium nuclei

4.3 Detectors

Detectors in Nuclear- and High-Energy Physics are devices which detect micro-particles (electrons and other elementary particles, nuclei, ions and el. magnetic radiation) and measure their energy and momentum. For the detection of electrons, γ-radiation, atomic nuclei and their constituents mainly the excitation and ionization of atomic or molecular electron shells is used. This means: The incident particles generate in the detector optical (fluorescence) or electrical (ion generation) signals, which are amplified and quantitatively measured [21–26].

In order to identify the species of the incident particle and to measure its energy, one has to know the interaction strength of the particle with the detector material and its energy dependence. Therefore this interaction has been discussed in Sect. 4.2 in more detail.

With trace detectors the track of particles can be made visible. If such detectors are placed in an external magnetic field, their spurs are curved. From the radius of curvature of their track, energy and momentum of the particle can be inferred. Such momentum measurements represent an important method for the determination of unknown particles, produced by collisions of two counter-propagating high energy particles.

Without modern particle detectors which consist of a complex combination of different detection techniques the experimental results of the sophisticated and expensive large colliders could not be interpreted and understood.

It is therefore worthwhile to study in more detail the different detector types [21–25].

The different devices for the detection and measurement of neutral and charged particles can be divided into the following categories:

- Simple detectors which monitor the number of particles, reaching the detectors without determining the species of the particles and their energy
- Spur-detectors where the track of a particles becomes visible (photoplate, cloud chamber, bubble chamber or spark chamber).
- Detectors with energy resolution which measure the energy of incident particles with a resolution ΔE, depending on the type of detector (semi-conductor—or scintillation-detectors, calorimeter).
- Compound detectors which represent combination of the types above.

The most important characteristics of a detector are the following:

- Its **sensitivity** or particle yield

$$\eta = \frac{N_S}{N_0} \leq 1$$

which is defined as the ratio of N_s detected particles to the number N_0 of particles incident onto the detector. It depends on the detector type and on the species and energy of the incident particles.

- Its energy resolving power $E/\delta E$, where δE is the smallest still resolvable energy interval and the energy dependence of its exit signal $S(E)$. With the energy distribution $f(E)$ of the incident particles the measured energy distribution $S(E)$ of the exit signal with the resolvable energy interval δE is given by

$$S(E) = \int\limits_{E-\delta E/2}^{E+\delta E/2} S(E') \cdot f(E')\mathrm{d}E'$$

- Its time resolution $1/\Delta t$ gives the smallest time interval Δt which can be still resolved. This is important for coincidence measurements where several detectors receive short signal pulses from different sources and the time correlation between the different signals has to be determined.
- Its spatial resolution, which is important for spur detectors where the track of particles has to be measured with high accuracy.
- Its capability to distinguish between different species of particles, for instance by using the differing detection sensitivity η for different species.

4.3.1 Ionization Chamber; Proportional Counter; Geiger-Counter

Ionization detectors, filled with gas are the oldest particle detectors used in Nuclear Physics. Already 1896 Becquerel (Fig. 1.1) detected radio-active radiation with such an ionization chamber. It consists of two electrodes in a chamber filled with gas (Fig. 4.68), where a voltage is applied between the electrodes. A particle, incident into the chamber ionizes part of the gas molecules creating positive ions and negatively charged electrons, which are accelerated towards the electrodes. If the voltage is sufficiently large, the electrons can gain enough energy to ionize the gas molecules on their way to the positive electrode creating new ion–electron pairs. The electrons are accelerated to the positive electrode and the ions to the negative one. The charge Q accumulated on one electrode causes a voltage pulse $U(t) = Q(t)/C$ on the capacitor C with a pulse form that depends on the time regime of the current $I(t) = \mathrm{d}Q/\mathrm{d}t$ arriving at the electrode. The voltage pulse is amplified and a counter counts the pulse rate, which can be converted by a digital-analogue-converter into an analogue signal. The count rate can be therefore measured either as a digital rate or as an analogue current signal.

We have up to now not taken into account that the charge carriers on their way to the electrode can recombine with opposite charges of the gas in the chamber, resulting in neutral atoms or molecules. The recombination rate

$$\frac{\mathrm{d}n}{\mathrm{d}t} = \alpha \cdot n^+ \cdot n^- \tag{4.76}$$

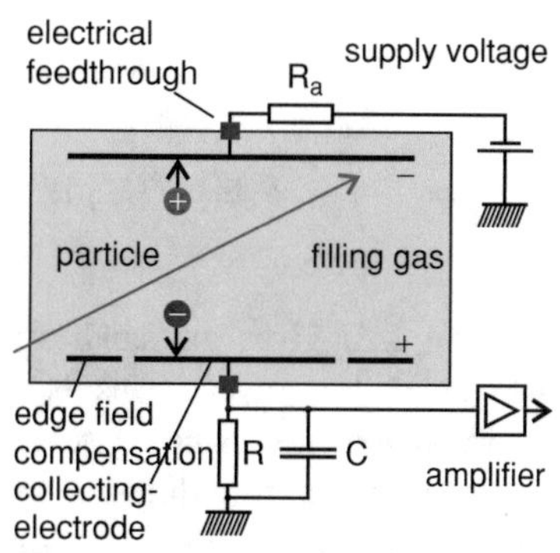

Fig. 4.68 Ionization Chamber

depends on the probability that the two charge carriers with opposite sign meet each other on the way to the electrodes. It is therefore proportional to the product $n^+ \cdot n^-$, on the creation rate of ion-pairs and on the drift velocity v_D of the charge carriers. Because of their much larger mass, v_D is smaller for ions than for electrons resulting in a larger concentration n^+. This leads to space charges.

The factor α in (4.76) is the **recombination coefficient** which depends on the type of ions. For the recombination of positive and negative ions in air at atmospheric pressure is $\alpha = 10^{-12}\ m^3 s^{-1}$. The recombination rate of positive ions with electrons is about 10^4-times smaller.

Such recombination processes in ionization detectors play a particular role if many ionizing particles hit the ionization detector from different directions. Then the probability that the produced charge carriers meet each other is larger and the recombination rate increases. Since not every incident particle creates an ion–electron pair, and some of these pairs recombine to neutral species the output signal increases less than linear with increasing incident rate and reaches finally a saturation value which does not further increase with increasing incident rate.

Because of the smaller recombination coefficient α for electron–ion-recombination and because of their shorter collection time generally the electrons collected on the positive electrode are used as detector signal.

The operation of an ionization chamber depends on the voltage between the electrodes. The current–voltage-characteristics can be divided into six ranges (Fig. 4.69): Increasing the voltage from zero to U_1 the current increases linear with U because the drift time and therefore also the recombination rate decreases with increasing voltage. In the range II from $U = U_1$ to $U = U_2$ all ions are collected and therefore the current does not further increase with U. This range is the operation range of ionization chambers because it gives a stable output signal proportional to the rate of incident particles.

With further increase of the voltage (range III) the secondary ionization starts, which results in a multiplication of the primary ion–electron pairs. In this proportionality range the output signal S is still proportional to the number of ion–electron pairs generated by the incident particles and therefore it is also a measure of the number of incident particles. The multiplication factor S/I_p of output signal to the primary current I_p can reach values up to 10^4. Since the specific ionization is larger for α-particles than for electrons with the same energy, the curve for α-particles in Fig. 4.69 lies above that for

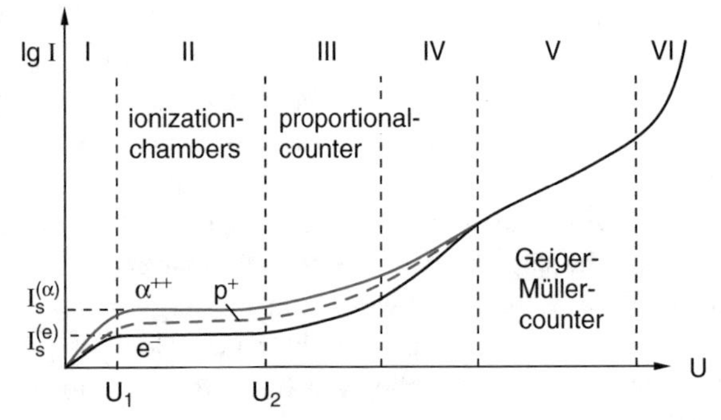

Fig. 4.69 Current–voltage characteristics of a gas discharge, indicating the different ranges of different gas-ionization detectors

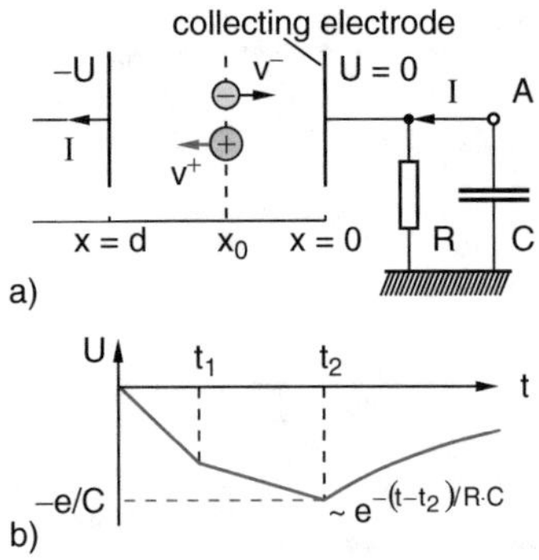

Fig. 4.70 Circuit for the formation of output pulses in the ionization chamber. **a** Drift of charge carriers. **b** Time curve of output pulses

electrons. It is therefore possible to distinguish between different particle species after calibration of the detector.

With increasing voltage in the range IV the density of ions formed by secondary ionization increases. The resulting space charge diminishes the effective value of the voltage U. Therefore the curves $I_S(U)$ for α-particles and for electrons approach each other at the end of range IV.

In the range V the amplification factor becomes so large that the output signal becomes independent of the input rate. Each incident particle generates an output pulse with a height which is independent of the species of incident particles. In this range **the Geiger-Müller Counte**r, the **spark chamber** and the different kinds of electronic trigger devices are operated.

Further increase of the voltage causes a self-sustained discharge independent of incident particles. This range is therefore not useful for particle detectors.

The ionization chamber operates in the range II. The two electrodes are realized as plates of a plane capacitor with a homogeneous electric field. The collection electrode is often surrounded by a guard ring (Fig. 4.68) in order to assure that only ionizing particles are collected out of a well-defined volume.

We will discuss the time response of the exit pulse. Generally the collection time t_1 for the electrons and $t_2 > t_1$ for the ions are both small compared to the time constant $\tau = R \cdot C$ of the amplifier input. Assume, that a particle flying through the ionization chamber, collides at the location x_0 with a gas atom and produces an ion–electron pair (Fig. 4.70a). In the electric field of the ionization chamber ions and electrons are separated. The electron is accelerated towards the positive electrode while the ion migrates into the opposite direction. During this migration time the capacitance C of the collection electrode receives by influence the charge $Q(t) = q^+(t) + q^-(t)$.

At the exit the voltage

$$U(t) = \frac{1}{C}[q^+(t) + q^-(t)] \qquad (4.77)$$

appears, where $q^+(t)$ and $q^-(t)$ depend on the distance between electrode and ion, resp. electron. They have the same sign as the influencing charges (see Vol. 2, Sect. 1.5) When the electron has reached the electrode at time t_1 it is $q^-(t_1) = -e$ while $q^+(t)$ still decreases until the ion reaches the negative electrode where $q^+(t_2) = 0$ (Fig. 4.70b).

Only until both charge carriers have reached the electrode, the full signal $U = -e/C$ appears at the exit, which then decreases with the time constant $\tau = R \cdot C$. Here C is the capacitance at the exit of the detector which should be as small as possible in order to optimize the signal height. The rise time of the exit pulse depends on the collection time of the charge carriers, but the decay time on the time constant of the detector output.

For electric field strengths above 10^6 V/m the proportional range III in Fig. 4.69 starts. Here each primary ion pair produces through secondary ionization a large exit pulse with an amplitude which is still proportional to the number of primary ions. This is the range of **proportionality counter tubes** (Fig. 4.71). They consist of a cylindrical tube with radius a at earth potential and a concentric thin wire with radius b at positive potential which serves as collecting electrode for electrons. The electric field strength at the radius r from the central axis is (see Vol. 2, Sect. 1.3)

$$\mathbf{E}(r) = \frac{U}{r \cdot \ln(a/b)}\hat{r}. \qquad (4.78)$$

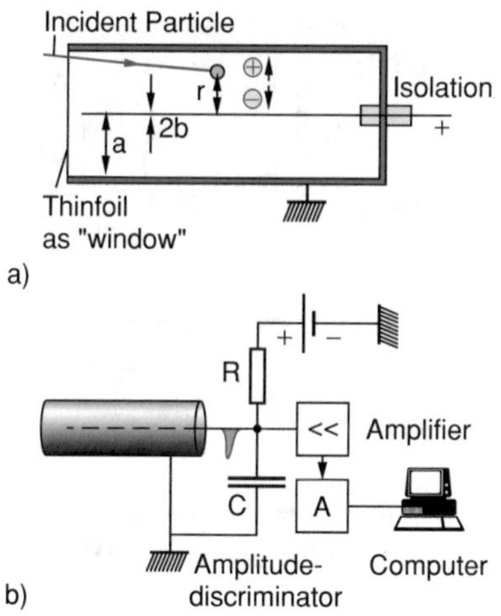

Fig. 4.71 Proportional counter **a** design **b** schematic circuit

The kinetic energy of electrons generated at the distance r_1 from the central axis is at the distance r

$$E_{\text{kin}}(r) = -e \cdot \int_{r_1}^{r} E(r)\mathrm{d}r = e \cdot U \cdot \frac{\ln(r_1/r)}{\ln(a/b)}. \qquad (4.79)$$

If E_{kin} is larger than the ionization energy of the atoms or molecules of the filling gas in the tube the primary electrons can produce by collisional ionization new electron–ion pairs.

For a properly chosen voltage U the necessary field strength for the secondary ionization is only reached in a small annular ring with $r \leq r_0$ around the central wire.

Example
$a = 10$ mm, $b = 0.1$ mm, $U = 1$ kV => $E(r \leq r_0) > -10^6$ V/m for $r_0 = 0.2$ mm.

Under these conditions ion-multiplication occurs only in the small volume around the central wire and the multiplication factor becomes independent of the localization where the primary ion-pair was produced. Since the number N of the primary generated ion pairs is proportional to the energy E_0 of the incident particle the amplitude U_A of the exit pulse

$$U_A = N \cdot k \cdot e/C \propto E_0 \cdot k/C \qquad (4.80)$$

is determined by the multiplication factor k which depends on the applied voltage.

The energy resolution of the *proportional counter tube* is about $\Delta E/E \approx 0.1$. A schematic circuit diagram is shown in Fig. 4.71b.

Neutrons can be detected by a counter tube filled with Boron-Trifluoride gas BF_3. The neutrons are absorbed by the Boron nuclei according to the reaction $^{10}B(n,\alpha)^7Li$. The binding energy of the neutron excites the nucleus and results in the emission of an α-particle, which induces the primary ionization of the filling gas. Using a BF_3-gas at a pressure of about 0.1 bar enriched with the boron isotope ^{10}B a detection sensitivity of about 20% of the incident neutrons can be achieved. The capture cross section of neutrons by boron nuclei decreases with increasing neutron energy (Fig. 4.66). Fast neutrons can be decelerated by an envelope of paraffin around the counter tube, which slows them down to thermal energies and therefore enhances their detection efficiency.

The device of the **Geiger-Müller-Counter** is identical to that of the proportional counter tube. The difference is the higher voltage U resulting in a self-sustained discharge (range V in Fig. 4.69). The amplitude of the exit pulse is then independent of species and energy of the incident particles. Therefore the Geiger-Müller Counter only measures the number of incident particles but not their energy. By choosing different thicknesses of the entrance window one can within certain limits distinguish between different particles. For example, α-particles with the energy $E_0 = 5$ MeV cannot penetrate a 50 μm aluminum foil.

The voltage is fed to the counter through a large resistance R. The gas discharge unloads the capacitance C to a voltage below the burning voltage of the counter tube and the discharge ends. The external voltage supply recharges the capacitor through the resistor R with the time constant $\tau = C \cdot R$. This time τ is also called the dead time of the counter, because the detector is not sensitive during this time. The dead-time must be longer than the collection time of the ions ($\approx 10^{-4}$ s) which is equal to the rise time of the exit pulse.

Because of the large multiplication factor k one does not need an amplifier, but can transfer the exit pulses directly to a counter or to a digital-analogue converter for obtaining a dc-signal. Often an additional acoustic gauge is used in order to warn people of an exceedingly large radioactive dose.

Example $R = 10^8\ \Omega$, $C = 10$ pF, $\Rightarrow \tau\ \tau = 10^{-3}$ s. With a multiplication factor $k = 10^5$ an incident particle generates about 10^3 primary ion pairs in the counter, which causes a charge input of $Q = 1.6 \cdot 10^{-11}$C onto the collecting wire, which produces at the capacitor $C = 10$ pF a voltage pulse of $U_a = 1.6$ V.

Instead of the electric pulses one can also use the light induced by the discharge for monitoring the incident particle. This is utilized in the **spark chamber** [24].

4.3.2 Scintillation Counters

In the scintillation counter the incident particles excite in an appropriate scintillation material atoms or molecules which then emit light pulses, which can be detected by photo-multipliers (Fig. 4.72). As scintillation materials can be used solids, liquids or gaseous media. In particular organic crystals doped with specific activator atoms, which enhance the light yield, are a good choice. Examples are NaI(Tl^+), i,e, sodium iodide, doped with thallium ions, or CsI(Tl) or ZnS (Ag).

Examples for organic scintillators are molecular crystals, such as Stilbene (1,2-diphenyl Ethen) or Anthracene ($C_{14}H_{10}$), amorphous polymers, such as Polystyrene or Polyvinyl-Toluene or other solutions of organic compound such as p-Terphenyle.

An incident particle with kinetic energy E_{kin} which is completely absorbed in the scintillator produces

$$N_{ph} = \delta \cdot E_{kin}/hv \tag{4.81}$$

photons $h \cdot v$, where the factor $\delta < 1$ describes the quantum efficiency of the scintillator. A fraction of these photons is lost by absorption in the scintillator or by incomplete reflection at the boundary surfaces of the crystal. Only the fraction $\beta \cdot N_{ph}$ reaches the photocathode of the photo-multiplier, which has the quantum efficiency η. The photocathode emits than $\eta \cdot \beta \cdot N_{ph\pi}$ photo-electrons which produce for a multiplication factor M across the capacitance C at the exit of the photo-multiplier a voltage pulse

$$\begin{aligned} U &= M \cdot \eta \cdot \beta \cdot N_{ph} \cdot e/C \\ &= M \cdot \eta \cdot \beta \cdot \delta \cdot e \cdot E_{kin}/(h \cdot v \cdot C) \end{aligned} \tag{4.82}$$

The voltage U is proportional to the energy E_{kin} of the incident particle. The exit pulses are sent to a pulse-height-analyser with a subsequent computer which gives the energy spectrum of the detected particles.

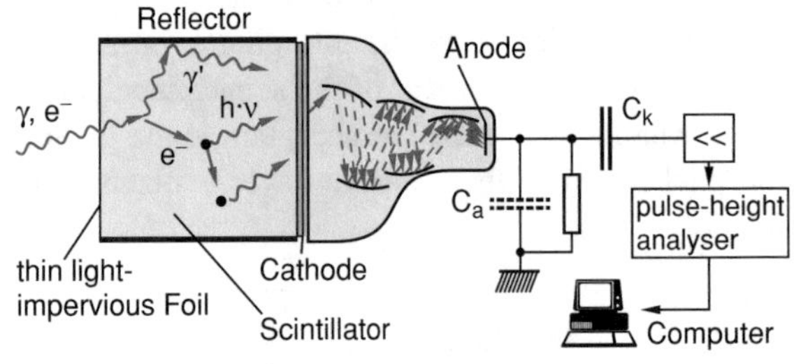

Fig. 4.72 Scintillation detector

The energy resolution depends on several factors:

- The scintillator must be sufficiently large to assure that the incident particle is completely absorbed within the scintillator.
- The imaging of the emitted light onto the photo-cathode must have the same efficiency for all points of the scintillator.
- The intensity of the emitted light must be low enough to avoid saturation of the photomultiplier.

The time profile of the output pulse can be estimated as follows:

The excitation time of the scintillator atoms is equal to the stopping time $T \approx 10^{-10}$ s of the incident particles. This is short compared with the mean lifetime τ of the excited atoms, which is in the range $T = 10^{-5} - 10^{-9}$ s. The organic scintillators (e.g. Polystyrene) have much shorter lifetimes than the activated inorganic crystals. The light emission of the excited scintillators follows the time behavior

$$I(t) = I_0 \cdot e^{-t/\tau}. \tag{4.83}$$

The photomultiplier (PM) delivers, even for an infinitively short input pulse, an exit pulse with finite width. The rise time of the exit pulse is determined by the time of flight spread of the photoelectrons on their way from the photo-cathode to the anode, which varies, dependent on the photomultiplier type, from 0.3 ns to 20 ns.

For an input pulse with a finite width the rise time of the exit pulse is given by the convolution of the input pulse time profile and the time spread in the PM. The decay time of the exit pulse is given by the convolution of the input pulse profile and the time constant $T_{MP} = R \cdot C$ of the multiplier exit, resp. the input time constant of the following amplifier and varies within wide limits.

For a statistical particle input the maximum counting rate R, that can be still safely detected for an exit pulse width ΔT should not exceed $R = 1/(3\Delta T)$.

Besides the detection of heavy particles or electrons also γ-quanta with the energy $h \cdot v$ can be detected. They are absorbed due to the photo-effect ($\propto Z^5$!!) or the Compton effect and produce secondary electrons which excite the scintillator atoms and cause the subsequent light emission.

For the detection of γ-quanta one has to use scintillator materials with a large nuclear charge number Z. For the Compton-effect (see Vol. 3, Sect. 3.1) a scattered γ-quantum $h \cdot v' < h \cdot v_\gamma$ is generated besides the Compton electron with kinetic energy $E_{kin} < h \cdot v_\gamma$ where the energy conservation demands $h \cdot v' + E_{kin} = h \cdot v_\gamma$. The pulse height distribution measured by the detector shows a sharp peak at the energy $h \cdot v_\gamma$ and a broad distribution with $E < E_{kin}$ which

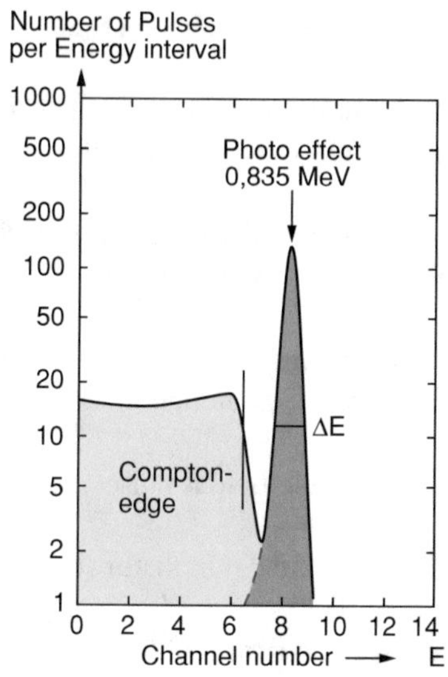

Fig. 4.73 Energy spectrum of mono-energetic incident γ-quanta with energy $E = 0.835$ MeV measured with a scintillation detector with a cylindrical Na(Tl) crystal (heights 7.6 cm, diameter 7.6 cm). The energya resolution is $\Delta E = 125$ keV

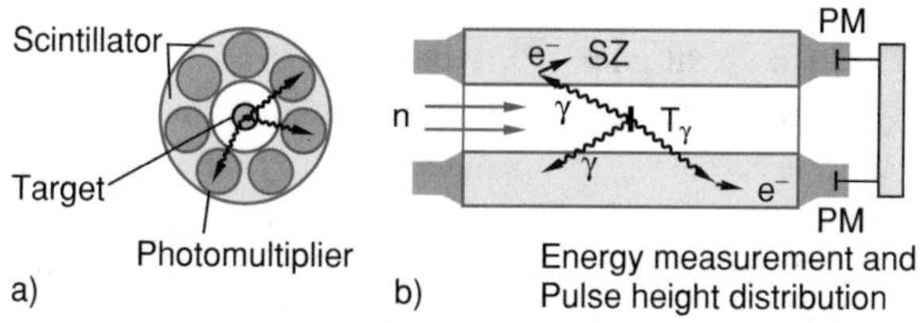

Fig. 4.74 Scintillation-Detector for detecting neutrons and for measuring their energy. **a** front view **b** side view

corresponds to the energy distribution of the Compton electrons (Fig. 4.73).

Also fast neutrons can be detected in scintillator-detectors, because they cause a recoil of protons in the scintillator which then can excite atoms with subsequent light emission. Another method uses the capture of neutrons by special nuclei which then emit γ-quanta which again can excite scintillator atoms with subsequent light emission.

In order to realize a maximum possible solid angle for the detection of the γ-quanta the neutron source is placed at the center of a large hollow cylinder which is filled with the scintillator liquid and is surrounded by many photomultipliers (Fig. 4.74).

4.3.3 Semiconductor Counters

A semiconductor counter is essentially a semiconductor diode operated in reverse direction. In the p-n transition layer the applied voltage causes a depletion of the charge carriers (Fig. 4.75a). If an incident particle is absorbed in this layer it

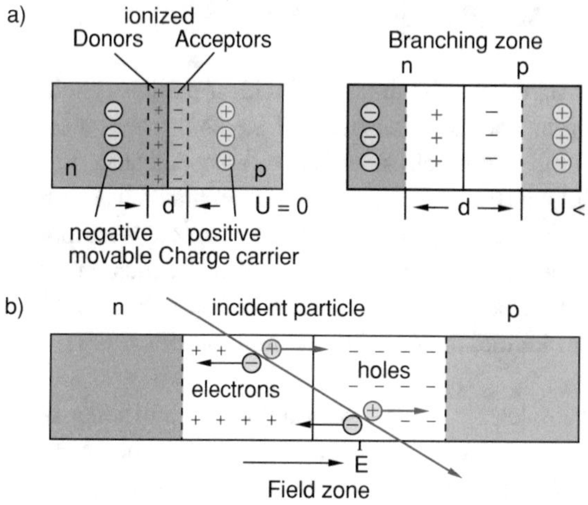

Fig. 4.75 Semi-conductor detector. **a** Depletion zone of charge carriers with and without external bias voltage. **b** Generation of electron–hole pairs by incident radiation or particles. The drift of electrons and holes to their electrodes is indicated by the arrows

produces many electron–hole-pairs, which are separated by the electric field in the layer and collected at the electrodes (Fig. 4.75b). The working principle is therefore similar to that of a gas-filled ionization detector. Compared with the ionization chamber the semiconductor detector has, however, the following advantages:

- Since the density of the solid material is larger by many orders of magnitude than that of a gas detector, incident particles with high kinetic energy can be already stopped in a much smaller volume of the detector.
- The production of an electron–hole-pair demands only the energy of the band gap around 1 eV. This is about 1/10 of the ionization energy of gas atoms. Therefore much more charge carriers are produced per energy loss ΔE than in the ionization chamber. This allows a higher energy resolution.
- The collection time for the electrons is only about 10–100 ns because of the very short distances to the electrodes. This is shorter by several orders of magnitude than in the ionization chamber. Semiconductor counters are therefore much faster and allow a better time resolution.

The material for semiconductor counters is generally silicon- or germanium crystals. Phosphorus or Antimony serve as donators in the n-part of the semiconductor, while Boron or Aluminum are used as acceptors in the p-part.

The thickness d of the transition layer is obtained from the relation

$$d \approx \sqrt{\frac{2\varepsilon\varepsilon_0}{q}\left(\frac{1}{n_D} + \frac{1}{n_A}\right) \cdot U} \qquad (4.84)$$

Example
For a reverse voltage $U = 500$ V, a concentration $n_D = 10^{17}\ m^{-3}$ of donor atoms and $n_A = 10^{22}\ m^{-3}$ of acceptor atoms d becomes $d = 1.3$ mm.

The capacity of the detector is

$$C_d \approx \varepsilon \cdot \frac{A}{4\pi \cdot d}. \tag{4.85}$$

Since area A and thickness d determine the active volume $V_{act} = A \cdot d$ it is advantageous for the detection of high energy particles to enlarge the active volume without increasing the capacity C, because C determines the response time of the detector.

In order to decrease the capacitance one therefore has to increase the thickness d of the transition layer. This can be achieved by generating a weakly doped intrinsic i-layer between the p- and the n-part of the semiconductor. One way to produce this layer is the diffusion of metallic lithium into one side of a weakly p-doped germanium crystal. The lithium atoms act as donators, forming a p-n structure with highly doped n-layer. By an external applied reverse voltage the Li^+-ions drift from the n-layer into the p-layer. This results in a depletion of the donor concentration in the n-layer and an enrichment in the p-region where the Li^+ ions recombine with the negative acceptor ions to form neutral stable dipole molecules. This charge compensation results in the formation of an intrinsic i-layer with equal donor and acceptor concentrations.

In this layer the diffusion voltage builds up a strong electric field, which causes a fast drift of the charge carriers, produced by the incident particles, to the electrodes.

In Fig. 4.76 a typical circuit of a semiconductor detector is shown. The achievable energy resolution is illustrated in Fig. 4.77 by the γ-spectrum of a radio-active dust particle which has been monitored by a Li-compensated Germanium detector [30].

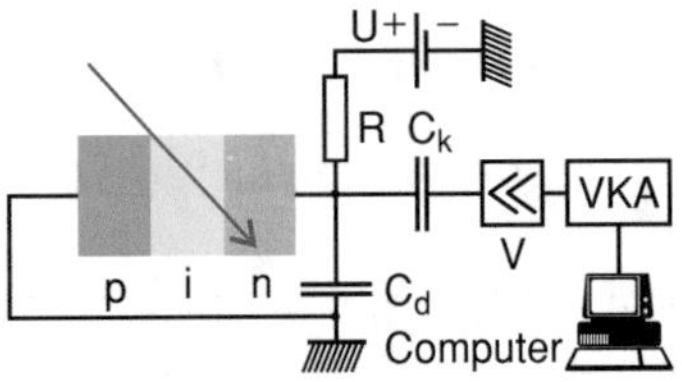

Fig. 4.76 Circuit of a semi-conductor detector with coupling capacitor C_k, amplifier V and multichannel analyzer VKA

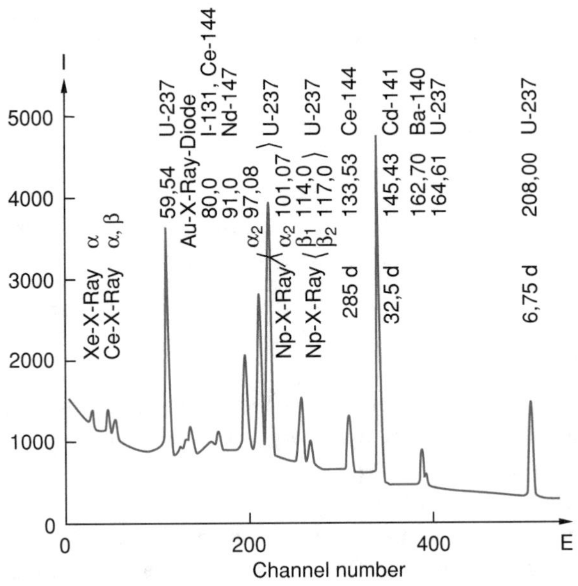

Fig. 4.77 Gamma spectrum of a sample of radio-active dust in the air after a Chinese hydrogen-bomb experiment, recorded with a lithium-compensated germanium counter. The number above the spectral lines give the energy $h \cdot \nu$ of the γ-quanta in keV and the mass numbers of the emitting atoms. (from E. Huber: Kernphysik. Ernst Reinhardt Verlag München 1972)

4.3.4 Trace Detectors

For many applications in nuclear—and high energy-physics it is very useful to detect the trace of an incident particle. Such trace detectors are

- Cloud chambers and bubble chambers. Here water droplets or steam bubbles are formed along the trace of the incident particle, which can be made visible by illuminating the chamber and observing the light scattered by the bubbles.
- Spark chambers and streamer chambers where the incident ionizing particle produces local spark discharges between two electrodes. Either the light emitted by these sparks or the electrical signals produced by the discharge is used for the trace detection of the incident particle.

When such a trace detector is placed in an external magnetic field B, the resultant radius of curvature r of the particle trace allows the determination of the momentum of the incident particle, according to

$$\frac{m \cdot v^2}{r} = q \cdot v \cdot B \Rightarrow m \cdot v = r \cdot q \cdot B$$

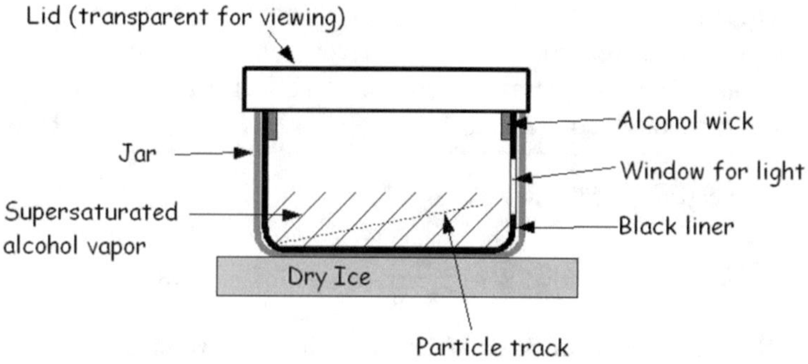

Fig. 4.77x Simple model of a cloud chamber operated with alcohol

The droplet density along the trace allows often the determination of kind and energy of the incident particle, if its momentum is known.

In the **Cloud Chamber** a super-saturated water vapor is generated by adiabatic expansion of a saturated gas–water-vapor mixture. The water vapor condenses around the ions, formed along the trace of the incident particle, and form small water droplets. If the volume of the chamber is changed adiabatically by a movable piston from V_1 to V_2 the equations for pressure and temperature are

$$\frac{p_2}{p_1} = \left(\frac{V_1}{V_2}\right)^{\kappa}; \quad \frac{T_2}{T_1} = \left(\frac{V_1}{V_2}\right)^{\kappa-1} \qquad \kappa = \frac{C_p}{C_V}. \tag{4.86}$$

After the expansion to the volume $V_2 = V_1 + \Delta V$ the temperature drops from T_1 to T_2 and the saturation vapor pressure $p_s(T_2) = a \cdot e^{-\Lambda/RT_2}$ (Λ = evaporation heat a = constant) becomes smaller than the surrounding pressure p_2 which causes condensation of the supersaturated water–vapor [31–32] (Fig. 4.77x).

Example

A cloud chamber filled with air of pressure $p_1 = 1$ bar and with saturated water vapor has at $T_1 = 10\ °\mathrm{C} = 283$ K a saturation pressure of 12 mbar. The adiabatic exponent κ of air is $\kappa = 1.4$. At the expansion to $V_2 = 1.3\ V_1$ the temperature drops, according to (4.86) to $T_2 = 254.8$ K($= -18.3$ °C) and the water vapor pressure decreases to $p_2 = 8.4$ mbar (assuming that the water vapor can be treated as ideal gas). The saturation vapor pressure is at -18.2 °C however, only 1.3 mbar, which means that a sixfold oversaturation exists, which causes a rapid condensation.

How are the droplets being formed?

Because of the surface tension σ an additional pressure $\Delta p = 2\sigma/r$ acts on the droplets with radius r (see Vol. 1, Sect. 6.4).The evaporation of the water droplets makes the radius r smaller and the energy gain is

$$\Delta W = 4\pi\sigma\left[r^2 - (r - \Delta r)^2\right] \approx 8\pi\sigma r \Delta r, \tag{4.87}$$

while the evaporation in the gas phase demands the energy Λ per mole. The evaporation in the gas phase is favored by the energy gain of the droplets. The vapor pressure above the surface of a droplet with radius r is higher than above a plane surface with $r = \infty$.

As has been already shown by Lord Kelvin the pressure as a function of the radius r is

$$p_s(r) = p_\infty \cdot e^{(2(\sigma/r)M)/(\varrho \cdot R \cdot T)} \tag{4.88}$$

where p_∞ is the saturation vapor pressure above a plane liquid surface ($r = \infty$, M = molar weight, R = gas constant, ρ = density of the liquid).

This shows that the vapor pressure above the surface of a neutral liquid droplet increases with decreasing radius r.

Condensation can only occur, if the vapor pressure $p_s(r)$ becomes smaller than the external pressure. This is the case for a minimal radius r_{min} of the droplet. Smaller droplets with $r < r_{min}$ just evaporate.

For a charged droplet the situation is different, because the electric field of the charged droplet attracts the water molecules with their electric dipole moment into the region of highest electric field, i.e. towards the droplet. Besides the surface tension there is now the repulsion between the charges of the droplet which results in a pressure $p_{el} = Z^2e^2/(8\pi\varepsilon_0 r^4)$ which acts in the outwards direction opposite to the pressure caused by surface tension. One obtains for the vapor pressure above the charged droplet.

$$p_s(r) = p_\infty \cdot e^{\left[2\sigma/r - Z^2e^2/\left(8\pi\varepsilon_0 r^4\right)\right]M/(\varrho \cdot R \cdot T)}. \tag{4.89}$$

Now the condition $p_s(r) < p$ can be fulfilled also for small droplets with $r < r_{min}$ and droplets develop, if charged ions act as condensation centers for droplet formation [29–35].

The cloud chamber is illuminated and the light scattered by the droplets (Mie-Scattering) is imaged onto a photo-plate or onto a CCD (charge coupled device) camera. For illustration is in Fig. 3.13 such a photograph shown, where a radioactive sample (Polonium as α-emitter) has been placed in the cloud chamber.

The **expansion cloud chamber** is only for a short time interval after the expansion in a condition where droplet formation is possible because the temperature changes rapidly due to heat exchange with the walls (Figs. 4.78 and 4.80X). For the demonstration of particle tracks a continuous cloud chamber is better suited, where a temperature gradient is maintained between a cooled wall and a heated wall on the opposite side (Fig. 4.79). As condensation medium often alcohol is used which evaporates at the higher temperature T_1, diffuses through the air at atmospheric pressure in the chamber and passes through a region where the condition for super-saturation are fulfilled, before it condenses at the cold wall.

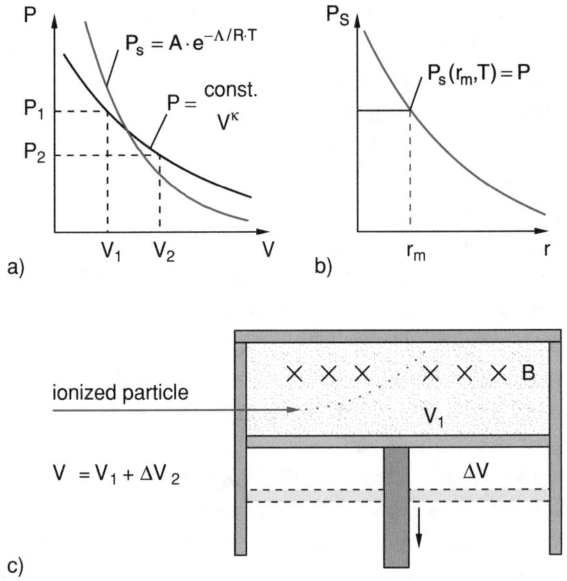

Fig. 4.78 Principle of the Expansion Cloud Chamber. **a** pressure p and vapor pressure p_s at the adiabatic expansion **b** Vapor pressure of a water droplet as a function of the droplet radius **c** schematic drawing of the expansion cloud chamber

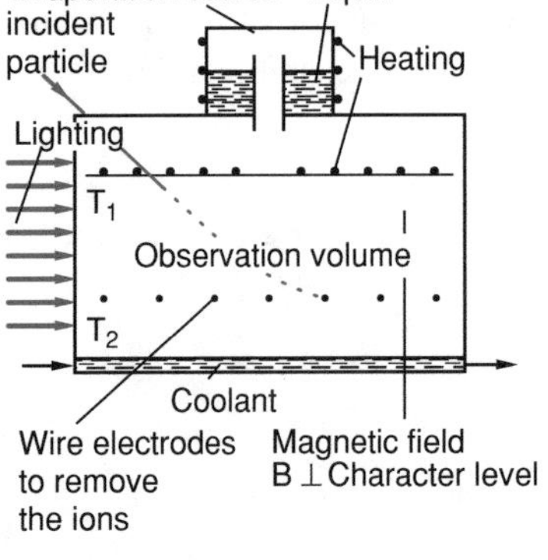

Fig. 4.79 Diffusion Cloud Chamber. The direction of observation is vertical to the drawing plane

The charged droplets can be removed by wire-electrodes which are charged for a short time in order to prepare the cloud chamber for new observations.

The detection of particles with high energy can be better realized by the bubble chamber, developed 1852 by *Donald Arthur Glaser*, (Nobelprize 1960). Here a liquid is kept at a temperature T and a pressure p which is only slightly higher than its vapor pressure $p_s(T)$. The liquid is therefore kept slightly below its boiling temperature (Fig. 4.80x).

When an ionizing particle passes through the bubble chamber it triggers the fast expansion of a piston, decreasing the pressure p for about 1 ms below the vapor pressure p_s. Thus the boiling temperature is exceeded and along the track of the ionizing particle occurs the formation of vapor bubbles. The radius of the bubbles increases until the expansion phase ends. The bubble track is recorded, as in the cloud chamber, by illumination with flash lamps through windows of the chamber and stereographic imaging of the light scattered by the bubbles. This allows the reconstruction of the three-dimensional trace of the particle (Fig. 4.80).

The bubble chamber is based on the inverse process used in the cloud chamber. Instead of the condensation of vapor here the evaporation of a liquid is utilized.

The advantage of the bubble chamber compared with the cloud chamber is the much larger density of the working material. Therefore the energy loss per cm of the ionizing incident particles is much larger and also the bubble density along the particle track is larger, which increases the energy resolution.

The bubble chamber is placed in the strong magnetic field of supra-fluid Helmholtz coils (up to 5 T). The curvature of

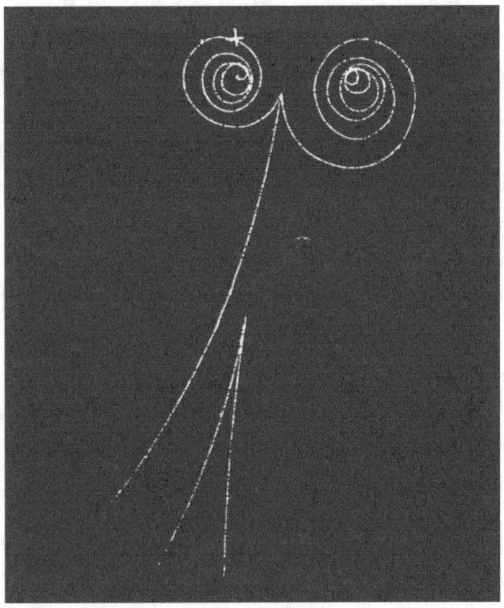

Fig. 4.80 Particle tracks in a bubble chamber. Gamma quanta, incident from above and not visible generate in the mid of the picture an electron–positron pair (Berkeley laboratories)

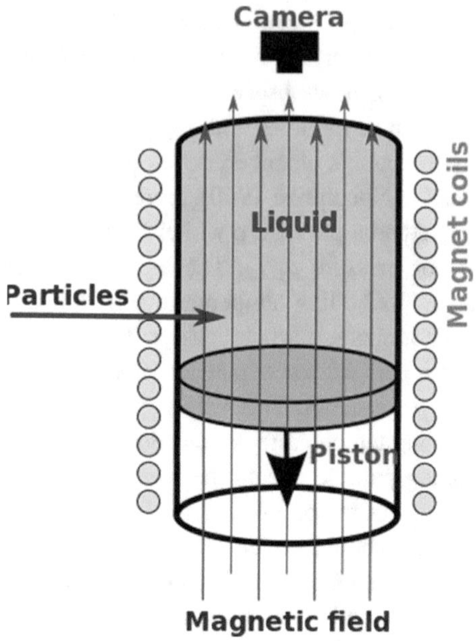

Fig. 4.80X Bubble chamber (Wikipedia)

the particle track gives information about the particle momentum. For velocities $v << c$ the mean energy loss per unit path length is proportional to $1/v^2$ (see (4.63)). This allows to deduce from the bubble density the velocity v of the particle and together with its momentum also its mass

$$m_0 = (p/v) \cdot \left(1 - \beta^2\right)^{1/2}$$

Depending on the special problem different liquids are used as working materials. Some examples are listed in Table 4.3. For the detection of protons liquid hydrogen is used, while for the detection of electrons or γ-radiation Xenon or Freon is a good choice because in these materials the radiation length is small (see Sect. 4.2.2).

In Fig. 4.80 the high spatial resolution and the good contrast ratio is illustrated by two events measured at the Berkeley laboratories: A γ- quantum, (not visible) enters the bubble chamber from below left. It generates an electron–positron pair (in the center of the picture), which proceeds to the bottom. Because of the high kinetic energy of electron and positron their path is only slightly curved in the magnetic field. In a second registered event the Compton effect is observed: A high energy γ-quant $h \cdot \nu$ produces in the upper part of the picture an electron with high kinetic energy E_{kin} which proceeds to the bottom, and a low energy photon with $h \cdot \nu' = h \cdot \nu - E_{kin}$. The photon creates an electron–positron pair. Because of its lower kinetic energy $E_{kin} = h \cdot \nu' - 2m_e c^2$ the tracks of both particles are much more curved in opposite directions in the magnetic field.

For the last years a new group of trace detectors have been developed which are based on spark production, when an ionizing particles passes through gases in strong electric fields. Detectors of this type are the **streamer chamber** (Fig. 4.81) and the **spark chamber** (Fig. 4.82).

One example of such spark detectors is the **Streamer-Chamber**, where a gas volume between two electrodes is traversed by an ionizing particle (Fig. 4.81). The passage of this particle is monitored by two detectors before and behind the streamer chamber, which deliver two signals that are measured in coincidence.

These signals trigger a high voltage pulse to the electrodes. The resultant high electric field ($E > 30$ kV/cm) accelerates the primary electrons and ions produced by the incident particle which gain sufficient energy to ionize the gas atoms between the electrodes. This results in the generation of a discharge channel (streamer) where the excited gas atoms emit fluorescence light which is monitored by a photo-multiplier. If the high voltage pulse is sufficiently short (about 1 ns) the discharge ends fast and the streamer are very short. The light emission is then restricted to the trace of the incident part.

In the streamer chamber (Fig. 4.81) the incident ionizing particle traverses a gas volume between two electrodes. The particle is monitored by two detectors before and after the chamber, delivering two signals in coincidence which trigger a high voltage pulse to the electrodes. The resultant high electric field strength ($E > 30$ kV/cm) accelerates the primary electrons, generated by the ionizing incident radiation.

These detectors have proved to be very effective. Their advantage is that the signals are electric pulses which can be directly processed by a computer.

Spark Chambers consist of a large number of parallel thin metal plates in a chamber filled with a noble gas

Table 4.3 Characteristic data of bubble chambers filled with different liquids. Liquid Temperature/K pressure/bar hadronic absorp- el.magn. radiation length/m. action length/m Hydrogen Deuterium Neon Propane CF_3Br

Liquid	Temperatur/K	Vapor pressure/bar	Hadronic absorption length/m	El.-magn. radiation length/m
Hydrogen	26	4.0	8.9	10
Deuterium	30	4.5	4.0	9
Neon	36	7.7	0.9	0.27
Propane	330	20	1.8	1.1
CF_3Br	300	18	0.7	0.1

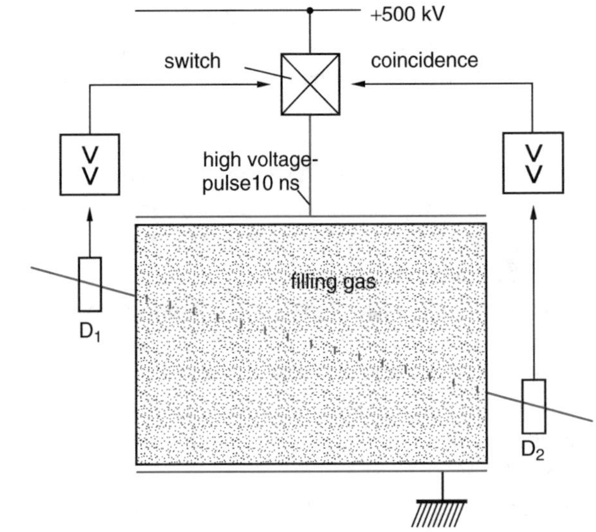

Fig. 4.81 Principle of a Streamer Chamber. The detectors D_1 and D_2 trigger (in coincidence) a high voltage switch, generating a short high voltage pulse between the electrodes. (From: St. Weinberg: Teile des Unteilbaren, Spektrum Weinheim 1984)

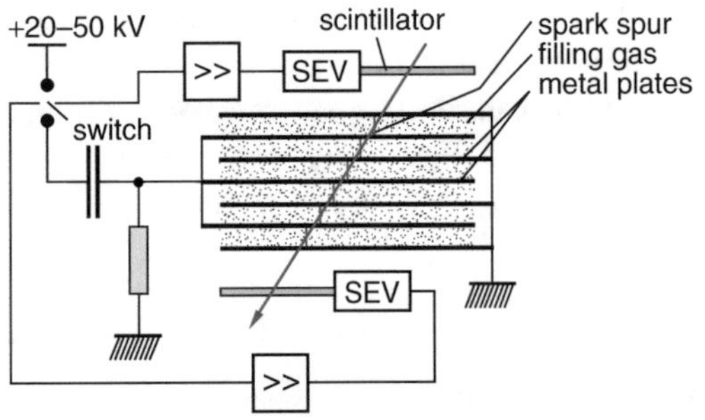

Fig. 4.82 Spark Chamber

(Fig. 4.82). Between the plates a high voltage is produced for a short time, triggered by the incident particle. This strong electric field produces a discharge channel along the trace of the ionizing incident particle, parallel to the electric field direction (Fig. 4.82x).

The light emission along the discharge channel is monitored and images the trace of the particle onto the detector.

The light emission along the discharge channel is monitored and images the trace of the particle onto the detector.

Another kind of spark chambers uses wire meshes instead of the metal plates, where each of the wires has its own voltage supply. The breakdown of the voltage by the discharge is measured selectively for each wire which allows the electric registration of the trace of the incident particle. The advantage is that the analysis of the trace is directly possible without the indirection over a photographic image or a CCD-detector (*charged coupled device*) as in the cloud- or bubble-chamber.

4.3.5 Cerenkov-Detector

When a charge particle passes through a dielectric (electric isolating material) it causes a temporary polarization of the atoms close to the particle path and induces a dipole moment changing in time, which emits electro-magnetic waves. The waves emitted by these dipoles superimpose each other with phase shifts that depend on the ratio $v/(c/n)$ of particle velocity v and light velocity $(c/n) = c_0/n$ in the medium with refractive index n. For $v < c_0/n$ the different partial waves cannot superimpose with equal phases and the superposition with random phases causes altogether destructive interference and no significant light emission occurs [4.29].

The situation changes if the particle velocity v becomes larger than the velocity of light c/n in the dielectric medium i.e. (c/n). Now there exists an angle ϑ_c against the propagation direction of the particle where all partial waves emitted from the atoms excited by the incident particle overlap in phase causing constructive interference and a strong total wave propagating into this direction. According to Fig. 4.83 we see that

$$\sin\vartheta_c = (c/n)/v = 1/(\beta\cdot n) \quad \text{with} \quad \beta = v/c.$$

Fig. 4.82x Spark Chamber

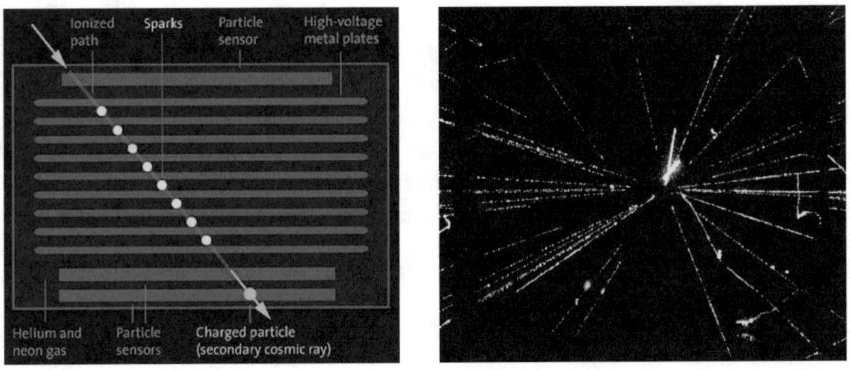

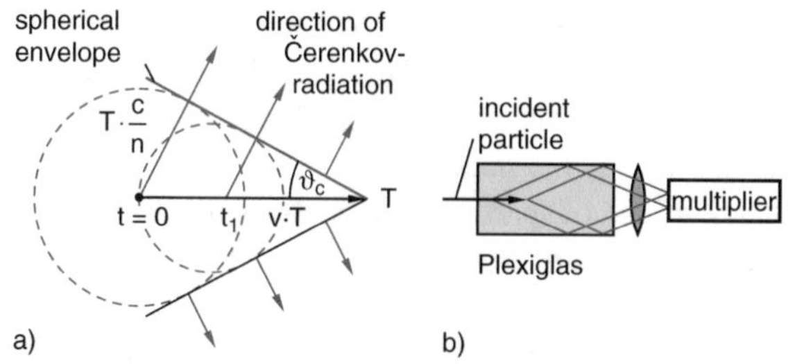

Fig. 4.83 **a** Cerenkov Radiation **b** Experimental setup, SEV = photon multiplier

The light emission into this direction is named **Cerenkov-radiation.**

Only particles with $\beta \cdot n > 1$ can induce Cerenkov radiation. Therefore Cerenkov detectors can be used for the discrimination against slower particles (threshold detectors) which are sensitive only for particles with $v > c/n$. in Fig. 4.83b the schematic design of a Cerenkov-detector is shown.

4.3.6 Detectors in High Energy Physics

When two high energy particles, travelling into opposite directions in storage rings, collide, their whole kinetic energy can be converted into the production of new particles. In order to measure all relevant data of these particles, such as their identity, their mass, their energy and their spin, complex detector systems have been developed, which consist of a combination of several different specific types of detectors. They have to meet the following requirements:

- They should embrace the collision zone and accept the total solid angel $\Omega = 4\pi$ to avoid that particles can escape the reaction volume without being detected.
- They should be large enough to ensure that even particles with very high energy are stopped in the detector or have at least sufficient long paths through the detector to determine their characteristic properties.

In Fig. 4.84 a cut through the cylindrical spark chamber is shown which surrounds one of the collision zones in the electron–positron storage ring at DESY in Hamburg. It consists of several hundred parallel thin gold-plated tungsten wires (∅ ≈ 20 μm), which are arranged on cylindrical shells with radius r_i. Between the wires which are kept on a high electric potential thin cathode cylinders are placed. To each of the wires the cylindrical coordinates (r_i, φ_i) can be assigned. When an ionizing incident particle passes through the spark chamber it generates, similar to the proportional counter, an electron avalanche to those wires where it flies past. This determines the coordinates (r, φ) of the particle path. The z-coordinate can be obtained from the charge ratio at the two ends of the wire. The spatial resolution can be better than half of the distance between the wires, if the time duration of the electron avalanche between start and arrival at the wire is measured.

If the detector system is placed inside an external axial magnetic field, the velocity components $v_\perp$ perpendicular to the z-direction are deflected. The curvature of the particle path gives information about the momentum component $p_\perp$. The spark density along the particle path is a measure of the particle velocity.

The spark proportional detector is surrounded by scintillation counters, streamer detectors and specific muon-detectors. The whole detector system has a weight of several thousand tons.

Calorimeters measure the total energy of a particle which is completely stopped in the calorimeter. Because of the different stopping powers for electrons and protons (see Sect. 4.2 and Fig. 4.51) different materials are used for the detection of electrons and hadrons [30, 31].

Electro-magnetic calorimeters determine the energy of electrons, positrons and photons by measuring the bremsstrahlung $e^- + \mathrm{N} + E_{\mathrm{kin}} \rightarrow e^- + \mathrm{N} + \gamma - \Delta E_{\mathrm{kin}}$ or the pair production

$$\gamma + \mathrm{N} \rightarrow \mathrm{e}^+ + \mathrm{e}^- + \mathrm{N}$$

where N is a nucleus of the stopping material.

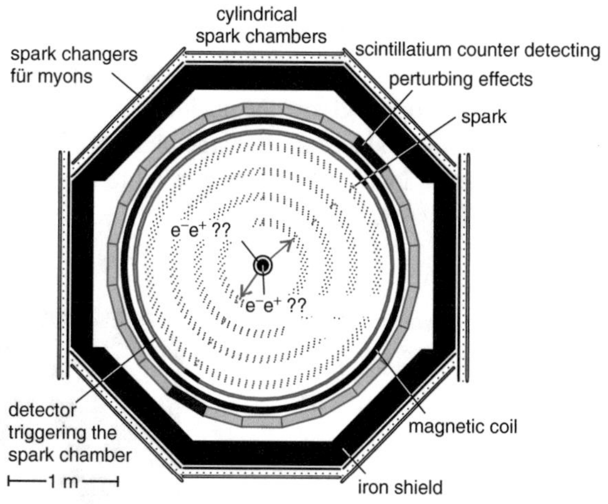

Fig. 4.84 Cylindrical proportional spark chamber, which surrounds the collision zone of an electron–positron collider at DESY Hamburg

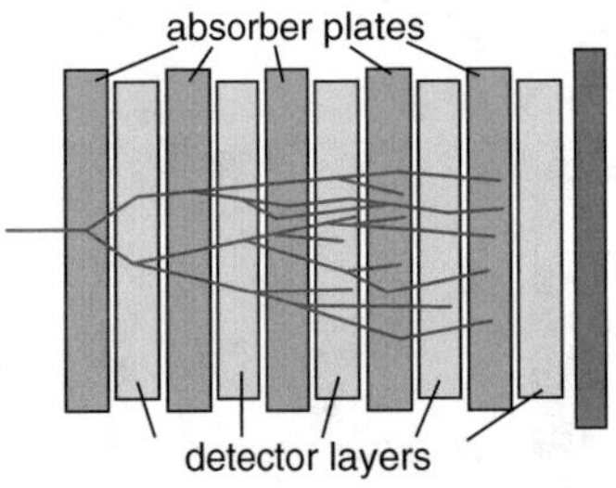

Fig. 4.85 Principle of a Calorimeter. The detectors can be Ionization Chambers, Spark Chambers or Scintillation Counters

An incident high energy electron produces a cascade of such processes (also called shower) until its energy is completely converted into a large number of electrons, positrons and photons (Fig. 4.85).

An incident electron with $E_{kin} = 10$ GeV produces, for example, about 10^3 secondary electrons with a mean energy of 10 MeV which can in turn generate many electron–positron pairs in the absorber material and can in addition excite many atoms or molecules. Since the bremsstrahlung increases with the square Z^2 of the absorber atoms, heavy elements such as lead with $Z = 82$ are preferentially used as absorber material. The fluorescence light emitted by the excited atoms is monitored by scintillation counters.

The kinetic energy of high energy hadrons (protons, π-mesons, neutrons etc.) is, due to their strong interaction, generally converted into the production of new particles, which can fly into all directions and produce in the detectors a large number of ionizing particles. Also γ-quanta can be produced which then can generate an electro-magnetic shower.

Hadron calorimeters (Fig. 4.85) consist of a large number of parallel iron plates (for the stopping of the incident particles) and plastic scintillators (for the detection of the produced γ-quanta) (see Sect. 4.2). The relative energy resolution of such calorimeters is about

$$\frac{\Delta E}{E} \approx \frac{1}{\sqrt{E/\text{GeV}}}.$$

Example

For incident protons with a kinetic energy of 30 GeV the relative energy resolution is $E/\Delta E = \sqrt{30} = 5.5$. The absolute energy resolution is then $\Delta E = 30/5.5 = 5.4$ GeV.

The Figs. 4.50 and 4.87 illustrate the large size and the complexity of such detectors. Figure 4.87 shows the H1-detector which surrounds one of the crossing points in the storage ring HERA (high energy ring accelerator) at DESY (Fig. 4.86) in Hamburg. Besides the central trace detectors a calorimeter with liquid argon is used which includes sections of lead as absorber of electrons and γ-quanta and stainless steel plates as absorbers of hadronic particles. The calorimeter is optimized in such a way, that it can distinguish between hadrons and electrons and measures the energy accurately for both types of particles [30, 35].

Fig. 4.86 View of the HERA tunnel in Hamburg. The large container in the foreground is the storage vessel for liquid helium which supplies the superconducting magnetic fields around the proton ring. The electron ring is just below the proton ring

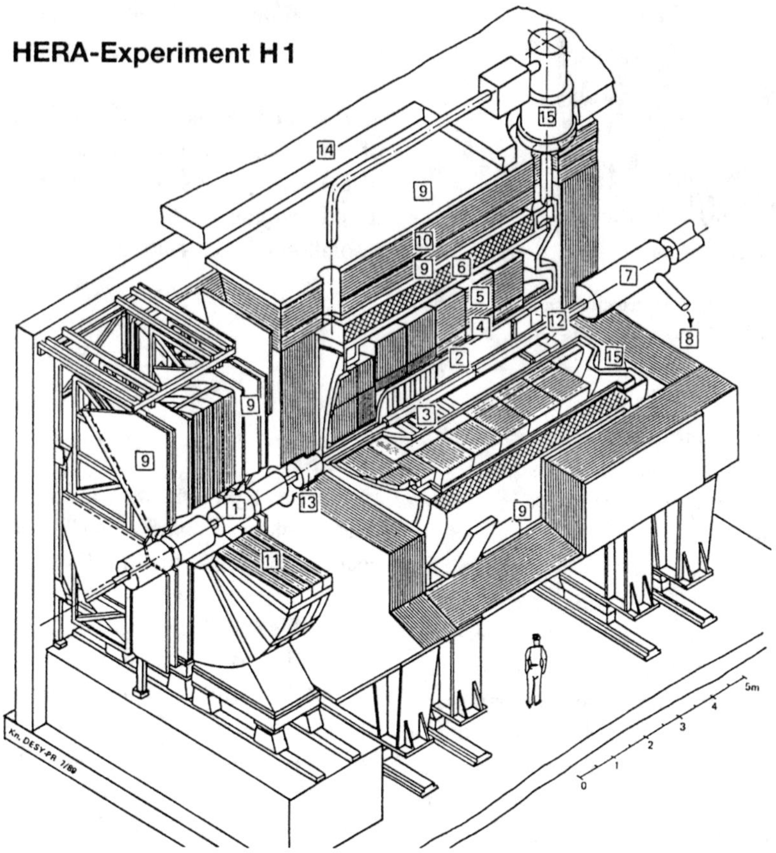

Fig. 4.87 Perspective view of the H_1-detector of at the HERA collider in Hamburg. 1 = beam tube and magnets, 2 = central spur- chamber 3 = foreword spur chamber 4 = electromagnetic calorimeter (lead) 5 = hadronic calorimeter (stainless steel) 4 and 5 are filled with liquid argon, 6 = superconducting magnetic field coil (1.2 T), 7 = compensation magnet 8 = liquid helium production 9 = Muon.Chamber 10 = iron sheets 11 = Muon toroidal magnet, 12 = warm electro-magnetic calorimeter, 13 = forward calorimeter 14 = Concrete shielding 15 = liquid argon cryostat. The total size of the H1-detector is 12 m × 10 m × 15 m, Total weight 2800 tons, (with kind permission of Prof. Eisele, DESY)

The largest detector installed at the large hadron collider LHC at CERN is the ATLAS-detector (**A** **T**oroidal **LHC** Apparatu**S**). It is 25 m high 45 m long and its weight is 7 000 tons (this is equal to the weight of 100 Jumbo jets 747) It consists of 4 components (Figs. 4.88):

- An internal trace detector, which monitors the curvature of the particle trace in the magnetic field and therefor measure the momentum of the particle.
- A calorimeter, which measures the energy of the particle
- A Muon-detector
- A strong magnetic field which is produced by superfluid coils.

A very important part of the detector is the computer-aided analysis, which filters about 100 relevant events out of 10^9 measured collisions, which are then analyzed in more detail and evaluated.

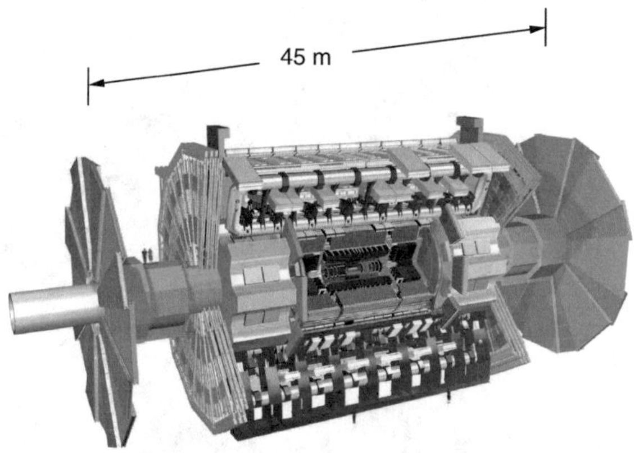

Fig. 4.88 Computer drawing of the ATLAS.detector (with kind permission of CERN)

A more detailed description of modern detectors can be found in the recommended book by *K.Kleinknecht* [22].

4.4 Scattering Experiments

In scattering experiments the deflection of a particle A is measured after its collision with a target B. As has been shown in Sect. 2.3 for the example of Coulomb scattering, measurements of the differential cross section $d\sigma/d\Omega$ allow the determination of the parameters of an assumed model potential $V(r)$ for the interaction between A and B. In particular the dependence of $V(r)$ on the distance r between A and B can be obtained, if $d\sigma/d\Omega$ is measured for different kinetic energies. In Sect. 2.2 it was discussed, that the deviation of the measured scattered intensity distribution $(dN(\theta)/d\theta)/N_0$ from the Rutherford's scattering law allows the determination of the distance r_{min} where the nuclear forces become noticeable.

For the *inelastic scattering* one has to determine besides the scattering angle also the energy loss of the scattered particles A and the excitation energy of the target particles B. For the *reactive scattering* new particles are produced and their identity, mass, energy and momentum can be measured.

For the experimental realization of scattering experiments the incident particles are formed into a collimated parallel beam in order to measure unambiguously their deflection angle at the scattering against the direction of the incident particles. Experimental techniques allow the realization of such collimated beams for electrons, protons neutrons, ions of different elements, mesons or neutrinos with energies ranging from 1 meV (cold neutrons) up to 10^{12} eV.(high energy protons, electrons or other charged particles).

As target particles B either particles at rest in solid, liquid or gaseous media are used or particles in collimated beams in colliders travelling into opposite directions as the particles A (Fig. 4.89b). The momentum of the scattered particles can be obtained by the curvature of their path in an external magnetic field and their energy either by their deflection in electric fields or by energy-selective detectors (See Sect. 4.3).

For collinear collisions the path and energy of the scattered particles or of newly produced particles are measured by trace detectors in magnetic fields. For particles with spin in addition the dependence of the scattering cross section on the relative spin orientation of the collision partners can be measured. This demands the total or preferential spin orientation of one of the collision partners.

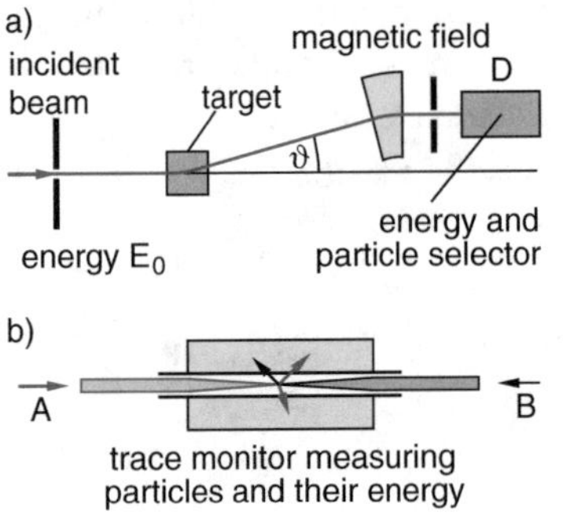

Fig. 4.89 Schematic setup of a scattering experiment **a** with a target at rest, **b** Scattering of two anti-collinear particle beams

4.4.1 Basics of Relativistic Kinematics

As has been shown in Vol. 1, Chap. 4, the total energy of a system and the total momentum is preserved for all possible collision processes if for inelastic collisions the internal energy of the collision partners is taken into account. As long as the kinetic energy of the collision partners is small compared to the rest mass energy m_0c^2 the description of the collision process can be based on the nonrelativistic Newtonian mechanics, which has been already discussed in Vol. 1, Sect. 4.2.

At relativistic energies ($E_{kin} > m_0c^2$) the relations between velocity, momentum and energy must be obtained from relativistic kinematics. Here the dependence of mass on the velocity has to be taken into account. However, as in classical mechanics here also conservation laws for momentum and energy are valid. They can be most clearly formulated if instead of the three-dimensional position vectors $r = \{x,y,z\}$ and momentum vectors $p = \{p_x, p_y, p_z\}$ the four-dimensional vectors

$$\mathcal{R} = \{c \cdot t, x, y, z\} = \{c \cdot t, \mathbf{r}\}, \tag{4.90a}$$

$$\mathcal{P} = \{E/c, p_x, p_y, p_z\} = \{E/c, \mathbf{p}\} \tag{4.90b}$$

are introduced. The metric of the non-Euclidian space, where these four-dimensional vectors are defined, is chosen in such a way, that the scalar products of two space vectors and that of two momentum vectors become

$$\mathcal{R}_a \cdot \mathcal{R}_b = c^2 t_a \cdot t_b - r_a \cdot r_b \tag{4.91a}$$

$$\mathcal{P}_a \cdot \mathcal{P}_b = E_a \cdot E_b/c^2 - p_a \cdot p_b \tag{4.91b}$$

The absolute squares of the 4-vectors are

$$|\mathcal{R}|^2 = \mathcal{R}^2 = c^2 t^2 - r^2, \tag{4.92a}$$

$$|\mathcal{P}|^2 = \mathcal{P}^2 = E^2/c^2 - p^2. \tag{4.92b}$$

These absolute squares are invariant against Lorentz-transformations i.e. they do not change under these transformations (see Problem 4.14).

Using the relativistic relation between energy and momentum (see Vol. 1, Sect. 4.4.3)

$$E = \sqrt{(m_0c^2)^2 + (cp)^2} = E_0 + E_{\text{kin}} \tag{4.93a}$$

We obtain the equation

$$E^2 - (cp)^2 = \left(m_0c^2\right)^2 \tag{4.93b}$$

This gives with (4.92b)

$$\begin{aligned} \mathcal{P}^2 &= \frac{1}{c^2}\left(m_0c^2\right)^2 \\ \Rightarrow |\mathcal{P}| &= \frac{1}{c} m_0 \cdot c^2 = m_0c \end{aligned} \tag{4.94}$$

The amount of the 4-momentum vector of a particle is proportional to its rest-energy!

Since for all collisions the total energy and the total momentum is conserved we arrive at the statement.

> For arbitrary collision processes the amount and all 4 components of the 4-vector are conserved.

This is a generalized summary of momentum- and energy conservation even for collisions where the mass of the collision partners changes.

We will illustrated this by two examples:

Examples

1. Scattering of an electron with the 4-vector $p_e = \{E_e/c, \boldsymbol{p}_e\}$ at a nucleus with $p = \{E_n/c, \boldsymbol{p}_n\}$.
   We assign all quantities after the collision with a dash and obtain

$$\begin{aligned} \mathcal{P}_e + \mathcal{P}_K &= \mathcal{P}'_e + \mathcal{P}'_K \\ \Rightarrow \mathcal{P}_e^2 + \mathcal{P}_K^2 &+ 2\mathcal{P}_e \cdot \mathcal{P}_K \end{aligned} \tag{4.95a}$$

$$= \mathcal{P}_e'^2 + \mathcal{P}_K'^2 + 2\mathcal{P}'_e \cdot \mathcal{P}'_K. \tag{4.95b}$$

For the elastic scattering the rest masses of the particles are conserved, which means

$$\mathcal{P}_e^2 = \mathcal{P}_e'^2 \text{ and } \mathcal{P}_K^2 = \mathcal{P}_K'^2,$$

From (4.95b) it therefore follows:

$$\mathcal{P}_e \cdot \mathcal{P}_K = \mathcal{P}'_e \cdot \mathcal{P}'_K \tag{4.96}$$

If the nucleus is at rest before the collision we have

$$\mathcal{P}_K = \{E_K/c, 0\} = \{m_0c, 0\},$$

And we get from (4.96) with (4.95a)

$$\begin{aligned} \mathcal{P}_e \cdot \mathcal{P}_K = \mathcal{P}'_e \cdot \left(\mathcal{P}_e + \mathcal{P}_K - \mathcal{P}'_e\right) &\Rightarrow E_e \cdot m_0c^2 \\ &= E'_e \cdot E_e - \mathbf{p}_e \cdot \mathbf{p}'_e \cdot c^2 \\ &+ E'_e \cdot m_0c^2 - \left(m_0c^2\right)^2 \end{aligned} \tag{4.97}$$

For high energies ($E_e >> m_0c^2$) we can neglect the term m_0c^2 and obtain for a scattering angle ϑ

$$\begin{aligned} p_e \cdot p'_e \cdot c^2 = p_e \cdot c^2 \cdot p'_e \cdot \cos\vartheta &= E_e \cdot E'_e \cos\vartheta E_e \cdot m_0c^2 \\ E'_e \cdot E_e(1 - \cos\vartheta) &+ E'_e \cdot m_0c^2 \end{aligned}$$

This gives the relation

$$E'_e = \frac{E_e}{1 + (E_e/m_0c^2)(1 - \cos\vartheta)} \tag{4.98}$$

between the energy of the electron E_e before and E_e' after the collision for a given scattering angle ϑ. For $\vartheta = 90°$ is $\cos\vartheta = 0$ and we get

$$E'_e = \frac{E_e}{1 + E_e/(m_0c^2)}.$$

For $E_e = m_0c^2$ is $E_e' = ½E_e$. Half of the initial energy is transferred to the recoil energy of the nucleus.

2. Reactive Scattering of two particles A and B according to the scheme

$$\text{A} + \text{B} \rightarrow \text{C} + \text{D}. \tag{4.99}$$

This process, where a projectile A hits a target nucleus B und converts it into a particle C while a particle D is emitted, can be in a short notation written as

B(*A D*)*C*. Because of the constant square $\mathcal{P}^2$ of the 4-vector it is

$$\begin{aligned} \mathcal{P}^2 &= (\mathcal{P}_A + \mathcal{P}_B)^2 = (\mathcal{P}_C + \mathcal{P}_D)^2 = \text{const.} \\ \Rightarrow c^2\mathcal{P}^2 &= (E_A + E_B)^2 - c^2(\mathbf{p}_A + \mathbf{p}_B)^2 \\ &= (E_C + E_D)^2 - c^2(\mathbf{p}_C + \mathbf{p}_D)^2. \end{aligned} \tag{4.100}$$

In the center-of mass system is $p_A + p_B = p_C + p_D = 0$. With the energies E_A^* and E_B^* in the center of mass system we get the invariant total energy

$$c^2\mathcal{P}^2 = \left(E_A^* + E_B^*\right)^2 = \left(E_C^* + E_D^*\right)^2 = \text{const.} \tag{4.101}$$

This energy can be completely converted into internal energy. E.g. for the production of new particles. Since $\mathcal{P}^2$ is invariant, we obtain for particles B at rest in the lab system ($\boldsymbol{p_B} = \boldsymbol{0}$) with (4.100)

$$\begin{aligned} c^2\mathcal{P}^2 &= \left(E_A + m_B c^2\right)^2 - p_A^2 c^2 \\ &= E_A^2 + \left(m_B c^2\right)^2 + 2E_A m_B c^2 - (p_A c)^2 \\ &= \left(m_A c^2\right)^2 + \left(m_B c^2\right)^2 + 2E_A \cdot m_B c^2. \end{aligned}$$

For the total energy in the center of mass system we therefore get

$$\begin{aligned} E_S &= \sqrt{c^2\mathcal{P}^2} \\ &= \sqrt{(m_A c^2)^2 + (m_B c^2)^2 + 2E_A \cdot m_B c^2} \\ &\approx \sqrt{2E_A \cdot m_B c^2} \end{aligned} \tag{4.102}$$

For $E_A >> m_A c^2$, $m_B c^2$. If, for instance, two protons with $E_A = E_B = 500$ GeV with opposite momenta collide, the total energy can be converted into excitation energy

$$E_S = (500 + 500)\text{GeV} = 10^3\,\text{GeV}.$$

In order to obtain the same excitation energy when a proton collides with a target particle B at rest the energy

$$E_A = \frac{E_S^2}{2m_B c^2} = \frac{10^6}{2 \cdot 0{,}93}\text{GeV} \approx 5{,}3 \cdot 10^5\,\text{GeV}$$

is required.

One can therefore achieve with collider-experiments the required excitation energy for the production of new particles at much lower energies than in experiments with the target at rest. The only disadvantage is the much lower density of the anti-collinear collision partners than with one collision partner at rest in liquid or solid targets.

4.4.2 Elastic Scattering

As has been shown in Vol. 1, Sect. 4.3 the elastic scattering of two particles can be described in the center- of mass-system as the motion of one particles with the reduced mass $M = M_A \cdot M_B/(M_A + M_B)$ in a potential that is determined by the mutual interaction between the two particles.

The quantum treatment replaces the particles by their wave packets. The scattering process is then described by moving wave-packets which pass through the potential range, while spherical symmetric wave-packets are leaving the potential range. Since the mathematical treatment of the time-dependent scattering is rather difficult one simplifies the situation: A particle moving into the z-direction with exactly defined momentum p_z while the other components $p_x = p_y = 0$, is described by the space-part $\exp(ikz)$ of the plane wave $\exp(i(kz\text{-}\omega t)$. Its extension into the z-direction is so large that the scattering process can be regarded as a stationary process (see [32, 33]). In this stationary description the wave-function of the scattering process is

$$\psi(r) = A \cdot \left[e^{ikz} + f(\vartheta) \cdot \frac{1}{r} e^{ikr}\right] \tag{4.103}$$

It consists of an incident plane wave and an outgoing spherical wave with an amplitude $A \cdot f(\vartheta)/r$ which depends on the scattering angle ϑ (Fig. 4.90).The detector with a sensitive area dF at the distance r from the scattering center measures all particles scattered into the solid angle $\text{d}\Omega = \text{d}F/r^2$. Its distance $D = r \cdot \sin\vartheta$ from the z-axis should be larger than half of the diameter d of the incident particle beam collimated by apertures. Otherwise part of the incident particles would be also detected. With the particle velocity v_a the particle flux hitting the detector is then

$$\begin{aligned} j_a \text{d}F &= v_a |\psi_a|^2 \text{d}F = v_a \left| A \cdot f(\vartheta) \cdot \frac{1}{r} e^{ikr} \right|^2 \text{d}F \\ &= v_a \cdot A^2 \cdot |f(\vartheta)|^2 \cdot \text{d}\Omega \end{aligned} \tag{4.104}$$

The differential scattering cross section $\text{d}\sigma/\text{d}\Omega$ is defined as the ratio $j_a \cdot \text{d}F/j_i$ of scattered particle flux to the incident particle flux. The comparison with (4.104) gives for the elastic scattering ($v_e = v_a$)

$$\frac{\text{d}\sigma}{\text{d}\Omega} = \frac{v_a \cdot A^2 \cdot |f(\vartheta)|^2}{v_e A^2} = |f(\vartheta)|^2. \tag{4.105}$$

The differential scattering cross section of the elastic scattering is equal to the absolute square of the scattering amplitude.

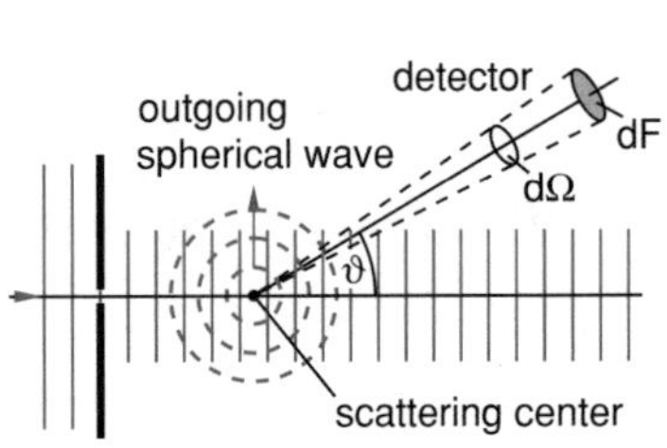

Fig. 4.90 Stationary description of scattering

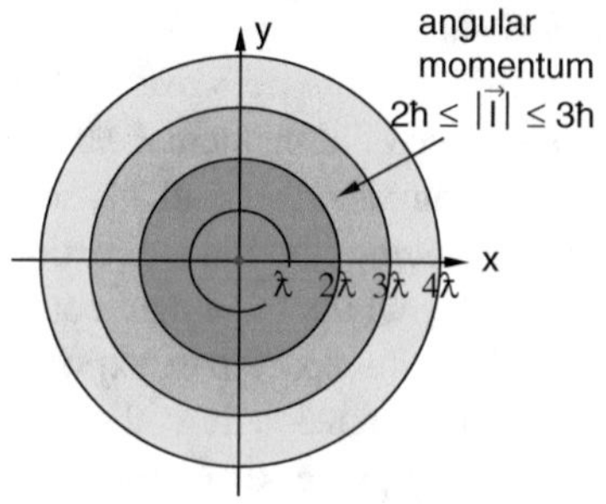

Fig. 4.91 The ring zones corresponding to the orbital angular momenta $n \cdot \hbar$ ($n = 1; 2; 3; \ldots$) around the scattering center

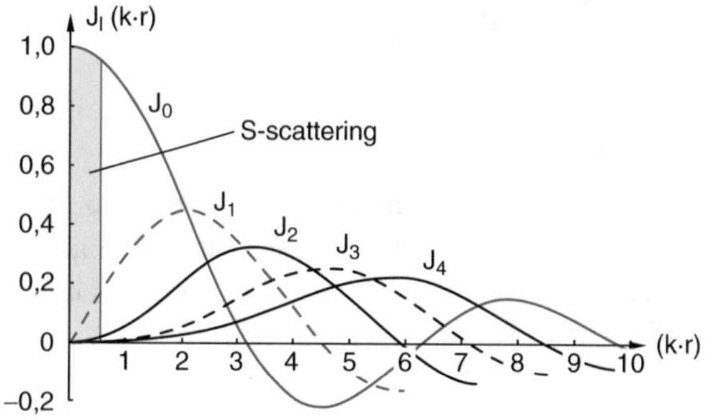

Fig. 4.92 Bessel-.functions $J_l(k \cdot r)$ with $l = 0; 1; 2; 3; 4$)

If the scattering amplitude f(ϑ) can be calculated for a given scattering potential, the differential scattering cross section can be obtained without experimental determination. However, this calculation is generally only approximately possible. One of these approximations is the partition of the incident wave into a sum of partial waves ψ_n with angular momentum $n \cdot \hbar$ with regard to the scattering center.

For incident particles with the momentum $\boldsymbol{p}$ we can divide the beam cross section into circular ring zones with radius $\rho_n = n \cdot \lambda_{dB} = n \cdot \hbar/p$ (Fig. 4.91). A particle with impact parameter $b = n \cdot \lambda/2\pi$ and momentum p has the angular momentum $\mathbf{l} = \mathbf{r} \times \mathbf{p}$ with $|\mathbf{l}| = l \cdot \hbar$ (0, 1, 2, 3, …).

The integral cross section for all particles with the orbital angular momentum $l \cdot \hbar$ is equal to the area of the l-th circular zone:

$$\begin{aligned}\sigma_l &= \pi(l+1)^2 ƛ^2 - \pi \cdot l^2 ƛ^2 \\ &= (2l+1)\pi \cdot ƛ\end{aligned} \tag{4.106a}$$

And the total scattering cross section is then

$$\sigma_{\text{tot}} = \sum_{l=0}^{l_{\max}} (2l+1)\pi ƛ^2. \tag{4.106b}$$

For a given energy E_{kin} and therefore also a given momentum p the maximum angular momentum of the scattered particles depends on the range of the interaction potential. The wave-function ψ of the incident wave can be expanded into Legendre polynomials P_l which gives the partial wave representation (see textbooks on quantum mechanics)

$$\begin{aligned}e^{ikz} &= e^{ikr\cdot\cos\vartheta} \\ &= \sum_{l=0}^{\infty} (2l+1)i^l \cdot J_l(kr) \cdot P_l(\cos\vartheta)\end{aligned} \tag{4.107}$$

where the functions $J_l(k \cdot r)$ are the spherical Bessel functions. Figure 4.92 illustrates that for $k \cdot r < 1$, the first term of the expansion (4.107) gives the major contribution, the larger $k \cdot r$ becomes the more partial waves participate in the scattering.

If only the first term with $l = 0$ gives a noticeable contribution, we speak of *S*-scattering where the angular distribution of the scattered particles is spherical symmetric. The scattering amplitude f is independent of ϑ.

In this case the scattering cross section $\sigma = \pi \cdot ƛ^2$ depends solely on the reduced de Broglie wavelength $\lambda_{dB}/2\pi = \hbar/p$ i.e. on the momentum p of the incident particle but not on the scattering angle ϑ. Its measurement does not bring new information on the interaction potential, besides the fact, that its range does not exceed $\lambda/2\pi$.

Example

For $1/k = \lambda = \hbar/(m \cdot v) = 10^{-14}$ m is $k \cdot r < 1$ for impact parameters $b < 10^{-14}$ m. If the range of the scattering center (for example the volume of the nucleus) is smaller than 10^{-14} m we get for $\lambda/2\pi = 10^{-14}$ m only S-scattering.

For large distances from the scattering center (i.e. $k \cdot r \cdot >> 1$) the Bessel function merges into the sine-function:

$$J_l(kr) \to \frac{1}{kr}\sin\left(kr - \frac{1}{2}\pi\right). \tag{4.108}$$

In this case we can describe the incident wave by

$$\begin{aligned}\Psi_e = e^{ikz} &= \frac{1}{2kr}\sum_{l=0}^{\infty}(2l+1)i^{l+1} \\ &\cdot \left[e^{-i(kr-(l/2)\pi)} - e^{+i(kr-(l/2)\pi)}\right] \cdot P_l(\cos\vartheta).\end{aligned} \tag{4.109}$$

In this representation the scattering process is described by the sum of incident plane wave and outgoing spherical wave with an amplitude f(ϑ) which is given by the Legendre Polynom $P_l(\cos\vartheta)$ with angular momentum l.

Equation (4.109) describes the incident plane particle wave which would be present if no potential exists. The potential changes amplitude and phases of the outgoing wave, because the passage through the potential area changes the de Broglie wavelength and therefore the phase, while the amplitude decreases by the factor $\alpha < 1$.

The resultant phase change is

$$\Delta\varphi = 2\pi \int \left(\frac{1}{\lambda_0} - \frac{1}{\lambda}\right) ds = \frac{2\pi}{h} \int \left[\sqrt{2mE} - \sqrt{2m(E - E_{pot})}\right] ds \quad (4.110)$$

where λ_0 is the de Broglie wavelength in the potential-free space. Taking into account amplitude and phase-changes we obtain instead of (4.109) with the abbreviations $\eta = \alpha(l) \cdot e^{i\Delta\varphi}$ and $\delta_l = kr - ½\pi$ for the wave transmitted through the potential area the wave-function

$$\Psi_t = \frac{1}{2kr} \sum_{l=0}^{\infty} (2l+1) i^{l+1} \cdot \left[e^{-i\delta_l} - \eta_l e^{+i\delta_l}\right] P_l(\cos\vartheta). \quad (4.111)$$

According to (4.103) the total wave-function should be

$$\Psi = \Psi_t = \Psi_e + \Psi_{Str} = e^{ikz} + f(\vartheta) e^{ikr}/r,$$

With (4.109 and 4.111) it follows

$$\Psi_{Str} = f(\vartheta) \cdot e^{ikr}/r = \Psi_t - \Psi_e = \frac{1}{2kr} \sum_{l=0}^{\infty} (2l+1) i^{l+1} (1 - \eta_l) \cdot e^{i(kr-(l/2)\pi)} P_l(\cos\vartheta),$$

and we obtain the scattering amplitude

$$f(v) = \frac{i}{2k} \sum_{l=0}^{\infty} (2l+1) i^l (1 - \eta_l) \cdot e^{i(kr-(l/2)\pi)} P_l(\cos\vartheta). \quad (4.112)$$

and the differential scattering cross section

$$\frac{d\sigma}{d\Omega} = \frac{1}{4k^2} \left| \sum_{l=0}^{\infty} (2l+1)(1 - \eta_l) P_l(\cos\vartheta) \right|^2. \quad (4.113)$$

The contribution of the different partial waves to the differential cross section is determined by their phase shifts when passing through the potential area.

Since in (4.113) the amplitudes of the partial waves with their different phases are added, interference effects occur which cause non-symmetric angular distributions for the scattered waves.

Using the orthogonality of the Legendre polynoms

$$\int P_l(\cos\vartheta) \cdot P_l(\cos\vartheta) d\Omega = \frac{4\pi}{2l+1} \delta_{l,l'}$$

one obtains from (4.113), when integrating over all scattering angles ϑ, the integral scattering cross section

$$\sigma_{int} = \frac{\pi}{k^2} \sum_{l=0}^{\infty} (2l+1)(1 - \eta_l)^2. \quad (4.114)$$

Each partial wave contributes the share

$$\sigma_l = \pi \cdot \lambda^2 (2l+1)(1 - \eta_l)^2 \quad (4.115)$$

to the integral cross section. The potential modifies the share (4.106a) without potential by adding the phase shift $\eta_l = \exp(2i\delta_l)$.

For the elastic scattering is $|\eta_l| = 1$ and the scattering amplitude becomes according to (4.112)

$$f(v) = \frac{i}{2k} \sum_{l=0}^{\infty} (2l+1)\left(1 - e^{2i\delta_l}\right) P_l(\cos\vartheta) = \lambda \sum_l (2l+1) e^{i\delta_l} \sin\delta_l \cdot P_l(\cos\vartheta) \quad (4.116)$$

and for the integral cross section we obtain

$$\sigma_{int}^{el} = 4\lambda^2 \sum_{l=0}^{\infty} (2l+1) \sin^2\delta_l. \quad (4.117)$$

The influence of the potential onto the *l*th partial wave is described by the phase shift δ_l.

From (4.116) it follows for the forwards scattering (ϑ = 0) with $\pi_l(0) = 1$

$$\mathrm{Im}(f(0)) = \lambda \cdot \sum_l (2l+l) \sin^2\delta_l = \frac{1}{4\lambda} \cdot \sigma_{int}^{el} \quad (4.118)$$

This relation between the integral cross section and the imaginary part of the scattering amplitude in forward direction is also called the **optical theorem**.

4.4.3 What Do We Learn from Scattering Experiments?

As has been shown in the previous section the measured integral scattering cross section $\sigma_{\text{int}}^{\text{el}}$ for the elastic scattering depends on the phase shifts δ_l which the matter wave suffers when passing through the potential area (see Eq. 4.117). Measuring $\sigma_{\text{int}}^{\text{el}}$ as a function of the energy of the incident particles, for sufficiently small energies only the S-wave with $l = 0$ contributes to the integral cross section. With increasing energy more and more partial waves with $l > 0$ contribute to the scattering cross section. Measuring the energy dependence of the cross section the different scattering phases δ_l can be selectively determined which in turn allows the determination of the r-dependence of the potential.

A more direct access to the r-dependence of the potential $V(r)$ is offered by the measurement of the differential cross section $d\sigma(\vartheta)/d\Omega$ as a function of the scattering angle ϑ, which is related to the impact parameter b of the incident particles. Such measurements therefore directly sample the r-dependence of the potential $V(r)$. However, they do not give information about the angular dependence of the potential which demands additional measurements, such as the scattering of spin-orientated particles (electrons, protons neutrons). These additional measurements allow the determination of the potential dependence on the relative orientation of the collision partners.

Information about the internal energy structure (e.g. energy levels of nuclei) is obtained from measurements of the energy loss in inelastic collisions. When measuring the inelastic differential cross section $\sigma_{\text{inel}}(\vartheta)$ it can show pronounced maxima at certain angles ϑ (Fig. 2.13). This implies that the energy transfer, i.e. the excitation of energy levels of the target particle shows maximum probability at certain impact parameters.

Resonances also occur for the elastic scattering, when the energy of the incident particle corresponds to an energy level of the compound particle, which is virtually excited but gives this energy away at the end of the scattering process. The projectile therefore does not loose energy. The energy width of such resonances gives information about the lifetime of the virtual levels of the intermediate complex. The phase shift δ_l of the partial wave which mainly contributes to the resonance changes rapidly when tuning over the resonance. It has a maximum of $\pi/2$ at the center of the resonance. This is quite similar to mechanical resonances of forced oscillations (see Vol. 1, Sect. 10.5).

4.5 Nuclear Spectroscopy

Nuclear spectroscopy covers measurements of the energy of γ-quanta, electrons, positrons or neutrons which are emitted by energetically excited nuclei.

Such excited nuclear levels are produced at the decay of radio-active parent nuclei (Sect. 3.2), at nuclear fission (Sect. 6.5), where the fission products are generally excited, at the excitation of nuclei by γ-quanta, or under collisions of nuclei with other projectiles.

Nuclear spectroscopy gives information not only about the energy of the excited levels, but also about their angular momentum, their parity and about possible electric or magnetic moments.

Time resolved measurements allow the determination of the lifetimes of excited nuclear levels, which range from 10^{-12} s to 10^{+9} years.

In many aspects the goal of nuclear spectroscopy is quite similar to that of the spectroscopy of the atomic shell. The difference lies in the order of magnitude of the different characteristic features. The energy is larger by many orders of magnitude. While the life-times of atomic levels are of the order of 10^{-6}–10^{-9} s that of excited nuclei span a much larger range.

The excitation cross sections for nuclei is smaller by several orders of magnitude than that of excited atomic levels.

4.5.1 Gamma-Spectroscopy

For measurements of the energy $h \cdot \nu$ of γ-quanta several experimental methods have been developed. A good choice are semi-conductor detectors (Sect. 4.3.3) or scintillation detectors (Sect. 4.3.2). Also a crystal spectrometer can be used where the γ-radiation is wavelength-selectively reflected under the angles α_m against the atomic planes according to the Bragg condition

$$2d \cdot \sin\alpha_m = m \cdot \lambda \Rightarrow h \cdot \nu = m \cdot \frac{h \cdot c}{2d \sin\alpha_m} \qquad (4.119)$$

If this condition is fulfilled, constructive interference appears between the different partial waves reflected by the parallel atomic planes (see Vol. 3, Sect. 7.5). Since for higher γ-energies very small angles appear, which cannot easily been realized, often higher interference orders m are utilized.

Example

$\sin\alpha_m = 0.01$, $m = 5$; $d = 0.2$ mm $\rightarrow h \cdot \nu = 22.5 \cdot 10^{-14}$ $J = 1.4$ MeV.

The most important criteria for the selection of the optimum detector are its spectral resolution $E/\Delta E$ and its

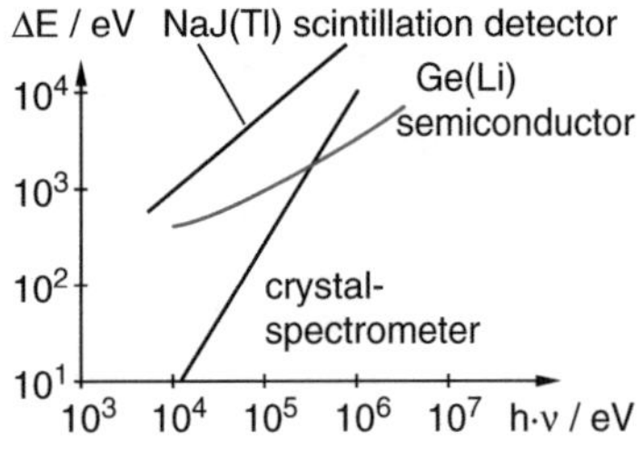

Fig. 4.93 Smallest still resolvable energy interval ΔE of different detectors for γ-radiation with quantum energy $h \cdot \nu$

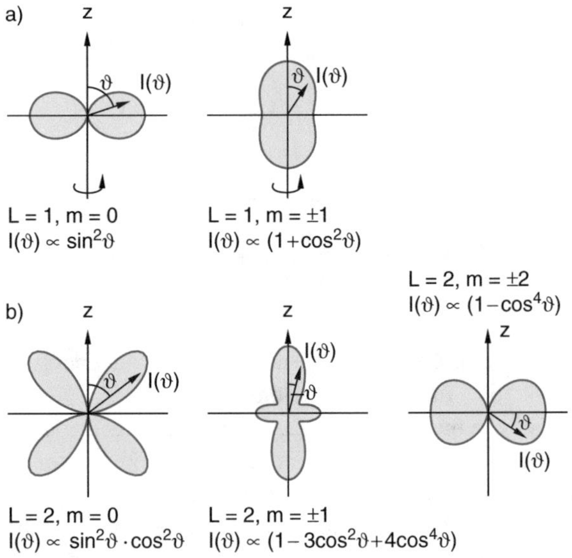

Fig. 4.94 Angular distribution of radiated γ-quanta for **a** dipole **b** quadrupole transitions

effectivity η which is defined as the ratio of the number of detected γ-quanta to the number of incident quanta.

The curves in Fig. 4.93 illustrate that the energy resolution depends on the energy $E = h \cdot \nu$ of the incident quanta. The crystal spectrometer has the highest energy resolution but the lowest efficiency, while scintillation detectors on the other side have the highest efficiency but the lowest energy resolution. The optimum compromise is offered by semi-conductor counters. They are therefore more and more applied.

The energy resolution of a germanium detector doped with Li is illustrated by the γ-spectrum of radio-active dust (Fig. 4.77) collected after a hydrogen-bomb explosion, where for $E = 100$ keV the energy resolution $\Delta E = 1$ keV could be achieved.

Scintillation detectors should have a sufficiently large sensitive volume to ensure that the γ-quanta are completely absorbed in the detector. This implies that also the secondary γ-quanta with lower energy generated by Compton-scattering are absorbed within the sensitive detector volume. Often the 4π-geometry shown in Fig. 4.74 is chosen where the γ-emitting sample is surrounded by scintillation detectors. Figure 3.32 shows the γ-spectrum of the excited nucleus ${}_{10}^{22}Ne$, measured with such an arrangement (see Sect. 3.5.1).

We had learned in Sect. 3.5.2 that excited nuclei can emit multipole-radiation of order 2^L depending on their angular momentum L and their parity, where $|L| = \sqrt{L \cdot (L+1)} \cdot$ is the amount of the angular momentum carried by the emitted γ-quantum and related to the center of the emitting nucleus. The angular distribution of the emitted γ-radiation depends on the order of the multipole radiation. Measurements of the number $N(\vartheta)$ of emitted γ-quanta as a function of the angle ϑ against the preferential direction (which we choose as the z-direction) therefore gives information about the character of the radiation. The angular intensity distribution of the multipole radiation is proportional to the square $|Y_L{}^m|^2$ of the spherical harmonics with the angular momentum quantum number L and the projection quantum number m.

For electric or magnetic dipole radiation is (Fig. 4.94)

$$\begin{aligned} I_1^0 &\propto \left|Y_1^0\right|^2 \propto \sin^2\vartheta \\ I_1^{\pm 1} &\propto \left|Y_1^{\pm 1}\right|^2 \propto (1+\cos^2\vartheta) \end{aligned} \tag{4.120a}$$

Where the quadrupole radiation gives the contribution

$$\begin{aligned} I_2^0(\vartheta) &\propto \sin^2\vartheta\cos^2\vartheta \\ I_2^{\pm 1}(\vartheta) &\propto \left(1-3\cos^2\vartheta+4\cos^4\vartheta\right) \\ I_2^{\pm 2}(\vartheta) &\propto \left(1-\cos^4\vartheta\right) \end{aligned} \tag{4.120b}$$

Equation (4.120) shows that $I(\vartheta) = I(\pi - \vartheta)$. It is therefore only necessary to measure the angular range of $\pi/2$, e.g. the range $90° \leq \vartheta \leq 180°$. The reason for this symmetry is the conservation of parity of the total system nucleus + γ-quantum for electro-magnetic interaction.

An anisotropic intensity distribution can be only expected, if the emitting nucleus is orientated which can be defined by the nuclear spin I with $I \geq 1$ (see Sect. 2.5). At thermal equilibrium the spins of a sample of many nuclei are without external magnetic field uniformly distributed, which means that the sublevels $|m\rangle$ of a rotational level $\langle J|$ all have the same energy, the spins of the statistical average have therefore no preferential direction. In this case even for nuclei with $I \neq 0$ the emitted γ-radiation is isotropic. In order to measure the angular distribution of the multipole radiation the spins have to be at least partially orientated. This can be realized by an external magnetic field, which defines a preferential direction.

Another method, which provides a preferential direction even without magnetic field is based on cascade transitions

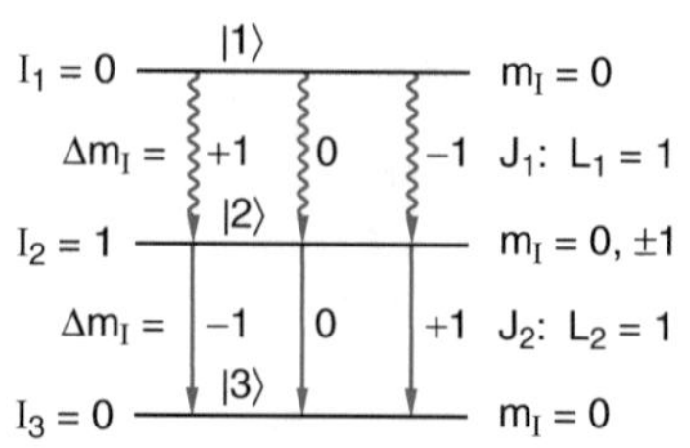

Fig. 4.95 Cascade γ-transitions between energy levels of a nucleus with $\Delta\ell = \pm 1$ and $\Delta m = 0, \pm 1$ (dipole–dipole cascade)

in nuclei (Fig. 4.95). Assume a γ-quantum with energy $h \cdot v$ and angular momentum quantum numbers L and m is emitted by a nucleus in an excited level $|1\rangle$. Resulting in the population of an intermediate level $|2\rangle$ which again can decay by emission of a γ-quantum ($h \cdot v_2$, L_2 m_2) into a lower level $|3\rangle$. The multipole character of γ_2 can be obtained by measuring the angular correlation of γ_2 The γ-quantum γ_1 defines the preferential direction and a delayed coincidence measurement (Fig. 4.96) gives the probability that after the quantum $\gamma_1(\vartheta)$ is emitted a second quantum γ_2 is observed under the angle ϑ against the direction of γ_1. The measurement of the energies $h \cdot v_1$ and $h \cdot v_2$ allows the determination of the nuclear levels $2\rangle$ and $|3\rangle$. In Fig. 4.96b the angular correlation $I(\gamma_1, \gamma_2, \vartheta)$ is shown. The intermediate level with nuclear spin $I_2 = 1$ has the energetically degenerate sublevels $m_l = 0; \pm 1$. If γ_1 corresponds to a transition with $\Delta m = +1$, the intermediate level with $m_l = +1$ is occupied and the transition $|2\rangle \rightarrow |3\rangle$ must obey $\Delta m = -1$. An analogue condition is valid for the cascade $\gamma_1(\Delta m = -1) \rightarrow \gamma_2(\Delta m = +1)$. The angular correlation therefore follows according to (4.120a) the angular distribution $|Y_1^{\pm 1}|^2 \propto \frac{1}{2}(1 + \cos^2\vartheta)$.

Because electric and magnetic multipole transition have opposite parity (see Sect. 3.5.2) but the same angular distribution the multipole character cannot be obtained solely from measurements of the angular correlation. However, the additional measurement of the polarization of the γ-radiation gives the direction of the electric field vector of the electro-magnetic wave (Fig. 4.96). Measurements of the polarization can use all effects that depend on the polarization of the wave; One example is the Compton effect, where the Compton scattering cross section not only depends on the angle φ between incident wave and scattered wave but also on the polarization direction of the scattered wave against that of the incident wave (see Vol. 3, Sect. 3.1.3). It is therefore possible to obtain with the arrangement of Fig. 4.97 which allows the measurement of polarization-scattering-correlations, the parity of the levels involved in the cascade transitions.

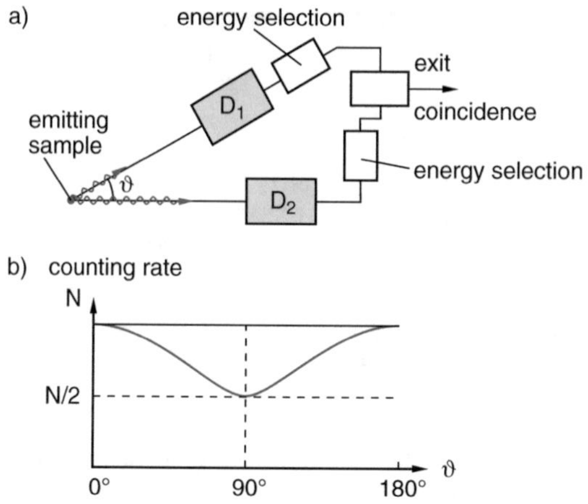

Fig. 4.96 **a** Experimental setup for the measurement of angular correlations of γ-quanta. **b** correlation counting rate $N(\vartheta)$ for a dipole–dipole cascade

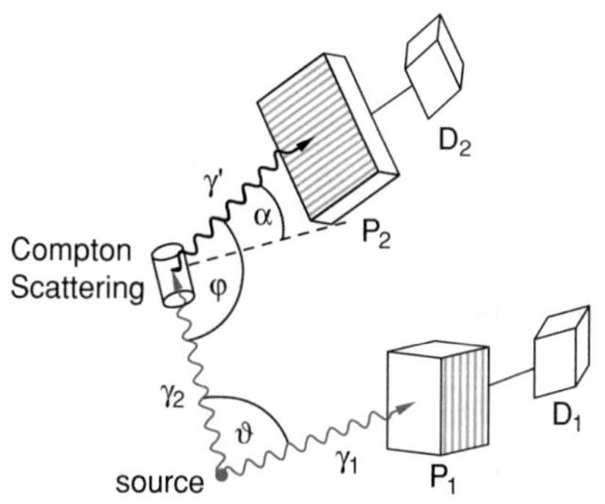

Fig. 4.97 Arrangement for measuring the polarization-dependent angular correlation using Compton scattering. P_i = D polarizer, D_i = D detector

4.5.2 Beta-Spectrometer

Measurements of the energy of electrons or positrons emitted by unstable nuclei (see Sect. 3.4) can be performed with Scintillation detectors or with semiconductor detectors. The best energy resolution is achieved with magnetic spectrometers. A basic device with a homogeneous magnetic field is shown in Fig. 3.26, where the focusing properties of a 180° magnetic sector field is utilized (see Sect. 3.4.4). A larger transmission is achieved with a specially formed magnetic axial field. Here all electrons emitted by the same point into all directions $|\alpha| < 90°$ are focused onto an axial point at a distance.

$$Z_f = \frac{2\pi}{e \cdot \int B\mathrm{d}z} \sqrt{2mE}$$

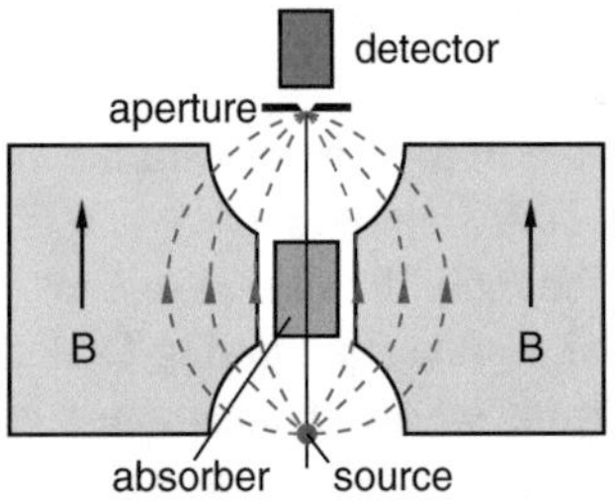

Fig. 4.98 Beta-Spectrometer with axial symmetric magnetic field acting as lens for the electrons

here $B(z)$ is the magnetic field along the electron path. A possible experimental design is shown in Fig. 4.98. The axial-symmetric magnetic field is produced by a magnetic coil with a z-dependent cross section which focusses all electrons emitted by a point source under the angle $|\alpha| < 90°$ onto a circular aperture in front of the detector. In order to prevent electrons emitted directly into the z-direction (which are not energy analyzed) to reach the detector a β—absorber is placed into the direct electron path. By variation of the magnetic field strength the energy spectrum of the electrons can be measured.

The assignment of β-transitions in β-γ-cascades is possible if the output signals of a γ-detector and a β-detector are measured in coincidence. Similar to the γ-γ coincidence measurements of the angular correlation here the coincidence rates between the two detectors as a function of the angle α between the directions of β-emission and γ-emission. Such measurements give information about the spins of the nuclear levels involved.

Summary

- Charged particles can be accelerated to high energies, either by passing once a high voltage difference (electrostatic accelerator) or many times an acceleration path with lower voltage difference where the energy is accumulated (ring accelerators)
- In circular accelerators the particles are kept on a circular orbit by a properly formed magnetic field. Their energy is successively increased by synchronously controlled high frequency acceleration sections.
- In the cyclotron the magnetic field remains constant in time. The radius of the particle orbits therefore increases with increasing particle energy. In the Betatron and the synchrotron the orbital radius remains constant. The magnetic field must therefore increase with increasing particle energy. For all circular accelerators the maximum achievable magnetic field constitutes the reachable upper limit for the kinetic energy of the particles.
- The particles in the accelerator can be injected into storage rings where they can be stored at constant energy for long times. Electric currents of the charged particles of up to 10 A can be realized. In order to keep the energy constant the radiation losses have to be compensated by energy supply in acceleration sections.
- The stability of the particle orbits is achieved by suitably formed magnetic fields and by electric quadrupole lenses.
- The detection of micro-particles (electrons, protons neutrons, mesons) is based on the particle type and on the energy dependent interaction of these particles with matter, where the kinetic energy of the particles in converted either into excitation- or ionization energy of the atoms or molecules of the detector or in electro-magnetic radiation energy..
- Energy and momentum of the particles can be inferred from measurements of their penetration depth in matter. A more accurate technique uses the deflection in electric or magnetic fields.
- Ionization chambers, Geiger counters and spark chambers convert the ionization produced by the incident particles into electric output pulses. In scintillation detectors the induced light emission caused by excitation of the detector atoms by the incident particles serves as detection signal.
- Electrons with velocities larger than the phase velocity of light in the transparent detector material emit Cerenkov-radiation, which is used as detector signal.
- In trace detectors (cloud or bubble or spark-chamber), the path of a particle produces a visible trace which is either photographed or detected by video cameras or by diode arrays. The curvature of this trace in a magnetic field yields the momentum of the particle.
- Detector systems in high energy physics are complicated and clever designed systems of various detector types which can monitor the type of a particle, its energy and momentum.
- Measurements of the differential collision cross section for the elastic scattering of particles give information about the dependence of the interaction potential on the inter-nuclear distance. For the determination of non-spherical potentials aligned particles (e.g. spin-orientated nuclei) are used. Generally a model potential with free parameters is set up, where the parameters are fitted to the experimental results.
- For inelastic or reactive collisions only the energy in the center-of mass system can be transferred into internal energy of the collision partners. The residual energy remains as translational energy of the collision partners. The maximum energy for transfer into internal energy is achieved in central collisions. They are realized in colliders where the collision partners fly before the collision into opposite directions

- For all collisions the components of the momentum-four-vector are preserved.
- In the stationary quantum mechanical treatment of elastic collisions the differential collision cross section is described by the absolute square $|f(\vartheta)|^2$ of the scattering amplitude $f(\vartheta)$ [32, 33]

Problems

4.1 An Electron, a Proton and an α-particle are each accelerated to the Kinetic Energy $e \cdot U = 1$ GeV. How large are the velocities and the total energy?

4.2 In a Betatron electrons shall be accelerated within 10 ms to a kinetic energy of 1 MeV. Calculate their total energy and their velocity. They start with the energy $E = 0$.What is the maximum magnetic flux Φ at the end of the acceleration period and how large is the magnetic field then at the electron circular path with radius $r = 1$ m?

4.3 In a Proton Synchrotron with a Radius $r = 10^3$ m Protons should be accelerated up to 400 GeV. How large is the necessary magnetic field?

4.4 A parallel proton beam with kinetic energy $E_{kin} =$ 100 MeV should be focused by a longitudinal magnetic field. What is the magnetic field for a focal lens $f = 10$ m?

4.5 What is the energy loss of an electron per round trip in the LEP storage ring with radius $r = 4000$ m due to radiation losses at the kinetic energy $E_{kin} = 50$ GeV? How large is the total energy loss at the electron current $I = 0.1$ A?

4.6 Show, that in a linear accelerator the particle acceleration is proportional to the energy gain dE/dx per length unit. How large is dE/dx in the Stanford linear accelerator SLAC, where the electrons are accelerated to the energy $E_{kin} = 50$ GeV along the distance $L = 3200$ m. How large must be dE/dx to make the radiation losses comparable to those at the LEP for the same final energy?

4.7 What is the maximum energy available for the production of new particles when a particle with mass m_1, energy E_1 and momentum p_1 collides with a particle m_2 that rests in the lab system? Calculate the threshold energy of a proton beam for the production of anti-protons in the reaction

$$p + p \rightarrow p + p + p + \bar{p} + p$$

(a) For a target at rest
(b) If the two protons with $p_1 = -p_2$ suffer a central collision.

4.8 Calculate the Luminosity for the Following Arrangements:
(a) A beam of α-particles with the electric current $I_1 = 1$ μA impinges onto a $6^{12}C$ foil with the surface mass density 64 μg/cm^2.
(b) In the double storage ring DORIS electrons and positrons with velocities $v \approx c$ collide under the angel $\alpha = 5°$. The corresponding currents are $I_1 = I_2 = 2$ A and both beams have a homogeneous density over the cross section $b \cdot h$ with $b_1 = b_2 = 1$ mm and $h_1 = h_2 = 4$ mm.
(c) In a storage ring currents $I_1 = I_2 = 3.2$ A of electrons and positrons are compressed into k bunches per circumference which collide with the frequency f ($b = h = 1$ mm, $k = 8$, $f = 10$ MHz).

4.9 The attenuation coefficient for X-rays with energy 100 keV in lead is $\mu = 5 \cdot 10^3\ m^{-1}$. How thick must a lead plate be in order to attenuate incident X-rays down to 1‰?

4.10 The momentum of an incident particles is measured in an experiment as $p = 367$ MeV/c. These particles produce in heavy flint-glas ($n = 1.70$) Cerenkov radiation under the angle $\alpha = 51°$. Which kind of particles is it? Which minimum energy is necessary for these particles in order to produce Cerenkow radiation? What is the dependence of the intensity of the radiation per unit path length on the angle α?

4.11 Protons with 1877 MeV kinetic energy impinge onto a sheet of Al and Pb resp. with thickness 1 g/cm^2. What is the attenuation factor for the two materials? In which material is the attenuation larger? (Note: insert into (4.63) $\langle E_B \rangle = 10 \cdot Z_2 \cdot eV$ as the mean binding energy of the target electrons). How does the result change if equal geometric thickness is assumed for both materials ($\rho_{Al} = 2.70$ g/cm^2 and $\rho_{Pb} = 11.34$ g/cm^2).

4.12 The cross section for neutron capture is for slow neutrons ($E_{kin} = 0.025$ eV) about $\sigma = 10^{-24}$–10^{-23} – m^2 although the nuclear radius is only $R = 1.2 \cdot A^{1/3}$. Can you explain this fact? Explain also why the capture cross section decreases as $1/v$ with increasing velocity v of the neutrons.

4.13 Justify Why
(a) Neutrons with kinetic energy $E_{kin} < 1$ MeV are scattered into all directions when colliding with nuclei with mass number $A = 20$.
(b) Show that the average relative energy loss of a neutron with energy E per collision is $\Delta E/E = 2A/(1 + A^2)$.

4.14 Show that the Squares R^2 and P^2 of the Four-Vectors (4.92) for Energy and Momentum are invariant under a Lorentz Transformation.

References

1. H.Wiedemann: Particle Accelerator Physics (Springer, Berlin Heuideklberg 1993)
2. E.J.N.Wilson: An Introduction to Particle Accelearators (Oxford Univ. Press 2001)
3. R.Jayakumar: Particle Accelerators, Colliders and the Story of High Energy Physics. (Springer Berlin, Heidelberg 2012)
4. S.Y.Lee: Accelerator Physics (world Scientific, Singapore 1999)
5. J. Wenz et al. Dual-energy electron beams from a compact laser-driven accelerator, *Nature Photonics* (2019). https://doi.org/10.1038/s41566-019-0356-z
6. M.Dunne et.al. Laser-Driven Particle Accelerators. Science Vol. 312, Issue 5772, p.374 (April 2006)
7. P. Ginter, Franzobel: LHC The large Hadron Collider (Edison Lammerhuber 2011)
8. Don Lincoln: The Large Hadron Collider (John Hopkins Univ. Press 2009)
9. E.Lyndon: The large Hadron Collider (EPFL-Press 2009)
10. J.Hein, R.Sauerbrey:generation of ultrahigh intensities and relativiticilaaser-matter interaction. (Springer handbook of Lasers and Optics Heidelberg 2007 page 287
11. J.Breuer, P. Hommelhoff: Laser-based Acceleration of nonrelativitic electronsat a dielectric structure. Phys. Rev. Lett. 27, Sept. 2013. https://doi.org/10.1103/Phs.Rev.Lett 111,134803
12. T.Plettner P.P Lu R.L Byer proposed few cycle laser-driven particle accelerator structure. Phys. Rev. Spec.Top 9 111301 (2006)
13. Mike Dunne: Laser-Driven Particle Accelerators. Science 312, 21st April 2006 p. 374376
14. S.van der Meer: Stochastic Cooling and the Accumulation of Antiprotons. Rev Mod. Phys. 57, 689 (1985)
15. S. Ebashi, M. Koch E Rubenstein: Handbook on Synchrotron Radiation (Springer, Heidelberg 1995)
16. W.Eberhardt (ed). Applications of Synchrotron Radiation (Springer, Heidelberg 1995)
17. https:de.wilipedia.org.wili/Higgs-Boson
18. M.Campenelli: inside CERN's Large Hadron Collider(World Scientific 2015)
19. P.Ginter, Franzhobel: LHC Large Hadron Collider (Edition Lammerhuber 2015)
20. St. Tabernier: Experimental Techniques in Nuclear and Particle Physics (Springer Heidelberg 2010)
21. H. Heitler: The Quantum Theory of radiation 3. Ed. Dover, Oxford 1984)
22. Paul Dirac: A theory of electrons and positrons, Proceedings of the Royal Society, Volume 126, 1930, 360
23. H.A.Bethe: Zur Theorie des Durchgangs schneller Korpuskular strahlung durh Materie. ANn. Phys. 397, 325 (1930)
24. N. Wermes, H.Kolanoski: Particvle Detectors: Fundamentals and Applications (Oxford Univ. Press 2020)
25. C. Henderson. Cloud and Bubble Chambers (Methuen London 1970)
26. K. Kleinknecht: Detectors, Particles and Radiation 2nd ed. (Cambridge Univ. Press 2008)
27. Claus Grupen. Particle Detectors (Cambridge Monographs on Particle Physics 2011)
28. Dan Green: The Physics of Particle Detectors (Cambridge Monographs on Particle Physics 2000)
29. https://en.wikipedia.org/wiki/Calorimeter_(particle_physics
30. https://en.wikipedia.org/wiki/Calorimeterhttps://en.wikipedia.org/wiki/Cherenkov_detector
31. U.Becker, U.Crowe: Complete scttering Experiments (Springer 2006 Heidelberg
32. H.Kleinpoppen, B.lohmann, Grzhimkailu: Perfect/Complete Scattering Experiments (Springer Heidelberg 2013)
33. https://www.desy.de/~beckerj/fsds/Detector_activities_FSDS.pdf

5 Nuclear Forces and Nuclear Models

From the results discussed in Chap. 2 we know, that the strong interaction between the nucleons in the nucleus is caused by attractive forces, which must be stronger than the repulsive Coulomb interaction between the charged protons. The homogeneous mass density inside the nucleus which is nearly independent of the size of the nucleus, tells us that these strong forces must have a short range and therefore act mainly between neighboring nucleons.

In this chapter we will win a more detailed insight into the physical nature of nuclear forces and we will learn about models for their adequate description. This will lead us to different models which are optimized to describe specific characteristics of nuclei.

The interaction between two nucleons can be best studied on isolated two-nucleon systems. The only existing stable two-nucleon system is the **deuteron** ${}^2_1\mathrm{H} = \mathrm{D}$ consisting of one proton and one neutron. Measurements of its binding energy, its nuclear spin, its magnetic dipole moment, electric quadrupole moment and its size can give us important information about nuclear forces and the substructure of nucleons. We will discuss this in Sect. 5.1.

The two other two-nucleon systems are p–p and n–n, which are both unstable. It is, however, useful to study the interaction between the nucleons in these systems and their dependence on the relative spin-orientation by scattering experiments. This will be explained in Sect. 5.2.

A simple mass balance proves that the deuteron nucleus must be stable. The sum of the masses of proton and neutron is $m(\mathrm{p}) + m(\mathrm{n}) = 1877.841\ \mathrm{MeV}/c^2$ while the mass of the deuteron is only $1875.613\,\mathrm{MeV}/c^2$. This mass difference must be caused by the binding-energy of the deuteron, which should be, according to this estimation, $E_\mathrm{B}({}^2\mathrm{D}) = -2.2\,\mathrm{MeV}$. Furthermore the β-decay of the neutron in the deuteron is not possible.

Because for the decay of the free neutron

$$\mathrm{n} \rightarrow \mathrm{p} + \mathrm{e}^- + \nu$$

the maximum kinetic energy of the electron is 0.78 MeV. If the decay should occur in the deuteron the binding energy of $E_\mathrm{B} = -2.2\,\mathrm{MeV}$ could not be overcome, because the transformation of the neutron into a proton would produce the p–p system with a positive binding energy (i.e. it is unstable).

5.1 The Deuteron

The binding energy of the deuteron can be measured by different methods:

- From the photo-induced fission of the deuteron with γ-quanta

$$ {}^2_1\mathrm{H} + h \cdot \nu \rightarrow \mathrm{n} + \mathrm{p} + E_\mathrm{kin} \qquad (5.1)$$

The kinetic energy of the fission products divides into equal parts for n and p, because they have nearly equal masses.

A deuteron target (e.g. heavy water D_2O) is irradiated with γ-radiation $h \cdot \nu$ of variable frequency ν, which is generated by bombardment of a target by electrons with variable energy $e \cdot U$ (Fig. 5.2). The maximum γ-energy $\mathrm{h} \cdot \nu_\mathrm{max}$ can be inferred from the electron energy $E_{kin} = e \cdot U = h \cdot \nu_\mathrm{max}$. A neutron detector measures the neutrons released by the fission. The proton energy can be determined either by the penetration through an opposing electric field, (Fig. 5.2) by the penetration depth into an absorbing material or by the deflection in a magnetic field. The binding energy of the deuteron is then

$$E_\mathrm{B} = [h \cdot \nu - E_\mathrm{kin}(\mathrm{p}) - E_\mathrm{kin}(\mathrm{n})]. \qquad (5.2)$$

W. Demtröder, *Nuclear and Particle Physics*, Undergraduate Lecture Notes in Physics,
https://doi.org/10.1007/978-3-030-58313-2_5

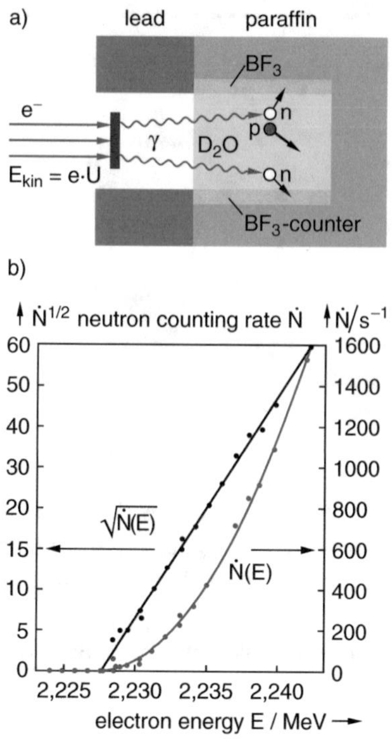

Fig. 5.1 Measurement of the Deuteron binding energy, based on the photo-induced fission. **a** Experimental setup **b** measured neutron rate $\dot{N}$ as a function of the electron energy (red curve) and $\sqrt{\dot{N}(E)}$ (black curve). (After R. C. Mobby, R. A. Laubenstein: Photo-Neutron Threshold of Beryllium and Deuteron. Phys, Rev. 80, 309 (1950))

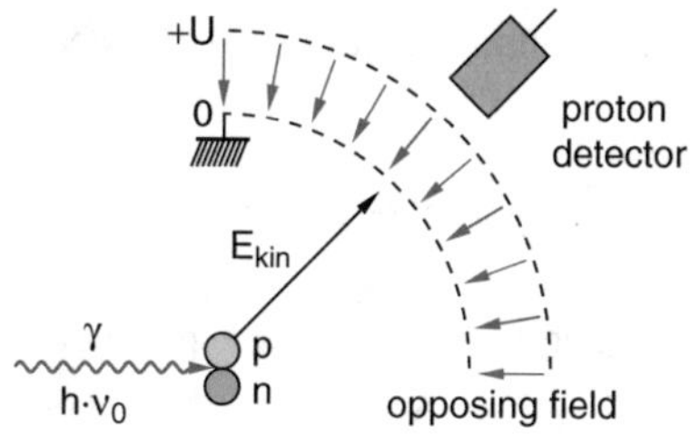

Fig. 5.2 Measurement of the kinetic energy of the proton released at the photo-fission of the deuteron with the opposing field technique

- From the recombination radiation emitted when a neutron attaches to a proton (for instance neutrons from a nuclear reactor are absorbed in paraffin or liquid hydrogen):

$$n + {}^1_1\mathrm{H} \rightarrow {}^2_1\mathrm{H} + \gamma \quad \text{with} \quad h \cdot \nu = E_\mathrm{B}. \tag{5.3}$$

Here the energy of the γ-quantum is measured with an energy-selective detector, e.g. a semiconductor detector (see Sect. 4.3.3) or by Bragg reflection at a crystal (Fig. 5.3).
The average value of the binding energy, obtained from the two methods is

$$E_\mathrm{B} = (2{,}224573 \pm 0{,}000040)\ \mathrm{MeV}$$

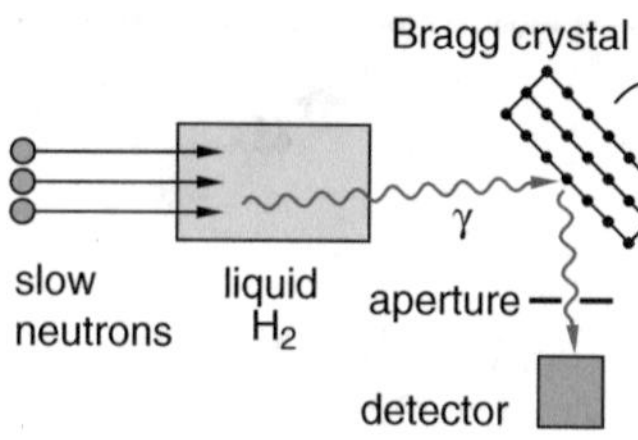

Fig. 5.3 Measurement of the γ-energy of the recombination radiation emitted at the accretion of slow neutrons with protons

It is much smaller than the average value $E_B/A \approx 8$ MeV per nucleon obtained for larger nuclei with A nucleons, which shows that the two nucleons in the deuteron nucleus must have a large kinetic energy.
This kinetic energy can be estimated using the uncertainty relation:
For a nuclear radius R the minimum momentum component is $p_i \approx \hbar/R (i = 1, 2, 3) \Rightarrow \ p^2 = \sum_{i=1}^3 p_i^2 = 3\hbar^2/R^2 \Rightarrow$ the kinetic energy

$$E_\mathrm{kin} \geq \frac{p^2}{2m} \approx \frac{3\hbar^2}{2mR^2}.$$

With $R = 2.0$ fm we get $E_\mathrm{kin} \geq 23$ MeV per nucleon and the total kinetic energy of 46 MeV/c^2. Since the binding-energy is the sum

$$-E_\mathrm{B} = -E_0 + E_\mathrm{kin}$$

of the negative potential- and the positive kinetic energy. According to this coarse estimation the depth of the potential box for the deuteron as a two-nucleon system is about −48 MeV (Fig. 5.4).
From measurements of the Hyperfine- structure and the Zeeman splitting of the ${}^2_1$H (Sect. 2.4) one can infer that the nuclear spin of the deuteron is $\mathrm{I} = 1 \cdot \hbar$, i.e. the spins of proton and neutron are parallel,
Very precise measurements of the splitting of the hfs-components (Fig. 5.5) give the magnetic moment of the deuteron nucleus

$$\mu_\mathrm{D} = (0{,}857348 \pm 0{,}00003)\mu_\mathrm{N}.$$

Fig. 5.4 Schematic representation of the potential depth E_0, kinetic energy E_kin and binding energy E_B of the two nucleons with parallel spin in the bound triplet state of the deuteron

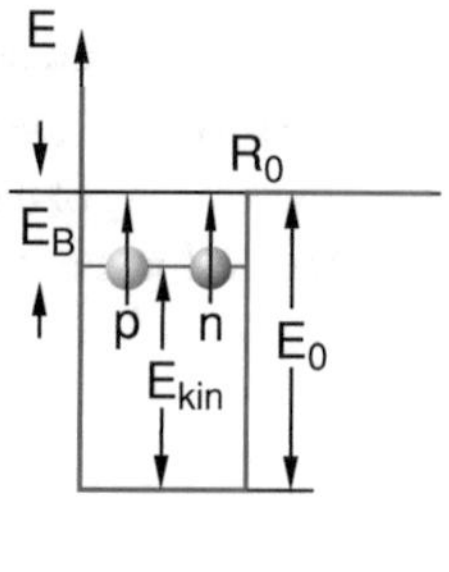

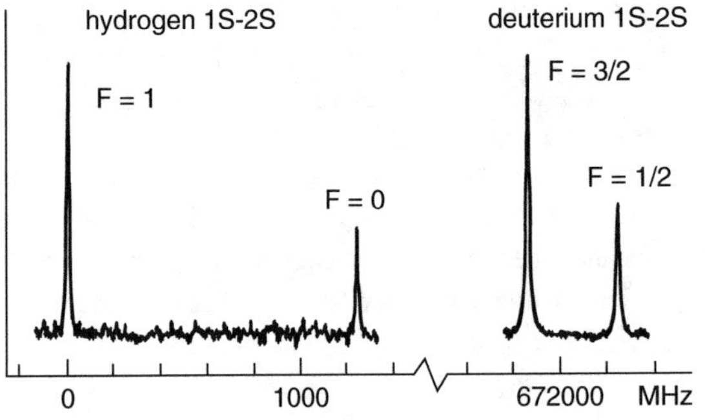

Fig. 5.5 Comparison of the hyperfine- splittings of the two-photon transitions $2S \leftarrow 1S$ for the two isotopes ${}^1_1\text{H}$ and ${}^2_1\text{H} = {}^2_1\text{D}$ (T. W. Hänsch; the hydrogen Atom. Ed. by G. F. Bassani, M. Inguscio and T. W. Hänsch (Springer, Heidelberg 1989)

Where μ_N is the nuclear magneton. It is only slightly smaller than the sum of the magnetic moments of proton and neutron, which is 0,87963 μ_N (see Table 2.4). This proves that the magnetic moment of the deuteron is essentially caused by the parallel spins of the two nucleons and that the orbital angular momentum L of the system should be zero.

This means that the deuteron has, in a first approximation an S-ground-state with $L = 0$ and spin quantum number $S = 1$ with a spherical symmetric position probability for the two nucleons. The ground-state of the deuteron is therefore labelled as ${}^3S$-state in analogy to the designation of states in the electron shell. It is the only bound state. There is no bound singlet state.

Experiments show, however, that the deuteron has an electric quadrupole moment

$$QM_D = (2{,}860 \pm 0{,}0003) \cdot 10^{-31}\text{cm}^2 \cdot e$$

This proves that its charge distribution could not be perfectly spherical symmetric.

Besides by the methods discussed in Sect. 2.5.2 the quadrupole moment can be also measured by the isotope shift of the spectral lines of ${}^2_1\text{H}$ against those of ${}^1_1\text{H}$ (Fig. 5.5), which is caused by the mass effect (different reduced masses $\mu = M_N \cdot m_e/(M_K + m_e)$), the magnetic hyperfine structure and the electric quadrupole moment (see Vol. 3, Sect. 5.6).

In order to determine the mean distance between proton and neutron in the deuteron nucleus we start in a first approximation with a spherical symmetric potential with the potential depth of E_0 and the radius R_0 where the proton and neutron move with their kinetic energy E_{kin} (Fig. 5.4). This two-body problem can be reduced to a one-body problem where one particle with the reduced mass

$$\mu = \frac{m_p \cdot m_n}{m_p + m_n}$$
$$\Rightarrow \mu \approx \frac{1}{2}m \quad \text{with} \quad m = \frac{1}{2}(m_p + m_n)$$

moves in a potential box

$$E_{pot} = \begin{cases} -E_0 & \text{for} \quad r < R_0 \\ 0 & \text{for} \quad r \geq R_0 \end{cases}.$$

As has been shown in Vol. 3, Sect. 4.2.4 the radial Schrödinger equation

$$\frac{d^2u}{dr^2} + \frac{m}{\hbar^2}\left[E - E_{pot}(r)\right]u = 0 \tag{5.4}$$

with the wave function $u(r) = r \cdot \psi(r)$ and the boundary conditions $u(0) = 0, u(\infty) = 0, E = -E_B < 0$ has the solutions

$$\begin{aligned} u_1 &= A_1 \sin(k_1 r) \quad \text{for} \quad r \leq R_0 \\ &\text{with} \quad k_1 = \frac{1}{\hbar}\sqrt{m(E_0 - E_B)}, \\ u_2 &= A_2 \cdot e^{-r/a} \quad \text{for} \quad r > R_0 \\ &\text{with} \quad \frac{1}{a} = \frac{1}{\hbar}\sqrt{m \cdot E_B}. \end{aligned} \tag{5.5}$$

Demanding that $u(r)$ and $u'(t)$ have to be continuous at $r = R_0$ we get

$$A_1 \sin(k_1 R_0) = A_2 e^{-R_0/a} \tag{5.6a}$$

$$k_1 A_1 \cos(k_1 R_0) = -\frac{A_2}{a} e^{-R_0/a}. \tag{5.6b}$$

Division of (5.6a) by (5.6b) yields

$$k_1 \cdot \cot(k_1 R_0) = -1/a, \tag{5.7}$$

This gives when inserting the expressions for k_1 and a

$$\cot\left[\frac{R_0 \cdot \sqrt{m(E_0 - E_B)}}{\hbar}\right] = -\sqrt{\frac{E_B}{E_0 - E_B}} \tag{5.8}$$

The radius R_0 of the potential box must correspond to the range of the nuclear forces, which is, according to the estimations in Sect. 2.3 around 1.5 Fermi.

Inserting $R = 1.5$ Fermi and the experimental value $E_B = -2.2$ MeV we obtain $E_0 = -45$ MeV.

The potential box is therefore much deeper than the only bound energy level $E = -E_B = -2{,}2$ MeV of the deuteron nucleus. The kinetic energy of proton and neutron (zero-point energy) is nearly as large as the absolute amount of the negative potential energy (Fig. 5.4).

With $E_0 \gg |E_B|$ we get for (5.8)

$$\cot \sqrt{\frac{R_0^2}{\hbar^2} m(E_0 - E_B)} \ll 1$$
$$\Rightarrow \frac{m}{\hbar^2} R_0^2 (E_0 - E_B) \approx \left(\frac{\pi}{2}\right)^2.$$

The depth of the potential box then becomes $E_0 - E_B \approx E_0$

$$E_0 = \left(\frac{\pi}{2}\right)^2 \frac{\hbar^2}{mR_0^2}. \quad (5.9)$$

This gives the relation between the depth E_0 and the radius R_0 of the potential box of the deuteron nucleus.

Also for $r > R_0$ the wave-function (5.5) gives a nonzero position probability for the two nucleons, which decays exponentially with increasing r. The probability $P(r)$ to find proton and neutron at a distance $r > R_0$ in the range from r to r + dr is

$$P(r) \cdot \mathrm{d}r = 4\pi r^2 |\psi(r)|^2 \mathrm{d}r = 4\pi |u_2(r)|^2 \mathrm{d}r$$
$$= 4\pi A_2^2 \mathrm{e}^{-2r/a}.$$

For $r = a$ the wave-function $u_2(r)$ drops to 1/e of its value at $r = 0$. The probability $P(r) \propto |u(r)|^2$ therefore drops to $1/\mathrm{e}^2$ (Fig. 5.6). The quantity a is called the radius of the deuteron nucleus. Inserting $E_B = -2{,}2$ MeV into (5.5) gives $a = 4.3$ fm, which is larger than the extension of the potential box R_0 by the factor 3.

The position probability for the proton and the neutron outside of the potential box ($r > R_0$) is

$$P(r > R_0) = 4\pi \int_{r=R_0}^{\infty} |u|^2 \mathrm{d}r = 3\pi a \cdot A_2^2 - \mathrm{e}^{-2R_0/a}. \quad (5.10)$$

For realistic values $R_0 = 1{,}5$ fm and $a = 4{,}3$ fm $P(r)$ becomes $P(r > R_0) \approx 2{,}15 \cdot \mathrm{e}^{-3/4{,}3} \approx 0{,}8$.

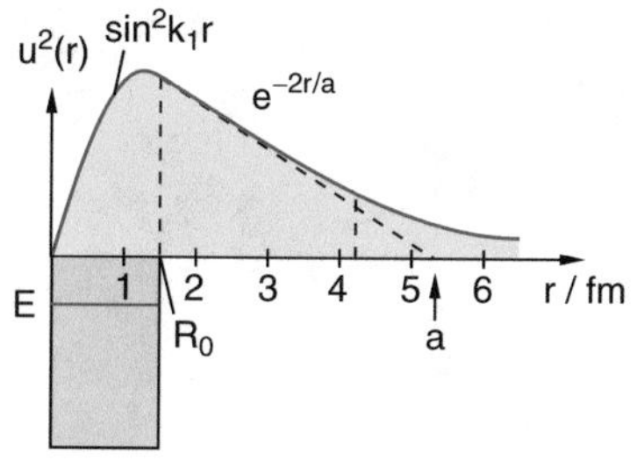

Fig. 5.6 Probability $4\pi r^2|\psi|^2 = 4\pi|u(r)|^2$ of finding the nucleons of the deuteron at the mutual distance r

The two nucleons spend during their relative motion about 80% of the time outside the potential box.

This is one of the reasons for the small binding energy. Heavy nuclei with $E_B/A \approx 8$ MeV per nucleon have a smaller mean distance between the nucleons and therefore a larger nucleon density and a stronger attraction between the nucleons.

We will now look for an explanation for

(a) the existence of the electric quadrupole moment,
(b) the small deviation of the magnetic moment from the vector sum $\mu_p + \mu_n$ and
(c) the nonexistence of á stable singlet state $I = 0$.

All three experimental facts can be explained if one assumes that the force between the nucleons contains besides the symmetric part an additional weaker part which depends on the relative spin orientation of the two nucleons. We therefore try the ansatz for the interaction potential between the two nucleons

$$V_{NN} = V_1(r) + V_2(\boldsymbol{I}_1, \boldsymbol{I}_2), \quad (5.11)$$

where the spin-dependent part

$$V_2(\boldsymbol{I}_1, \boldsymbol{I}_2) = a_I \cdot \left[3 \cdot \frac{(\boldsymbol{I}_1 \cdot \boldsymbol{r}) \cdot (\boldsymbol{I}_2 \cdot \boldsymbol{r})}{r^2} - \boldsymbol{I}_1 \cdot \boldsymbol{I}_2\right] \quad (5.12)$$

depends on the orientation of the two dipoles against the connecting axis (Fig. 5.7). This is similar to the interaction between two atomic dipoles which also depends on the relative orientation of the two dipoles. The last term in (5.12) becomes maximum if the two dipoles are parallel ($I_1 \| I_2$), which means for the triplet state.

The spin term can contribute to the binding energy, if $V_2 < 0$. This is realized for an array of proton and neutron stretched in the spin direction. The nucleon density of the deuteron nucleus shows a slight deviation from the spherical symmetry. The charge distribution corresponds to a slightly stretched prolate rotational ellipsoid and causes a positive electric quadrupole moment $QM > 0$ in accordance with the experimental results, which results in a lower energy and therefore a larger binding energy than the disc form (Fig. 5.8).

This can be also describe as follows:

The s-wave function with $L = 0$ is superimposed by a small contribution of a wave function with $L > 0$. Since the experiments have unambiguously proved that the parity of

Fig. 5.7 Spin-dependence of the nuclear interaction potential

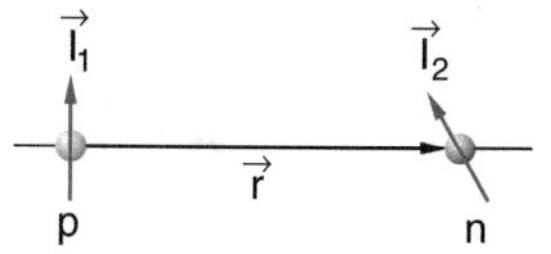

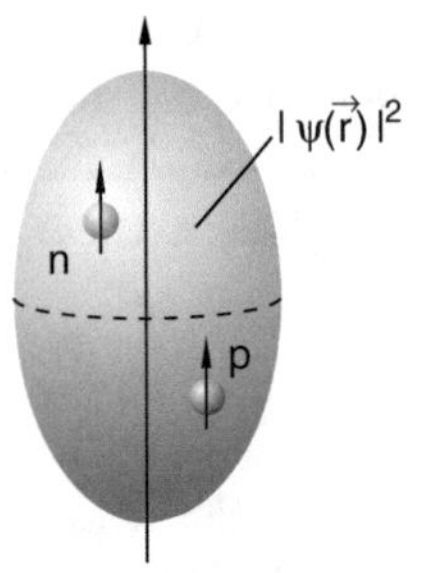

Fig. 5.8 Description of the mass- and charge distribution in a nucleid as a prolate rotational ellipsoid

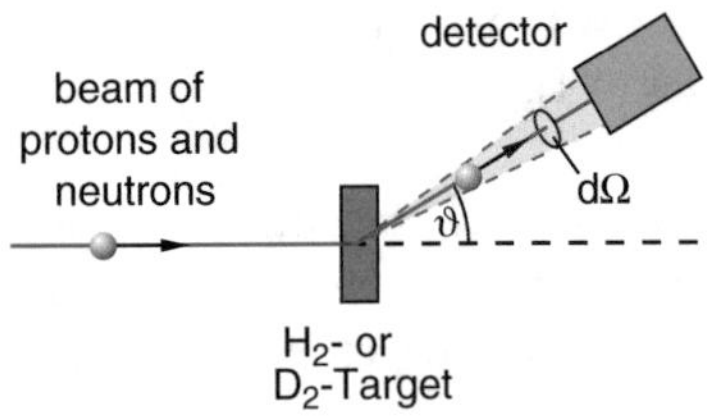

Fig. 5.9 Schematic experimental setup for measuring the differential cross section of nucleon-nucleon scattering

the deuteron is positive the admixture wave function must have an even value of L, for example a *d*-function with $L = 2$. Since an additional angular momentum contributes to the magnetic moment which, on the other side, deviates only slightly from the pure spin moment $\mu_p + \mu_n$, the admixture of the d-function can be only a few percent.

It should be emphasized that the spin-dependent part of the wave function is not equal to the interaction between two magnetic dipoles, which is much too small to explain the observed effects. The real reason is the specific characteristics of the strong interaction, namely the spin-dependent nuclear force.

This model of the spin-dependent nuclear force explains the fact, that only the triplet state of the deuteron nucleus is a bound state, and also the existence of the electric quadrupole moment and the small deviation of the magnetic moment from the vector sum of the spin moments of proton and neutron expected for a state with $L = 0$.

The dependence of the nucleon-nucleon interaction on the relative orientation of the spins is confirmed by experiments of nucleon-nucleon scattering, which we will now discuss.

5.2 Nucleon-Nucleon Scattering

Measurements of the total and the differential scattering cross sections for the elastic scattering of nucleons yield important information on the nucleon-nucleon interaction. The details of this information depend strongly on the relative energy of the collision partners, as will be discussed in the following:

While for the neutron-proton scattering only the strong interaction is active, (the magnetic interaction between the magnetic moments of the nucleons is completely negligible) in the proton-proton scattering also the Coulomb interaction must be considered.

5.2.1 Basic Fundamentals

The experimental design for such scattering experiments is shown in Fig. 5.9. A collimated beam of protons or neutrons with kinetic energy E_0 impinge on a target (liquid or gaseous H_2 or D_2) which contains protons or neutrons at rest. One measures the number of nucleons scattered at the angle ϑ into the solid angle dΩ.

As target generally liquid H_2 is used. Since the distance d $\approx 10^{-10}$ m between the two H-atoms is large compared to the proton radius $r_0 \approx 10^{-15}$ m the scattering at each of the two protons cam be regarded as independent. This means that there is no essential difference between the scattering at H-atoms or H_2-molecules. For the scattering at D_2-molecules the contributions to the scattering amplitudes of protons and neutrons just add.

The calculation of the scattering is simpler in the center-of-mass system, because then the two-body problem can be reduced to a single particle with reduced mass μ scattered by a potential with a fixed center (see Vol. 1, Sect. 4.1).The energy of the collision system is in the center-of-mass system at non-relativistic energies $(E_{kin} \ll mc^2)$

$$\mu \cdot v_{rel}^2/2 = E_0/2$$

Because $m_p \approx m_n \to \mu = m/2$ with $m = \frac{1}{2}(m_p + m_n)$.
The orbital angular momentum

$$\begin{aligned} &\boldsymbol{L} = \boldsymbol{r} \times \boldsymbol{p} \\ &\Rightarrow |\boldsymbol{L}| = b \cdot \mu \cdot v_0 = b \cdot \sqrt{mE_0/2} \end{aligned} \tag{5.13}$$

of the incident particle, referred to the scattering center $(r = 0)$ is small, because the impact parameter b must be smaller than the range r_0 of the nuclear forces in order to detect a measurable deflection of the incident particle. With

$$|\boldsymbol{L}_{max}| = l_{max} \cdot \hbar \lesssim 2R_0 \cdot \mu \cdot v_0 = 2R_0\hbar k$$

we obtain the maximum quantum number ℓ of the obital angular momentum

$$\ell_{max} \leq 2R_0 \cdot k = \frac{2R_0}{\hbar}\sqrt{m \cdot E_0/2}. \tag{5.14}$$

Example

For the proton-neutron scattering at $E_0/2 = 1$ MeV in the center of mass system we obtain for $r_0 = 1,5 \cdot 10^{-15}$ m $\to$ $\ell_{max} \leq 0{,}5$. At this energy mainly nucleons with $\ell = 0$ are scattered but only a few with $\ell = 1$ (*s*- and *p*-scattering). For $E_0 \leq 0.2$ MeV $\to$ $\ell_{max} \leq 0.16$. This gives pure *s*-

scattering with $\ell = 0$. i.e. the scattering is spherical symmetric (see Sect. 4.4).

The quantum–mechanical treatment of elastic scattering describes the incident particles by an incident plane matter wave (see Sect. 4.4.2) and the integral scattering cross section by

$$\sigma_{\mathrm{int}}^{\mathrm{el}} = 4\pi\lambda_{\mathrm{dB}}^2 \sum_{l=0}^{l_{\max}} (2l+1)\sin^2\delta_l, \tag{5.15}$$

where the scattering phases δ_ℓ describes the phase shift of the scattered wave with orbital angular momentum $|l| = \sqrt{l(l+1)}\hbar$ (see 4.117).

The information about the interaction potential is expressed by the phases $\delta(E_0)$ which depend on the energy E_0 of the incident particles.

The higher the energy E_0, the smaller becomes the de Broglie wavelength $\lambda_{\mathrm{dB}} = \mathrm{h/p}$ and the more angular momenta $\ell \cdot \mathrm{h}$ contribute to the scattered wave. Measuring the total cross section $\sigma_{\mathrm{tot}}^{\mathrm{el}}(E_0)$ as a function of the energy E_0, the interaction potential V(r) can be sampled as a function of the distance r between the colliding nucleons. As has been already shown in Sect. 2.2, the measurement of $\sigma_{\mathrm{tot}}^{\mathrm{el}}(E_0)$ does not directly give the potential progression V(r) because in (5.15) several phases contribute to the total cross section. Their number increases with increasing energy. It is, however, possible to assume a model potential with free parameters and calculate with this potential the phases $\delta(E_0)$ and the cross section $\sigma_{\mathrm{tot}}^{\mathrm{el}}(E_0)$ using (5.15). Now the calculated cross section is compared with the measurements and the free parameters in the model potential are varied until the agreement is satisfactory.

5.2.2 Spin-Dependence of the Nuclear Forces

The neutron-proton scattering at small energies ($E_0 < 10$ MeV) yields mainly scattered particles with zero orbital angular momentum ($\ell = 0$). If the particles are not polarized, i.e. their spins are not oriented, the spins of proton and neutron at the collision can be either parallel (triplet scattering with $S = 1$), or anti-parallel (singlet scattering with $S = 0$). If the interaction potential is spin-dependent, the scattering phases δ will differ for the triplet scattering from those for the singlet scattering. The quantization axis is the z-direction which is also the direction of the orbital angular momentum perpendicular to the scattering plane (x, y). For parallel spins ($\mathbf{I_1} \parallel \mathbf{I_2}$) of the colliding nucleons the total spin $\mathrm{I} = I_1 + I_2$ can point either into the $\pm\, z$-direction or perpendicular to the z-direction (i.e. in the (x, y)-plane), because the spin I with $|I| = \sqrt{(I \cdot (I+1)} \cdot \hbar$ has for $I = 1$ the three possible projections with $I_z = \pm 1$ and 0. Because there are three possible orientation of the spin for the triplet scattering, but only one for the singlet scattering, the total scattering cross section for unpolarized particles can be written as weighted average of singlet- and triplet-scattering

$$\sigma_{\mathrm{tot}}^{\mathrm{el}} = \frac{4\pi}{k^2}\left[\frac{3}{4}\sin^2\delta_0^{\mathrm{t}} + \frac{1}{4}\sin^2\delta_0^{\mathrm{s}}\right] \tag{5.16}$$

where the wavenumber k of the incident particles is $\mathrm{k} = 2\pi/\lambda_{\mathrm{dB}}$.

For the neutron-proton scattering there are no Coulomb forces active and the pure nuclear forces can be studied. The experimental result for the neutron-proton scattering at small neutron energies ($E_{\mathrm{kin}} \approx 10$ eV) is

$$\sigma_{\mathrm{tot}}^{\mathrm{el}} = 20{,}3 \cdot 10^{-24}\,\mathrm{cm}^2 = 20{,}3\,\mathrm{barn}.$$

How is this value partioned into singlet-and triplet scattering?

In the foregoing section the potential depth of the deuteron was found as $E_0 = -45$ MeV and the potential width was $R_0 = 1{,}5$ fm.The triplet state has a binding energy of $E_{\mathrm{B}} = -2{,}2$ MeV. For the singlet state, on the other side no bound state exists but a state for the scattering has a positive energy $E > 0$. The Schrödinger Eq. (5.4) yields for $E > 0$ the solutions

$$\begin{aligned} u_1 &= A_1 \sin k_1 r \quad \text{for} \quad r < R_0 \\ &\text{with } k_1 = \sqrt{2\mu(E_{\mathrm{kin}} + E_0)/\lambda_{\mathrm{dB}}^2} \\ u_2 &= A_2 \sin\left(k_2 r + \delta_0^{\mathrm{t}}\right) \quad \text{for} \quad r > R_0 \\ &\text{with } k_2 = \sqrt{2\mu E_{\mathrm{kin}}/\lambda_{\mathrm{dB}}^2} \end{aligned} \tag{5.17}$$

The continuity condition for $r = R_0$ gives, analogue to (5.7) the condition

$$k_1 \cot(k_1 R_0) = k_2 \cot\left(k_2 R_0 + \delta_0^{\mathrm{t}}\right). \tag{5.18}$$

For very small incident energies $(\{E_0 \to 0)$ is $\{k_2\{R_0\}\} \ll \delta_0^{\mathrm{t}}$ and we obtain for $\mathrm{E}_0 \to 0$ from (5.18)

$$\cot(k_1 R_0) = \frac{k_2^0}{k_1}\cot\delta_0^{\mathrm{t}}. \tag{5.19}$$

For the total triplet cross section one gets for the limiting case of small incident energies

$$\sigma_{\mathrm{tot}}^{\mathrm{t}} = \frac{4\pi}{k_2^2}\sin^2\delta_0^{\mathrm{t}} = \frac{4\pi}{k_2^2}\frac{1}{1+\cot^2\delta_0^{\mathrm{t}}}, \tag{5.20a}$$

Because of $\cot^2\delta_0 \gg 1$ we get with (5.18)

$$\lim_{\substack{E_{kin}\to 0 \\ (k_2 \to 0)}} \left(\sigma_{\mathrm{tot}}^{t}\right) = \frac{4\pi}{k_{10}^2 \cot^2(k_{10}R_0)} = 4\pi a^{*2} \tag{5.20b}$$

with

$$a^* = \lim_{E_{\text{kin}}\to 0} = a(0) = \lim_{k\to 0}[(k\cdot\cot\delta_0(k))]^{-1} \quad (5.20c)$$

where $k_{10}^2 = m\cdot E_0/\hbar^2$. The total triplet scattering cross section $\sigma_{\text{tot}}^{\text{t}} = 4\pi a^2$ is determined by the parameter a* which gives that distance from the center $r = 0$ where the local probability of the nucleons has decreased to 1/e of its maximum value. The parameter a* is called the **scattering length**. Inserting the value a* = 5.5 fm, which has been obtained from the spectroscopy of the deuteron (Fig. 5.6) we get

$$\sigma_{\text{tot}}^{\text{t}} \approx 3{,}8\cdot 10^{-24}\,\text{cm}^2 = 3{,}8\ \text{barn}.$$

When this is compared with the value $\sigma_{\text{tot}} = 20.3$ barn for the total cross section, it follows from (5.16) that the singlet cross section σ^{s} has the substantially larger value

$$\sigma_{\text{tot}}^{\text{s}} = 4\cdot\sigma_{\text{tot}}^{\text{el}} - 3\cdot\sigma_{\text{tot}}^{\text{t}} \approx 70\cdot 10^{-28}\,\text{m}^2 = 70\ \text{barn}$$

The singlet scattering cross section is essentially larger than the triplet scattering cross section.

The nuclear forces must be therefore spin-dependent and the interaction potential for the singlet scattering must have a larger range that that for the triplet scattering.

The scattering length a* can be descriptively visualized by Fig. 5.10. The solution *u(r)* of the Schrödinger Eq. (5.4) approaches for $E = 0$ and $r > R_0$ a straight line ($u_2''0 = 0$), which is drawn in Fig. 5.10 as dashed line for the two cases $E < -E_{\text{pot}}$ (bound state) and $E > -E_{\text{pot}}$ which are the tangents to the curve u(r), (scattering state). The intersection of these straight lines with the abscissa $u(r) = 0$ corresponds to the scattering length $a^* = a(0)$, which is positive for the bound state and negative for the unbound scattering state.

With the scattering length the total scattering cross section can be written in a common formula valid for singlet as well as for triplet scattering. Using (5.19) and (5.20a, 5.20b) we get

$$\sigma_{\text{total}}^{\text{el}} = \frac{4\pi}{k^2 + \left[\frac{1}{a^*} + \frac{1}{2}k^2R_0\right]^2}. \quad (5.21)$$

this implies:

The measured scattering cross section at low energies allow only the determination of the two parameters a^* and R_0, but not of the radial profile of the interaction potential.

In Table 5.1 the scattering length for the 3 systems pp. nn and np are compiled. In order to learn more about the radial dependence of the potential *V(r)* one has to use fast neutrons with small De Broglie wavelengths λ_{dB}, where, however, larger orbital angular momenta contribute to the scattering. This means, one has to measure the energy-dependence of the

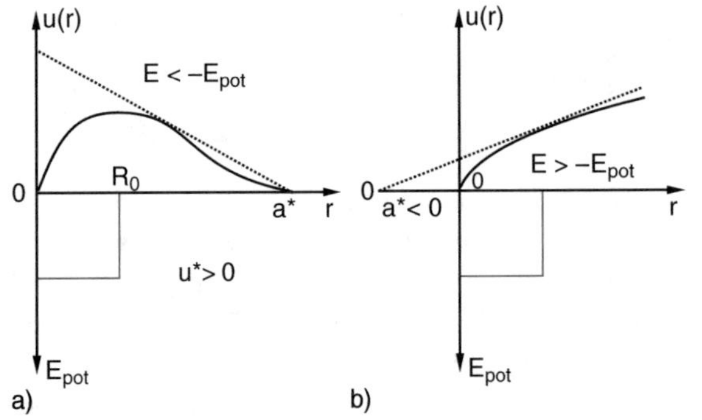

Fig. 5.10 Scattering length a* for bound states with $(E < -E_{\text{pot}}, a^* > 0)$ **b** for positive scattering states with $(E > -E_{\text{pot}}, a^* < 0)$

Table 5.1 Scattering lengths and effective range of nuclear forces for the systems np; pp and nn

Collision partner s	Scattering length	Effective range
np	$a_s(0) = -23,7\,\text{fm}$	$r_{0s} = 2,75\text{fm}$
	$a_t(0) = +5,4\,\text{fm}$	$r_{0t} = 1,76\text{fm}$
pp	$a_s(0) = -7,8\,\text{fm}$	$r_{0s} = 2,77\,\text{fm}$
nn	$a_s(0) = -16,7\,\text{fm}$	$r_{0s} = 2,85\,\text{fm}$

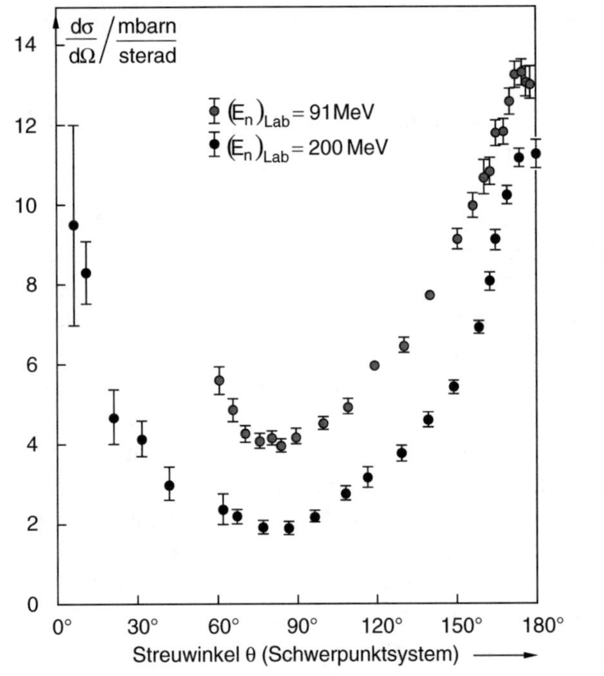

Fig. 5.11 Elastic differential cross section of the scattering of fast protons at neutrons for two different kinetic energies (After A. Bohr, B. Mottelson: Struktur der Atomkerne Bd 1 und 2 Akademie-Verlag Leipzig 1975)

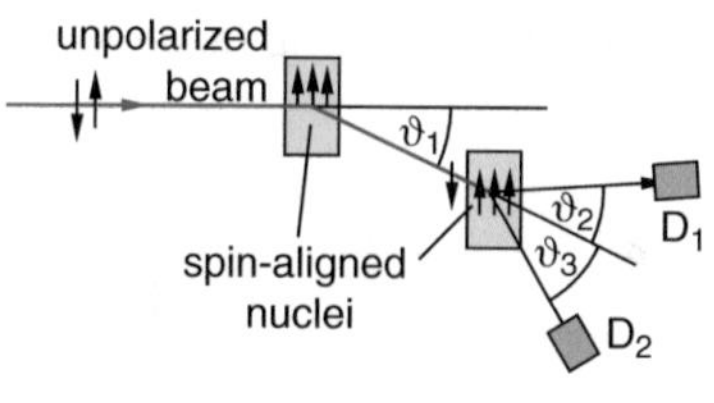

Fig. 5.12 Measurement of the spin dependence of nuclear forces using spin-polarized particles produced by a double scattering arrangement

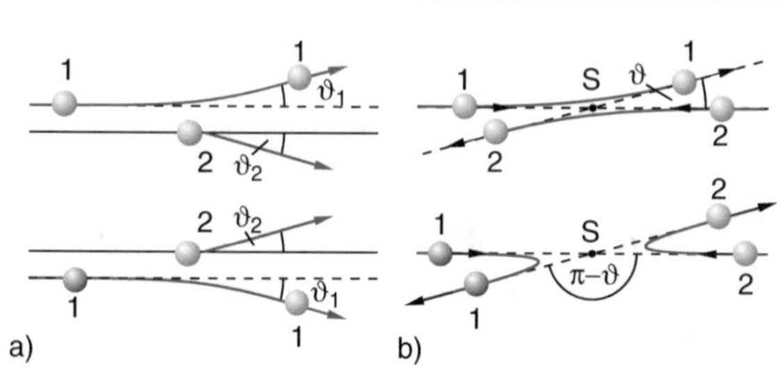

Fig. 5.13 Undistinguishable collision processes at the scattering of identical particles particles **a** in the lab-system **b** in the center of mass system

differential scattering cross section. The result of such measurements is shown in Fig. 5.11 for two different energies.

The most detailed information about the spin-dependence of nuclear forces is gained from scattering experiments with spin-polarized nucleons. Such experiments are more difficult because they demand a double scattering (Fig. 5.12).

The incident unpolarized particle beam is scattered by a spin-polarized target. The target nucleons can be, for example, oriented at low temperatures in an external magnetic field (**nuclear spin polarization**). Since the scattering cross section depends on the relative spin orientation of the two collision partners, more particles are scattered into the angle ϑ_1 if their spins are anti-parallel to that of the target nucleon than if they are parallel. The total beam scattered into the direction of ϑ_1 is therefore partially polarized. This means that the first scattering center acts as polarizer. The scattered particles are now again scattered in a second target of polarized nucleons. Measurements of the differential scattering cross section $\sigma(\vartheta_2)$ as a function of the scattering angle ϑ_2 for parallel and $\sigma(\vartheta_3)$ for anti-parallel spins are performed when orientating alternately the spin direction of the target particles into opposite directions by reversing the magnetic field.

5.2.3 The Charge-Independence of the Nuclear Forces

In order to investigate whether the nuclear forces depend on the electric charge one has to compare the scattering cross sections for n–p scattering with those for p–p scattering and n–n scattering. This comparison has, however, to take into account the following fact:

For the p–p- and n–n scattering identical particles are scattered. As illustrated in Fig. 5.13 one cannot distinguish whether the scattered particle is the incident or the target particle. This implies in the center of mass system that the scattered particle can be found at the angles ϑ or $\pi - \vartheta$. Therefore the total scattering amplitude must be written as the sum of the two different scattering amplitudes. This is completely analogue to the double slit experiment in optics where the light amplitude observed on a screen behind the slits is the sum of the partial amplitudes transmitted by the two slits. The differential cross section for identical particles is therefore, instead of (4.105), given by

$$\left(\frac{d\sigma}{d\Omega}\right)_{el} = |f(\vartheta) \pm f(\pi - \vartheta)|^2. \qquad (5.22)$$

For sufficiently slow collision partners the orbital angular momentum is $l = 0$. In this case the spatial wave-function is spherical symmetric and it is $f(\vartheta) = f(\pi - \vartheta)$. This implies that the wave-function is also symmetric against the exchange of the two particles. However, for Fermions (half-integer spin) the total wave-function (product of spatial part times spin function) must be antisymmetric (see Vol.3 Sect. 6.1) Therefore the spins of the two nucleons must be anti-parallel.

The scattering of protons on protons or of neutrons on neutrons at low energies occurs exclusively as singlet scattering.

One therefore has to compare the scattering cross sections of the p–p- or the n–n-scattering with the singlet cross section of the p–n-scattering.

When comparing the p–p cross sections with those of the n–n-scattering one has to take into account that for the p–p-scattering besides the nuclear forces also the Coulomb forces contribute. However, the Coulomb contribution can be exactly calculated and therefore it can be subtracted. When doing this the experimental results confirm the fact, that the cross sections due to nuclear forces are equal for p–p-scattering and for n–n-scattering.

This means: The nuclear forces are independent of the nucleon charge.

The comparison of the results of the n-p scattering with those of the p–p– or n–n-scattering shows, however, that the cross sections of the n–p-scattering are slightly larger than those of the n–n- or p–p scattering. This implies that the nuclear forces between proton and neutron are about 1.5% stronger than those between equal nucleons. The reason for this small difference will be explained in Sect. 5.4.

5.3 Isospin Formalism

The charge-independence of the nuclear forces indicates that with regard to the strong interaction proton and neutron can be regarded as identical particles (called *nucleons*) which differ only by their charge, i.e. with regard to the Coulomb interaction. Therefore *Heisenberg* had already 1932 proposed, proton and neutron to regard as equal particles which exist in two different charge states $q = 1 \cdot e$ and $q = 0 \cdot e$.

The two charge states of the nucleon can be characterized analogue to the description of the electron in its two spin states $m_s = +\frac{1}{2}$ and $m_s = -\frac{1}{2}$ which are described by the two vectors for the two spin orientations

$$\psi_e = \psi(r) \cdot \chi^{\pm}$$

$$\chi^+ = \begin{pmatrix} 1 \\ 0 \end{pmatrix} \text{ for } \uparrow \qquad \chi^- = \begin{pmatrix} 0 \\ 1 \end{pmatrix} \text{ for } \downarrow \tag{5.23}$$

The wave-function of the electron

$$\psi_e = \psi(\mathrm{r}) \cdot \chi^s$$

can be written as the product of spatial wave-function and spin function. In a similar way the wave-function of the nucleon is written as the product

$$\psi_p = \psi_N(\boldsymbol{r}) \cdot \pi \quad \text{with} \quad \pi = \begin{pmatrix} 1 \\ 0 \end{pmatrix} \quad \text{for the Proton,}$$

$$\psi_n = \psi_N(\boldsymbol{r}) \cdot \nu \quad \text{with} \quad \nu = \begin{pmatrix} 0 \\ 1 \end{pmatrix} \quad \text{for the Neutron}$$

The electron is characterized by its spin with the spin component $S_z = \pm m_s \hbar$, while the nucleon can be characterized by a vector $\boldsymbol{\tau} = \{\tau_1, \tau_2, \tau_3\}$ which is called **isospin** and which has all properties of an angular momentum. Its components can be represented, exactly as the Pauli-matrices for the electron by two-column matrices

$$\tau_1 = \frac{1}{2}\begin{pmatrix} 0 & 1 \\ 1 & 0 \end{pmatrix}, \quad \tau_2 = \frac{1}{2}\begin{pmatrix} 0 & -\mathrm{i} \\ \mathrm{i} & 0 \end{pmatrix} \qquad \tau_3 = \frac{1}{2}\begin{pmatrix} 1 & 0 \\ 0 & -1 \end{pmatrix} \tag{5.24}$$

The third component of this isospin applied to the functions π and ν give

$$\tau_3\pi = \frac{1}{2}\begin{pmatrix} 1 & 0 \\ 0 & -1 \end{pmatrix}\begin{pmatrix} 1 \\ 0 \end{pmatrix} = \frac{1}{2}\begin{pmatrix} 1 \\ 0 \end{pmatrix} = \frac{1}{2}\pi, \qquad \tau_3\nu = \frac{1}{2}\begin{pmatrix} 1 & 0 \\ 0 & -1 \end{pmatrix}\begin{pmatrix} 0 \\ 1 \end{pmatrix} = -\frac{1}{2}\begin{pmatrix} 0 \\ 1 \end{pmatrix} = -\frac{1}{2}\nu. \tag{5.25}$$

The proton has the isospin-component $\tau_3 = +\frac{1}{2}$, the neutron $\tau_3 = -\frac{1}{2}$.

Just as the total spin S of a many-electron system is the vector sum $\boldsymbol{S} = \sum s$ of the individual electron spins, the total isospin T of a nucleus is the vector sum of the individual isospins $\boldsymbol{\tau}_i$ of the nucleons.

$$\boldsymbol{T} = \sum \tau_k. \tag{5.26}$$

For the isospin $T = |\boldsymbol{T}|$ there are 2T + 1 possible values of the component $T_3 = \sum \tau_{3k}$.

The relevance of the isospin, where the components are **not** components in a position space but in an abstract isospin space, shall be illustrated by the following example:

We regard at first two mirror-nuclei, which blend into each other if the numbers of protons and neutrons are exchanged. The nuclei $^{11}_{6}C$ (6 protons and 5 neutrons) and $^{11}_{5}B$ (5 protons and 6 neutrons), for example, form a pair of mirror nuclei (Fig. 5.14a). Without the Coulomb repulsion the two nuclei should have the same wave-functions and the same energy levels. Because of the Coulomb repulsion between the protons the energy levels of $^{11}_{6}C$ are slightly higher. If this energy difference ΔE_C is larger than $(m_n - m_p)c^2$ the nucleus $^{11}_{6}C$ can spontaneously transform by β^+-emission (conversion of a proton into a neutron) into the $^{11}_{5}B$-nucleus (Fig. 5.14b).

Due to the Pauli Principle (see Sect. 2.6) the nucleons occupy the lowest energy levels where each level can be only occupied by two nucleons with opposite spin. For these fully occupied levels the isospin component is always $T_3 = 0$.

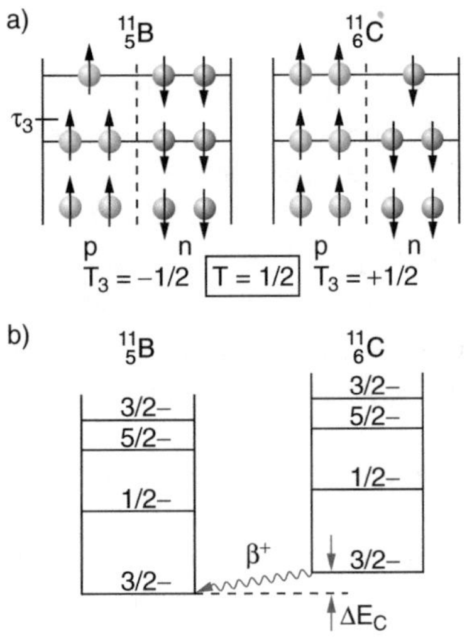

Fig. 5.14 **a** The two mirror nuclei $^{11}_{5}B$ and $^{11}_{6}C$ with the Isospin components τ_3 of the nucleons (the arrows symbolize τ_3, not the nuclear spin!) **b** Ground state and the lowest excited states of the mirror nuclei with nuclear spin and parity.

Table 5.2 Isospin component $T_3 = -\frac{1}{2}(Z-N)$ for some nuclei with odd numbers of protons or neutrons

Nucleus	$^{45}_{20}Ca$	$^{45}_{21}C$	$^{45}_{21}Ti$	$^{45}_{23}V$	$^{45}_{24}Cr$	$^{45}_{25}Mn$
T_3	−5/2	−3/2	−1/2	+ 1/2	+ 3/2	+ 5/2

Only the unpaired nucleon can contribute to T_3. For $^{11}_{6}C$ for example, is $T_3 = +\frac{1}{2}$, because the unpaired nucleon is a proton, while for $^{11}_{5}B$ is $T_3 = -\frac{1}{2}$ because the unpaired nucleon is a neutron. The two mirror-nuclei differ by the direction of the isospin in the isospin configuration space. The two nuclei represent an **Isospin-Dublet**.

The β^+-decay of $^{11}_{6}C$ transforms a proton into a neutron. In the isospin formalism it leads to a transition from $T_3 = +\frac{1}{2}$ to $T_3 = -\frac{1}{2}$. This transition does not alter the binding conditions due to the nuclear forces, it only changes the Coulomb repulsion between the protons, which is, however, small compared to the strong interaction.

For the general case of a nucleus with Z protons and N neutrons the isospin component is (Table 5.2)

$$T_3 = \sum_{k=1}^{A} \tau_{3k} = \frac{1}{2}(Z-N). \tag{5.27}$$

The third component $T_3 = ½(Z{-}N) = ½(N_p - N_n)$ of the isospin gives half of the proton excess of a nucleus, i.e. half of the difference $N_p{-}N_n$ of proton number and neutron number.

The component T_3 defines a definite nucleus within an isobar-series. Therefore the vector T is often called **Isobar-spin** or in the short form **isospin**.

We will now look at the isospin of the simplest nuclei (Fig. 5.15):

- The unbound system n–n (di-neutron) has $T_3 = -1$
- The deuteron n-p has $T_3 = +1$ in the bound triplet state and $T_3 = 0$ in the unbound singlet state, where the spins of proton and neutron are anti-parallel. Since the spin of both particles can be up or down it should be described by the wave-function $(1/\sqrt{2})(\uparrow\downarrow - \downarrow\uparrow)$. The unbound singlet state can be only detected by scattering experiments).
- The unbound system p-p has $T_3 = +1$

The only stable system of these three two-nucleon systems pp; nn; np is the deuteron np.

Analogue to the spin function of the Helium atom with 2 electrons we get for the two-nucleon systems the

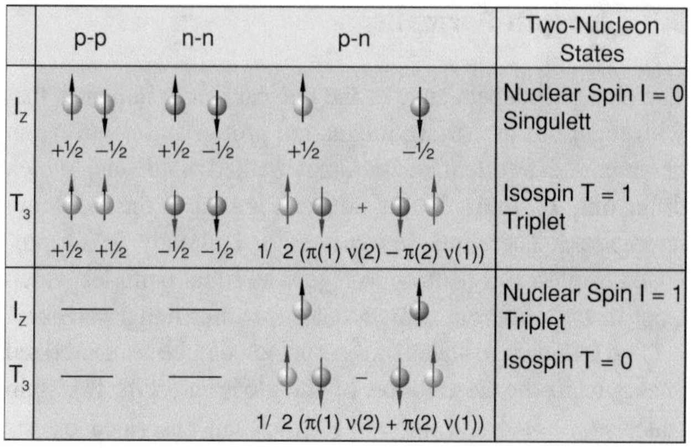

Fig. 5.15 Possible orientations of nuclear spin I and Isospin τ_3 of the three two-nucleon states. only the deuteron with $I = 1$ and $T = 0$ is stable. The nuclear spins I are drawn black, the isospins T red. (p = red, n = black)

Table 5.3 Isospin T and its component T_3 for the two-nucleon system and its isospin function

T	T_3	Isospin function
1	1	$\varphi_1^1 = \pi(1)\cdot\pi(2)$ Isospin triplet
1	0	$\varphi_0^1 = \frac{1}{\sqrt{2}}[\pi(1)\nu(2) + \pi(2)\nu(1)]$ Isospin triplet
1	−1	$\varphi_{-1}^1 = \nu(1)\cdot\nu(2)$ Isospin triplet
0	0	$\varphi_0^0 = \frac{1}{\sqrt{2}}[\pi(1)\nu(2) - \pi(2)\nu(1)]$ Isospin singlet

isospin-wave-functions in Table 5.3. The functions of the isospin triplet are symmetric against exchange of two nucleons, while those of the isospin singlet are antisymmetric (Fig. 5.15).

5.4 Meson-Exchange Model of Nuclear Forces

When two electrons are scattered by each other, electro-magnetic radiation is emitted because during the collision process electric charges are accelerated or retarded.

$$e^- + e^- \rightarrow e^- + e^- - \Delta E_{kin} + h\cdot\nu \quad \text{with} \quad \Delta E_{kin} = h\cdot\nu \tag{5.28a}$$

Another way to describe this is: The electro-magnetic interaction between two charges causes the emission of a photon $h\cdot\nu$ (Fig. 5.16a). If the scattering occurs in a strong electro-magnetic field (e.g. in a laser beam) also the inverse process can happen:

$$e^- + e^- + h \cdot \nu \rightarrow e^- + e^- + \Delta E_{kin} \quad (5.28b)$$

(**inverse bremsstrahlung**), where a photon $h \cdot \nu$ is absorbed and the electrons gain the corresponding energy $\Delta E_{kin} = h \cdot \nu$ (Fig. 5.16b).

Analogous to the interaction between two H-atoms, where the exchange of the two electrons contribute to the binding in the H_2-molecule (see Vol. 3. Sect. 9.3) the electro-magnetic interaction between two particles can be in a general way described by the exchange of *virtual photons*, where one particle emits a photon and the other particle absorbs it (Fig. 5.16).

This can be graphical visualized by *Feynman diagrams*, (Fig. 5.17), where the ordinate is the time axis and the space direction is symbolized by the *x*-axis. The exchange of virtual photons causes the interaction between the two colliding electrons and induces the deflection of the electrons during the collision. In Fig. 5.17b the Feynman diagram of the scattering of electrons by protons is illustrated.

Since the generation of a photon demands the additional energy $E = h \cdot \nu$, this photon can exist only for a short span of time

$$\Delta t \leq \hbar / h \cdot \nu = 1/(2\pi\nu)$$

in order to fulfill the *Heisenberg Uncertainty Relation*

$$\Delta E \cdot \Delta t \geq \hbar$$

for the fluctuation $\Delta E = h \cdot \nu$ of the total energy E by the photon energy $h \cdot \nu$.

The span of time $\Delta t = r/c$ depends on the distance between the two interacting particles. With increasing distance *r* the energy $h \cdot \nu$ of the virtual photons must decrease, because

$$h \cdot \nu \leq \hbar / \Delta t = \hbar \cdot c/r$$

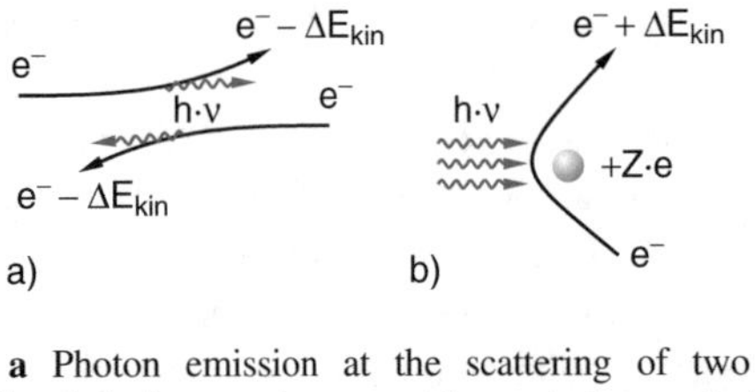

Fig. 5.16 **a** Photon emission at the scattering of two electrons (bremsstrahlung) **b** inverse bremsstrahlung at the scattering of an electrons in the Coulomb field when the scattered electron is irradiated by an electro-magnetic wave

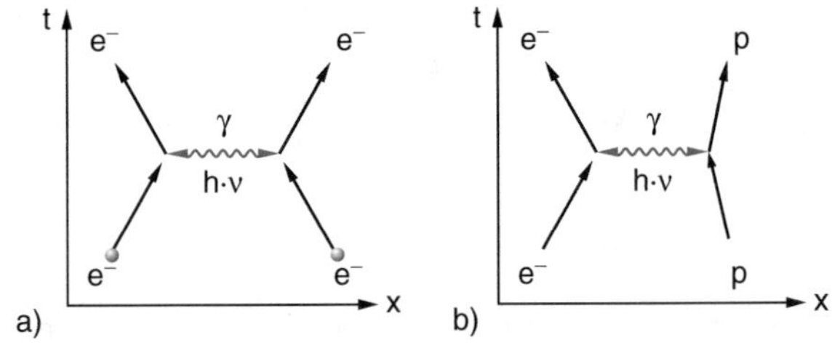

Fig. 5.17 Feynman diagram as the schematic representation for the exchange of virtual photons at the Coulomb interaction between two charges. **a** electron–electron scattering **b** electron-proton scattering

The interaction energy between the two particles is proportional to the product of virtual photon energy and the inverse exchange time, which are both proportional to *1/r*. The interaction energy is then proportional to $1/r^2$ which is the *Coulomb Law*.

When measuring the scattering of nucleons by each other one observes above a threshold energy $E_{kin} > 200$ MeV the generation of new particles which later on were identify as π-Mesons (see Chap. 7). The different inelastic scattering processes can be described, similar to (5.28a) by the reaction equations

$$\begin{aligned} p + p &\Rightarrow p + p - \Delta E_{kin} + \pi^0 \\ &\rightarrow p + n - \Delta E_{kin} + \pi^+ \end{aligned} \quad (5.29a)$$

$$\begin{aligned} p + n &\Rightarrow p + n - \Delta E_{kin} + \pi^0 \\ &\rightarrow p + p - \Delta E_{kin} + \pi^- \\ &\rightarrow n + n - \Delta E_{kin} + \pi^+, \end{aligned} \quad (5.29b)$$

where $\Delta E_{kin} = E_{kin} - E'_{kin}$ is the difference of the kinetic energy E_{kin} of the incident particles and E'_{kin} of the reaction products. This shows that by collisions between nucleons where the nuclear force is acting, π-mesons can be produced.

It is therefore tempting to describe the strong interaction by the exchange of virtual π-mesons analogue to the description of the electro-magnetic interaction by the exchange of virtual photons.

In the corresponding *Feynman diagrams* which can be regarded as short hand descriptions of the reaction processes, the coordinate axes are often omitted (Fig. 5.18).

This meson-exchange model of nuclear forces has been already proposed in 1936 by the Japanese Physicist *Hideki Yukawa* (1907–1981, Fig. 5.19) long before the π-mesons

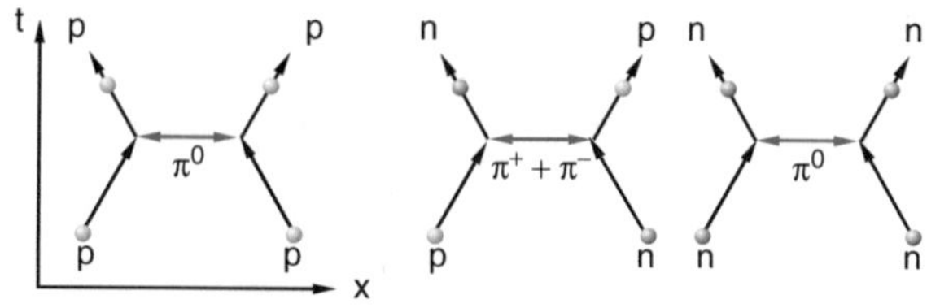

Fig. 5.18 Feynman diagrams of the strong interaction between p–p; p-n; and n–n nucleons caused by the exchange of virtual π-mesons

were discovered. Yukawa could even predict the mass of the π-meson from the known range of the nuclear forces.

The binding energy of the strong interaction between two nucleons is proportional to the mass energy $E = m \cdot c^2$ of the exchange particle with mass m, whereas the range of the interaction is proportional to 1/m. While for the electro-. magnetic interaction photons with rest mass zero are responsible and the interaction potential $V(r) \propto 1/r$ extends to $r = \infty$, the meson-exchange demands, because of the finite mass m_π, a finite range

$$\Delta E \cdot \Delta t \geq \hbar \quad \text{with} \quad \Delta E = m_\pi c^2, \quad r \leq c \cdot \Delta t$$

$$r \leq r_0 = \frac{\hbar}{m_\pi \cdot c}, \tag{5.30}$$

because the π-meson cannot fly with a velocity $v > c$. Inserting the numerical value $m_\pi \cdot c^2 = 139\,\text{MeV}$ for the rest mass of the π-meson we obtain the range $r_0 = 1.4 \cdot 10^{-15}\,\text{m}$, which is only slightly larger than the radius of the proton.

Yukawa proposed for the nuclear forces the model potential

$$V(r) = \frac{g}{r} \cdot e^{-(m_\pi \cdot c/\hbar) \cdot r} \tag{5.31}$$

Fig. 5.19 Hideki Yukawa (bing.com images)

with a strength given by the coupling constant g. At the distance $r = r_0 = \hbar/(m_\pi c)$ the interaction drops to 1/e of its maximum value at $r = 0$.

Note: With this definition the range of the nuclear forces is equal to the Compton-Wavelength (see Vol. 3. Sect. 3.1.3)

$$r_0 = \lambda_{c\pi} = \frac{\hbar}{m_\pi c} \tag{5.32}$$

of the π-meson. This short range of the nuclear forces effects that for nuclei with many nucleons the nuclear force acts only between neighboring nucleons. Therefore the nuclear volume is proportional to the number of nucleons and the nuclear radius is $r_0 = r_0 \cdot A^{1/3}$ (see Sect. 2.2).

This exchange model also explains why the strong interaction is slightly larger for the p-n-system than for the p–p- or n–n-systems. While for the p–p and n–n systems π^0. mesons are exchanged, for the p-n nuclei in addition π^+ and π^- mesons can be exchanged which have a slightly larger mass than the π^0-mesons.

5.5 Different Models of the Nucleus

In order to explain the experimental results discussed in the previous sections, such as binding energies, magnetic and electric moments, instability of radioactive nuclei, different models of atomic nuclei have been developed. Each of these models explains some of the observations but not all of them. The real nuclei are more complex than the models presented here. Only the advancement of more detailed models, which combine the different aspects of the crude models and which take into account that the nucleons are composed of quarks (see Sect. 7.4) allows the adequate description of real nuclei. Such more complex models are nowadays capable of satisfactorily describing and explaining all observed characteristics of real nuclei. Their complete description exceeds, however, the level of this textbook.

There are essentially two different approaches to describe nuclei as many-particle systems. The first starts from a single particle model where a single nucleon is regarded which moves in the nuclear potential generated by the time-average of the interaction with all other nucleons in the nucleus. This is completely analogue to the *Hartree–Fock* treatment for the calculation of the atomic electron shell (see Vol. 3, Sect. 6.4). It leads in its simple form to the Fermi-model of free particles and its advancement represents the shell model of nuclei.

There exist a couple of phenomena (e.g. vibrations and rotations or the fission of nuclei) where the interaction between the nucleons is essential and can no longer be treated as small perturbation of the single particle motion but has an important influence. Here collective models are required which take into account the collective movement of the nucleons in the nucleus.

In this section we will shortly discuss some of these different models. For more details the reader is referred to the literature [1–4].

5.5.1 Nucleons as Fermi-Gas.

We have seen in Chap. 2, that the nucleon density in the nucleus is approximately constant and only around the nuclear radius R_N in a narrow range between $R_N - \Delta r/2$ and $R_N + \Delta r/2$ it rapidly approaches zero. This experimental finding can be explained by a model where the nucleons are kept in a potential well *V(r)* with a flat bottom and a steep edge. This potential well can be simplified by a potential box which has a vertical jump from $V = -V_0$ to $V = 0$ at $r = a$, if we neglect the Coulomb force. This potential is due to the strong interaction of the nuclear force.

We will now find out which energy levels exist in this potential and which are occupied by nucleons. In particular we will find out, which characteristic properties of the nucleus can be explained by this model.

At first we assume that each nucleon can freely move in the potential box, which means, we neglect collisions between the nucleons. This assumption is justified by the fact that the diameter $2R_0$ of the repulsive hard core of a nucleon is small compared with the mean distance ($\approx$ 2.8 fm) between the nucleons in the nucleus.

The potential V*(r)* is obtained by the average interaction which an arbitrary nucleon N_i experiences due to the attraction by all other nucleons. Inside the nucleus the vector sum of all forces acting on the nucleon N_i by all other nucleons add up to zero, because of the homogeneous distribution of the nucleons. This implies that the potential inside the nucleus must be constant. At the surface $r = a$ of the nucleus this is no longer true, because of the missing nucleons for $r > a$. Here a strong inward directed attractive force $F = -\mathbf{grad}\, V(r)$ is present which results, because of the short range of the nuclear forces), in a steep increase of the potential. This is completely analogue to the situation in a liquid drop.

According to our model each nucleon moves freely in the potential box with the potential energy

$$E_{pot}(r) = \begin{cases} -E_0 & \text{for} \quad r < a \\ 0 & \text{for} \quad r \geq a \end{cases}, \tag{5.33}$$

which acts like a spherical shell enclosing the nucleons.

We will now determine the possible energy levels and their occupation with protons and neutrons. In Vol. 3. Chap. 4 the wave-functions and their Eigen-values, which give the energy levels, have been calculated for a two-dimensional rectangular potential box. In an analogue way the Schrödinger-equation for a particle with mass m in a three-dimensional cuboid potential box

$$\frac{-\hbar^2}{2m}\Delta\psi - E_0\psi = E\psi \tag{5.34}$$

can be solved where $E_{pot} = -E_0$ is the depth of the potential box. We try again the product ansatz

$$\psi(x,y,z) = \psi_1(x)\cdot\psi_2(y)\cdot\psi_3(z) \tag{5.35}$$

which gives the three equations

$$\begin{aligned} \frac{-\hbar^2}{2m}\frac{\partial^2\psi_1}{\partial^2 x^2} &= E_x\psi_1, \\ \frac{-\hbar^2}{2m}\frac{\partial^2\psi_2}{\partial^2 y^2} &= E_y\psi_2, \\ \frac{-\hbar^2}{2m}\frac{\partial^2\psi_3}{\partial^2 z^2} &= E_z\psi_3 \end{aligned} \tag{5.36}$$

The energy values follow the condition

$$E_x + E_y + E_z = E + E_0$$

where $E - E_{pot} = E_{kin}$ is the kinetic energy and $E_{pot} = -E_0$ is the potential energy of a nucleon. Introducing the components of the wave vector **k** as $k_i = \frac{1}{\hbar}\sqrt{2mE_i}$ with i = *x*, *y*, *z* we can simplify the Eqs. (5.36) as

$$\begin{aligned} \frac{\partial^2\psi_1}{\partial^2 x^2} &= -k_x^2\psi_1 \\ \frac{\partial^2\psi_2}{\partial^2 y^2} &= -k_y^2\psi_2 \\ \frac{\partial^2\psi_3}{\partial^2 z^2} &= -k_z^2\psi_3 \end{aligned} \tag{5.36a}$$

Since the wave-functions ψ_i must vanish at the edges $x = a$, $y = b$, $z = c$ of the potential box the solutions have the form

$$\begin{aligned} \psi_1 &= A_1 \sin k_x x \quad \text{with} \quad k_x = n_x \cdot \pi/a, \\ \psi_2 &= A_2 \sin k_y y \quad \text{with} \quad k_y = n_y \cdot \pi/a, \\ \psi_3 &= A_3 \sin k_z z \quad \text{with} \quad k_z = n_z \cdot \pi/a, \end{aligned} \tag{5.37}$$

where $n_x, n_y, n_z = 1, 2, 3, \ldots$ are integers (Fig. 5.20). For the cubic potential box is $a = b = c$. The corresponding energy levels are then

$$E_i = \frac{\hbar^2}{2m}\frac{\pi^2}{a^2}n_i^2, \quad i = x, y, z, \tag{5.38}$$

and for the total kinetic energy the energy values are

$$E + E_0 = \frac{\hbar^2}{2m}\frac{\pi^2}{a^2}\left(n_x^2 + n_y^2 + n_z^2\right). \tag{5.39}$$

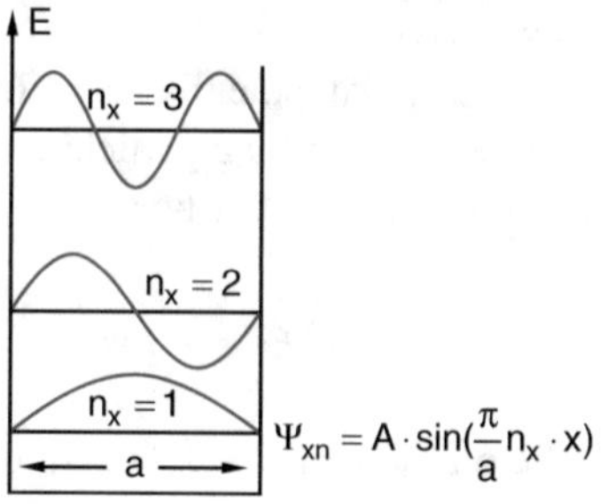

Fig. 5.20 Wave-function of a free nucleon in a one-dimensional potential box

In a coordinate system in k-space with the intercepts $\hbar\pi/(a \cdot \sqrt{2m})$ of the x,-y,-and z-axes each possible energy value corresponds to a point (Fig. 5.21) in a grid.

The number n of possible energy values $E \leq E_{\max} - E_0 < 0$ in the potential box with depth E_0 is equal to the number n of grid points in the octant $(k_x > 0, k_y > 0, k_z > 0)$ within the sphere in k-space with radius

$$k_{\max} = \frac{a}{\pi} \cdot p_{\max}/\hbar = \frac{a}{\hbar\pi}\sqrt{2mE_{\max}}. \tag{5.40}$$

This gives

$$n = \frac{1}{8} \cdot \frac{4}{3}\pi k_{\max}^3 = \frac{a^3}{6\pi^2\hbar^3}(2m \cdot E_{\max})^{3/2} \tag{5.41}$$

possible energy levels. The $A = N + Z$ nucleons in the nucleus with mass number A are Fermions and obey the Pauli-Principle, which demands that each of these levels can be occupied at most with two protons and two neutrons with opposite spins. They occupy at first the lowest energy levels with $n \leq n_F$ until the Fermi energy E_F is reached. For Z = N these are $n_F = A/4$ levels, for $N > Z$ is $n_F = Z/2 + (N-Z)/2$.

The number of occupied levels per volume unit is with $V = a^3$ according to (5.41)

$$\frac{n_F}{V} = \frac{1}{6\pi^2} \cdot \left(\frac{2mE_F}{\hbar^2}\right)^{3/2}. \tag{5.42}$$

It therefore depends solely on the Fermi-energy not on the volume of the potential box. The Fermi-energy can be deduced from (5.42) as

$$E_F = \frac{\hbar^2}{2m}\left(\frac{6n_F\pi^2}{V}\right)^{2/3}. \tag{5.43}$$

The density of states in the three-dimensional potential box increases with $\sqrt{E}$ (Fig. 5.22) which implies that the energy distance between neighboring states decreases with increasing energy.

$$\frac{dn}{dE} = \frac{a^3}{4\pi^2}\left(\frac{2m}{\hbar^2}\right)^{3/2} \cdot \sqrt{E}. \tag{5.44}$$

For all stable nuclei is $E_F < E_{\max}$, i.e. $E < 0$. All occupied levels lie below the potential limit $E_{\text{kin}} = -E_0$ and are therefore bound levels.

The mass density distribution $\varrho(r)$ of the nucleons (see Sect. 2.3) gives the mean nucleon density

$$\varrho_N \approx \frac{n_F}{V} \approx 0{,}15\ \text{Nukleonen/fm}^3.$$

This gives a Fermi-energy of about 35 MeV. Since the mean binding energy of nucleons in nuclei is smaller than 10 MeV the potential depth is about

$$-E_0 = 40 - 50\ \text{MeV},$$

independent of the size of the nucleus.

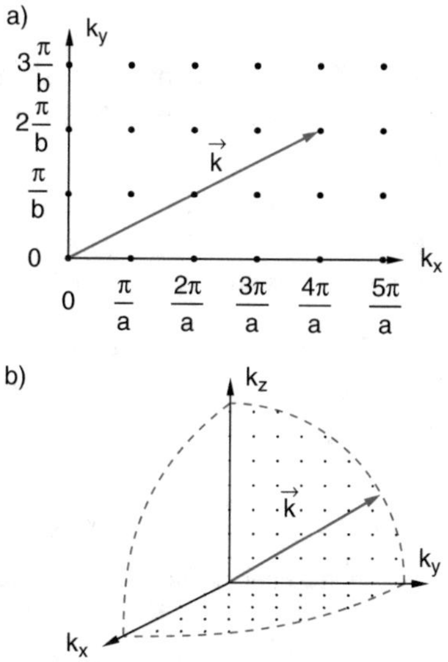

Fig. 5.21 **a** Representation of the possible k-vectors allowed by the boundary conditions as points in the two-dimensional grid in *k*-space. **b** Spherical octant in the three-dimensional *k*-space.

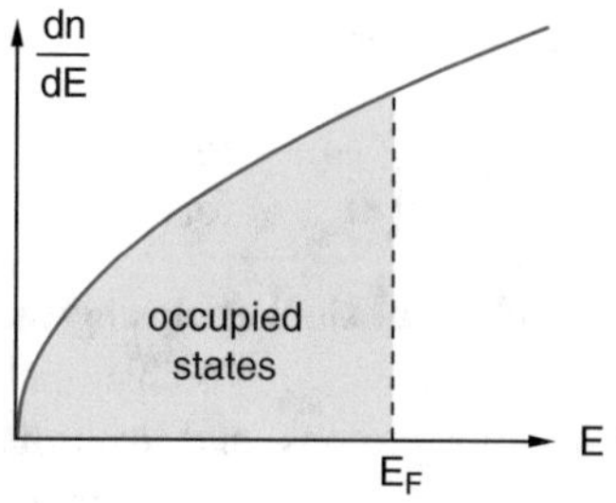

Fig. 5.22 Density of states dn/dE in the three-dimensional potential box

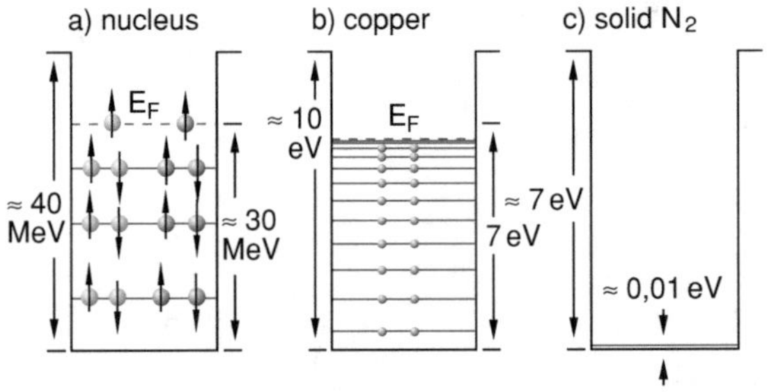

Fig. 5.23 Comparison of the potential depth and Fermi energy in different areas of physics: **a** Nucleons in a potential box **b** Electron gas in a solid **c** Electrons in solid nitrogen (after Mayer-Kuckuk, Kernphysik, Teubner Stuttgart 1993)

Remark

The treatment of nucleons as Fermi-gas is completely analogue to that of an electrons gas in metals (see Vol. 3. Sect. 13.1).

It is instructive to compare the ratio of Fermi-energy to the potential depth for different systems (Fig. 5.23).

The ratio E_F/E_0 is for electrons in the conduction band in copper with $E_F = 7\,\text{eV} E_0 = 10\,\text{eV} \rightarrow E_F/E_0 = 0.7$ and has therefore about the same value as for nuclei $(E_F = 35\,\text{MeV}, E_0 = 45\,\text{MeV} \rightarrow E_F/E_0 = 0.78)$.

Although the absolute values of E_F and E_0 in the nuclear potential box are larger by 6 orders of magnitude.

The situation is different for molecules. For example for nitrogen N_2 the three valence electrons are bound in the configuration $N \equiv N$. The Fermi energy of solid nitrogen at low temperatures is with $E_F = 10^{-2}$ eV very small compared to the binding energy $E_B = 7$ eV of the two N-atoms. The ratio $E_F/E_B = 1.3 \cdot 10^{-3}$ is then smaller by nearly 3 orders of magnitude compared to the other examples.

In order to adjust this simple model of the free nucleon gas to the real situation, the Coulomb repulsion between the protons has to be taken into account. In this improved model we assume two different potential boxes for protons and neutrons, where the potential for protons is slightly higher than that for neutrons (Fig. 5.24) and furthermore has a

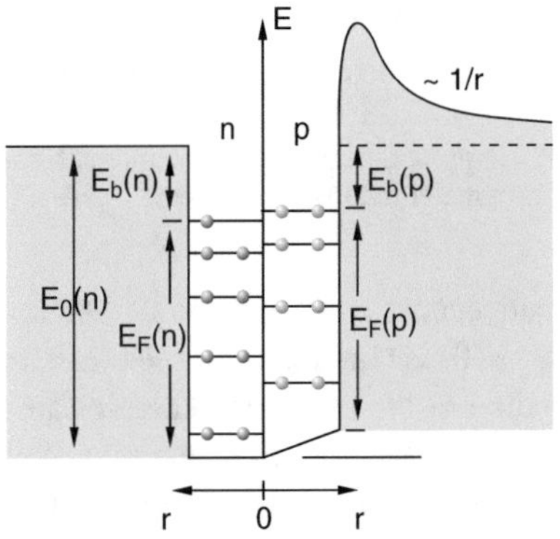

Fig. 5.24 Separated potential boxes for neutrons and protons

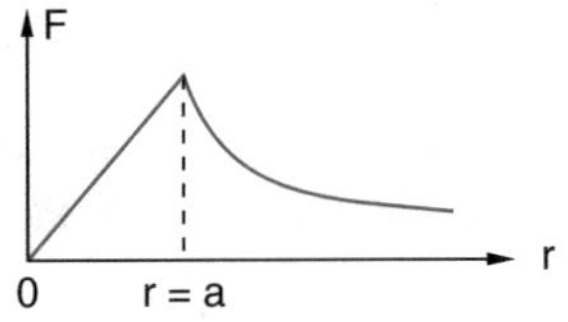

Fig. 5.25 Radial dependence of the Coulomb force inside and outside the potential box

Coulomb barrier at the potential edge. Strictly speaking the bottom of the potential box is no longer flat, because the Coulomb potential $V_C(r)$ has the r-dependence shown in Fig. 5.25 (see Vol. 2. Sect. 1.3).

The energy levels in both potential boxes are occupied with two nucleons until the Fermi-energy is reached. This illustrates that nuclei with a given atomic mass number A but a larger neutron excess have a smaller total energy than those with Z = N.

For smaller nuclei the spin dependence of the nuclear forces is responsible for the fact that the potential of the n−p-configuration (deuteron) has a lower energy than the di-neutron n–n (see Sect. 5.2).

The model of the potential box has the advantage of easy calculation of the energy levels. However, it represents only a crude approximation to the real potential. Another choice would be the harmonic oscillator potential

$$V(r) = \begin{cases} -V_0\left[1 - (r/R_m)^2\right] & \text{for} \quad r < R_m \\ 0 & \text{for} \quad r \geq R_m \end{cases} \tag{5.45}$$

where the energy levels can be readily calculated from the Schrödinger equation. A more realistic potential is the **Woods-Saxon-Potential**

$$V(r) = -\frac{V_0}{1 + e^{(r-R_0)/b}}, \tag{5.46}$$

where the fit-parameter b is a measure for the width of the potential edge, i.e. for the range Δr where the potential changes rapidly from 90% to 10% of its maximum value (Fig. 5.26). The radial dependence of the real potential lies between the potential box and the harmonic potential.

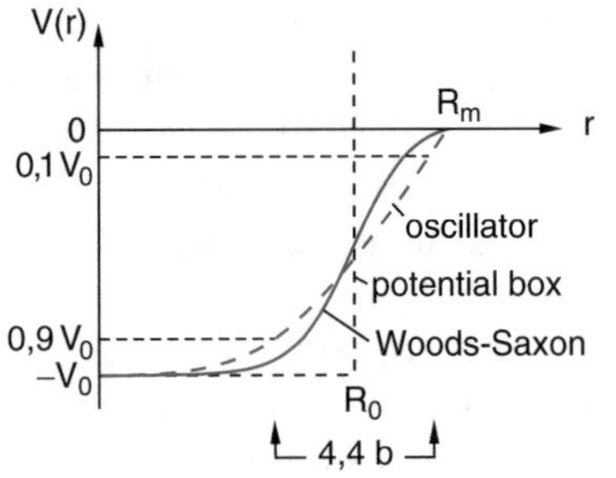

Fig. 5.26 Comparison between potential box, parabolic potential of the harmonic oscillator and Woods-Saxon potential

5.5.2 Shell-Model of Nuclei

The energy E_B which has to be expended to detach a proton or neutron from the nucleus is equal to the difference

$$-E_B = E_{kin}^{max} + E_{pot} = E_F - E_0 \quad (5.47a)$$

of the maximum kinetic energy in the highest occupied energy level $E_{kin}^{max} = E_F$ and the potential energy $E_{pot} = -E_0$ (Fig. 5.24). It is also called the separation energy of the proton $E_s(p)$ resp. $E_s(n)$ = of the neutron. It can be determined from the difference of the binding-energies of parent and daughter nucleus.

$$\begin{aligned} E_s(p) &= E_B(N,Z) - E_B(N,Z-1) \\ E_s(n) &= E_B(N,Z) - E_B(N-1,Z). \end{aligned} \quad (5.47b)$$

If two protons or two neutrons with opposite spin recombine to a pair, the binding energy of the nucleus becomes generally larger. The corresponding pair-energy is defined as

$$\begin{aligned} E_p(p) &= 2[E_B(Z,N) - E_B(Z-1,N)] \\ &\quad - [E_B(Z,N) - E_B(Z-2,N)] \\ E_p(n) &= 2[E_B(Z,N) - E_B(Z,N-1)] \\ &\quad - [E_B(Z,N) - E_B(Z,N-2)]. \end{aligned} \quad (5.47c)$$

Note: Although the free di-nucleons pp and nn are not stable, inside a nucleus the binding to pairs increases the binding energy, i.e. decreases the energy of the di-nucleons. This is due to the fact, that two protons resp. two neutrons with opposite spins can occupy the same energy level (Pauli principle) with negative potential energy and the binding energies of the two protons resp neutrons nearly add up.

The Fermi-gas model states that this energy is (at least for larger nuclei) independent of the mass number A, since E_F depends solely on the nucleon density n_F/V which is nearly independent of the mass number A.

However, experiments show, that the separation energies $E_s(Z, N)$ show for protons distinct maxima when plotting E_s as a function of the proton number Z as well as for neutrons as a function of the neutron number N. In Fig. 5.27 the separation energies $E_s(n, N)$ and the pair energies $E_P(n, N)$ of a neutron pair are plotted for the different isotopes of Calcium ($Z = 20$) as a function of the neutron number N. The maxima of the pair energies occur at even neutron numbers, which illustrates that the binding energy has maxima, if the nucleons occur in pairs.

The pair energies have distinct maxima at the "*magic numbers*" 2, 8, 20, 28, 50, 82, and 126 (Fig. 5.28). These numbers were called "magic" because at that time there was no explanation for these maxima.

Also the excitation energies for protons or neutrons from the ground state into the first excited state, which can be measured by absorption of γ-quanta show maxima at the magic numbers (Fig. 5.29) and even the abundance of stable isotopes or isotones (Fig. 5.30) is maximum for the magic numbers.

These maxima of the separation energies are reminiscent of similar maxima of the ionization energies of atoms, which could be completely explained by the shell structure of the

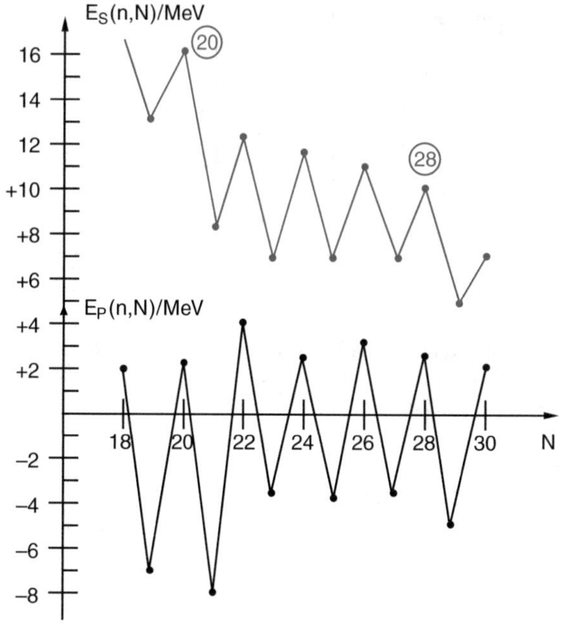

Fig. 5.27 Separation energy $E_s(n, N)$ of.a neutron and pair energy $E_p(n, N)$ of a neutron pair for the Calcium isotopes ${}^{A}_{20}Ca(N)$ with $A = 20 + N$ as a function of the neutron number N

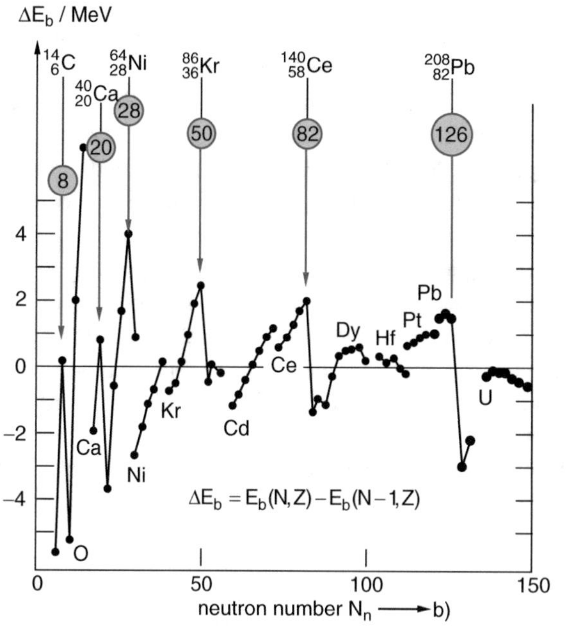

Fig. 5.28 Pair energy of a neutron pair for different isotopes of some elements with maxima at the magic numbers

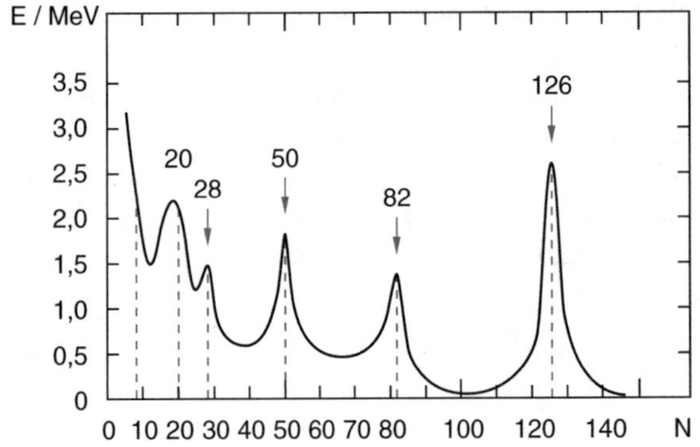

Fig. 5.29 Energy of the first excited states of g–g nuclei as a function of the neutron number N

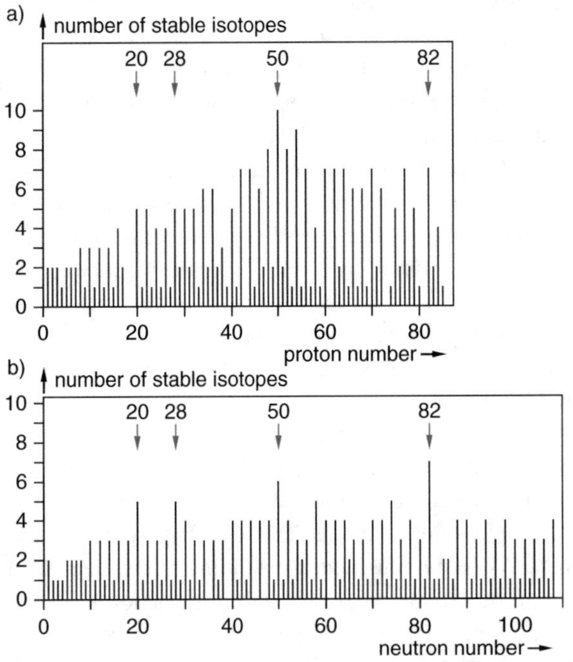

Fig. 5.30 Abundance of the stable isotopes as a function **a** of the proton number A and **b** of the neutron number N. (After K. Bethge: Kernphysik Springer Heidelberg 1996)

atomic electrons (see Vol. 3, Sect. 6.2). Always when a completely filled shell was reached (noble gases) the ionization energy shows distinct maxima, while for the alkali atoms where the buildup of a new shell starts a minimum of the ionization energy is observed.

It suggests itself to try this shell model also for nuclei, in order to explain the magic numbers This has indeed been successfully done and we will explain in the following the basic idea of this nuclear shell model:

In a spherical symmetric potential the Schrödinger equation can be separated into the product of an angular factor and a radial function (see Vol. 3, Sect. 4.3.2). The solutions of the angular part are for every spherical symmetric potential the spherical harmonics Y_l^m, independent of its radial dependence. These functions therefore depend solely on the quantum numbers ℓ of the orbital angular momentum and m of its z-component.

In order to obtain the solutions $R(r)$ for the radial part of the wave function $\psi(\mathrm{r}, \vartheta, \varphi)$ we solve the Schrödinger equation for the function $u(r) = r \cdot R(r)$

$$\frac{\mathrm{d}^2 u}{\mathrm{d}r^2} + \frac{2m}{\hbar^2}\left[E - V(r) - \frac{l(l+1)\hbar^2}{2mr^2}\right]u = 0 \qquad (5.48)$$

the wave-functions are obtained by inserting the specific potential $V(r)$ (see Vol. 3, Sect. 5.1). Before this is discussed, we will try at first a more descriptive way:

The real nuclear potential $V(r)$ has to be between the two limiting model potentials: The potential box and the harmonic potential (Fig. 5.26). While the potential box (5.33) with vertical walls is too steep, the harmonic potential (parabolic potential (5.45) has a too shallow edge.

The solutions and energy eigenvalues for the potential box have been already presented in Sect. 5.5.1. For the three-dimensional potential

$$V(r) = c \cdot r^2 + d, \quad (d = -V_0) \qquad (5.49a)$$

of the three-dimensional harmonic oscillator, which can be rearranged as

$$(V(r) - d) \propto r^2 = x^2 + y^2 + z^2 \qquad (5.49b)$$

Similar to the solutions of the potential box, the solutions $\psi(r)$ for the parabolic potential can be written for $\ell = 0$ as product of three functions of one variable

$$\psi(r) = \psi_x(x) \cdot \psi_y(y) \cdot \psi_z(z) \qquad (5.50)$$

Each of these functions is a solution of the one-dimensional harmonic oscillator (see Vol. 3, Chap. 4).

The total energy is then

$$\begin{aligned} E_n &= \hbar\omega\left(n_x + \frac{1}{2}\right) + \hbar\omega\left(n_y + \frac{1}{2}\right) + \hbar\omega\left(n_z + \frac{1}{2}\right) \\ &= \hbar\omega\left(n + \frac{3}{2}\right) \\ &\text{with} \quad n = n_x + n_y + n_z = 0, 1, \ldots \end{aligned} \qquad (5.51a)$$

There are $q = (n+1) \cdot \left(\frac{n}{2} + 1\right)$ different combinations of n_x, n_y, n_z, which result in the same value of n and therefore have the same energy (see Problem 5.6). The energy levels are therefore q-fold degenerate (Table 5.4).

We call all q degenerate energy levels with the same energy (i.e. the same value of n) an energy shell, analogue to the situation in the atomic electron shell. If each of these shells is now occupied with $2q$ protons and $2q$ neutrons (obeying of course the Pauli-principle, where one has to keep in mind that levels with the same energy but different values of n_x, n_y, n_z, must be regarded as different levels

Table 5.4 Degree of degeneration $q(n)$ for the three-dimensional oscillator

n	n_x	n_y	n_z	q
0	0	0	0	1
1	1	0	0	3
	0	1	0	
	0	0	1	
2	2	0	0	6
	0	2	0	
	0	0	2	
	1	1	0	
	0	1	1	
	1	0	1	
3	3	0	0	10
	0	3	0	
	0	0	3	
	2	1	0	
	2	0	1	
	1	2	0	
	1	0	2	
	0	1	2	
	0	2	1	
	1	1	1	

because their spatial wave-functions (5.50) are different. These states are therefore in spite of the same energy not identical quantum states.

Regarding these remarks one obtains the occupation numbers $(N_p + N_n)_n$ of a nuclear shell.

The occupation numbers $\sum_0^n N(p,n)$ reproduce in fact up to $n = 2$ the magic numbers. However, for $n > 2$ they deviate from them (Table 5.5). The reason for this deviation will be discussed later on.

The meaning of the quantum numbers can be illustrated when we regard the wave-function

$$\psi(r,\vartheta,\varphi) = R_n(r) \cdot Y_l^m(\vartheta,\varphi)$$

Table 5.5 Maximum occupation number of the eigen-states in the three-dimensional oscillator potential

n	0	1	2	3	4	5	
q	1	3	6	10	15	21	
N_p, N_n	2	6	12	20	30	42	
$\sum_{i=0}^{n} N_i(\mathrm{p,n})$	2	8	20	40	70	112	
Magical numbers	2	8	20	28	50	82	126

as product of its radial and its angular part. The radial part defines the quantum number n of the total energy, the angular part the quantum number ℓ of the orbital angular momentum and m_l of its z-projection. Solving the radial Eq. (5.48) for the three-dimensional oscillator potential, including the centrifugal term, one obtains, analogue to (5.51a) the equidistant energy Eigen-values

$$E(n,l) = \hbar\omega\left(2n + l + \frac{3}{2}\right). \tag{5.52}$$

with $n = 0, 1, 2, \ldots$ and $l = 0, 1, 2, \ldots$.

They do not depend on m_ℓ because of the spherical symmetry. The possible values of ℓ belonging to the lowest n-values are listed in Fig. 5.31. One can see that to each energy level E_n with $n \geq 2$ several ℓ-values contribute.

Passing from the oscillator potential to the potential box the degenerate levels with different ℓ-values split (Fig. 5.31). The splitting is, however, very small. One therefore observes for each n a narrow group of levels. The occupation numbers in Table 5.5 do not change between oscillator potential and potential box. According to Fig. 5.26 one would expect that the energy levels of the real nucleus lie between those of the two model potentials (Fig. 5.32).. However, the experimental results disagree, at least for larger values of n, with this assumption.

Otto Haxel and *Johannes Hans Daniel Jensen* and independently *Maria Goeppert-Mayer* (Fig. 5.33, Nobel-prize 1963) recognized 1948, that this discrepancy is caused by the spin–orbit coupling. This coupling splits the energy levels. While the fine-structure splitting in the electron shell is small due to the weak spin–orbit-interaction (see Vol. 3. Chap. 5), in nuclei this coupling is much larger due to the strong nuclear forces. It results in a large splitting which might be even larger than the energy difference between successive shells.

In Sect. 5.2 we have seen that the interaction between two nuclei depends on the mutual orientation of their spins. In analogy to the treatment in the electron shell M. Goeppert-Mayer, Jensen and Haxel added to the spherical symmetric potential V(r) a spin-dependent term, thus introducing the effective potential

$$V_i(r) = V(r) + V_{ls}(r) \cdot \boldsymbol{l} \cdot \boldsymbol{s} \tag{5.53}$$

The expectation value of $\boldsymbol{l} \cdot \boldsymbol{s}$ is (see Vol. 3. Sect. 5.6.1)

$$\langle \boldsymbol{l} \cdot \boldsymbol{s} \rangle = \frac{1}{2}[j(j+1) - l(l+1) - s(s+1)] \cdot \hbar^2. \tag{5.54}$$

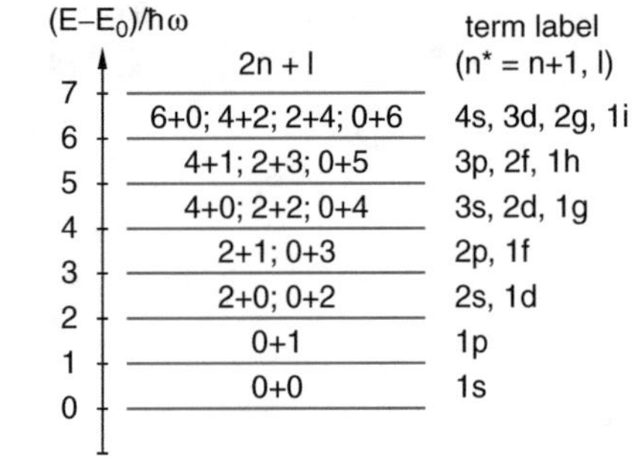

Fig. 5.31 Equidistant energy levels $E(n, \ell)$ of the three-dimensional harmonic oscillator with the term labels (n, ℓ) analogue to that in the atomic electron shell

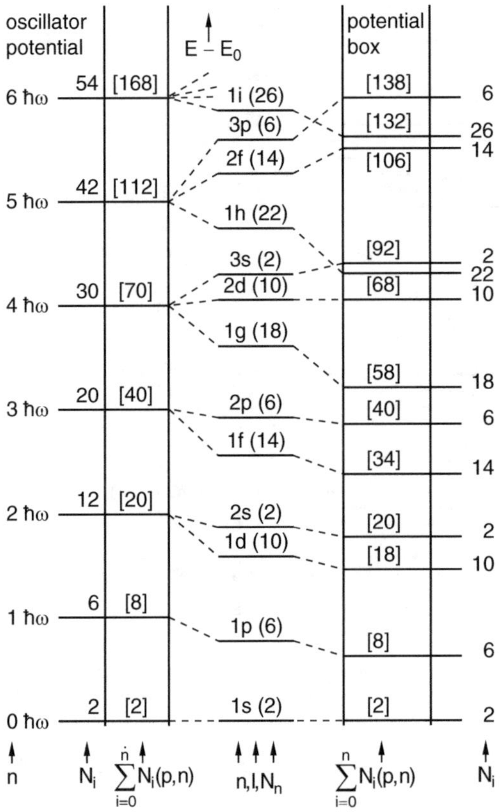

Fig. 5.32 Comparison of the energy levels of particles in a three-dimensional potential box and in the potential of a three-dimensional harmonic oscillator. (M. Goeppert-Mayer, J. H. D Jensen: Elementary Theory of Nuclear Shell Structure (Wiley, New York 1955)

With $s = \pm 1/2$ is $j = l \pm 1/2$ and we obtain for $\langle l \cdot s \rangle$ the two values $(l/2) \cdot \hbar^2$ and $-(1/2)(l+1) \cdot \hbar^2$ and therefore the two effective potentials

$$V_i^{\text{eff}}(r) = V(r) + \frac{1}{2} V_{ls}(r) \cdot l \cdot \hbar^2 \quad \text{for} \quad j = l + 1/2, \tag{5.55a}$$

Fig. 5.33 Maria Goeppert-.Mayer (Nobelpreis 1963). From; Die Nobel-.Preisträger der Physik (Heinz Moos Verlag München 1964)

$$V_i^{\text{eff}}(r) = V(r) - \frac{1}{2} V_{ls}(r) \cdot (l+1) \cdot \hbar^2 \quad \text{for} \quad j = l - 1/2, \tag{5.55b}$$

These two potentials result in an energy splitting

$$\Delta E = (2l+1) \cdot \frac{1}{2} V_{ls} \cdot \hbar^2 \tag{5.56}$$

of the two levels which is proportional to $(2\,l + 1)$. The amount of this splitting is of the same order of magnitude as the energy separation of the levels (5.51a) with different n-values.

It turns out that $V_{ls}(r)$ and $V(r)$ are both negative. Therefore the levels with $j = 1 - 1/2$ have a higher energy than those with $j = l + 1/2$.

The maximum possible number of nucleons in each level (n, l, j) is obtained, when we determine all allowed combinations of n, l, and the resulting values of j (Table 5.6). Since there are for every value of j $(2j + 1)$ possible orientations with the same energy, one gets the nucleon numbers listed in the last column of Table 5.6 populating the levels with increasing energy up to the last occupied level.

The energetic order and the absolute energies are only obtained by an extensive calculation.

Plotting the energetic order of all levels, obtained from such calculations with their occupation numbers in Fig. 5.34, one obtains indeed the experimentally found magic numbers, representing the sum of all proton- and neutron numbers which can occupy all levels up to that level $E_{n,j}$ above which a particular large energy gap to the next higher level occurs. This is even more pronounced in Fig. 5.35 where the energy levels in the same shell are drawn as hatched red bands, which makes the energy gaps more obvious. Nuclei where the proton numbers as well as the neutron numbers are both magic numbers are called **double magic nuclei**. They have a particular high stability. Examples are ${}^4_2\text{He}(N_p = 2, N_n = 2), {}^{16}_8\text{O}(N_p = 8, N_n = 8),$

Table 5.6 Energetic level sequence and occupation number of levels *(n, l, j)* according to the nuclear shell model including spin–orbit coupling

n, l, j	1, 0, 1/2	1, 1, 3/2	1, 1, 1/2	1, 2, 5/2	2, 0, 1/2	1, 2, 3/2	1, 3, 7/2
Designation	$1s_{1/2}$	$1p_{3/2}$	$1p_{1/2}$	$1d_{5/2}$	$2s_{1/2}$	$1d_{3/2}$	$1f_{7/2}$
2j + 1	2	4	2	6	2	4	8
$\sum_j 2j+1$	2	6	8	14	16	20	28

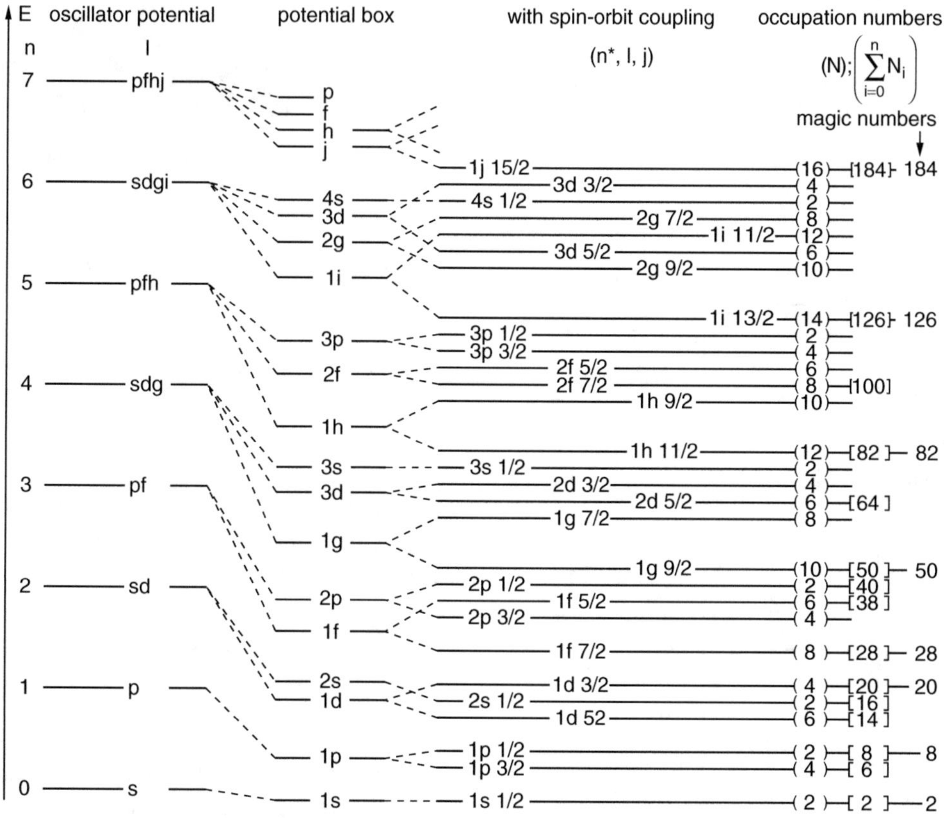

Fig. 5.34 One-particle energy levels according to the Shell Model including spin–orbit coupling. (After H. Bucka: Nukleonen-Physik, de Gruyter, berlin 1981)

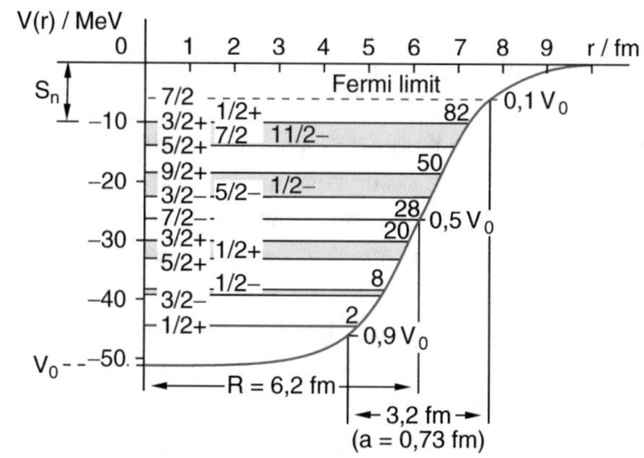

Fig. 5.35 Neutron energy levels of a nucleus with N = 82 in the Wood-Saxon potential with spin–orbit coupling. (after T. Mayer-Kuckuk: Kernphysik, Teubner Stuttgart 1993)

${}^{40}_{20}\mathrm{Ca}(N_p = 20, N_n = 8), {}^{48}_{20}\mathrm{Ca}(N_p = 20, N_n = 28), {}^{208}_{82}\mathrm{Pb}$ $(N_p = 82, N_n = 126)$.

The shell model explains, besides the physical meaning of the magic numbers, additional experimental facts: When a shell (i.e. all levels occupied by nucleons with a given total angular momentum *j*) is fully occupied with nucleons, the total angular momentum $j = \Sigma j_i$ of the nucleus ***I***, called the nuclear spin, has to be zero, because all sublevels m_j with – j $\leq m_j \leq j + 1$ are occupied.

Nuclei with closed occupied shells must be spherical symmetric, have the nuclear spin *I* = 0 and possess no electric quadrupole moment.

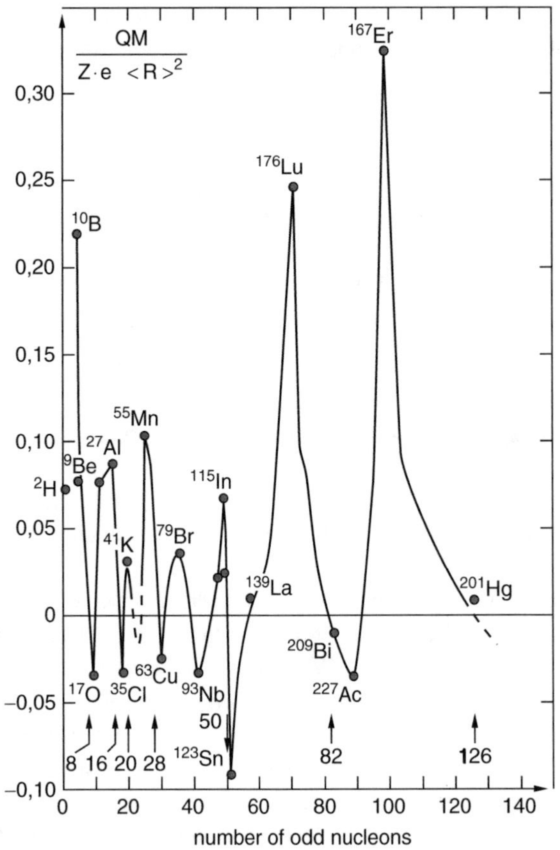

Fig. 5.36 Nuclear electric quadrupole moments as a function of the number of odd nucleons. (after Povh et.al. Teilchen und Kerne. Springer, Heidelberg 1994)

This is exactly what has been observed (Fig. 5.36).

When adding another nucleon N_i to a nucleus with closed shell, this nucleon has to occupy a level in the next higher shell. The nuclear spin

$$\boldsymbol{I} = j = l_i + s_i$$

is now determined by the orbital angular momentum l_i and the spin s_i of this unpaired nucleon. The quantum numbers l_i, s_i and j_i are the quantum number of the open shell just above the highest closed shell. The situation is analogues to that of the electron shell of the alkali atoms.

In case of two nucleons in incompletely filled shells the situation becomes more complicated, because now more possibilities for the coupling of angular momenta exist. Since the spin–orbit coupling between ℓ_i and s_i of the same nucleon is stronger than the coupling between $\ell_i\ell_k$ resp. s_i, s_k of two different nucleons, we can assume j-j-coupling, i.e. the angular momenta $j_i = \ell_i + s_i$ of the different nucleons couple to a total angular momentum

$$j = I = \sum_i j_i.$$

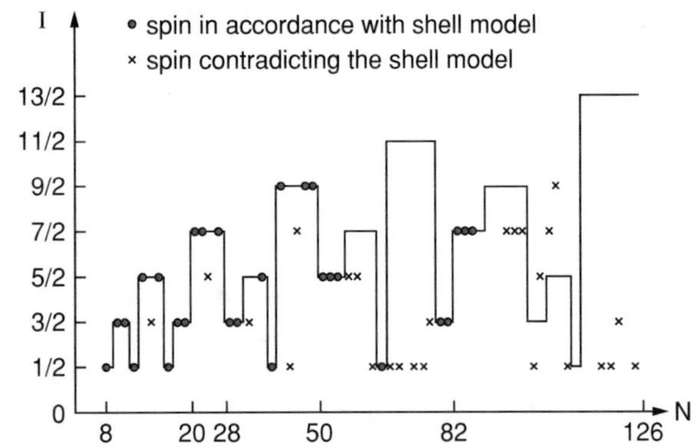

Fig. 5.37 Experimental values of the nuclear spin as a function of the neutron number N. (from: P. Marmier, E.Sheldon,: Physics of Nuclei and Particles, Academic Press, New York 1969)

which is the spin I of the nucleus, because the filled shells have $I = 0$ and therefore do not contribute to the spin I.

Note: Nuclei in an excited state can have a different spin I than in the ground- state, if ℓ_i and ℓ_k differ.

While the predictions of the shell model agree well with the experimental results for small and medium nuclei, discrepancies arise for heavy nuclei (Fig. 5.37). They can be explained by the following facts:

- For heavy nuclei the deviations from the spherical symmetry become larger. Also the average potential used in the single-particle approximation for the motion of a nucleon deviates with increasing nuclear mass more and more from a pure radial potential. One reason is that the single-particle approximation becomes worse because of the residual interactions (e.g. the pair correlation between two nucleons which have an angular part) which increase with increasing size of the nucleus. Another reason is the increasing Coulomb repulsion between the protons. This results in a minimum of the potential energy for ellipsoidal distribution of the protons (see Sect. 5.6) and breaks the spherical symmetry.
- In a non-spherical potential the angular momentum is no longer constant in time and therefore also the total angular momentum of a closed shell might not necessary be zero. In addition those deformed nuclei can rotate and add another angular momentum to the intrinsic momenta.

This last point will be discussed in more detail in the next section.

5.6 Rotation and Vibration of Nuclei

Up to now we have regarded a single nucleon moving in the time-averaged spherical potential which is generated by the interaction between all nucleons. The energy levels calculated in this single-particle model are occupied by protons and neutrons up to the Fermi-energy obeying the Pauli-principle. The interaction between the nucleons was only indirectly and blankly taken into account by the averaged potential.

With increasing number of nucleons the deviations from a spherical symmetric potential become more severe. The necessary corrections of the central field approximation can be incorporated in a collective model of *A.Bohr* and *B. Mottelson*, which represent a refinement and extension of the **liquid drop model** (Sect. 2.6.3). This collective model explains furthermore many properties of heavy nuclei, such as collective rotations or vibrations of nuclei. This collective behavior of nuclei appears in particular for deformed nuclei. Spherical symmetric nuclei cannot be induced to rotations while deformed nuclei experience during collisions with other nuclei a torque and can therefore change their angular momentum.

5.6.1 Deformed Nuclei

A nucleon with angular momentum $\ell > 0$ outside a closed shell can deform the otherwise spherical symmetric nuclear body by interaction with the nucleons in the closed shell. This is completely analogue to the polarization of the electron shell of atoms by an electron outside of closed shells. (Fig. 5.38).

The nuclear deformation into an ellipsoid with rotational symmetry can be described for $\beta \ll 1$ by the polar representation

$$R(\vartheta) = R_0\left[1 + \beta \cdot Y_2^0(\cos\vartheta)\right] \quad (5.57)$$

of the angular dependent nuclear radius $R(\vartheta)$, where R_0 is the mean radius, $Y_2{}^0$ the spherical harmonic function and

$$\beta = \frac{4}{3}\sqrt{\frac{\pi}{5}} \cdot \frac{\Delta R}{R_0} \ll 1 \quad (5.58)$$

the **deformation parameter** which is a measure of the deviation

$$\Delta R = a - b \quad (5.59a)$$

of the ellipsoid with the large semi-axis a and the small one b from the spherical configuration with $a = b$. Often the normalized quantity

$$\delta = \frac{a-b}{R_0} = 0,946\beta \quad (5.59b)$$

is used as deformation parameter.

With increasing deformation the surface of the ellipsoid increases and with it the surface energy. Instead of (2.40b) for a spherical nucleus we get the surface energy.

$$E_S = E_0\left(1 + \beta^2 + \ldots..\right) \text{ with } E_0 = a_S \cdot A^{2/3} \quad (5.59c)$$

Nuclei with $\beta > 0$, (i.e. a > b) form a prolate symmetric top (an ellipsoid prolonged in the direction of the symmetry axis) (Fig. 5.39a) while nuclei with $\beta < 0$ (a < b) are oblate symmetric tops (ellipsoid squeezed in the direction of the symmetry axis) (Fig. 5.39b).

Plotting the potential energy a s function of the deformation parameter β $E_{pot}(\beta)$ increases monotonically for nuclei with closed shells (Fig. 5.40). With increasing number of nucleons outside closed shells the increase is slower, turns into a decrease and $E_{pot}(\beta)$ reaches a minimum for $\beta \neq 0$. This means that for such nucleon-configurations the stable nucleus exists only as deformed ellipsoid and not as a sphere. This is illustrated in Fig. 5.41 for the Samarium isotopes, where the curves $E_{pot}(\beta)$ are shown for different neutron numbers.

Nuclei, for which the potential energy $E_{pot}(\beta)$ has a minimum for $\beta \neq 0$ are deformed in their ground state because of energetic reasons. Their form can be described by a prolate or oblate rotational ellipsoid., depending on the sign of the deformation parameter β.

In such deformed nuclei the proton distribution is no longer spherical symmetric. The nuclei therefore possess an electric quadrupole moment, which increases with the deformation parameter β.

Due to this polarization of the nuclear trunk with closed shells the total angular momentum can deviate from zero because the potential is no longer spherical symmetric. The nuclear spin is no longer solely caused by the nucleons outside closed shells but also by the nucleons in closed shells. This explains the observed large spin values of heavy nuclei.

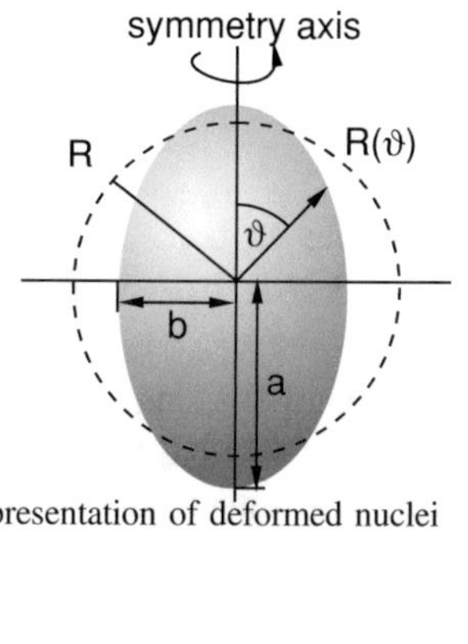

Fig. 5.38 Polar representation of deformed nuclei

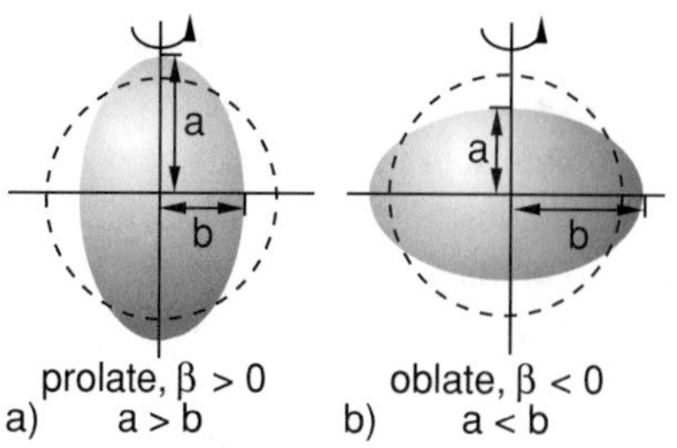

Fig. 5.39 **a** Prolate, **b** oblate form of the rotational ellipsoid of a deformed nucleus.

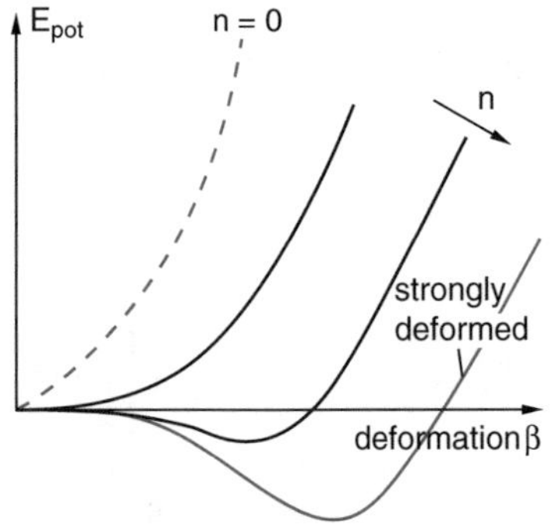

Fig. 5.40 Schematic dependence of the deformation energy as a function of the deformation parameter ß for nuclei with increasing number of nucleons outside of closed shells

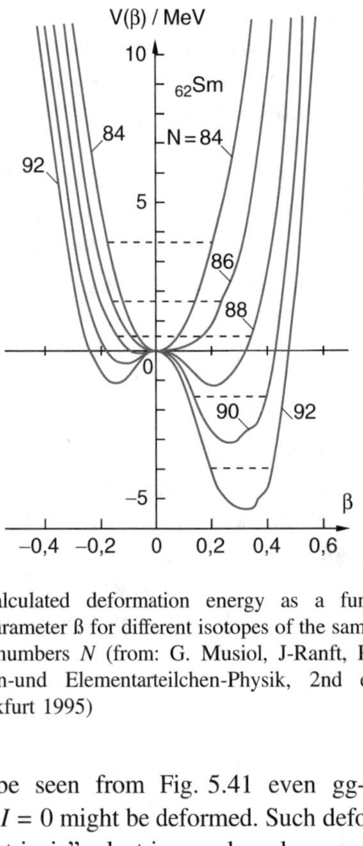

Fig. 5.41 Calculated deformation energy as a function of the deformation parameter ß for different isotopes of the samarium nucleus with neutron numbers N (from: G. Musiol, J-Ranft, R. Reifand D. Seeliger: Kern-und Elementarteilchen-Physik, 2nd edition Harry Deutsch, Frankfurt 1995)

As can be seen from Fig. 5.41 even gg-nuclei with nuclear spin $I = 0$ might be deformed. Such deformed nuclei have an "intrinsic" electric quadrupole moment, which, however, cannot be observed because for I = 0 all directions of the symmetry axis are randomly distributed into all directions, and therefore the observable average quadrupole moment is zero (see next Section).

5.6.2 Rotating Nuclei

Nuclei without spherical symmetry can gain large angular momenta when colliding with other heavy particles (protons, α-particles or other nuclei), if the colliding particles transfer

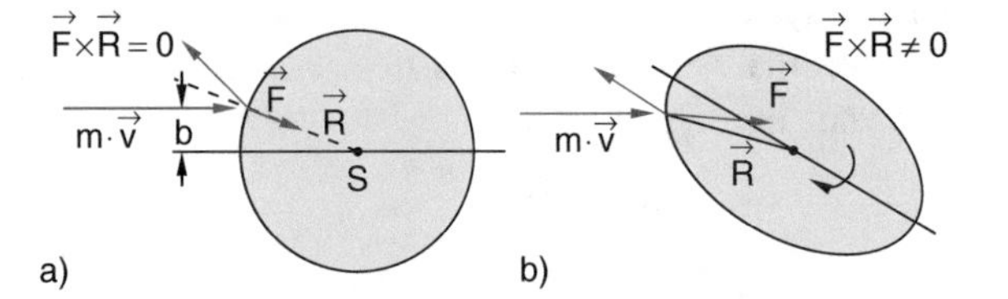

Fig. 5.42 **a** only a radial force acts but no torque. Therefore no transfer of angular momentum is possible. **b** Transfer of nuclear angular momentum onto a deformed nucleus by collisions. For spherical symmetric nuclei

a torque to the nucleus in question. As illustrated in Fig. 5.42b such a transfer of torque occurs because the force on opposite points of the ellipsoid are not equal. For a sphere they are symmetric and therefore no torque is transferred (Fig. 5.42a) and the angular momentum transfer is zero. Such a transfer of orbital angular momentum $L = b \cdot \mu \cdot v$ of the collision pair with reduced mass μ and relative velocity v into the nuclear spin I of one of the collision partners can occur already at larger collision parameters b, where the short range nuclear force is no longer present but the long range Coulomb forces cause the transfer.

The nucleus, which has gained angular momentum can rotate as a whole (collective rotation of all nucleons). The rotating nucleus loses its rotational energy by radiation, analogue to a rotating molecule. While for molecules this radiation occurs in the microwave range, one observes for rotating nuclei γ-radiation in the energy range 0.1–10 MeV, i.e. at about 10^8–10^{10} times higher energies.

We will now look at the rotational levels of rotating nuclei in more detail. We will at first regard those nuclei, which have before their excitation the nuclear spin $I = 0$. For rotational ellipsoids rotations about the symmetry axis cannot be excited since at collisions the transferred torque is zero (Fig. 5.42a). The rotational angular momentum **R** must be therefore perpendicular to the symmetry axis (see Vol. 1, Chap. 5). The rotational energy of such a symmetric top is for a rotation about an arbitrary axis perpendicular to the symmetry axis.

$$E_{\mathrm{rot}} = \frac{R^2}{2\Theta}, \tag{5.60}$$

where Θ is the moment of inertia, which has been labelled by the letter Θ instead of I to avoid any possible confusion with the nuclear spin I.

The different nucleons in the nucleus have their own angular momenta $j_i = \ell_i + s_i$. The vector sum $I = \sum j_i$ of all j_i is for non-closed shells generally not zero. The total angular momentum of the nucleus in the lowest rotational level, where no energy has been transferred from outside, is $I_0 = \sum j_i$. This total angular momentum is not constant if measured in the nuclear coordinate frame, because the force field is not spherical symmetric. The nuclear spin I precesses about the symmetry axis, so that only its projection $\Omega = \sum \Omega i$ onto the symmetry axis is nonzero, while the other components average to zero.

$$\Omega = \sum_i \Omega_i = K \cdot \hbar. \qquad (5.61)$$

If a nucleus with $I_0 \neq 0$ is induced by collisions to collective rotation with angular momentum $R \perp \hat{e}_z$, the two angular momenta couple to a total nuclear angular momentum (Fig. 5.43).

$$\boldsymbol{I} = \boldsymbol{I}_0 + \boldsymbol{R}$$

where I is no longer perpendicular to the symmetry axis, which we choose as the z-axis.

Since $\boldsymbol{R}$ is perpendicular to the z-axis, the projection $I_z = K \cdot h$ of the total angular momentum $\boldsymbol{I}$ onto the z-axis is independent of $\boldsymbol{R}$.

The total rotational energy excited by collisions is

$$E_{\text{rot}} = \frac{I_x^2}{2\Theta_x} + \frac{I_y^2}{2\Theta_y}, \qquad (5.62a)$$

Since $\Theta_x = \Theta_y = \Theta$ and $I_x^2 + I_y^2 = I^2 - I_z^2$ with $I_z^2 = K^2\hbar^2$ this can be written as

$$E_{\text{rot}} = \frac{[I(I+1) - K^2]\hbar^2}{2\Theta}. \qquad (5.62b)$$

The quantum number K of the projection of I onto the symmetry axis of the nucleus can have all values 0, 1, 2,….. I for integer I or 1/2, 3/2, 5/2,….I for half-integer I.

One therefore obtains for each value of I an energy ladder $E(I, K)$ of all allowed values of K (Fig. 5.44). This is completely analogue to the symmetric top molecule which rotates about an axis through the center of gravity inclined against the symmetry axis (Vol. 3, Chap. 9).

Before we discuss the general case of rotation- spectra of nuclei we will start with the more simple case with $I_0 = 0 \Rightarrow K = 0$. In this case the nuclear angular momentum $I = R$ is always perpendicular to the symmetry axis. The wave-function of such a rotational level is symmetric with respect to a rotation about the direction of R.

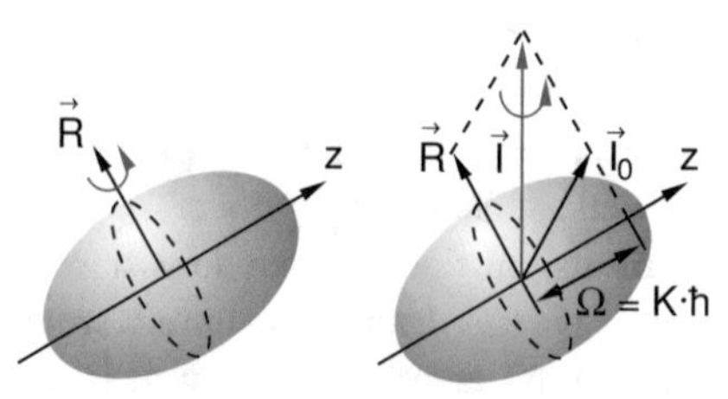

Fig. 5.43 Nuclear spin I and collective angular momentum R in the body-fixed coordinate system of the nucleus.

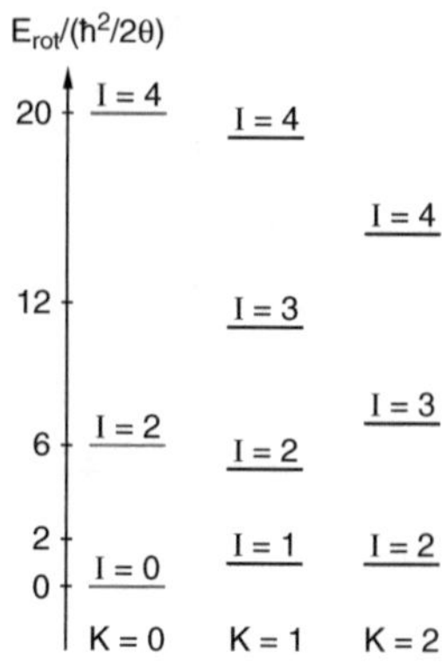

Fig. 5.44 Rotational energy levels of an oblate nucleus for different projection quantum number K

Because of parity conservation for transitions between rotational levels and the defined parity of the wave-functions $Y_I^m(\vartheta, \varphi)$ only transition from a level with I = 0 into levels with even values of I can be excited. In Fig. 5.45 the possible rotational levels are listed with their energies, which are compared with measured rotational transitions of the $^{238}_{92}U$-nucleus.

For downwards transitions with $\Delta I = 2$ between rotational levels $I = I + 2 \rightarrow I = 0$, a γ-quantum is emitted with the energy

$$h \cdot \nu = \frac{(I+2) \cdot (I+3) - I(I+1)}{2\Theta}\hbar^2 = \frac{2I+3}{\Theta}\hbar^2. \quad (5.63)$$

according to (5.62b) for $\Delta K = 0$.

Measurements of the γ-energy allows the determination of the moment of inertia Θ of the nucleus and therefore its mass distribution.

Nuclear models where all nucleons perform synchronous motions (they all rotate or vibrate in phase) are called **collective models**.

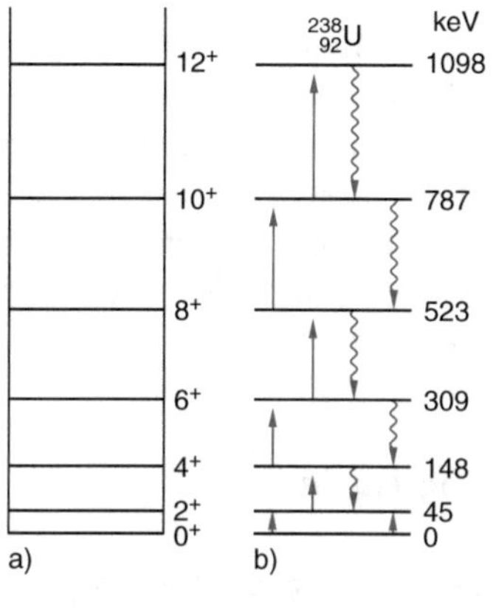

Fig. 5.45 **a** Excitable rotational energy levels in a deformed gg-nucleus **b** observed rotational transitions in $^{238}_{92}U$, which had been excited by electron impact.

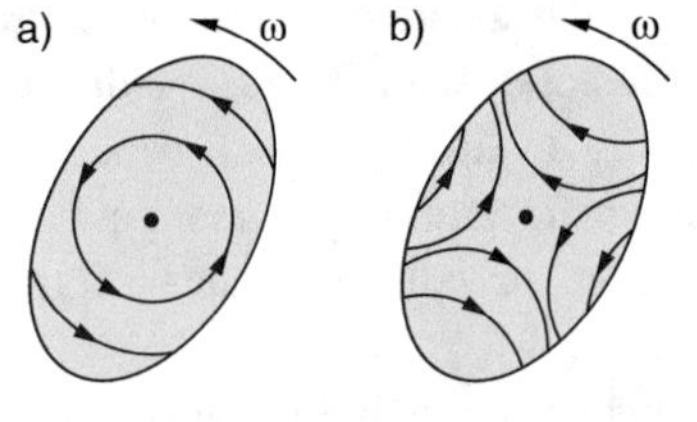

Fig. 5.46 Schematic representation of the rotation about an axis perpendicular to the drawing plane **a** of a rigid nucleus **b** a non- rigid nucleus

However, when comparing the experimental results with those obtained from nuclear collective models assuming a rigid rotating nucleus, discrepancies occur where the experiments give smaller values of the moment of inertia than the expression

$$\Theta_s = \frac{2}{5} M R_0^2 = \frac{2}{5} A \cdot m_N \cdot R_0^2 \qquad (5.64)$$

for the rigid rotating nucleus with the mass number A. The conclusion for solving this discrepancy is the assumption that not all nucleons participate in the rotation, but only those in the edge region of the nucleus.

According to this model the nucleus rotates differentially like a viscose liquid in a rotating cylinder, i.e. the angular velocity ω depends on the distance from the center.

Such a non-rigid rotation can be illustrated by the phenomenon of the tides (see Vol. 1. Sect. 6.6). Due to the gravitational attraction by the moon and the rotation of the earth two elevations (high tide) and two depressions (low tide) travel per day around the earth, Similar the outer nucleons of a differentially rotating nucleus travel around the kernel of the nucleus, Such a differential rotation can be composed of a rotation and a deformation-vibration which results in an angular velocity which increases with the distance r from the center of mass. In Fig. 5.46b the nucleus is regarded as a viscous liquid where the nucleons perform individual motions (indicated by the arrows), where the directions of these motions can be arbitrary.

5.6.3 Nuclear Vibrations

While rotations of nuclei can be only excited for deformed nuclei, nuclear vibrations can occur in a manifold way and even spherical symmetric nuclei can be vibrational excited. The situation for nuclear vibrations can be illustrated by the example of a liquid drop. The following vibrational modes can occur:

- Radial compression vibrations (Fig. 5.47a), where the density of the nucleus changes periodically and no angular momentum is excited (**monopole vibrations**).

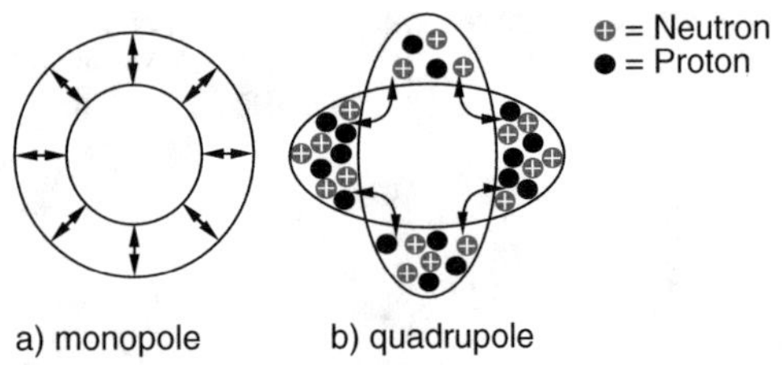

Fig. 5.47 **a** radial compression-vibration **b** Deformation vibration

Since nuclei are nearly incompressible, such radial vibrations demand a high excitation energy.

- Surface vibrations where the volume of the nucleus is not changed but the nucleus is deformed. Here the spherical nucleus can be alternately deformed into a prolate or oblate rotational ellipsoid (Fig. 5.47b).

The manifold of possible surface vibrations can be quite generally described by the displacement of an arbitrary point $(R_0, \vartheta, \varphi)$ at the surface of the nucleus. The probability of such a displacement can be expressed by the absolute square of the function

$$R(\vartheta, \varphi, t) = R_0 \left[1 + \sum_{l=0}^{\infty} \sum_{m=-l}^{+l} a_{l,m}(t) Y_l^m(\vartheta, \varphi) \right] \qquad (5.65)$$

and the time-dependent coefficients $a_{l,m}(t)$. For $l = 0$ we obtain the radial vibration (monopole vibration) for $l = 1$ the dipole vibration is described (which can however, not be excited because there are only positive charges in the nucleus). For $l = 2$ the quadrupole vibration occurs, which is described according to (5.56) by.

$$R(\vartheta, \varphi, t) = R_0 [1 + \beta(t) \cos\gamma(t) \cdot Y_2^0(\vartheta, \varphi) + \frac{1}{\sqrt{2}} \beta(t) \sin\gamma(t) \cdot \left(Y_2^2(\vartheta, \varphi) + Y_2^{-2}(\vartheta, \varphi) \right)], \qquad (5.66)$$

where the time-dependent deformation-parameter $\beta(t)$ is defined by (5.59). While $\beta(t)$ gives the magnitude of the deformation, the parameter γ describes its form. The comparison with (5.65) shows that

$$a_{20} = \beta \cos\gamma, \quad a_{22} = \frac{\beta}{\sqrt{2}} \sin\gamma. \qquad (5.67)$$

If only $\beta(t)$ changes with time but γ remains constant the rotational symmetry of the nucleus remains unchanged, while for a time variation of $\gamma(t)$ the rotational symmetry of the nucleus is not preserved. In Fig. 5.48 some vibrational modes are illustrated.

Measurements of the vibrational frequencies give information about the forces between the nucleons and their dependence on the direction of the elongation and on the nucleon spin.

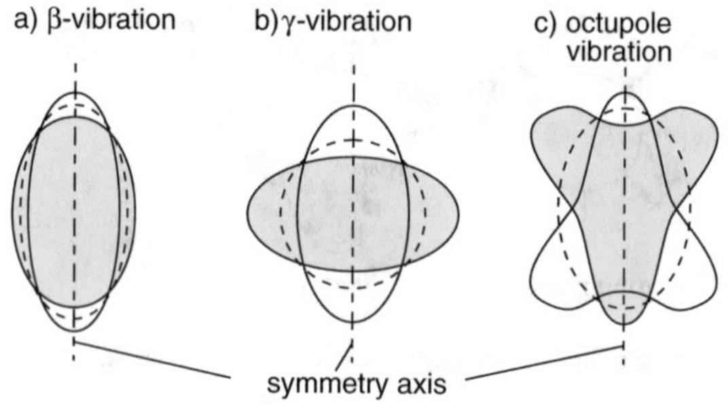

Fig. 5.48 Possible vibrational modes of nuclei. The dashed curve gives the equilibrium structure of the nucleus

Only those vibrations can be excited where the center of mass does not change. Since the functions Y_l^m with odd values of l result in a vibration of the center of mass they cannot be excited. Besides the radial vibration with l = 0 the next higher vibrational mode is the quadrupole vibration with $l = 2$.

For sufficiently small amplitudes the restoring forces are proportional to the elongations (Hooke's law) and we obtain as solutions for the vibrational energy levels the eigenvalues of the harmonic oscillator

$$E_{\text{vib}} = \hbar\omega_{l,m}\left(n + \frac{1}{2}\right), \tag{5.68}$$

where the vibrational frequencies $\omega_{l,m}$ depend on the form of the vibrations and are generally different for the different deformations $Y_{l,m}$.

5.7 Experimental Detection of Excited Nuclear Rotational- and Vibrational States

When a fast charged particle with sufficient kinetic energy passes close to a nucleus, the resultant electric induction pulse transferred to the nucleus can result in the excitation of the nucleus (Coulomb excitation). The projectile loses kinetic energy (inelastic collision). Such a Coulomb excitation can be best observed when the impact parameter is larger than the range of the nuclear force. This is always the case when the kinetic energy of the projectile is sufficiently small so that it cannot penetrate through the repulsive Coulomb barrier (Fig. 3.16).

The cross section for the Coulomb excitation of a nuclear level with the multipole order ℓ by a projectile with charge $Z_1 \cdot$ e and mass M_1

$$\sigma_{\text{CE}} \propto \left(\frac{Z_2 e}{\hbar v_0}\right)^2 \cdot a^{-(2l+2)} \cdot f(Z_1, Z_2, v_0, v_e) \tag{5.69}$$

is proportional to the square of the charge $Z_2 \cdot$ e of the target nucleus and decreases with increasing distance a between projectile and target. The function f, which depends on the charges of projectile and target and on the velocities v_0 before and v_e after the collision, can be found in tables of nuclear physics [2].

Nuclear levels with very high angular momentum can be produced in nuclear fusion reactions where heavy ions collide with the target nucleus and get stuck. In this case the total orbital angular momentum of the projectile, referred to the center of mass of the target, is transferred into the internal angular momentum of the compound nucleus (Fig. 5.49).

When collective vibrations of a nucleus are excited by absorption of γ-quanta, one observes strong resonances in the absorption cross section $\sigma(E_\gamma)$. One example is the reaction

$${}^{A}_{Z}\text{N} + \gamma \;\rightarrow\; {}^{A-\nu}_{Z}\text{N}^* + \nu \cdot n, \tag{5.70}$$

where the γ-absorption creates an excited nucleus which emits ν neutrons (Fig. 5.50).

Such a "giant-resonance" corresponds to collective radial vibration where all protons of the nucleus oscillate against the neutrons without changing the form of the nucleus. Such "segregation-vibrations" can occur as monopole-vibration ($\ell = 0$), dipole oscillation (when the oscillation occurs in one direction and $\ell = 1$) or even quadrupole vibration with $\ell = 2$. These vibrational modes are generally accompanied by changes of the nuclear spin I and Isospin T (Fig. 5.51).

If protons and neutrons oscillate concordant i.e. with the same phase, there is no segregation of protons and neutrons, the Isospin does not change ($\Delta T = 0$). If, however, i.e. protons and neutrons oscillate against each other, a periodic change of the difference $\Delta = N_p – N_n$ occurs in each volume element ΔV. The isospin component $T_3 = 1/2(N_p - N_n)$

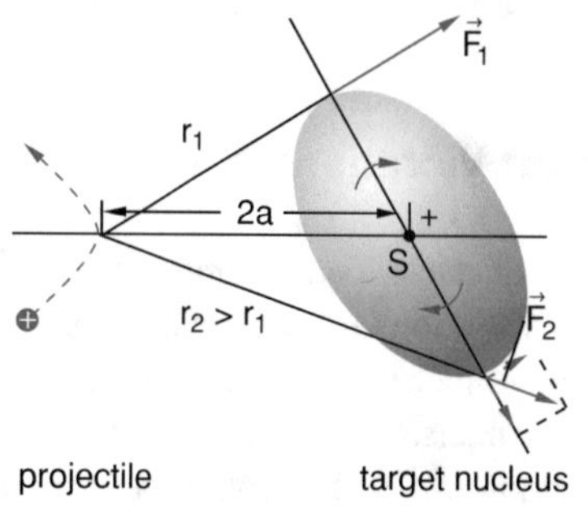

Fig. 5.49 Coulomb-excitation of nuclear rotation

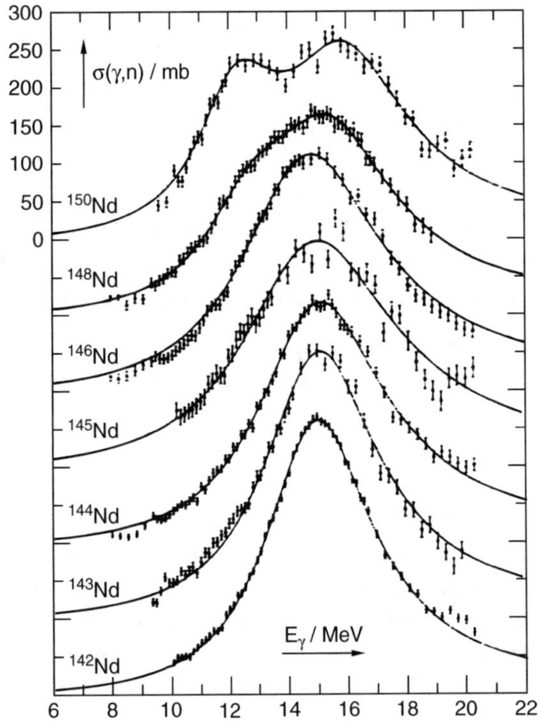

Fig. 5.50 Giant resonance in the cross section $\sigma(\gamma, n)$ for the γ-induced neutron emission after the excitation of different Neodymium-isotopes. (after B. L. Berman, C. S. Fultz. Rev. Mod. Physics **47**,713 (1975))

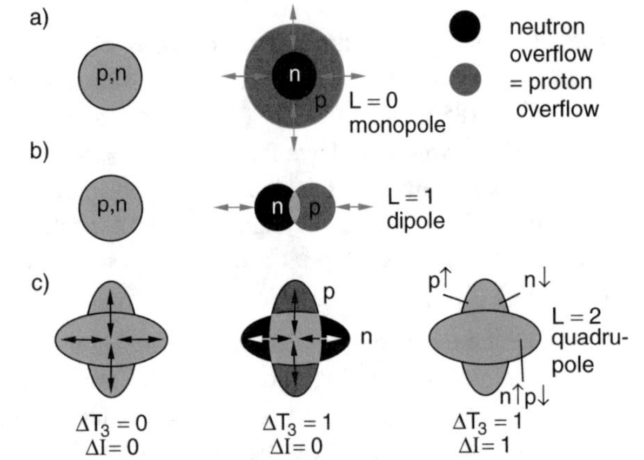

Fig. 5.51 Picturesque drawing of the excitation of giant resonances as collective segregational vibrations of different multimode orders. **a** radial segregation vibration **b** dipole oscillation **c** quadrupole vibrations without and with segregation and with segregation of nuclear spins

therefore oscillates for $\Delta T_3 = 1$ between + ½ and −½. Such an oscillation with $\Delta T_3 = 1$ is called **iso-vectorial resonance**.

Summary

- The interaction between two nucleons can be best studied on two-nucleon systems. The only stable system is the deuteron d = (p, n) with parallel spins of the two nucleons.
- The binding energy of the deuteron is $E_B = -2.2$ MeV, but the potential depth, on the other hand is $E_{pot} = -45$ MeV. The magnetic dipole moment is $\mu_D = 0.857\ \mu_N$ where $\mu_N = 5.0500783 \times 10^{-27}$J/T. is the nuclear magnetic moment. The electric quadrupole moment is $Q_D = 2.86 \cdot 10^{-27}\text{cm}^2$. The mean distance between proton and neutron in the bound state of the deuteron is 4.3 fm and therefore much larger than the radius $R_0 = 1.5$ fm of the potential box of the attractive nuclear force.
- All other two-nucleon systems (p, p and n.n) are not bound. They can be studied by scattering experiments. Detailed information can be obtained from scattering experiments with spin-polarized nucleons. They allow investigations of the spin-dependent part of the nuclear forces. Such experiments are, however, difficult.
- The strong interaction between nucleons is independent of their electric charge, but it depends on the relative spin orientation of the nucleons.
- In the model of the isospin-formalism proton and neutron can be regarded as two isospin-components of a nucleon. The isospin can be treated as a vector with three components in an abstract space. The three components are represented by two-row quadratic matrices. The isospin T of a nucleus is the vector sum of the isospins $\boldsymbol{T} = \sum_k \boldsymbol{\tau}_k$ of the nucleons. The amount of the third component of the isospin for a nucleus with N_p = Z protons and N_n neutrons is $T_3 = 1/2(Z-N)$.
- The strong interaction can be formally described by the exchange of virtual π-mesons. Its range r_0 is equal to the reduced Compton wavelength $r_0 = \bar{\lambda}_C/2\pi = \hbar/(m_\pi c)$ of the π-meson with mass m_π.
- Many observed properties of nuclei can be explained by the single-particle model of free nucleons in a mean potential. This model calculates the mean interaction of an arbitrary nucleon N_i with all other nucleons. The most realistic model uses the Woods-Saxon-potential with a radial dependence between the potential box and the parabolic potential.
- The nuclear shell model was developed in analogy to that of the atomic electron shell. To each principal quantum number n there are $(n-1)$ quantum numbers ℓ of the orbital angular momentum and $(2\ell + 1)$ quantum numbers m_ℓ of the projection of ℓ. All sublevels (n, ℓ, m_ℓ) with equal n form a nuclear shell. Nuclei, where all sublevels

are occupied with nucleons, are called closed shells. Nuclei with closed shells of nucleons have a larger stability.

- Only the consideration of the spin–orbit coupling gives the correct energetic order of the closed shells and explains the magic numbers for protons and neutrons, where the nuclei have maxima of their stability.
- The nuclear spin I_0 of a nucleus in its ground-state is the vector sum $j = \sum j_i$ of the spins of the nucleons. If the orbital angular momentum is zero it is $I = J$.
- For deformed nuclei with nuclear spin I_0 a collective rotation of all nuclei with angular momentum R around the center of mass can be excited by collisions or by absorption of γ-quanta. The total nuclear angular momentum is then $I = I_0 + R$.
- Deformed nuclei can be described by a rotational ellipsoid. The deformation parameter $\delta = (a-b)/R_0$ gives the relative magnitude of the deformation of a sphere with radius R_0.
- Nuclei excited into higher rotational levels can radiate their excitation energy as γ-quanta (Multipole-radiation).
- Nuclei can be excited into different vibrational states. Besides radial vibrations which preserve the spherical symmetry of nuclei, deformation-vibrations can be excited, where the magnitude and form of the deformation oscillates with the vibrational period.
- The excitation of rotations and vibrations of nuclei appear as resonances in the inelastic cross section $\sigma(E)$.

Problems

5.1

a) Which levels of the deuteron with angular momentum quantum number $J = 1$ are possible?
b) Which of these levels have positive parity?
c) Estimate from the magnetic dipole moment and the electric quadrupole moment of the deuteron the relative share of the d-wave-function.

5.2 Measurements with a mass spectrometer give the mass differences

$$M(^2D_3^+) - \frac{1}{2}M(^{12}C^{++}) = 42,306 \cdot 10^{-3}u,$$
$$M(^1H_2^+) - M(^2D^+) = 1,548 \cdot 10^{-3}u,$$

where D_3^+ is the triatomic molecular deuterium ion. Determine with these data.

1. The mass excess of ^{1}H, ^{2}H, and their absolute masses, using the atomic mass unit

$$\frac{1}{12}M(^{12}C) = 1\text{AMU} = 931,4943\,\text{MeV}/c^2,$$

2. The masses m_p of the proton, m_D of the deuteron and (using the binding energy of the deuteron) the mass m_n of the neutron.

5.3 A rotational ellipsoid with the half-axes $a = R \cdot (1 + \varepsilon)$, $b = R \cdot (1 + \varepsilon)^{-1/2}$ and $c = b$ has the volume $V = (4/3)\pi \cdot a \cdot b \cdot c = (4/3) \cdot \pi \cdot R^3$ and the surface

$$A_s = 2\pi ab\left(\frac{b}{a} + \frac{1}{x}\arcsin x\right)$$
$$\text{with} \quad x = \frac{(a^2 - b^2)^{1/2}}{a}$$

If a spherical symmetric nucleus with radius $R = r_0 A^{1/3}$ is deformed into such an ellipsoid, the energy term for the volume energy in the total binding-energy remains, according to the Bethe-Weizsäcker formula, constant but the surface energy term changes by the factor $(1 + 2/5 \cdot \varepsilon^2 \ldots)$ and the Coulomb energy by the factor $(1 - 1/5 \cdot \varepsilon^2 + \ldots)$.

1. Show, that for small values of ε the expression (5.59c) for E_s is correct by expanding the formula for the surface A_s up to terns of ϵ^2.
2. Prove that the dependence of the Coulomb energy on ε has the correct form.

5.4 The ^{6}Li nucleus has three excited states above the ground state $J^p = 1^+$ with α-particle substructure $J^p = 3^+(2.185\text{MeV})$, $J^p = 2^+(4.32\text{MeV})$ and $J^p = 1^+(5.7\text{MeV})$ which all have the isospin $T = 0$. Give an explanation for this energetic order and determine from the splitting the sign and the expectation value of the spin–orbit coupling interaction $V_{ls} = a \cdot \sum l_i \cdot s_i$.

5.5 Calculate the magnetic moment of a g-u or u-g nucleus using the vector-coupling model for the two cases $I = j = l \pm ½$ ($g_s = 5.586$ for p and $g_s = -3.826$ for n). Which magnetic moment can be expected for ^{7}Li ($J = 3/2$), ^{13}C ($J = ½$) and ^{17}O ($J = 5/2$)?

5.6 How many degenerate energy levels exist for the three-dimensional harmonic oscillator for the principal quantum number n?

5.7 Explain why there are only a few stable u-u nuclei using the Bethe-Weizsäcker formula and the Fermi-gas model.

5.8 Justify qualitatively in the framework of the shell model, why the nuclear spin quantum number I in the ground state of the nucleus $^{14}_{7}$N is $I = 1$.

References

1. W. Greiner. J. Maruhn: Nuclear Models (Springer Heidelberg 1996)
2. Giant Resonances in nuclei. Rep. Progr. Phys. **44**, 719 (1981)
3. M. Zeeman: Inner Models and Lartge Cardinals. (de Gruyter, Berlin 2002)
4. K.L.G.Heyde: The Nuclear Shell Model. (Springer, Heidelberg 2013)
5. M. Goeppert-Mayer, J.H.D. Jensen: Elementary Theory of Nuclear Shell Structure. (Wiley, New York 1955)

6 Nuclear Reactions

Nuclear reactions are induced by inelastic collisions, which can excite nuclei into higher energy states, or they may transfer the initial nucleus into other nuclei, or they can cause the fission of nuclei or they even may create new particles. For the investigation of such reactions generally projectile particle with kinetic energy E_{kin} are shot onto target nuclei. Projectile particles may be electrons, positrons protons neutrons or mesons, but also atomic nuclei, such as alpha particles or C_6^+-charged carbon nuclei.

For sufficiently high kinetic energies of the projectile particles collisions with the target nuclei can result in the generation of new particles. For example collisions between two protons with kinetic energies $E_{kin} > 300$ MeV produces a neutron and a π^+-meson according to the reaction

$$p + p \rightarrow n + p + \pi^+ .$$

Such high energy reactions where new elementary particles are created, will be treated in Chap. 7.

In the present chapter we will discuss the basic foundations and experimental techniques for the investigation of nuclear reaction in the "middle energy range", where nuclei are excited or split (nuclear fission) are converted into other nuclei, or fused to larger nuclei (nuclear fusion). (Fig. 6.1)

6.1 Basic Foundations

A nuclear reaction where a particle a hits a nucleus X and converts it into a nucleus Y and a particle b is emitted, can be described by the reaction equation

$$a + X \rightarrow Y + b \tag{6.1a}$$

Or written in a short notation as

$$X(a, b)Y \tag{6.1b}$$

Such nuclear reactions can be grouped into different categories:

6.1.1 Inelastic Scatterring with Nuclear Excitation

Inelastic collisions with nuclear excitation can be written as Fig. 6.1.

$$a(E) + X \rightarrow X^* + a(E - \Delta E), \tag{6.2a}$$

where the part $\Delta E = \Delta E_1 + E_{recoil}(X^*)$ of the kinetic energy E_{kin} of the incident projectile a is converted into excitation energy ΔE_1 and recoil energy of the target nucleus X which was at rest before the collision. The excitation energy can be rotational or vibrational energy of the target nucleus X*. The excited nucleus can release its "internal" energy by emission of γ-quanta.

$$X^* \rightarrow X + \gamma \tag{6.2b}$$

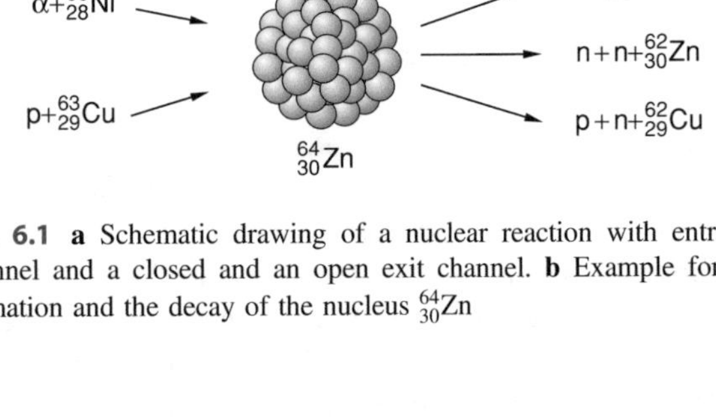

Fig. 6.1 **a** Schematic drawing of a nuclear reaction with entrance channel and a closed and an open exit channel. **b** Example for the formation and the decay of the nucleus $^{64}_{30}Zn$

W. Demtröder, *Nuclear and Particle Physics*, Undergraduate Lecture Notes in Physics,
https://doi.org/10.1007/978-3-030-58313-2_6

The excited nucleus X* either returns into its initial state, or it converts by emission of electrons e^-, positrons e^+ or α-particles into other nuclei.

6.1.2 Reactive Scattering

Here the target nucleus X is converted into another nucleus Y

$$a + X \rightarrow Y + b. \tag{6.3}$$

And a particle b (sometimes even several particles b_1, $b_{2...}$) is emitted (conversion reaction). Examples are

$$p + {}^7_3\text{Li} \begin{cases} \nearrow {}^7_4\text{Be} + n, \\ \searrow {}^4_2\text{He} + {}^3_1\text{H} + p. \end{cases} \tag{6.4}$$

6.1.3 Collision-Induced Nuclear Fission

If a projectile *a* collides with the target nucleus X *a* can be captured by X and a compound nucleus (*a*X) is formed which is excited. The excited compound nucleus (*a*X)* can decay into 2 fragments Y_1 and Y_2 and in addition several neutrons can be released.

$$a + X \rightarrow (aX)^* \rightarrow Y_1 + Y_2 + \nu \cdot n, \tag{6.5}$$

The kinetic energy of the projectile *a* and its binding energy in the compound nucleus results in the internal excitation of the compound nucleus. If the excitation energy E_{exc} is sufficiently high the compound can decay according to (6.5).

One example is the fission of the uranium nucleus induced by fast neutrons

$$\begin{aligned} n(E) + {}^{238}_{92}\text{U} &\rightarrow {}^{239}_{92}\text{U}^* \\ &\rightarrow {}^{A_1}_{Z_1}Y_1 + {}^{A_2}_{Z_2}Y_2 + \nu \cdot n \end{aligned} \tag{6.6a}$$

with $A_1 + A_2 = 239 - \nu$ and $Z_1 + Z_2 = 92$, The reaction starts for $E \geq 1.5$ MeV and releases, besides the fission products Y_1 and Y_2 also ν neutrons (see Sect. 6.5).

6.1.4 Energy Threshold

The probability of the reactions presented above, depends on the collision partners a and X but in particular on the collision energy. Many reactions start only above an energy threshold E_{th}, specific for the reaction considered. For positively charged projectiles (protons, α-particles, C^{6+}nuclei) the kinetic energy must be large enough to overcome the Coulomb barrier in order to start the reaction.

The energy balance (6.1) can be also written as mass balance

$$M(a) + M(X) = [M(b) + M(Y)] + Q/c^2.$$

The quantity Q is the **reaction heat.** It equals the difference $\Delta M\ c^2$ between the energies in the entrance- and the exit channel. For $Q > 0$ the reaction is **exothermic.** It starts already at small energies where the Coulomb barrier can be overcome. Q gives then the kinetic energy and excitation energy of the reaction products.

For $Q < 0$ the reaction is **endothermic.** In the entrance channel at least the kinetic energy $E_{kin} > |Q|$ has to be provided in order to reach the reaction threshold.

Many nuclear reactions can be described by a compound model, proposed by *Niels Bohr* (Fig. 6.2X).

According to this model the two collision partners *a* and *X* stick together and form a compound nucleus (aX)* which is in an excited state, because of the kinetic energy of a and its binding energy of the compound (aX). This excited compound nucleus rapidly decays into fragments. Since the excitation energy of (aX)* is distributed among many degrees of freedom of the nucleons in (aX)* the excited compound nucleus "forgets" the details of its formation, this means that the exit channel depends on the energy **but not** on the specific entrance channel.

> The Compound nucleus model divides the reaction into two steps: The fusion of the two reactants and the decay of the compound nucleus.

Examples of such fusion reactions are

$$\begin{aligned} d + d \rightarrow {}^4_2\text{He}^* &\rightarrow {}^3_2\text{He} + n + 3.25\ \text{MeV} \\ &\rightarrow {}^3_1\text{H} + p + 4.0\ \text{MeV} \end{aligned} \tag{6.7a}$$

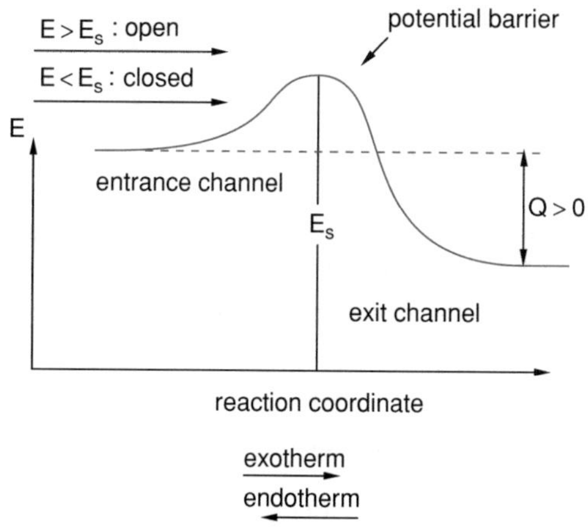

Fig. 6.2 Schematic progression of the potential $E_{pot}(r)$ for exothermic and endothermic reactions

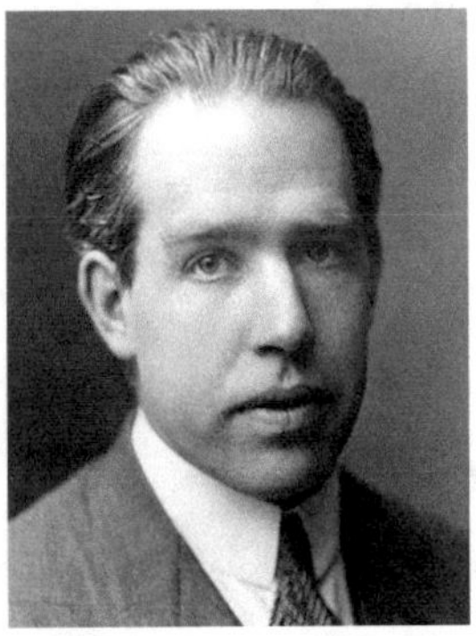

Fig. 6.2X Niels Bohr (Nobelprize 1922)

$$\begin{aligned} p + {}^{6}_{3}\mathrm{Li} &\rightarrow {}^{7}_{4}\mathrm{Be}^{*} \\ &\rightarrow {}^{4}_{2}\mathrm{He} + {}^{3}_{2}\mathrm{He} + 22.4\,\mathrm{MeV}. \end{aligned} \tag{6.7b}$$

If the energy in the entrance channel surpasses the threshold energy $E_S(k)$ for a specific reaction (k) one says that the exit channel is open for the reaction (k) (Fig. 6.2), otherwise it is closed.

The kinetic energy $E_{kin} = p_a^2/(2m_a)$ of the projectile a must be large enough to provide the reaction heat Q and the kinetic energy of the center of mass motion. For a target nucleus X at rest the velocity of the center of mass is

$$v_s = \frac{m_a \cdot v_a}{m_a + m_X} \tag{6.8}$$

which gives the kinetic energy of the center of mass motion

$$E_{kin}(S) = \frac{(m_a \cdot v_a)^2}{2(m_a + m_X)}. \tag{6.9}$$

And the threshold energy

$$E_S = Q + \frac{p_a^2}{2(m_a + m_X)} = E_{kin}^{min}. \tag{6.10a}$$

With $p_a^2/2m_a = E_{kin}(a)$ we obtain

$$\begin{aligned} E_{kin}\left(1 - \frac{m_a}{m_a + m_X}\right) &= Q \\ \Rightarrow E_{kin}^{min} &= Q\left(1 + \frac{m_a}{m_X}\right). \end{aligned} \tag{6.10b}$$

Examples

1. For the reaction $\alpha + {}^{4}_{2}\mathrm{He} \rightarrow {}^{7}_{3}\mathrm{Li} + p$ (collisions of two α-particles) the reaction heat is $Q = -17$ MeV. The threshold energy for this reaction is then

$$E_S = E_{kin}^{min} = 17 \cdot (1 + 4/4)\,\mathrm{MeV} = 34\,\mathrm{MeV}$$

which is twice as high as the reaction heat Q.

2. For the excitation of the first excited state in the iron nucleus at $E_1 = 0.87$ MeV by inelastic collisions with a neutron, according to

$$n + {}^{56}\mathrm{Fe} \rightarrow {}^{56}\mathrm{Fe}^{*} + n - \Delta E_{kin}$$

the threshold energy is

$$E_S = 0.87(1 + 1/56)\,\mathrm{MeV} = 0.88\,\mathrm{MeV}$$

and therefore only slightly higher than the excitation energy.

Applying energy- and momentum- conservation laws one can obtain details of the kinematics of reactive collisions. For the target nucleus X at rest momentum conservation gives

$$\begin{aligned} m_a v_a &= m_b v_b \cos\vartheta + m_y v_y \cos\varphi \\ 0 &= m_b v_b \sin\vartheta - m_y v_y \sin\varphi \end{aligned}. \tag{6.11a}$$

Elimination of φ taking into account the relation. $(\cos^2\varphi + \sin^2\varphi) = 1$ gives the equation

$$m_y^2 v_y^2 = m_b^2 v_b^2 + m_a^2 v_a^2 - 2 m_a m_b v_a v_b \cdot \cos\vartheta. \tag{6.11b}$$

The non-relativistic energy law yields

$$\frac{1}{2} m_a v_a^2 + Q = \frac{1}{2} m_b v_b^2 + \frac{1}{2} m_y v_y^2. \tag{6.11c}$$

Inserting $m_b v_b^2$ from (6.11b) into (6.11c) yields the relation

$$\begin{aligned} Q = \left(\frac{m_a}{m_y} - 1\right) E_a - \left(\frac{m_b}{m_y} - 1\right) E_b \\ -2\frac{\sqrt{m_a m_b}}{m_y}\sqrt{E_a \cdot E_b} \cdot \cos\vartheta. \end{aligned} \tag{6.12}$$

The solution of this quadratic equation for $\sqrt{E_b}$ is

$$\sqrt{E_b} = \frac{\sqrt{m_a \cdot m_b}}{m_y + m_a} \cdot \sqrt{E_a} \cdot \left[\cos\vartheta \pm \sqrt{\cos^2\vartheta + \frac{m_y(m_b + m_y)}{m_a \cdot m_b}\left(\frac{Q}{E_a} + 1 - \frac{m_a}{m_y}\right)}\right]. \tag{6.13}$$

Only solutions with real positive values are physically realistic. This implies for E_b that the expression in brackets [] in (6.13) must be positive and real. If

$$Q \geq \left(\frac{m_a}{m_y} - 1\right) \cdot E_a \tag{6.14}$$

the square root has a value larger than $\cos\vartheta$ and therefore only the + sign before the square root can be taken.

This means: For each scattering angle ϑ a particle with uniquely defined energy E_b exists.

For $Q < \left(\frac{m_a}{m_y} - 1\right) E_a$ the Eq. (6.13) has two real solutions, i.e. particles with two different energies can be scattered into the same angle ϑ.

The minimum excitation energy is

$$E_a^{\min} = -\frac{m_y}{m_y - m_a} \cdot Q. \qquad (6.15)$$

6.1.5 Reaction Cross Section

The probability that a reaction in a specific entrance channel (i) passes to a selected exit channel (k) depends on the cross section $\sigma_{ik}(E)$ for this reaction (Fig. 6.4a). It generally depends on the scattering angle ϑ of the particle in the exit channel against the direction of the incident projectile (Fig. 6.3).

The rate dN_k/dt of reactions that result in particles in the exit channel (k) depends on the incident projectile flux (number of projectiles per sec and per cm^2 in the entrance channel (i)) and on the target particle density n_X in the reaction volume V. It is

$$\dot{N}_k = \sigma_{ik}(E) \cdot n_X \cdot V \cdot \Phi_a(i). \qquad (6.16)$$

where $\sigma_{ik}(E)$ is the energy-dependent integral cross section for the transition from the entrance channel (i) into the exit channel (k), integrated over all scattering angles ϑ. The measurement of the integral cross section therefore gives no information about the specific scattering direction.

In order to measure the particles scattered into a specific direction the **differential cross section** has to be determined. It is defined as.

$$\frac{d}{d\Omega}(\sigma_{ik}(\vartheta, E)) = \frac{1}{n_X \cdot V \cdot \Phi_a(i)} \cdot \frac{d\dot{N}_k(\vartheta)}{d\Omega}. \qquad (6.17)$$

The energy dependence $\sigma_{ik}(E)$ of the total cross section is called the *excitation function* of the reaction with the entrance channel (i) and the exit channel (k). In Fig. 6.5 the excitation functions of the reactions.

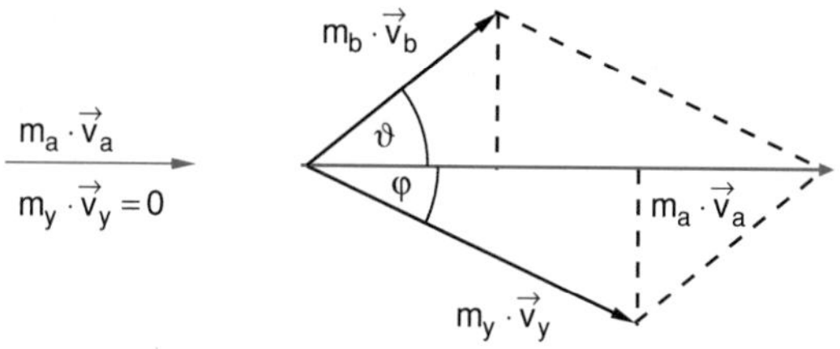

Fig. 6.3 Momentum conservation for a reactive collision X(a,b)Y when a nucleus X hits a nucleus Y at rest

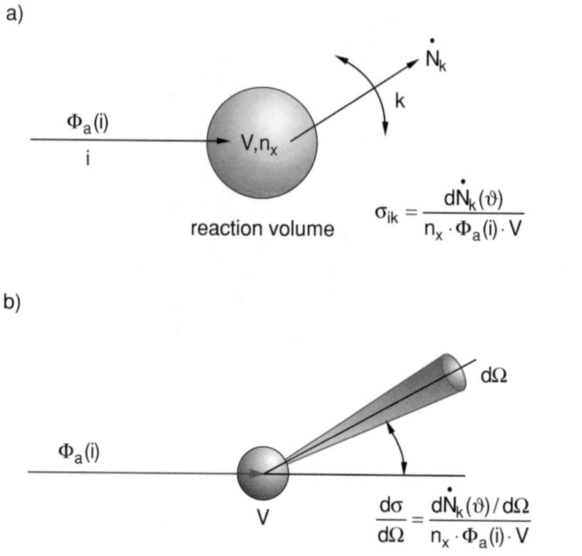

Fig. 6.4 **a** Cross section of a nuclear reaction **b** Definition of the differential cross section

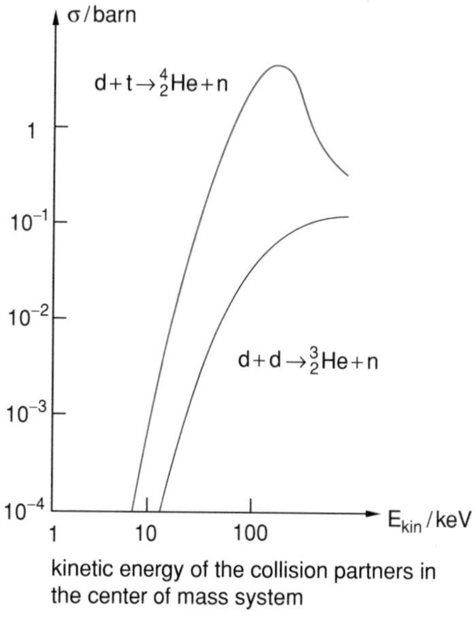

Fig. 6.5 Excitation functions of the two fusion reactions d + t → $^2_4$He + n and d + d → $^3_2$He + n

$$^2_1\text{H} + ^3_1\text{H} \rightarrow ^4_2\text{He} + \text{n}$$
$$^2_1\text{H} + ^2_1\text{H} \rightarrow ^3_2\text{He} + \text{n}$$

are shown, which illustrates that the energy dependence can be very drastically (note the logarithmic ordinate) and that it can differ considerably for equal projectiles but different target nuclei.

6.2 Conservation Laws

For all nuclear reactions, investigated so far, it was always found that energy and momentum are preserved. We will now discuss whether there are additional quantities which are also conserved during nuclear reactions.

6.2.1 Conservation of the Nucleon Number

For energies below the threshold energy E_S where new particles are generated, the number of nucleons remains for all reactions conserved. This means that for the reaction between nucleons A

$$A_1 + A_2 \rightarrow A_3 + A_4. \tag{6.18}$$

the nucleon number on both sides are the same.

6.2.2 Conservation of the Electric Charge

For all reactions of the kind (6.1) it was always found that

$$Z_1 + Z_2 = Z_3 + Z_4 \tag{6.19}$$

which means that the total charge is preserved.

6.2.3 Conservation of Angular Momentum

For the reaction (6.3) the total angular momentum J is the sum of all individual angular momenta:

$$\begin{aligned} \boldsymbol{J}(\mathrm{a}+\mathrm{X}) &= \boldsymbol{I}_\mathrm{a} + \boldsymbol{I}_\mathrm{X} + \boldsymbol{L}_\mathrm{aX} \\ &= \boldsymbol{J}(\mathrm{b}+\mathrm{Y}) = \boldsymbol{I}_\mathrm{b} + \boldsymbol{I}_\mathrm{Y} + \boldsymbol{L}_\mathrm{bY}, \end{aligned} \tag{6.20}$$

where $\boldsymbol{L}_{aX}$ is the orbital angular momentum of the projectile with reference to the target $\boldsymbol{X}$ and $\boldsymbol{I}_i$ are the nuclear spins of the particles involved in the reaction. With the momentum $\boldsymbol{p}_a$ of the projectile a and the impact parameter b_a the orbital angular momentum is $|L_{aX}| = p_a \cdot b_a$.

The maximum orbital angular momentum L_{aX}^{max} of a nuclear reaction is determined by the kinetic energy E_{kin} of the projectile and the maximum range R of the interaction energy, where the reaction can be still initiated. With the impact parameter b we obtain in the laboratory system

$$\begin{aligned} &|L_{aX}| = p_a \cdot b < L_{aX}^{max} = p_a \cdot R \\ &\text{with } L = \hbar\sqrt{l(l+1)} \text{ and } p_a = \sqrt{2m_a \cdot E_{kin}}; \\ &\Rightarrow L_{aX}^{max} = \hbar\sqrt{l_{max}(l_{max}+1)} \le R \cdot \sqrt{2m_a \cdot E_{kin}} \\ &\Rightarrow l_{max} \le \frac{R}{\hbar}\sqrt{2m_a \cdot E_{kin}}. \end{aligned} \tag{6.21}$$

In the center of mass system the mass m_a is replaced by the reduced mass $\mu = m_a \cdot m_X/(m_a + m_X)$. The orbital angular momentum is then referred to the center of mass.

For the nucleon-nucleon scattering (a = n, X = p) Is $R = r_p + r_n \approx \sim 2.6 \cdot 10^{-15}$ m, $m_a = 1.67 \cdot 10^{-27}$ kg. Inserting these numerical values into (6.21) gives

$$l_{max}^{lab} \le 1.5 \cdot 10^6 \sqrt{E_{kin}/J} \approx 0.6 \cdot \sqrt{E_{kin}/\mathrm{MeV}}$$

In the lab-system and

$$l_{max}^{S} \le 0.4\sqrt{E_{kin}^{S}/\mathrm{MeV}}$$

In the center of mass system.

Example

1. With $R = 2.6$ fm, $E_{kin} = 1$ MeV $\rightarrow l_{max} \le 0.6$, which means that only S-scattering (with $l = 0$) can occur. For $E_{kin} = 100$ MeV we obtain $l_{max} = 6$.
2. For lead target nuclei the range of the nuclear forces is $R = r_0 \cdot (1 + A^{1/3}) \approx 9.5$ fm and the maximum angular momentum quantum number l becomes $l_{max} = 7$ for $E = 10$ MeV.

6.2.4 Conservation of Parity

The parity P describes the behavior of the wave-function under mirror imaging of all coordinates at the origin.

In a spherical symmetric potential the wave-function can be split into a radial part and an angular part

$$\psi(r, \vartheta, \varphi) = R(r)\, Y(\vartheta, \varphi).$$

Since $R(r)$ is invariant against mirror imaging at the origin the parity is solely determined by the angular part $Y_l^m(\vartheta, \varphi)$. It is

$$Y_l^m(\pi - \vartheta, \varphi + \pi) = (-1)^l Y_l^m(\vartheta, \varphi).$$

States with even l have therefore even parity P, those with odd l have odd parity.

For all processes induced by the strong force the parity remains preserved. It is

$$\begin{aligned} P_{a+X} &= P_a \cdot P_X \cdot (-1)^{l_{aX}} \\ &= P_{b+Y} = P_b \cdot P_Y \cdot (-1)^{l_{bY}}, \end{aligned} \tag{6.22}$$

where P_a and P_X are the "internal" parities of the particles involved in the reaction, which depend on the spins of the particles.

Example Elastic scattering of protons at neutrons ($I_a = I_X = ½$). Since the nature of the particles is not changed in the reaction, it must be.

$$P_a = P_X = P_b = P_Y,$$

and (6.22) becomes

$$(-1)^{l_{aX}} = (-1)^{l_{bY}}$$
$$\Rightarrow \Delta l = l_{aX} - l_{bY} = \text{gerade}$$

Only such processes are possible where the quantum number l of the angular momentum changes by even numbers, i.e. the parity remains constant.

Since for a spin flip of a particle the spin quantum number changes by$\Delta I = 1$, such processes with spin flip are forbidden by parity conservation rules. Because the total angular momentum must be conserved (see (6.20)) only $\Delta l = 0$ and $\Delta I = 0$ or $\Delta l = 2$ and $\Delta I = -2$ or $\Delta l = -2$ and $\Delta I = +2$ can occur. However, since for $\Delta I = \pm 2$ the spins of both nuclei must flip, this process is very unlikely.

All of these conservation laws restrict nuclear reactions which would be allowed by energetic reasons.

6.3 Special Collision-Induced Nuclear Reactions

We will now illustrate the above selection rules and the energy balance by some specific examples.

6.3.1 The (α, p)-Reaction

The historical first artificial nuclear conversion of the type

$$^4_2\mathrm{He} + ^A_Z\mathrm{X} \rightarrow ^{A+4}_{Z+2}\mathrm{Y}^* \rightarrow ^{A+3}_{Z+1}\mathrm{Y} + ^1_1\mathrm{H} + Q \qquad (6.23)$$

was discovered by *E. Rutherford*, when he bombarded nitrogen atoms in a cloud chamber by α-particles.

$$\alpha + ^{14}_7\mathrm{N} \rightarrow ^{17}_8\mathrm{O} + \mathrm{p} \qquad (6.23a)$$

In this experiment a nitrogen nucleus is converted into an oxygen nucleus. This historical cloud chamber photograph is shown in Fig. 6.6. The quantity

$$Q = E_{kin}(\alpha) + [M(\alpha) + M(X) - M(Y) - M(p)]c^2$$

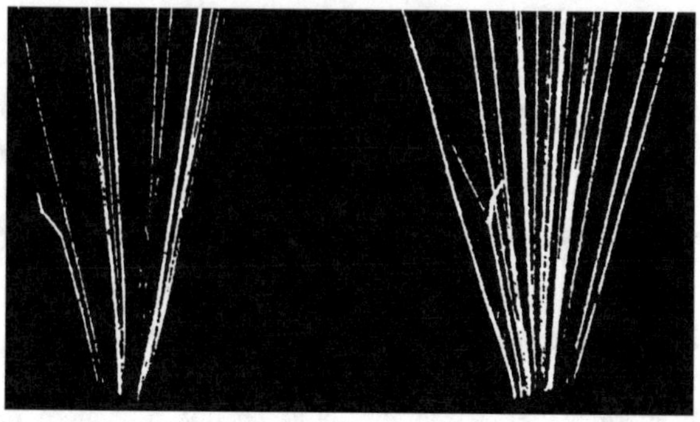

Fig. 6.6 Two stereographic cloud chamber photos of the first by Rutherford discovered nuclear conversion at the reaction $\alpha + ^{14}_7\mathrm{N} \rightarrow ^{17}_8\mathrm{O} + \mathrm{p}$. The α-particles enter the cloud chamber from below. The thin spur is due to the proton, the thick spur is caused by the recoil of the O-nucleus

gives the energy balance of the reaction.

Further examples of (α, p)-reactions are:

$$\begin{aligned} &^4_2\mathrm{He} + ^{10}_5\mathrm{B} \rightarrow ^{14}_7\mathrm{N}^* \\ &\quad \rightarrow ^{13}_6\mathrm{C} + ^1_1\mathrm{H} + 4.04\,\mathrm{MeV}, \\ &^4_2\mathrm{He} + ^{27}_{13}\mathrm{Al} \rightarrow ^{31}_{15}\mathrm{P}^* \\ &\quad \rightarrow ^{30}_{14}\mathrm{Si} + ^1_1\mathrm{H} + 2.26\,\mathrm{MeV}, \\ &^4_2\mathrm{He} + ^{32}_{16}\mathrm{S} \rightarrow ^{36}_{18}\mathrm{Ar}^* \\ &\quad \rightarrow ^{35}_{17}\mathrm{Cl} + ^1_1\mathrm{H} - 2.10\,\mathrm{MeV}. \end{aligned} \qquad (6.24)$$

The reaction energy Q can appear as kinetic energy or as excitation energy of the reaction products. This is illustrated in Fig. 6.7 for the reaction.

$$\alpha + ^{27}_{13}\mathrm{Al} \rightarrow ^{31}_{15}\mathrm{P}^* \rightarrow ^{30}_{14}\mathrm{Si} + \mathrm{p}$$

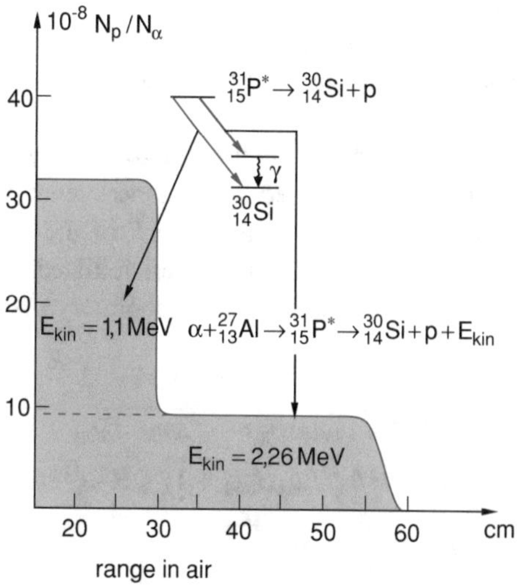

Fig. 6.7 Energy distribution of the protons (measured by their range in air) for the reaction $\alpha + ^{27}_{13}\mathrm{Al} \rightarrow ^{30}_{14}\mathrm{Si} + \mathrm{p}$.

where the kinetic energy of the protons is deduced from the range of the protons in air at atmospheric pressure. One can recognize two different groups of protons with ranges of 28 and 58 cm which correspond to the energies

$$E_{kin_1} = 1.1\ \text{MeV} \quad \text{and} \quad E_{kin_2} = 2.26\ \text{MeV}.$$

This proves that the silicon nucleus is produced in this reaction in at least two different energetic levels.

The excitation energy $E_a = E_{kin_2} - E_{kin_1}$ can be emitted as γ-quantum with the energy $h \cdot \nu = E_a$. This was also observed experimentally.

6.3.2 The (α, n)-Reaction.

If beryllium-nuclei are bombarded by α-particles, neutrons are released according to the reaction

$$\alpha + {}^9_4\text{Be} \to {}^{13}_6\text{C}^* \to {}^{12}_6\text{C} + \text{n}. \tag{6.25}$$

For many of such (α, n)- reactions the nuclei are formed in excited states. The emitted neutrons therefore appear in groups with different kinetic energies. In Fig. 6.8 the energy distribution of neutrons is shown which is observed when ${}^9_4$Be-nuclei are bombarded by α-particles from a radioactive polonium source. The neutron energies are deduced from the measured recoil energies of the protons during the deceleration of the neutrons in paraffin.

Further examples of (α, n)-reactions are

$$\begin{aligned}
&\alpha + {}^7_3\text{Li} \to {}^{11}_5\text{B}^* \to {}^{10}_5\text{B} + \text{n},\\
&\alpha + {}^{11}_5\text{B} \to {}^{15}_7\text{N}^* \to {}^{14}_7\text{N} + \text{n},\\
&\alpha + {}^{19}_9\text{F} \to {}^{23}_{11}\text{Na}^* \to {}^{22}_{11}\text{Na} + \text{n},\\
&\alpha + {}^{27}_{13}\text{Al} \to {}^{31}_{15}\text{P}^* \to {}^{30}_{15}\text{P} + \text{n}.
\end{aligned} \tag{6.26}$$

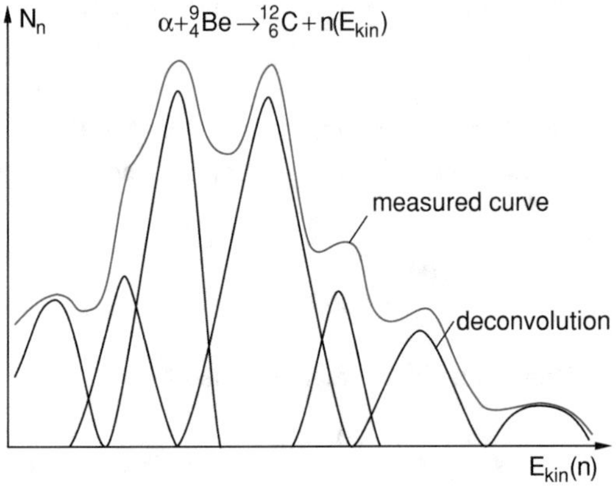

Fig. 6.8 Energy groups of neutrons for the reaction ${}^9_4\text{Be}(\alpha,\text{n}){}^{12}_6\text{C}$ (after Whitmore and Baker Phys. Rev. 78, 799 (1950)

6.4 Collision-Induced Radioactivity

The collision-induced (artificial) radioactivity was first discovered 1934 by Irene Curie and Frederic Joliot-Curie, when they bombarded light nuclei by α-particles. They found that the bombarded substances emit β^+ and β^-- particles and γ-quanta. This emission continues also after the end of the bombardment. This phenomenon was found for the following reactions:

$$\begin{aligned}
&\alpha + {}^{10}_5\text{B} \to {}^{14}_7\text{N}^* \to {}^{13}_7\text{N} + \text{n}\\
&{}^{13}_7\text{N} \to {}^{13}_6\text{C} + \beta^+ + \nu(\tau = 9.96\ \text{min}),
\end{aligned} \tag{6.27a}$$

$$\begin{aligned}
&\alpha + {}^{27}_{13}\text{Al} \to {}^{31}_{15}\text{P}^* \to {}^{30}_{15}\text{P} + \text{n}\\
&{}^{30}_{15}\text{P} \to {}^{30}_{14}\text{Si} + \beta^+ + \nu(\tau = 2.5\ \text{min}).
\end{aligned} \tag{6.27b}$$

The nuclei formed by α-bombardment are unstable and decay by emission of β^- or β^+ particles.

To distinguish the *natural radioactive substances* existing in nature from these collision-induced radioactive products the latter are called *artificial radioactive substances.* Meanwhile there are a large number of artificial radioactive substances which are used in medicine, biology and for the solution of technical problems.

Most of the artificial radioactive nuclei are formed in nuclear reactors as fission products (see Sect. 6.5) or by neutron bombardment of samples which are brought into the reactor kernel in specially designed tubes.

For example many (n.γ) reactions where thermal neutrons ($E_{kin} \approx 0.03$ eV) are captured by nuclei brought into the reactor core with a high flux of neutrons, produce artificial radioactive nuclei.

One example is the reaction

$$\begin{aligned}
&{}^{23}_{11}\text{Na} + \text{n} \xrightarrow[\sigma=53\ \text{fm}^2]{} {}^{24}_{11}\text{Na}^*\\
&\underset{\gamma}{\to} {}^{24}_{11}\text{Na} \xrightarrow[1,39\text{MeV}]{\beta-} {}^{24}_{12}\text{Mg}^* \to {}^{24}_{12}\text{Mg} + \gamma,
\end{aligned} \tag{6.28a}$$

The β^--emitting nuclide ${}^{24}_{11}$Na has a half-lifetime of 14.96 h (Fig. 6.9).

For cancer therapy the radioactive isotope ${}^{60}_{27}$Co is used which is produced by bombardement with neutrons according to the reaction

$$\begin{aligned}
&{}^{59}_{27}\text{Co} + \text{n} \xrightarrow[\sigma=3700\ \text{fm}^2]{} {}^{60}_{27}\text{Co}^*\\
&\xrightarrow{\gamma_1,\beta-} \frac{60}{28}\text{Ni}^* \to {}^{60}_{28}\text{Ni} + \gamma_2 + \gamma_3
\end{aligned} \tag{6.28b}$$

Further examples are the reactions ${}^{14}_7$N (n, p) ${}^{14}_6$C or ${}^{36}_{17}$Cl (n, p) ${}^{35}_{16}$S, following the sequence

$$\begin{aligned}
&{}^{14}_7\text{N} + \text{n} \xrightarrow{\sigma=181\ \text{fm}^2} {}^{14}_6\text{C}^* + \text{p}\\
&{}^{14}_6\text{C}^* \xrightarrow[\tau=5730\ \text{a}^{\beta^-(0.15\ \text{MeV})}\ {}^{14}_7\text{N},]{} .
\end{aligned} \tag{6.29a}$$

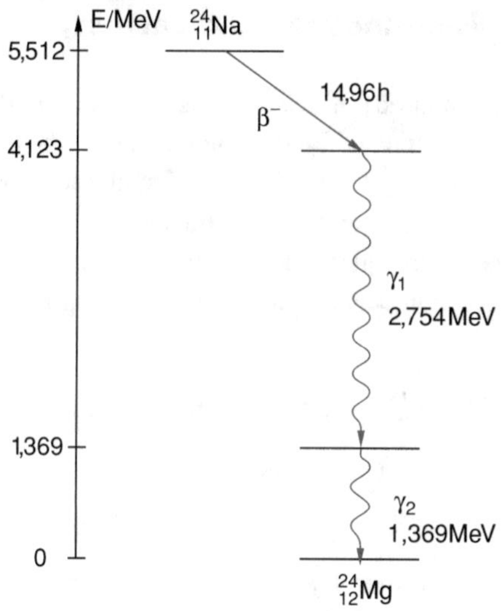

Fig. 6.9 Decay scheme of the excited radioactive isotope $^{24}_{11}\mathrm{Na}^*$

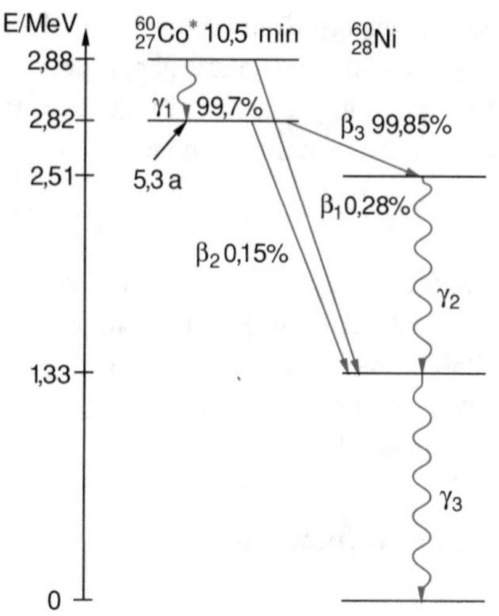

Fig. 6.10 Decay scheme of the radio-active nucleus $^{60}_{27}\mathrm{Co}$ produced by the reaction (6.28b)

$$^{35}_{17}\mathrm{Cl} + \mathrm{n} \xrightarrow{\sigma=7.8\,\mathrm{fm}^2} {}^{35}_{16}\mathrm{S}^* + \mathrm{p} \tag{6.29b}$$

$$^{35}_{16}\mathrm{S}^* \xrightarrow[\tau=87.5\,\mathrm{d}]{\beta^-(0.167\,\mathrm{MeV})} {}^{35}_{17}\mathrm{Cl}. \tag{6.29c}$$

Under bombardment by α-particles not only the (α, n) or (α, p) reactions can occur but the excited compound nucleus can also decay by γ-emission and return into its ground state. This stabilizes the compound nucleus, with the result that besides the γ-quanta no other particles are emitted in the exit channel.

One example is the reaction

$$\alpha + {}^{7}_{3}\mathrm{Li} \rightarrow {}^{11}_{5}\mathrm{B}^* \rightarrow {}^{11}_{5}\mathrm{B} + \gamma. \tag{6.30}$$

Figure 6.11 shows the yield of γ-quanta as a function of the kinetic energy of the α-particles. At sharply defined energies (0.4, 0.82, and 0.96 MeV) a steep increase of the yield is observed accompanied by the emission of γ-quanta with the corresponding energies (Fig. 6.11b). This is due to the resonance capture of the α-particles by the Li-nucleus. At these resonance energies E_r observed in the reaction $\alpha + {}^{7}_{3}\mathrm{Li} \rightarrow {}^{11}_{5}\mathrm{B}^*(E_r)$ energy levels of the compound nucleus $^{11}_{5}\mathrm{B}^*$ are excited where the capture- cross section of the α-particles reaches a maximum.

The activity of the artificial radio-nuclides can be obtained by the following considerations:

With the neutron flux density Φ in the reaction volume and the excitation cross section σ_a the creation rate dN/dt of the radio-nuclides from N_0 stable parent nuclides is

$$\frac{dN}{dt} = \sigma_a \cdot \Phi \cdot N_0. \tag{6.31a}$$

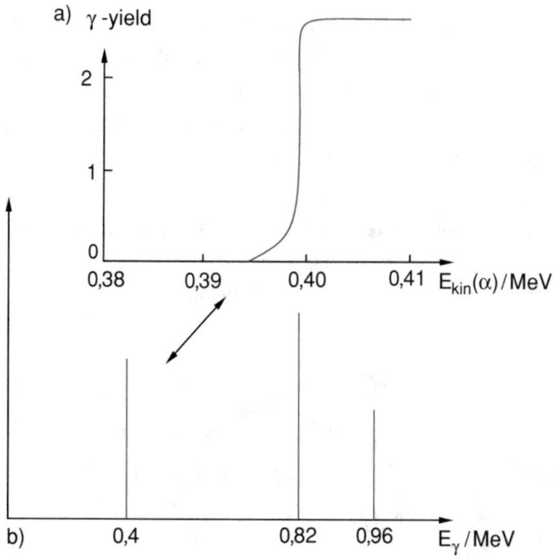

Fig. 6.11 **a** Yield of γ-quanta in the reaction $^{7}_{3}\mathrm{Li}(\alpha,\beta)\ {}^{11}_{5}\mathrm{B}$ as a function of the kinetic energy of the α-particles **b** γ-spectrum of the excited $^{11}_{5}\mathrm{B}^*$ nucleus for the energy E_{kin} = 1.5 MeV of the α-particles

The created unstable radio-nuclides decay with the decay constant λ (see Sect. 3.2). The true decay rate of the parent nuclides is then

$$\frac{dN}{dt} = \sigma_a \Phi N_0 - \lambda \cdot N. \tag{6.31b}$$

Integration over the irradiation time t_B gives

$$N(t_B) = \frac{\sigma_a \cdot \Phi \cdot N_0}{\lambda}\left(1 - e^{-\lambda \cdot t_B}\right). \tag{6.32}$$

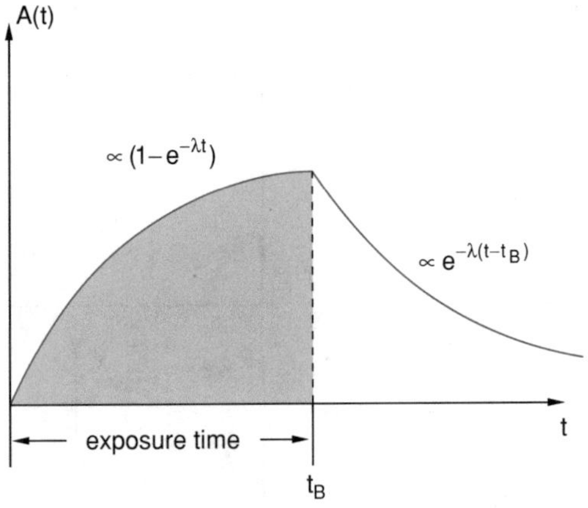

Fig. 6.12 Time progression of the activity *A(t)* of artificial radio-active nuclei during and after the irradiation

The activity $A = \lambda \cdot N$ of the irradiated sample is then

$$A(t_B) = \sigma_a \cdot \Phi \cdot N_0\left(1 - e^{-\lambda t_B}\right). \quad (6.33)$$

After the end of the irradiation the activity of the sample decays exponentially with the decay constant λ (Fig. 6.12).

6.5 Nuclear Fission

By collisions with suitable projectiles nuclei can be split into two or more fragments. For very heavy nuclei also spontaneous fission, i.e. without external energy supply, has been observed. However, such spontaneous fission can be nowadays only observed, if the fission probability is very small, because otherwise the radio-active elements formed at the origin of our solar system would have been already disappeared by fission.

6.5.1 Spontaneous Nuclear Fission

Most nuclei are nearly spherical symmetric. They can only split if their size is deformed into a non-spherical elliptical configuration. This deformation demands the supply of energy. In Fig. 6.13 the potential energy is plotted schematically as a function of the deformation parameter ε (see Sect. 5.6.1 and Problems 5.3 and 6.5).

The energy balance can be illustrated by the droplet model of nuclei (Sect. 2.6.3). The two essential parts of the total binding energy (2.45) which change with the nuclear deformation are the surface energy and the Coulomb energy.

When the spherical size of a nucleus with radius R is changed into a rotational elliptical form with the axes

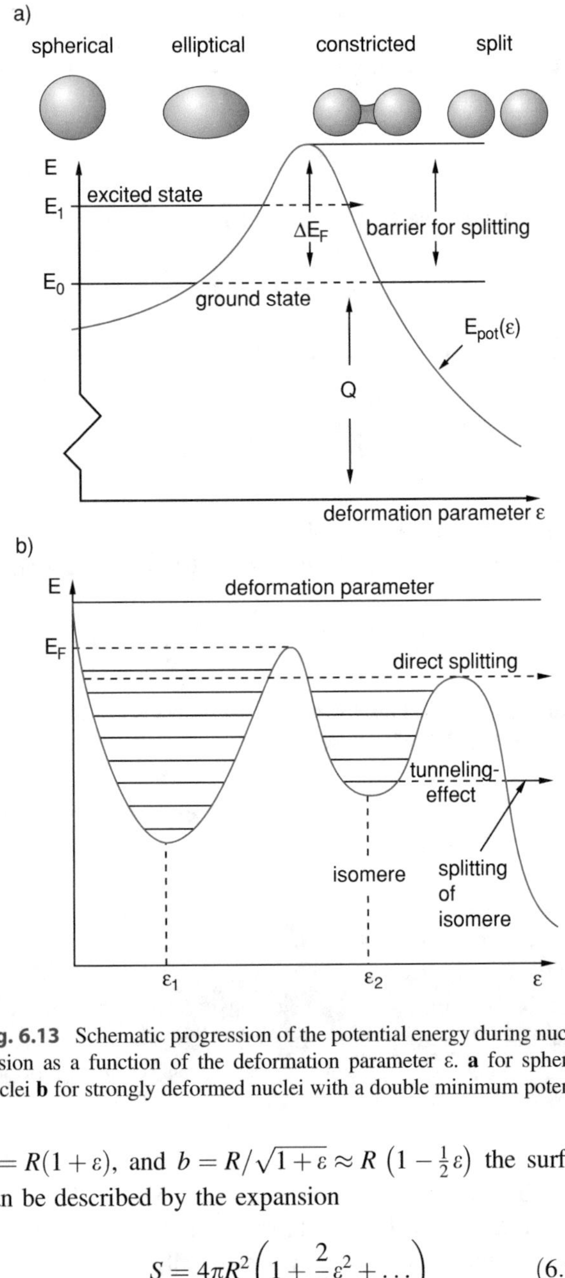

Fig. 6.13 Schematic progression of the potential energy during nuclear fission as a function of the deformation parameter ε. **a** for spherical nuclei **b** for strongly deformed nuclei with a double minimum potential

$a = R(1+\varepsilon)$, and $b = R/\sqrt{1+\varepsilon} \approx R\left(1 - \frac{1}{2}\varepsilon\right)$ the surface can be described by the expansion

$$S = 4\pi R^2\left(1 + \frac{2}{5}\varepsilon^2 + \ldots\right) \quad (6.34)$$

The surface energy therefore increases during the deformation from E_0^S to

$$E_e^S = E_0^S\left(1 + \frac{2}{5}\varepsilon^2\right) = E_0^S + \Delta E^S \quad (6.35)$$

The Coulomb energy, on the other hand, decreases during the deformation (Problem 6.5) to

$$E_e^C = E_0^C\left(1 - \frac{1}{5}\varepsilon^2\right) = E_0^C - \Delta E^C. \quad (6.36)$$

The nucleus remains stable if for small deformations

$$\Delta E^{\mathrm{C}} \leq \Delta E^{\mathrm{S}} \tag{6.37a}$$

With $\Delta E^{\mathrm{C}} = \frac{1}{5}\varepsilon^2 E_0^{\mathrm{C}}$ and $\Delta E^{\mathrm{S}} = \frac{2}{5}\varepsilon^2 E_0^{\mathrm{S}}$ this gives the condition for spontaneous fission

$$E_0^{\mathrm{C}} \geq 2E_0^{\mathrm{S}}. \tag{6.37b}$$

Introducing the fission parameter

$$X_{\mathrm{S}} = \frac{1}{2} E_0^{\mathrm{C}} / E_0^{\mathrm{S}} \tag{6.38}$$

we can formulate the condition for fission as:

Nuclei with $X_{\mathrm{S}} \geq 1$ can undergo spontaneous fission.

With the numerical values of surface- and Coulomb energy given in Eq. (2.46) the fission parameter becomes

$$X_S = \frac{a_C \cdot Z^2/A^{1/3}}{2a_S \cdot A^{2/3}} = \frac{a_C}{2a_S}\frac{Z^2}{A}. \tag{6.39}$$

Inserting the values from (2.46) with a_C = 0.714 MeV/c^2 and a_s = 18.33 MeV/c^2 we obtain

$$X_S \geq 1 \quad \text{for} \quad \frac{Z^2}{A} \geq 51. \tag{6.40}$$

Nuclei with $Z^2/A > 51$ split spontaneously and are therefore nowadays no longer present. They can be, however, formed by collisions between small nuclei (artificial fusion), but have a short lifetime. Nuclei with $Z^2/A < 51$ can split if one supplies the necessary energy ΔE_{F}.

Note: Because of the tunnel effect (see Vol. 3, Sect. 3.4) nuclei can spontaneously split even for $X_S < 1$ (Fig. 6.14). However, the probability decreases rapidly with decreasing values of X_S. because the fragments have large masses and therefore the tunnel probability is very small. In such cases the probability for α-emission will be often much larger than that for fission. In Table 6.1 some numerical values for spontaneous fission and for α-decay are listed.

6.5.2 Collision-Induced Fission of Light Nuclei

The first fission of light nuclei was discovered 1932 by *Cockcroft* and *Walton* when they bombarded $^7_3$Li with protons ($E_{\mathrm{kin}} \leq$ 0.5 MeV) and observed α-particles according to the reaction

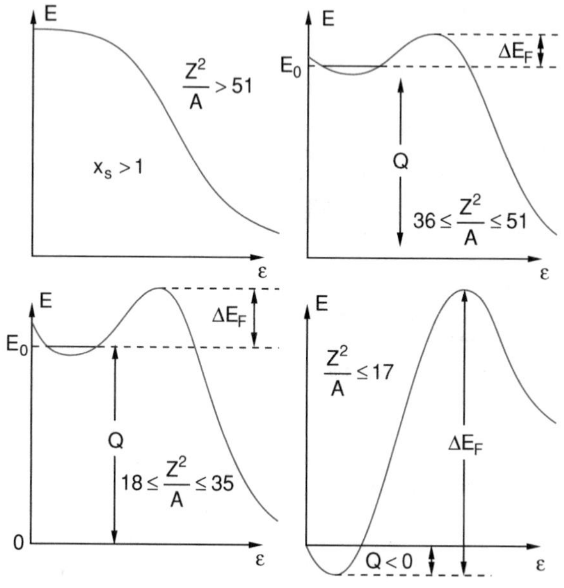

Fig. 6.14 Potential barriers ΔE_f for nuclear fission for different values of the ratio Z^2/A as a function of the deformation paramter

Table 6.1 Fission parameters and half-lifetimes ½ for the spontaneous fission and the α-decay of some nuclei

Nuclide	X_s	$T_{½}$ (fission)	$T_{½}$ (α-decay)
$^{232}_{90}$Th	0.68	>10^{19} a	1.4 · 10^{10} a
$^{235}_{92}U$	0.70	~10^{17} a	7 · 10^8 a
$^{238}_{92}U$	0.693	~10^{16} a	4 · 10^9 a
$^{242}_{94}$Pu	0.71	~10^{11} a	~4 · 10^5 a
$^{252}_{98}$Cf	0.74	6.10^1 a	2.2 a
$^{254}_{100}$Fm	0.76	246 d	3.4 h
$^{255}_{102}$No	0.80	?	180 s

$$\mathrm{p} + {}^7_3\mathrm{Li} \rightarrow {}^8_4\mathrm{Be}^* \rightarrow \alpha + \alpha + Q. \tag{6.41}$$

The two α-particles had a range of 8.3 cm in air corresponding to a kinetic energy of 8.63 MeV. The reaction heat Q of this reaction is therefore

$$Q = 17.26\ \mathrm{MeV}.$$

With sufficiently large kinetic energy of charged projectiles also heavier nuclei can be split. Examples are the fission- reactions

$${}^{63}_{29}\mathrm{Cu} + \mathrm{p} \begin{array}{l} \nearrow {}^{38}_{17}\mathrm{Cl} + {}^{25}_{13}\mathrm{Al} + \mathrm{n} \\ \searrow {}^{24}_{11}\mathrm{Na} + {}^{39}_{19}\mathrm{K} + \mathrm{n}, \end{array} \tag{6.42}$$

induced by fast protons. The cross sections of these reactions are plotted in Fig. 6.15 as a function of the proton energy. The

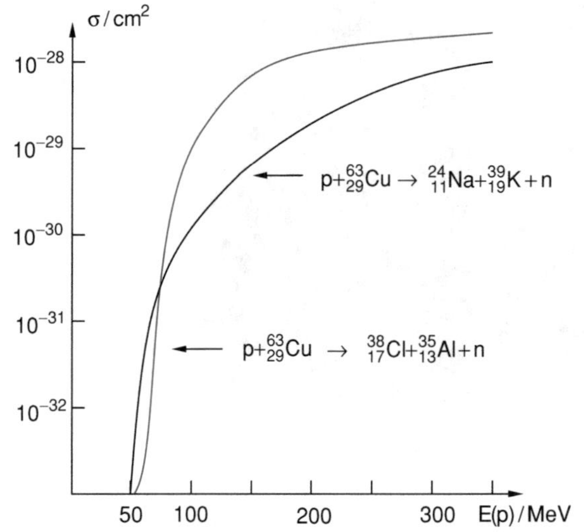

Fig. 6.15 Cross section σ(E) for the fission of Cu-nuclei by fast protons with energy E. (Batzel, Seaborg Phys. Rev. 82, 609 1952)

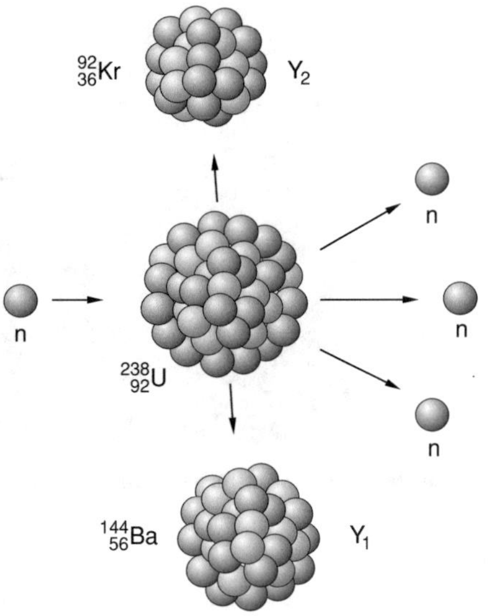

Fig. 6.16 Schematic vivid representation of the neutron-induced nuclear fission

curves show that the threshold energy is about 50–60 MeV, because the protons have to overcome the Coulomb barrier.

Particularly effective projectiles for the collision-induced fission are neutrons because for them there is no Coulomb barrier. Therefore also slow neutrons can induce nuclear fission.

Examples are the reactions

$$n + {}^{6}_{3}\mathrm{Li} \rightarrow {}^{7}_{3}\mathrm{Li}^{*} \rightarrow {}^{3}_{1}\mathrm{H} + {}^{4}_{2}\mathrm{He} \qquad (6.43a)$$

$$n + {}^{10}_{5}\mathrm{B} \rightarrow {}^{11}_{5}\mathrm{B}^{*} \rightarrow {}^{7}_{3}\mathrm{Li} + {}^{4}_{2}\mathrm{He}. \qquad (6.43b)$$

The (n, α) boron reaction has a very large cross section and is therefore often used as detection technique for neutrons, or also for the controlled absorption of neutrons in nuclear reactors, where it is used for the control of the fission rate.

6.5.3 Collision-Induced Fission of Heavy Nuclei

Based on the results of many experiments *Enrico Fermi* (1901–1954) and his coworkers tried to generate new elements by bombardment of Uranium with neutrons. They found β-active radioactive reaction products which they attributed to trans-Uranium elements with $Z \geq 93$.

Very careful chemical investigations of such reaction products 1939 by *Otto Hahn* (1879–1968) and *Fritz Straßman* (1902–1980) revealed, however, that the element Barium was undoubtedly one of the reaction products Y in the reaction

$$n + {}^{238}_{92}\mathrm{U} \rightarrow \mathrm{Y}_1 + \mathrm{Y}_2 + \nu \cdot n \qquad (6.44)$$

Lise Meitner (1878–1968) first recognized that the reaction (6.44) described the fission of the uranium nucleus by neutrons into two fragments Y_1 and Y_2 with comparable masses and which emit β^-particles (electrons) because of the large neutron surplus.

The fission reaction proceeds according to the scheme (Fig. 6.16).

$$n + {}^{238}_{92}\mathrm{U} \rightarrow \left({}^{239}_{92}\mathrm{U}^{*}\right) \rightarrow \mathrm{Y}_1^{*} + \mathrm{Y}_2 + \nu \cdot n \qquad (6.45)$$

It starts at the threshold energy $E_{\mathrm{kin}} \geq 1$ MeV (kinetic energy of the neutrons).

The Uranium isotope ${}^{235}_{92}\mathrm{U}$ can be already split by slow neutrons with a very much larger cross section (Fig. 6.17). The reason for this large cross section is the following:

At the collision-induced nuclear fission an excited intermediate nucleus is formed, where the excitation energy is provided by the kinetic energy of the projectile and by its binding energy in the compound nucleus. Nuclear fission can occur if this excitation energy is larger than the critical energy $E_c = \Delta E_F$, which is necessary to overcome the potential barrier ΔE_F in Fig. 6.13a. In Table 6.2 these critical energies are listed for some fissionable nuclei.

Example

1. With the attachment of a neutron by the ${}^{239}_{92}\mathrm{U}$ nucleus the g-u- nucleus ${}^{239}_{92}\mathrm{U}$ is formed with a binding energy of the neutron which is smaller than for g-g nuclei (see Sect. 2.6), The binding energy can be deduced from the mass difference between initial and final nuclei.

$$E_B = \left[m\left({}^{238}_{92}\mathrm{U}\right) + m_n - m\left({}^{239}_{92}\mathrm{U}\right)\right]c^2 = 5.2\ \mathrm{MeV}.$$

This is not enough to cause the fission of ${}^{239}_{92}\mathrm{U}$ because the critical minimum energy is $E_c = 5.9$ MeV. The

Fig. 6.16X Otto Hahn, Fritz Strassmann and Lise Meitner

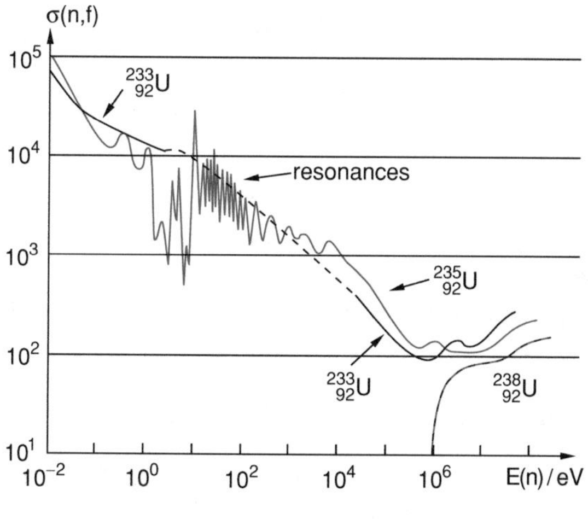

Fig. 6.17 Fission cross section $\sigma_f(U, n, E)$ as a function of the neutron energy for the uranium isotopes ${}^{238}_{92}U$; ${}^{135}_{92}U$; and ${}^{233}_{92}U$

Table 6.2 Critical energy E_c (heights of the potential barrier), neutron binding energy E_b in the compound nucleus and fission threshold energy $\Delta E_F = E_c–E_b$ of the kinetic energy of the neutrons inducing the fission

Targetkern X	Compoundkern X + n	E_c (MeV)	E_b (MeV)	$E_c–E_b$ (MeV)
${}^{233}{}_{92}U$	${}^{234}{}_{92}U$	5,8	7,0	−1,2
${}^{235}{}_{92}U$	${}^{236}{}_{92}U$	5,3	6,4	−1,1
${}^{234}{}_{92}U$	${}^{235}{}_{92}U$	5,8	5,3	+ 0,5
${}^{238}{}_{92}U$	${}^{239}{}_{92}U$	6,1	5,0	+ 1,1
${}^{231}{}_{91}$ Pa	${}^{232}{}_{91}$ Pa	6,2	5,5	+ 0,7
${}^{232}{}_{90}Th$	${}^{233}{}_{92}Th$	6,8	5,5	+ 1,3

neutrons must therefore have a minimum kinetic energy of 0.7 MeV in order to split the ${}^{238}_{92}U$ nucleus.

2. The integration of a neutron by the nucleus ${}^{235}_{92}U$ forms the g-g nucleus ${}^{236}_{92}U$ with a larger binding energy $(E_B = 6.4\ \text{MeV} > E_c = 5.3\ \text{MeV})$ of the neutron. Therefore the capture of slow neutrons by the ${}^{236}_{92}UU$ nucleus leads already to the fission of the ${}^{236}_{92}U$ nucleus. The cross section for this neutron capture with subsequent fission increases steeply with decreasing neutron energy. The fission cross section for low thermal neutrons is therefore for ${}^{235}_{92}U$ larger by about one order of magnitude than that for ${}^{233}_{92}U$ by neutrons with large kinetic energies. (see Sect. 8.3).

The mass distribution of the fission products shows that the fission occurs generally into two products with slightly different masses (Fig. 6.18). Of course the mass balance must obey the conservation of the total mass ${}^{A_1}_{Z_1}X_1, {}^{A_2}_{Z_2}X_2$ and follows the reaction.

$$A_1 + A_2 + \nu = A \quad \text{and} \quad Z_1 + Z_2 = Z. \tag{6.46}$$

One of the possible fission reactions is for example

$$\begin{aligned} n + {}^{235}_{92}U &\rightarrow {}^{236}_{92}U^* \\ &\rightarrow {}^{141}_{56}Ba + {}^{92}_{56}Kr + 3n + Q. \end{aligned} \tag{6.47}$$

Heavy nuclei can be also split by charged projectiles if their energy is sufficiently large to overcome the Coulomb barrier, because they must penetrate into the nucleus in order to initiate the fission. With increasing energy of the projectiles the mass distribution of the fission products becomes more symmetric (Fig. 6.19).

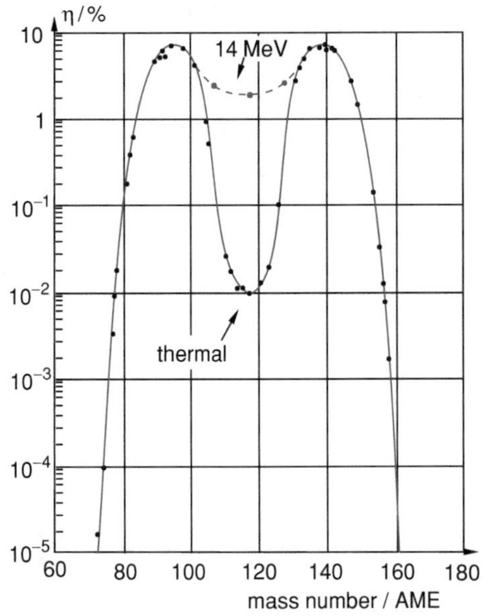

Fig. 6.18 Fission probability η in % as a function of the mass number of the fission products for the fission of $^{235}_{92}U$ by slow thermal neutrons and by fast (E_{kin} = 14 MeV) neutrons

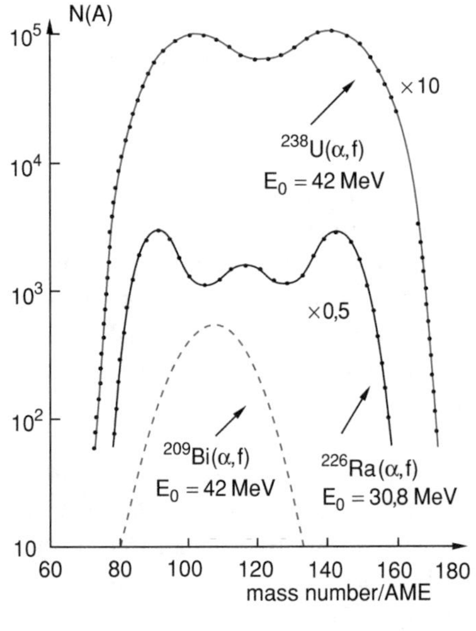

Fig. 6.19 Mass distribution of the fission products for some nuclear fission reactions induced by fast α-particles with the kinetic energy$E_0(\alpha)$. (after R. Vandenbosch, J. R. Huzenga: Nuclear Fission, Academic Press New York 1973)

Heavy nuclei can be split also by photons, (γ-quant $h \cdot \nu$) of sufficient energy $h \cdot \nu$, which can be generated by irradiation of tungsten with electrons from an electron synchrotron. Since the energy of the γ-quanta can be continuously varied by changing the electron energy in the synchrotron, the threshold energy for photon-induced nuclear fission can be determined very accurately.

6.5.4 Energy Balance of Nuclear Fission

From the mass difference between initial and final reaction partners in the reaction (6.47) the energy balance and the reaction heat Q can be deduced. This gives the very large value of Q = 180 MeV. The main part of this energy appears as kinetic energy of the fission products (167 MeV), a minor part as kinetic energy of the neutrons (6 MeV). The energy distribution $N_n(E_{kin})$ of the fission neutrons (Fig. 6.20) can be approximately described by the function

$$N(E) = C \cdot \sinh \sqrt{E_{kin}/\text{MeV}} \cdot e^{-E_{kin}/\text{MeV}}. \tag{6.48}$$

The fission products are generally excited and show a neutron excess. They give their excitation energy away by emission of γ-radiation and their neutron excess by emission of β^-radiation, i.e they emit electrons.

$${}^{A_X}_{Z_X}X^* \rightarrow {}^{A_X}_{Z_{X+1}}\text{Y} + \beta^- + \bar{\nu}. \tag{6.49}$$

This β^- decay is accompanied by emission of anti-neutrinos. which escape and carry their kinetic energy with them, because their absorption probability is vanishingly small.

Some fission products emit also neutrons.

According to the reaction.

$${}^{A_X}_{Z_X}\text{X}^* \rightarrow {}^{A_X-1}_{Z_X}\text{X} + \text{n}. \tag{6.50}$$

The decay time of this reaction ranges from a few milliseconds up to some days. This neutron emission is therefore delayed against the promped neutron emission during the direct fission of the parent nucleus. These delayed

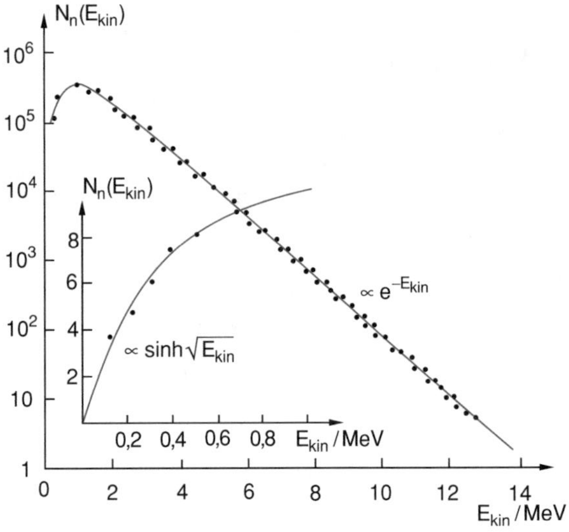

Fig. 6.20 Energy distribution of the fission neutrons

neutron play an important role for the control of nuclear reactors (see Sect. 8.3)

The total energy which is released at the fission of an uranium nucleus is

E_{kin} (fission products)	167 MeV
E_{kin} (fission neutrons)	6 MeV
Direct γ-radiation	7 MeV
Total energy per fission of one U-nucleus	180 MeV
γ -radiation of fission products	6 MeV
β^- radiaton of fission products	5 MeV
Anti-neutrino radiation	10 MeV
Total delayed energy release per nucleus	21 MeV

Altogether the energy release per uranium nucleus is therefore 201 MeV, from which the non-observable energy of the escaping anti-neutrinos has to be subtracted.

The kinetic energy of the fission products can be measured with a time-of-flight method (Fig. 6.21a). A thin foil of fissionable material (e.g. ${}^{235}_{92}UO_3$) is irradiated by thermal neutrons from a nuclear reactor. The two fragments produced during the nuclear fission fly into opposite directions (conservation of momentum). Those fragments flying parallel to the axis of the evacuated time-of-flight-tube are monitored by the detectors D_1 and D_2. (e.g. Scintillation counters Sect. 4.3.2) as short pulses. The distance between the source and D_1 is very small (about 1 cm) while the distance QD_2 is about 350 cm. The exit signal of D_1 starts a linear voltage ramp in a time-to pulse- height converter and that of D_2 stops it. In this way the time of flight is converted into a voltage output. Corresponding to the mass distribution of the fission products one observes two maxima in the time-of-flight distribution, where the lighter product has the higher velocity (Fig. 6.21b).

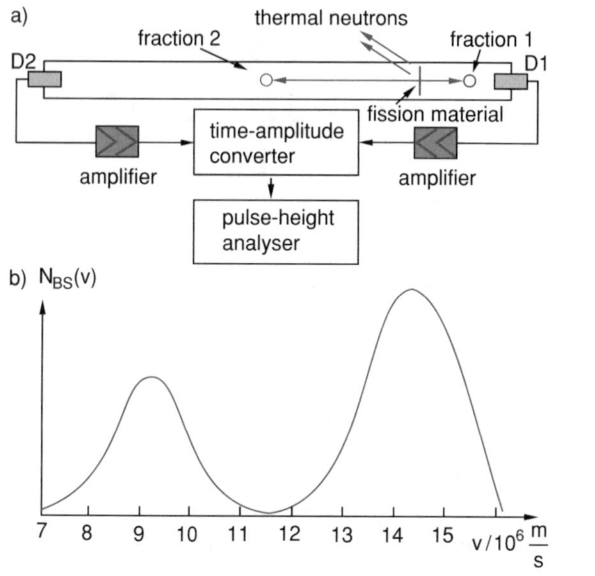

Fig. 6.21 **a** Experimental setup for the measurement of the velocity of the fission products **b** Velocity distribution N(v) of the fission fragments for the fission of ${}^{235}_{92}U$ by thermal neutrons, measured with the time of flight method of (**a**). (after R. B. Leachman Phys. Rev. **87,** 444 1952)

Example For the reaction.

$$ {}^{2}_{1}H + {}^{2}_{1}H \rightarrow {}^{3}_{1}H + p + 3.0\,\mathrm{MeV} $$

Is $a_1 = a_2 = 1.5 \cdot 10^{-15}\mathrm{m}, Z_1 = Z_2 = 1 \Rightarrow E_{kin} \geq \frac{0.5\mathrm{MeV}}{(m/2)v^2}$.

This corresponds to the mean thermal energy $(m/2)v^2$ at a temperature of $6 \cdot 10^9$ K.

Taking into account that the nuclear radius r increases with $A^{1/3}$ the minimum energy for overcoming the Coulomb-barrier increases with $E_{kin} \propto (Z_1 + Z_2)/A^{1/3}$.

Because of the tunnel-effect two nuclei can already fuse at lower energies.

6.6 Nuclear Fusion

The collision between two nuclei can lead to the fusion of the two nuclei into a heavier nucleus. The kinetic energy of the reactants must be sufficiently high in order to overcome the Coulomb barrier:

$$ E_{kin} \geq \frac{Z_1 \cdot Z_2 \cdot e^2}{4\pi\varepsilon_0 (a_1 + a_2)}, $$

where a_i is the range of the nuclear forces in the nucleus with charge $Z_1 \cdot e$.

The probability for fusion decreases exponentially with decreasing kinetic energy of the colliding particles because of the exponentially decreasing tunnel probability. Therefore the probability of fusion increases very steeply with increasing energy (Fig. 6.5).

These considerations show that nuclear fusion occurs mainly for light nuclei, because the Coulomb barrier increases proportional to $(Z_1 \times Z_2)^{2/3}$

The binding energy of nuclei increases with increasing number $N = A$ of nucleons until the iron nucleus at $N_{Fe} = 56$.

Therefore the mass of the fusion reactants $A_1 + A_2$ is always larger than the resultant nucleus with $A = A_1 + A_2$. One wins energy at the fusion of light nuclei with $A < 56$.

Examples of fusion reactions, which also play a role in the interior of stars (see Vol. 5) are

$$p + p \rightarrow {}^2_1H + e^+ + \nu_e + 1.19\ \text{MeV}, \tag{6.51a}$$

$$d + d \rightarrow {}^3_2He + n + 3.25\ \text{MeV}, \tag{6.51b}$$

$${}^3_1H + {}^2_1H \rightarrow {}^4_2He + n + 17.6\ \text{MeV}, \tag{6.51c}$$

$${}^3_1H + {}^3_1H \rightarrow {}^4_2He + 2n + 20.7\ \text{MeV}, \tag{6.51d}$$

$${}^6_3Li + {}^2_1H \rightarrow {}^4_2He + {}^4_2He + 22.4\ \text{MeV}. \tag{6.51e}$$

With the flux density of fusion reactants $n_1(v)$ and $n_2(v)$ in the velocity interval dv (s/m^4) and the cross section $\sigma(v)$ for fusion, which depends on the relative velocity of the collision partners, the number of fusion reactions dN/dt per sec in the volume dV is given by

$$\begin{aligned} \frac{dN}{dt} &= \iint n_1(v_1) \cdot n_2(v_2) \cdot \sigma(\mathbf{v_1} - \mathbf{v_2}) \\ &\cdot |v_1 - v_2| dv_1 dv_2 \\ &\approx \frac{n_2 \cdot n_1}{v_{W_1}^3 \cdot v_{W_2}^3} \cdot \iint v_1^2 v_2^2 e^{-m_1 v_1^2/m_1 v_1^2 (2k_B T)} \\ &\cdot e^{-m_2 v_2^2/m_2 v_2^2 (2k_B T)} \sigma(v) v dv_1 dv_2 \end{aligned} \tag{6.52}$$

where $\mathbf{v} = \mathbf{v_1} - \mathbf{v_2}$ is the relative velocity of the fusion partners.

The energy gain per fusion process of light nuclei is comparable or even larger than that for the fission of heavy nuclei. Therefore great efforts are undertaken in order to realize controlled nuclear fusion in specially designed fusion reactors. The main experimental difficulty is the heating of charged particles up to temperatures of above 10^7 K and to confine them for a sufficiently long time in a limited volume and prevent them of reaching the walls of the container (see Sect. 8.4).

In the hydrogen bomb, which is based on fusion of hydrogen, the necessary high temperature and density are reached by implosion of fusion material induced by a fission bomb. This techniques makes the probability of the fusion reaction (6.51d) very high for a very short time interval resulting in an explosion caused by the energy release of the fusion reaction.

6.7 Generation of Trans-Uranium Elements

We have shown in Chap. 2 that the stability of nuclei is determined by the difference between attractive nuclear forces between the nucleons and repulsive Coulomb force between the protons. With increasing number Z of protons the binding energy should therefore decrease. Although nuclei with $Z > 92$ are unstable against fission or α-emission, theoretical estimations, based on the nuclear shell model, suggest, that within a certain mass range with $Z > 117$ stable nuclei should be again possible (*islands of stability*), because in this range closed shells of nucleons could increase the stability.

There are quite a lot of experimental efforts to realize this prediction. One technique for the production of trans-uranium nuclei is based on high energy collisions between projectile nuclei with medium mass and heavy target nuclei [1–8]. Since the binding energy per nucleon for nuclei beyond iron decreases again, the reaction heat Q of such fusion reactions is negative, i.e. one has to supply more energy for the initiation of such fusion reactions than can be gained by the fusion. Such fusion experiments therefore do not aim at the realization of energy sources but only for winning information about trans-uranium nuclei.

Stable fusion products can be only formed, if the high kinetic energy necessary for overcoming the Coulomb barrier, can be dissipated before the excited compound nucleus decays again into fragments (Fig. 6.22). This implies that the excited compound nucleus must release its excitation energy fast enough by emission of γ-quanta or neutrons.

The first experiments for the generation of trans-uranium nuclei started with the bombardment of heavy nuclei with neutrons resulting in the following reactions:

$${}^{238}_{92}U + {}^1_0n \rightarrow {}^{239}_{92}U \xrightarrow[23.5\text{min}]{\beta^-} {}^{239}_{93}Np \xrightarrow[2.35\ d]{\beta^-} {}^{239}_{94}Pu \tag{6.53a}$$

Bombarding heavy nuclei with light nuclei one obtains for instance

$${}^{22}_{10}Ne + {}^{238}_{92}U \rightarrow {}^{260}_{102}No^* \rightarrow {}^{256}_{102}No + 4n, \tag{6.53b}$$

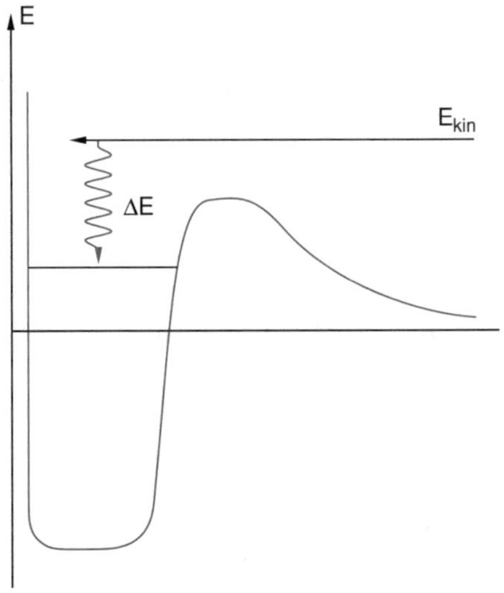

Fig. 6.22 Illustration of the formation and stabilization of transuranium nuclei

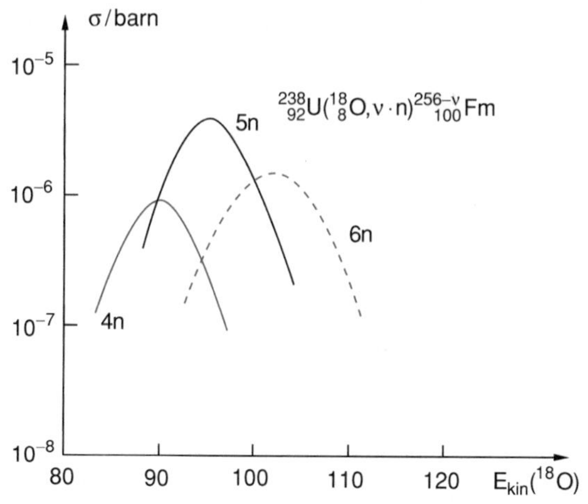

Fig. 6.23 Reaction cross sections for the formation of Fermium isotopes by collision of fast $^{18}_{8}$Oxygen nuclei with uranium targets as a function of the kinetic energy of the oxygen nuclei. (after Musiol: Kern- und Elementarteilchen Physik (Harry Deutsch Frankfurt 1995)

$$^{18}_{8}O + ^{243}_{95}Am \rightarrow ^{261}_{103}Lr^{*} \rightarrow ^{256}_{103}Lr + 5n. \qquad (6.53c)$$

The cross sections for these fusion reactions strongly depend on the kinetic energy of the collision partners. The maximum cross section depends on the number of emitted neutrons (Fig. 6.23). Note, that for all reactions only the energy in the center of mass system is available, because part of the energy is necessary for the kinetic energy of the product nucleus due to momentum conservation (see Problem 6.8).

The compound nucleus formed in the reaction rapidly transforms by β^--emission into nuclei with higher Z-numbers. The cross sections for the generation of super-heavy nuclei, i.e. for the formation of trans-uranium elements, are generally very small. They range from μ-barn to femto-barn (1 barn = 10^{-24} cm^2, 1 femtobarn = 10^{-39} cm^2) and they decrease rapidly with increasing number Z of protons in the trans-uranium nucleus. For the reaction

$$^{238}U + ^{48}Ca \rightarrow ^{283} 114 + 3n$$

they are about 10^{-36} cm^2. For trans-uranium elements with large Z-numbers the measuring time to produce even only a few elements sufficient to be detected, range up to several weeks. Their lifetimes range from milliseconds to a few seconds (Table 6.3).

In Fig. 6.26 the modern version of the periodic table is shown, including the transuranium elements, which are listed together with their the trans-Uranium elements are listed together with their abbreviations. Their names are generally chosen by the discoverer. Often the names of famous physicists are selected or the name of the city or institution where the discovery took place (Fig. 6.24).

Table 6.3 List of trans-uranium elements with Z > 109

Z	Name	Symbol	Lifetime
110	Darmstadtium	^{281}Ds	11 s
111	Roentgenium	^{280}Rg	3.6 s
112	Copernicium	^{285}Uub	29 s
113	Nihonium	^{284}Uut	0.49 s
114	Flerovium	^{289}Uuq	2.6 s
115	Moscovium	^{288}Uup	88 ms
116	Livermorium	^{293}Uuh	61 ms
117	Tennessine	Yet unknown	N/A
118	Organeson	^{294}Uuo	0.89 m

$$^{48}_{20}Ca + ^{242}_{94}Pu \rightarrow ^{A}_{Z}X + x\,n$$
$$^{48}_{20}Ca + ^{238}_{92}U \rightarrow ^{A}_{Z}X + x\,n$$
$$^{48}_{20}Ca + ^{244}_{94}Pu \rightarrow ^{289}_{114}X + 3n.$$

All artificially produced trans-uranium elements are radioactive. They decay by α- or ß-emission into other nuclei with mean lifetimes ranging from several μs up to many days The natural trans-uranium elements which can be still found today, have atomic numbers $Z < 95$ and half-lifetimes $\tau_{1/2} > 10^7$ years (e.g. $^{244}_{94}$Pu).

Theoretical estimations predict islands of stability for trans-uranium elements with atomic numbers around Z = 126, Z = 144 and Z = 164 corresponding to nuclei with closed nucleon shells, which should have a higher stability than nuclei with neighboring Z-values. Up to now, however, the search for such stable trans-uranium nuclei was not successful.

The different trans-uranium elements can be identified by their decay chains which contain already known elements. The lifetimes and the energies of the emitted α-particles are measured. For illustration Fig. 6.27 shows the production reaction of the element Copernicium $_{112}$Cn and its decay chain which ends here at the already known radioactive element fermium (^{253}Fm). This element Fm decays further by α- and ß- decay into elements with lower Z-number and ends finally in the natural occurring element Neptunium Np.

Up to now elements with atomic numbers $Z < 119$ have been produced and safely assigned [9]. The discovery of some elements with $Z > 119$ has been reported but not yet confirmed.

$$\begin{aligned} ^{240}_{96}Cm &\rightarrow ^{236}_{94}Pu + ^{4}_{2}He \quad (T = 26,8d\ ),\\ ^{243}_{97}Bk &\rightarrow ^{239}_{95}Am + ^{4}_{2}He \quad (T = 4,5\ h),\\ ^{244}_{98}Cf &\rightarrow ^{240}_{96}Cm + ^{4}_{2}He \quad (T = 25\ min),\\ ^{261}_{107}Ns &\rightarrow ^{257}_{105}Ha + ^{4}_{2}He \quad (T = 1\ ms). \end{aligned} \qquad (6.54)$$

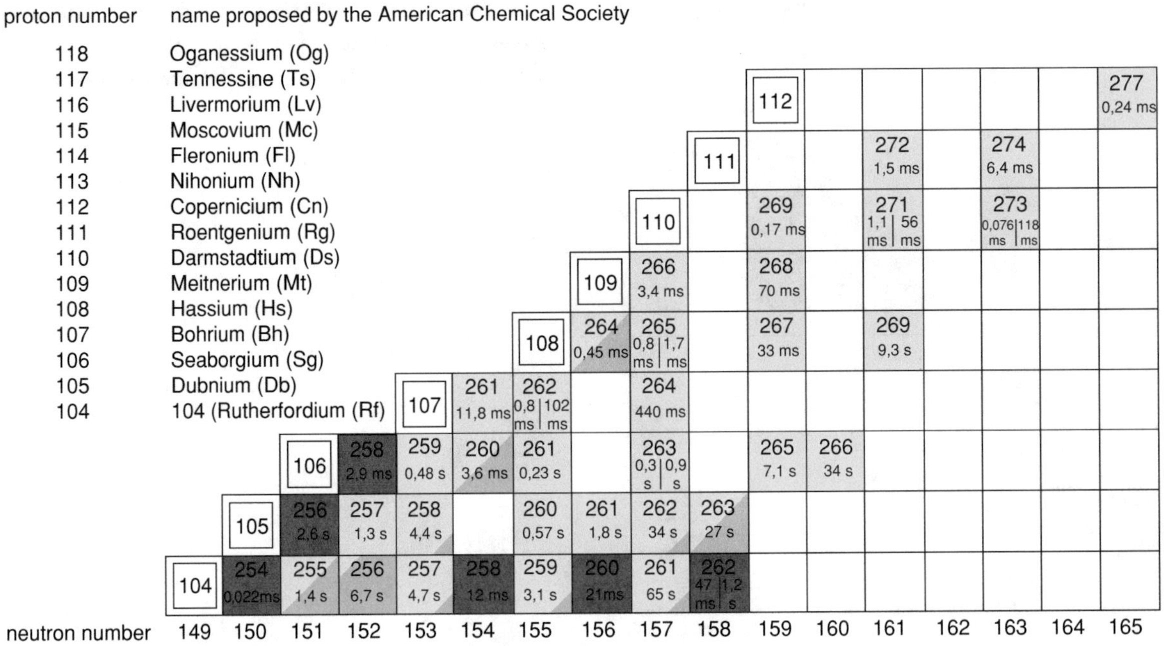

Fig. 6.24 Chart of the trans-uranium nuclei formed by heavy ion induced fusion with their mean lifetimes. Color code: bright red: α_decay; dark red; fission; (after Armbruster Spektrum der Wissenschaft Dec. 1996 S. Hofmann: Physik Journal July 2007 page 19)

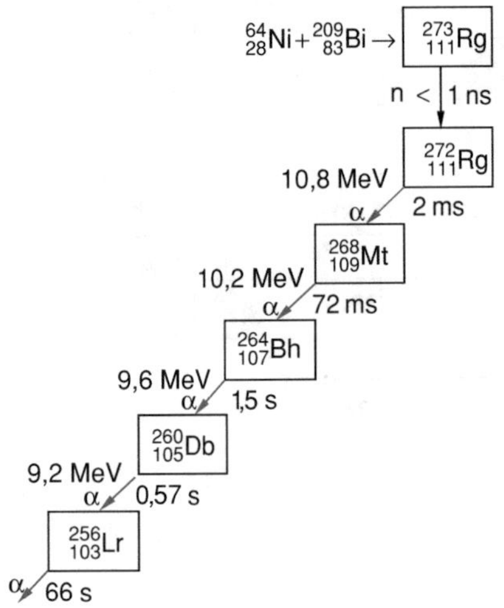

Fig. 6.25 Decay chain of the compound nucleus $^{273}_{111}$Rg formed by the collisions $^{64}_{28}$Ni+ $^{209}_{83}$Bi (after Armbruster, Spektrum der Wiss, Dec. 1996)

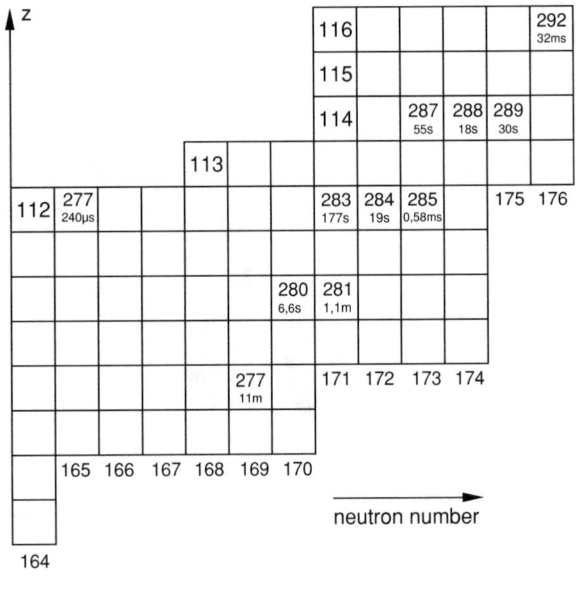

Fig. 6.26 Nuclide chart for the elements 112–116

In Fig. 6.28 some reactions for the production of trans-uranium elements are compiled. Most of these elements decay by α-emission of by fission into lighter elements [4–7].

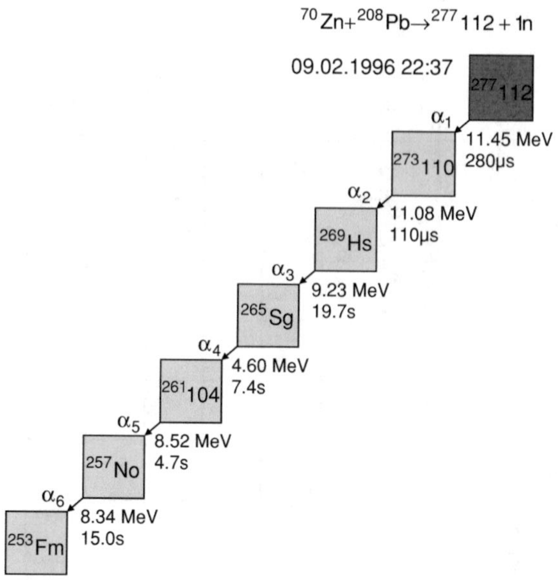

Fig. 6.27 Formation and decay chain of the trans-uranium element (Copernicum) 112 [4]

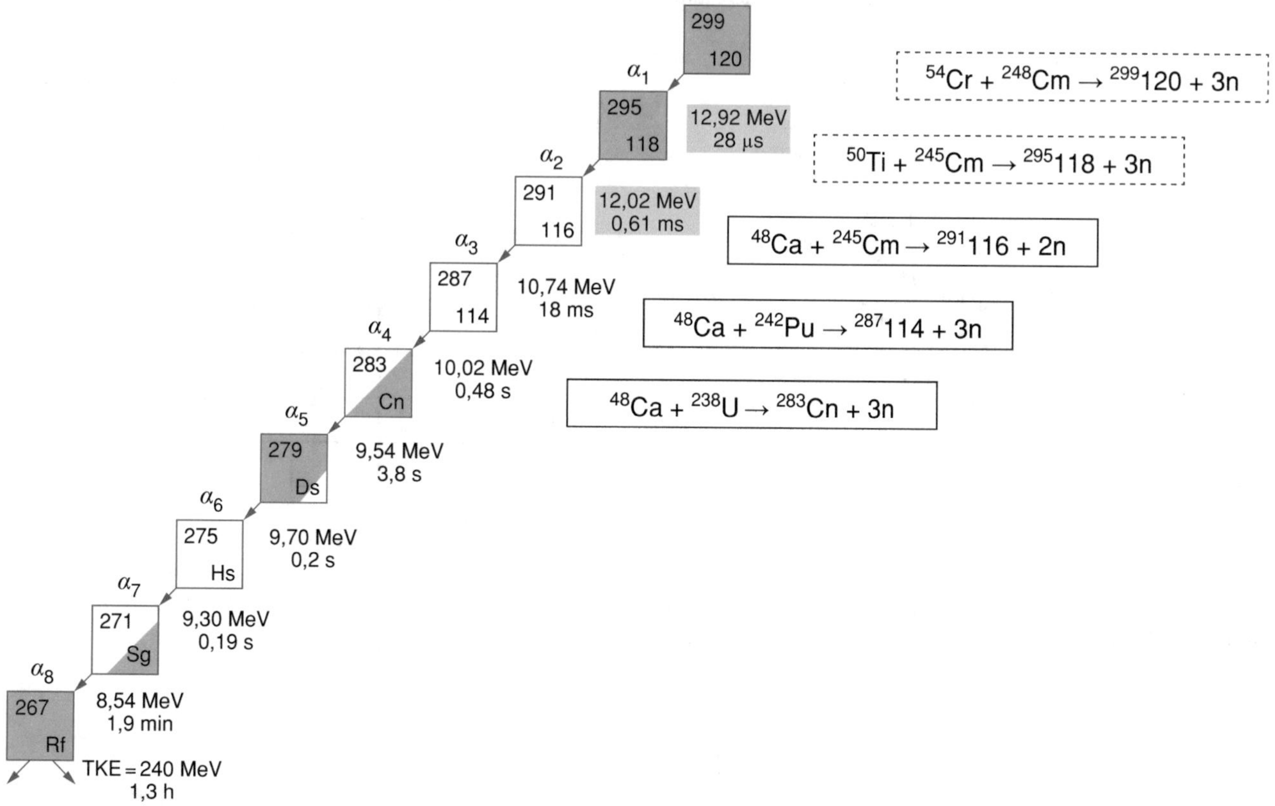

Fig. 6.28 Decay chain of eight consecutive-decays and spontaneous fission of ^{267}Rf at the end. The partially red marked isotopes ^{283}Cn, ^{279}Ds and ^{271}Sg show the probabilities of these nuclei versus '-decay (white) and spontaneous fission (red). For each decay of the nuclei from ^{291}116 to ^{271}Sg the measured? energy and the half-life are given, at ^{267}Rf the fission energy (TKE = Total Kinetic Energy). The elements 118 and 120 have not been observed so far. The theoretically calculated energies and half-lives are given here [5]

Summary

- Nuclear reactions include inelastic and reactive collisions of nuclei where energetically excited nuclei are generated or even new nuclei are produced. The collision-induced nuclear fission or fusion are special nuclear reactions.
- Many nuclear reactions can be described by the compound model where during the reaction a collision complex (compound nucleus) is formed which then can decay into several decay channels.
- Nuclear reactions are described by their reactivity (reaction heat) Q. They proceed only for energies $E > E_S$ above a threshold energy E_{th}. Reactions with $Q > 0$ are exothermic. They also may have to overcome a reaction barrier. However, for $Q > 0$ the energy in the exit channel is always larger than that in the entrance channel.
- For all nuclear reactions the total number of nucleons, the total charge, the angular momentum and the parity are conserved.
- By collision-induced nuclear reactions artificial radioactive elements can be produced which do not occur in nature.
- Nuclei ${}^A_Z X$ with $Z^2/A > 51$ can split into smaller nuclei even without tunnel-effect. For $Z^2/A < 51$ the fission is only possible due to the tunnel-effect. The smaller the ratio Z^2/A becomes the longer is the half lifetime for spontaneous fission.
- Before a spherical symmetric nucleus can undergo fission it has to be deformed into an elliptical shape with half axes a and b. The potential energy as a function of the deformation parameter $\delta = (a-b)/R_0$ is determined by the increase of the surface energy and the decrease of the Coulomb repulsion.
- For all nuclei that do not split spontaneously, the potential energy $E_{pot}(\varepsilon)$ has a maximum which lies above the highest occupied energy level.
- Many stable nuclei can be split by sufficient energy supply, which must be larger than the potential barrier ΔE_S for fission.
- For photon-induced fission the photon energy $h \cdot \nu > \Delta E_S$ must be larger than the potential barrier for fission
  - For the neutron-induced nuclear fission the sum of kinetic energy of the neutron and its binding energy in the compound nucleus contribute to the fission. It must be $E_{kin} + E_B > \Delta E_S$.
  - Collisions between nuclei ${}^{A1}_{Z1}X_1$ and ${}^{A2}_{Z2}X_2$ can result in the formation of heavier nuclei ${}^A_Z X$ with $A = A_1 + A_2$ (*nuclear fusion*). The kinetic energy of the reaction partners in the center of mass system must be larger than the Coulomb potential barrier. The fusion of nuclei smaller than the iron nucleus releases energy
- Collisions between nuclei with medium atomic number A (or of light nuclei colliding with heavy nuclei) can produce trans-uranium nuclei. The reaction is endothermic, i.e. one needs energy for its initiation. The produced trans-uranium nuclei with atomic numbers $92 < Z < 118$ are unstable. They decay by emission of α-or β-particles or by fission into lighter nuclei.
- Up to now the theoretically predicted stability islands at $Z \approx 126$, 144 and 164 which correspond to nuclei with closed shells could not be found experimentally.

Problems

6.1 A particle with mass m_1 and kinetic energy E_{kin} is captured by a nucleus at rest. The compound nucleus with mass M_0 emits a light particle m_3 in the direction perpendicular to the direction of the incident particle and transforms into a nucleus with mass M_2. What are the kinetic energies of m_3 and M_2, if the reaction heat Q is given by the mass difference between entrance and exit channel?

6.2 Use the results of problem 6.1 in order to determine the energy of the neutron which is emitted at the reaction

$$d(0.2\,\text{MeV}) + {}^3_1H(0\,\text{MeV}) \rightarrow {}^4_2He + n$$

under 90° against the direction of the incident deuteron.

6.3 A neutron with the kinetic energy E collides elastically with a nucleus with mass number A. Neutron and target nucleus should be regarded as hard spheres with radius $r = r_0 \sqrt[3]{A}$.

a. What is the relation between impact parameter b and scattering angle ϑ?
b. How is the energy loss related to the impact parameter b?
c. What is the mean energy loss, averaged over all impact parameters? Apply the general result to the specific cases with $A = 1$ (hydrogen) and $A = 12$ (carbon).

6.4 A nuclear reaction has in the lab system the differential cross section $d\sigma/d\Omega$ with $d\Omega = 2\pi \cdot \sin\theta \cdot d\theta$. Calculate with these data the differential cross section $d\sigma/d\Omega$ in the center of mass system.

6.5 Show that for small values of the deformation parameter the surface energy and the Coulomb energy ε are described by the equations (6.35) and (6.36).

6.6 Calculate the kinetic energy E_{kin} of the fission products for the fission of ${}^{235}_{92}U$ by thermal neutrons if the mass ratio of the products is 1.25:1 and $\nu = 3$ neutrons are emitted with a mean energy of 2 MeV and a γ-quant with $h \cdot \nu = 4.6$ MeV. How is the energy divided among the fragments if their mass ratio is $m_1/m_2 = 1.4$?

6.7 At the fission of $^{235}_{92}U$ the total kinetic energy of the fragments ($Z_1 = 35$, $A_1 = 72$) and ($Z_2 = 57$, $A_2 = 162$) is assumed to be 200 MeV. What was their distance from the outer part of the Coulomb barrier at the fission?

6.8 What is the energy of the relative motion in the center of mass system for the maximum of the reaction $^{238}_{92}U(^{18}_{8}O, 5n)^{251}_{100}Fm$ shown in Fig. 6.23?

6.9 Which exit channels are energetically possible, when an α-particle with the energy E_{kinx}=17 MeV hits a tritium nucleus $_1^3H$ at rest? Regard as possible final products the nuclei ^4He, ^5He, ^6He, ^6Li and ^7Li with masses given in the table at the end of this textbook.

6.10

a. Show that for the reaction $a + A \rightarrow b + B^*$ (Excitation energy E_B and $v_A = 0$) the scattering angle θ in the lab system cannot be larger than θ_{max} in the center of mass system with $\sin\theta_{max} = v_b/v_S = G$ and v_S = velocity of the center of mass
b. How large is G for a given kinetic energy $E_{kin}(a)$ and a given reaction heat Q?
c. Calculate the numerical value of G for the elastic scattering.

References

1. Eric Scerri, A Very Short Introduction to the Periodic Table, Oxford University Press, Oxford, 2011
2. *Silva, Robert J. (2006). "Fermium, Mendelevium, Nobelium and Lawrencium". In Morss, Lester R.; Edelstein, Norman M.; Fuger, Jean (eds.). The Chemistry of the Actinide and Transactinide Elements (Third ed.). Dordrecht, The Netherlands:* Springer Science+Business Media. SBN 978–1–4020–3555–5.
3. W.M.Gibson: The Physics of Nuclear Reactions (Pergamon Press Oxford 1980)
4. R.Brass Nuclear Reactions with heavy Ions (Springer . Heicdelberg 1980) https://www.britannica.com/science/transuranium-element
5. L.R.Morss, N-M-Edelstein J.Fuger: The Chemistry of the actinides and trans-actinides Elements. (Springer, Dordrecht 2006
6. M.Schädel: Chemistry of Superheavy Elements. Phil. Trans.Royal Soc. **A 373**, 019
7. Eliav, E.; Kaldor, U.; Borschevsky, A. (2018). Scott, R. A. (ed.). Electronic Structure of the Transactinide Atoms. Encyclopedia of Inorganic and Bioinorganic Chemistry. John Wiley & Sons.(2018)
8. Kragh, H. (2018). From Transuranic to Superheavy Elements: A Story of Dispute and Creation. Springer. ISBN 978–3–319–75813–8.
9. https://en.wikipedia.org/wiki/Transuranium_element

7 Physics of Elementary Particles

We have discussed in Chap. 5 that, according to the model of nuclear forces, proposed in 1935 by *H. Yukawa*, the strong interaction between nucleons can be described by the exchange of particles with a mass of about 140 MeV/c^2. This model had the big advantage that most experimental results could be correctly described, but the essential drawback, that such a particle had not been found. Therefore an intense search for the postulated particle was started. At that time no accelerators were available and the only source of high energy particles was the cosmic radiation, where the primary particles (p^+, e^-, γ) collide with the nuclei of atoms and molecules in the earth atmosphere and produce new particles (secondary radiation Fig. 7.1) [1–3].

7.1 The Discovery of Muons and Pions

Carl David Anderson (1905–1991) found 1937 indeed in his Cloud Chamber (see Sect. 4.3.4) traces of particles produced by the cosmic radiation with a mass approximately equal to the mass of the predicted Yukawa particle. Similar observations were performed independently by *Street* and *Stephenson*. Positively charged as well as negatively charged particles with equal masses were observed.

More detailed further experiments proved, however, that these particles could not be the looked for Yukawa particles, because they did not show strong interaction with other particles but rather a weak interaction and furthermore they decayed relatively slowly into leptons with a mean lifetime of 2 μs. These particles were named muons (myons). They decay according to the scheme

$$\mu^- \rightarrow e^- + \nu_\mu + \overline{\nu}_e, \tag{7.1a}$$

$$\mu^+ \rightarrow e^+ + \nu_e + \overline{\nu}_\mu \tag{7.1b}$$

into electrons or positrons and two different neutrinos: The electron neutrino and the muon-neutrino (see Sect. 7.3), where $\overline{\nu}$ is the anti-neutrino (Fig. 7.2).

Only 10 years later the searched Yukawa particle was found in 1947 by *C.F. Powel* and coworkers. They analyzed traces on photo-plates, which they had exposed to the cosmic rays on the *Jungfraujoch* in the Swiss Alpes (3000 *m* altitude) and in the *Andes* (5000 m) [4]. Later photo-plates exposures in a balloon at 10 000 m altitude confirmed, that the traces were caused by particles in the cosmic rays which occurred as positively as well as negatively charged particles with masses predicted by Yukawa. They were called **π-Mesons** or **pions**. They decay with a mean lifetime of $2.6 \cdot 10^{-8}$ s into muons and neutrinos according to the scheme

$$\pi^+ \rightarrow \mu^+ + \nu, \tag{7.2a}$$

$$\pi^- \rightarrow \mu^- + \overline{\nu} \tag{7.2b}$$

The muons decay further according to (7.1a, 7.1b) into electrons or positrons and neutrinos. The π-mesons interact in deed with nucleons. Experiments showed the following reactions:

$$\pi^+ + {}^A_Z\mathrm{N} \rightarrow p + {}^{A-1}_{Z}\mathrm{N}, \tag{7.3a}$$

$$\pi^- + {}^A_Z\mathrm{N} \rightarrow n + {}^{A-1}_{Z-1}\mathrm{N}. \tag{7.3b}$$

In these reactions the π^+-mesons convert a neutron into a proton, while the π^- mesons convert a proton into a neutron.

This proved that the particles postulated by Yukawa really do exist. It turned out much later, however, that the Yukawa theory is not sufficient to explain all characteristics of the strong interaction. Only the quark-model, developed much later (see Sect. 7.4), could elucidate all experimental observations.

W. Demtröder, *Nuclear and Particle Physics*, Undergraduate Lecture Notes in Physics,
https://doi.org/10.1007/978-3-030-58313-2_7

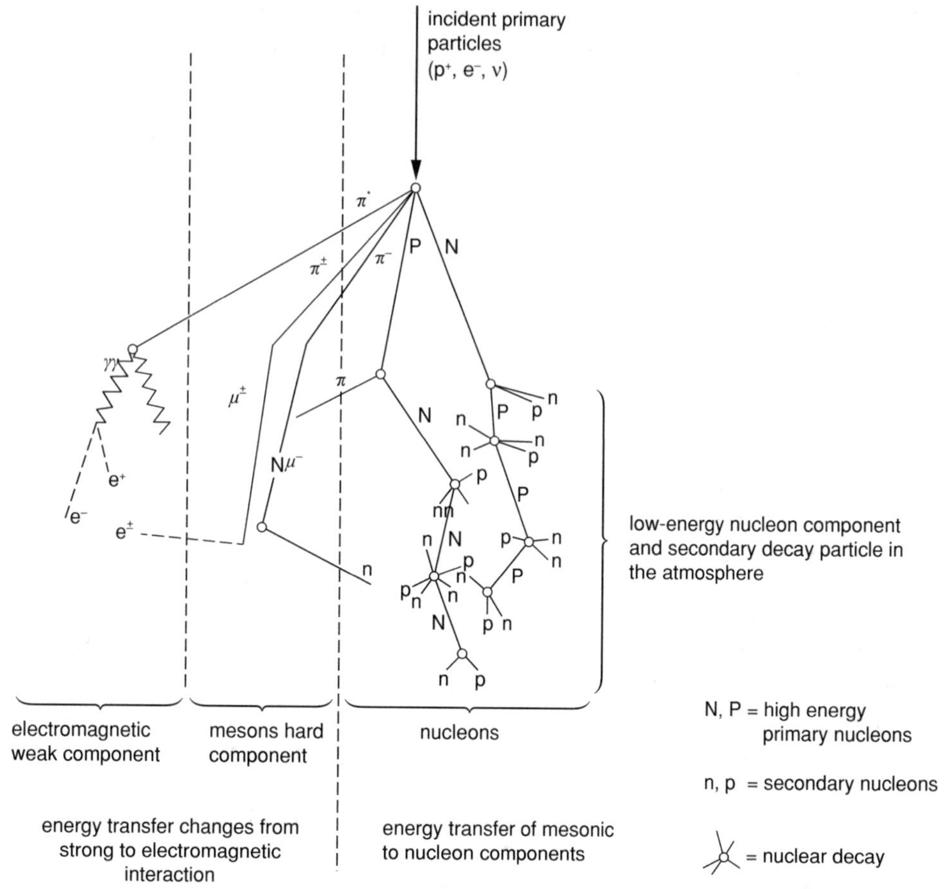

Fig. 7.1 Schematic illustration of cosmic rays and their secondary products in the atmosphere

Fig. 7.2 The tracks of π^+, μ^+, e^+ in a photo-emulsion, which show the decay $\pi^+ \to \mu^+ \to e^+$ of the π^+ -pion present in the cosmic rays. (From H, Brown et. al. Nature **163**, 47, (1949))

7.2 The Zoo of Elementary Particles

Until 1947 the only elementary particles known so far were the light particles (e^-, ν) and their anti-particles (e^+, and $\bar{\nu}$), the heavy particles (p^+, n, π^+) and the muons, μ^- and μ^+ with intermediate masses (see Table 7.1). In 1948 a new particle was found in the cosmic rays, the neutral Kaon K^0, which decays into two pions

$$K^0 \to \pi^+ + \pi^- \tag{7.4}$$

Due to the development of particle accelerators it became possible to generate large currents of high energy particles, which could create in collisions a large variety of new unknown particles. Now one did no longer rely on random and seldom events in the cosmic rays in order to observe new particles. These new opportunities opened the way for an intensive successful search for new elementary particles. During the next years 16 strong interacting particles with masses below the proton mass were found and even more than 100 particles with masses above the proton mass were discovered (Tables 7.1 and 7.2). However, all these particles are not stable but decay with lifetimes between 10^{-6} and 10^{-24} s into other particles [5].

This large number of particles destroyed for a certain time the hope of a reduction of all observable material substances onto a few really elementary building blocks. In order to achieve a certain order into this "Zoo" of different particles detailed measurements of all possible characteristics, as for instance mass, charge, spin, parity, lifetimes and possible decay channels have been performed, in order to find common features of the different particles [5]. Such a procedure

Table 7.1 Some stable and instable particles

Particle	Forecast	Experimental evidence	Mass ·c² in MeV	Spin in ħ	Charge e	Interaction*
Electron	–	1895	0,51	1/2	−1	e, sch, g
Proton	–	≈1905	938,28	1/2	+ 1	all
Neutron	1920	1932	939,57	1/2	0	s, sch, g
Positron	1928	1932	0,51	1/2	+ 1	e, sch, g
Antiproton	1928	1955	938,28	1/2	−1	all
Myon	–	1937	105,66	1/2	± 1	e, sch, g
π-Meson	1935	1947	134,96	0	0 ± 1	all
τ-Lepton	–	1975	1784	1/2	−1	e, sch, g
e-Neutrino	1930	1956	$< 10^{-6}$	1/2	0	sch, g?

*e—electromagnetic, s—strong, sch—weak, g—gravitational

Table 7.2 Characteristic properties of particles with lifetimes longer than 10^{-22} s

	Particle	Symbol	Baryon number B	Mass (MeV/c²)	Charge	Spin in ħ	Isospin T	Components T_3	Strangeness	Lifetime in s
	Photon	γ	0	0	0	1	0	0	0	∞
Leptons	Neutrino	$\nu_e, \bar{\nu}_e \nu_\mu, \bar{\nu}_\mu$	0	$< 10^{-6}$	0	1/2	0	0	0	∞
		$\nu_\tau, \bar{\nu}_\tau$	0	$< 10^{-6}$	0	1/2	0	0		∞
			0	?	0	1/2	0	0		∞
	Elektron	e^+, e^-	0	0,511	$\pm e$	1/2	0	0	0	∞
	Myon	μ^-, μ^+	0	105,66	$\pm e$	1/2	0	0	0	$2{,}199\text{-}\ 10^{-6}$
Mesons	Pionen	π^-, π^+	0	139,57	$\pm e$	0	1	± 1	0	$2{,}602 \cdot 10^{-8}$
		π^0	0	134,97	0	0	1	0	0	$8{,}4 \cdot 10^{-17}$
	Kaonen	K^+, K^-	0	493,7	$\pm e$	0	1/2	± 1/2	± 1, −1	$1{,}238\text{-}\ 10^{-8}$
		K^0_S	0	497,71	0	0	1/2	−1/2	± 1	$8{,}93 \cdot 10^{-8}$
		K^0_L	0	497,71	0	0	1/2	+ 1/2	−1	$5.2 \cdot 10^{-8}$
	Eta-	η	0	548.5	0	0	0	0	0	$2{,}5 \cdot 10^{-17}$
	Rho-	ϱ	0	768,5	0, $\pm e$	1	0	0	0	$3{,}3 \cdot 10^{-21}$
	Phi-	ϕ	0	1019	0	1	0	0	0	$1{,}5 \cdot 10^{-22}$
	Psi-	ψ	0	3095	0	1	0	0	0	10^{-20}
Baryons	Proton	p^+, p^-	1, −1	938.26	$\pm e$	1/2	1/2	± 1/2	0	∞
	Neutron	$n, \bar{n}$	1, −1	939.55	0	1/2	−1/2	∓ 1/2	0	887
	Lambda-	$\Lambda, \bar{\Lambda}$	1, −1	1115.68	0	1/2	0	0	-1, +1	$2.5 \cdot 10^{-10}$
	Sigma-	$\Sigma^+, \bar{\Sigma}^+$	1, −1	1189.4	$\pm e$	1/2	1	∓ 1	-1, +1	$8 \cdot 10^{-11}$
			1, −1	1192.5	0	1/2	1	± 1	-1, +1	$< 7 \cdot 10^{-20}$
	Sigma-	$\Sigma^0, \bar{\Sigma}^0$	1	1232	+ e, 0,	3/2	0	0	0	$5\text{–}10^{-24}$
	Delta-	$\Delta^+, \Delta^0, \Delta^-$	1	1314.8	$-e$	1/2	1/2	−1/2	−2	$2.9 \cdot 10^{-10}$
	Xi-	Ξ^0	+ 1	1672.4	0	3/2	0	0	−3	$1.3 \cdot 10^{-10}$
	Omega-	Ω^-			$-e$					

is very helpful, if a general theory of elementary particles is not yet available, because one may find certain conservation laws, or deviations from up to then accepted rules. This will be illustrated by some examples.

The **mass** of a particle can be determined by a combined measurement of momentum and energy. The relativistic relation

$$E_{\text{kin}} = \sqrt{(m_0c^2)^2 + (pc)^2} - m_0c^2 \tag{7.5}$$

between kinetic energy and momentum yields after resolving for the rest mass m_0

$$m_0 = \frac{p^2 - E_{\text{kin}}^2/c^2}{2E_{\text{kin}}}. \tag{7.6}$$

The momentum can be measured by the deflection of the particle track in a magnetic field $\boldsymbol{B}$ (see Vol. 3, Sect. 2.6). The energy can be obtained by the range of the particle in absorbing material (Sect. 4.2). Measurements of further characteristic properties of particles will be illustrated in the following sections by the example of the π-meson (pion).

7.2.1 Lifetime of the Pion

The experimental methods for measuring lifetimes depend on the order of magnitude of the time scale. The first measurements of the lifetime of the π^+-meson was performed by *O. Chamberlain* [6]: Electrons accelerated in a 340 MeV synchrotron were focused onto a target of heavy nuclei where their bremsstrahlung produced a directed beam of high energy photons (γ-radiation, Fig. 7.3a). These photons impinge onto protons in a paraffin block and induce the reaction

$$\gamma + \mathrm{p} \rightarrow \pi^+ + \mathrm{n}. \quad (7.7)$$

The pions are slowed down in a scintillation crystal by exciting and ionizing the scintillator molecules. The light intensity emitted by the excited molecules is a measure of the particle energy. The stopped pions ($E_{kin} \approx 0$) decay according to ($\pi^+ \rightarrow \mu^+ + \nu$) into muons and transfer their kinetic energy

$$E_{kin}(\mu^+) = \left(m_{\pi^+} - m_{\mu^+}\right)c^2 - E_{kin}(\nu) \quad (7.8)$$

to the scintillator molecules and decay further

$$\mu^+ \rightarrow \mathrm{e}^+ + \nu_e + \overline{\nu}_\mu, \quad (7.9)$$

where the positron delivers its kinetic energy again to the scintillator molecules before it recombines with an electron and delivers the annihilation radiation $e^+ + e^- \rightarrow 2\gamma$ in form of two γ-quanta emitted into opposite directions.

Altogether one observes therefore at the decay chain π^+ $\mu^+ \rightarrow e^+$ three light pulses which are generated by the decay of the three particles (Fig. 7.3b). The time interval Δt_1 between the first and second pulse corresponds to the lifetime of the π^+-pion, whereas the interval Δt_2 between the second and third pulse gives the much longer lifetime of the μ^+ Muon.

Note, that the decay of a particle is a random process, which means that not all π^+-pions decay after the same time interval. The number of decays within the time interval from t to $t + dt$ equals the decrease $-dN = -\lambda \cdot N$ of the original pion number N. Integration gives

$$\Rightarrow N(t) = N_0 \cdot \mathrm{e}^{-\lambda \cdot t} = N_0 \cdot \mathrm{e}^{-t/\tau}, \quad (7.10)$$

where $\tau = 1/\lambda$ is the mean lifetime, after which only $N(\tau) = N(0)/e$ pions are left.

It is therefore necessary to perform many of such measurements, in order to determine the random distribution of the time intervals Δt_1, This was done by Chamberlain and coworkers. Their result is shown in Fig. 7.4, where the measured decay numbers dN/dt are plotted on a logarithmic scale against the measured time intervals $t = \Delta t_1$. According to (7.10) the slope of the straight line $\ln(dN/dt) = -\lambda\, t$ gives the decay constant $\lambda = 1/\tau$ which equals the inverse mean lifetime $\tau = 16$ ns.

7.2.2 Spin of the Pion

When the pion was discovered, one of the open questions was whether its spin is an even or odd multiple of ħ/2, which implies the classification of the pion as Boson (as e.g. the photon) or Fermion (as the proton or electron). This can be decided by measuring the cross sections for the two inverse reactions

$$\mathrm{p} + \mathrm{p} \rightarrow \pi^+ + \mathrm{d} \quad (7.11a)$$

$$\pi^+ + \mathrm{d} \rightarrow \mathrm{p} + \mathrm{p} \quad (7.11b)$$

Since the spin of the proton is $I = 1/2\hbar$ and that of the deuteron $I = 1\hbar$, the spin of the pion must be $I = n \cdot \hbar$ with

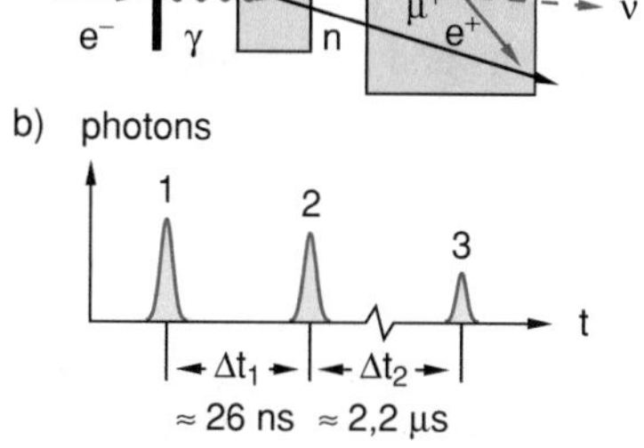

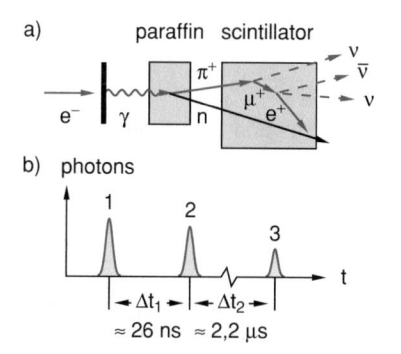

Fig. 7.3 **a** Experimental setup for measuring the lifetime of the π^+-meson **b** Time sequence of the light pulses in a scintillation crystal, where the decay $\pi^+ \rightarrow \mu^+ \rightarrow e^+$ takes place

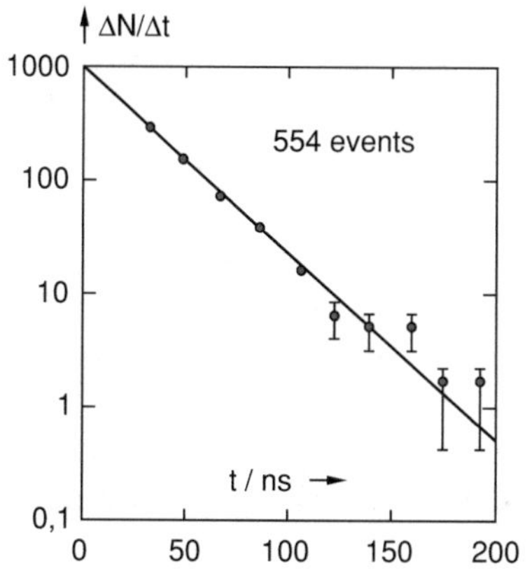

Fig. 7.4 Counting rate of the measured π^+ decay $\Delta N/\Delta t$ in a time interval $\Delta t = 18$ ns as a function of the time interval after the first light pulse (from O. Chamberlain, Phys. Rev 79, 394 (1950))

n = integer, if the reaction (7.11a) is observed, because the sum of the proton spins $(I_1 + I_2)/\hbar$ on the left side of (7.11a) is integer and an eventually present orbital angular momentum $L/\hbar$ is always integer.

The pions are therefore Bosons with integer spin I.

The magnitude of the spin can be obtained as follows.

Since the reaction cross section σ is proportional to the statistical weight of the final reaction products, it must be for reaction (7.11a) with unpolarized protons (random spin orientation)

$$\begin{aligned}\sigma(\mathrm{p}+\mathrm{p}\to\pi^{+}+\mathrm{d}) &\propto 2(2I_\pi+1)(2I_\mathrm{d}+1)\\ &= 2(2I_\pi+1)\cdot 3.\end{aligned} \tag{7.12}$$

For the inverse reaction we obtain

$$\begin{aligned}\sigma(\pi^{+}+\mathrm{d}\to\mathrm{p}+\mathrm{p}) &\propto 2\cdot\frac{1}{2}(2I_\mathrm{p}+1)(2I_\mathrm{p}+1)\\ &= 4,\end{aligned} \tag{7.13}$$

where the factor 1/2 in (7.13) takes into account that the two protons generated in the reaction (7.11b) cannot be identical in all quantum numbers, according to the Pauli-Principle. In the Eq. (7.13) therefore only half of the possible spin orientations can be realized. The ratio of the cross sections for the two reactions is therefore

$$\begin{aligned}\frac{\sigma(\mathrm{p}+\mathrm{p}\to\pi^{+}+\mathrm{d})}{\sigma(\pi^{+}+\mathrm{d}\to\mathrm{p}+\mathrm{p})} &= \frac{3}{2}(2I_\pi+1)\\ &= \begin{cases}3/2 & \text{für}\quad I_\pi=0,\\ 9/2 & \text{für}\quad I_\pi=1.\end{cases}\end{aligned} \tag{7.14}$$

Measurements of this ratio show unambiguously that $I_{\pi+} = 0$. Also for the negative pion π^- one obtains $I_{\pi-} = 0$.

π-mesons have the spin $I = 0$.

7.2.3 Parity of π-Mesons

As has been already discussed in Sect. 6.2.4 the parity of a particle state describes the behavior of its wave function under reflection of all coordinates at the origin. If $P\psi(x, y, z) = +\psi(-x, -y, -z)$ the parity of this state is positive whereas for $P\psi(x, y, z) = -\psi(-x, -y, -z)$ the parity is negative. The parity of a system of two particles A and B is defined by the product

$$P_\mathrm{AB} = P_\mathrm{A}\cdot P_\mathrm{B}\cdot(-1)^{l}, \tag{7.15}$$

where ℓ is the quantum number of the orbital angular momentum of the relative motion of A against B and P_A, P_B are the parities of A and B.

The parity of the nucleons is defined as $P_\mathrm{p} = P_\mathrm{n} = +1$. It turns out that for all reactions which occur due to the strong interaction the parity is preserved. This is different for the weak interaction, where parity violations have been observed (see Sect. 7.6).

For illustration we regard the reaction

$$\pi^{-}+\mathrm{d}\to\mathrm{n}+\mathrm{n}, \tag{7.16}$$

where a negative pion is absorbed in deuterium, captured by a deuterium nucleus (strong interaction) and two neutrons are produced.

Since the reaction occurs with pions nearly at rest, the orbital angular momentum is zero, i.e. $\ell = 0$. The total angular momentum of the left side of (7.16) is with $I_{\pi-} = 0$ and $I_\mathrm{d} = 1$; $\ell = 0$ with the quantum number $j = 1$.

Because the angular momentum must be preserved for the reaction (7.16) the total angular momentum of the reaction products on the right side of (7.16) must be also $|j| = \sqrt{j(j+1)}\hbar$ with $j = 1$.

Because of the different possibilities for the coupling of the neutron spin I_n to the total spin $\boldsymbol{I}$ and the possible orbital angular momentum ℓ of the fly away neutrons we obtain the total angular momentum $J = \mathrm{I} + \ell$ and for $\ell \leq 2$ the possible final state ($\ell = 0, I = 1$), ($\ell = 1, I = 0$), ($\ell = 1, I = 1$), ($\ell = 2, I = 1$). The Pauli-Principle restricts the possible final states. Since the two reaction products are identical fermions, the total wave function must be anti-symmetric against exchange of the two particles. The symmetry of the spatial part of the wave function is *gerade* for even values of ℓ and ungerade for odd values.

The spin part of the wave function is symmetric for $I = 1$ because both spins of the neutrons are parallel, it is antisymmetric for $I = 0$, where both spins are anti-parallel. Therefore combinations of even values of ℓ with $I = 1$ and odd values of ℓ with $I = 0$ are forbidden by the Pauli-Principle. From the different possibilities listed above only one is allowed with ($\ell = 1, I = 1$) where the total wave function is antisymmetric.

The parity of the right side of (7.16) is according to (7.15) $P = -1$. Because of parity conservation for all processes mediated by the strong interaction the parity of the exit channel on the right side of (7.16) must be also $P = -1$. The parity of the deuteron is $P_\mathrm{d} = +1$, because the deuteron is a bound state of proton and neutron with $I = 1$ and $\ell = 0$. Since the orbital angular momentum at the capture of the π^- pion is $\ell = 0$ the parity of the π-meson must be $P_\pi = -1$.

In a similar way one finds that also for the positive pion π^+ the parity is $P_{\pi+} = -1$.

The parity of π-mesons is $P_\pi = -1$.

7.2.4 The Discovery of More Particles

In collisions of high energy stable particles (e^-; p; n) a large number of new particles were discovered. Sometimes short living particles can be only observed as resonances in the excitation cross section or by detecting the long living particles into which they decay. This shall be illustrated by some examples.

In anti-collinear collisions of electrons with positrons in the crossing points of storage rings (Fig. 4.35) besides the reaction

$$e^- + e^+ \to \mu^- + \mu^+ \tag{7.17}$$

also heavy particles, the τ-leptons, were discovered

$$e^- + e^+ \to \tau^- + \tau^+, \tag{7.18a}$$

$$\tau^+ \to \mu^+ + \nu_\mu + \overline{\nu}_\tau \tag{7.18b}$$

$$\llcorner e^+ + \nu_e + \overline{\nu}_\tau, \tag{7.18c}$$

$$\tau^- \to \mu^- + \overline{\nu}_\mu + \nu_\tau, \tag{7.18d}$$

$$\llcorner e^- + \overline{\nu}_e + \nu_\tau, \tag{7.18e}$$

These τ-leptons have a mass of 1.777 GeV/c^2 and decay with a mean lifetime of $2.9 \cdot 10^{-13}$ s into muons, electrons and neutrinos.

When neutrinos were first discovered, it was assumed that only one sort of neutrino exists. However, meanwhile it has been proved that to each of the leptons e, μ, and τ a specific neutrino and its anti-neutrino exists, named as ν_e, ν_μ and ν_τ with their anti-particles $\overline{\nu}_e$, $\overline{\nu}_\mu$ and $\overline{\nu}_\tau$ (see Sect. 7.3).

The cross section for the production of lepton pairs decreases with increasing center of mass energy E_s of the collision partners as $1/E_S^2$. It has about the same magnitude for the two reactions (7.17) and (7.18a, b, c, d) (Fig. 7.5). It does not show any resonances.

The situation is different for the production of hadrons in $e^- + e^+$ collisions (Fig. 7.6). Here distinct maxima appear at certain center of mass collision energies which are called *resonances*. Their height is a measure of the production rate of new particles. The resonance width Γ = ΔE is proportional to the inverse lifetime of the created particles.

Broad resonances correspond to very short living particles, narrow resonances to long living particles.

Example The resonance at $E_S = 770$ MeV in Fig. 7.6, which corresponds to the generation of the ρ^0-particle has a width $\Delta E = 150$ MeV. According to the uncertainty relation $\Delta E \cdot \tau \geq \hbar$ this corresponds to a mean lifetime $\tau = 3 \cdot 10^{-24}$ s. The resonance at $E_S = 3$ GeV which corresponds to the long living ψ-particle is with $\Delta E \leq 87$ keV much narrower and gives a mean lifetime of $\tau = 7 \cdot 10^{-21}$ s horizontal line to indicate the end of the example.

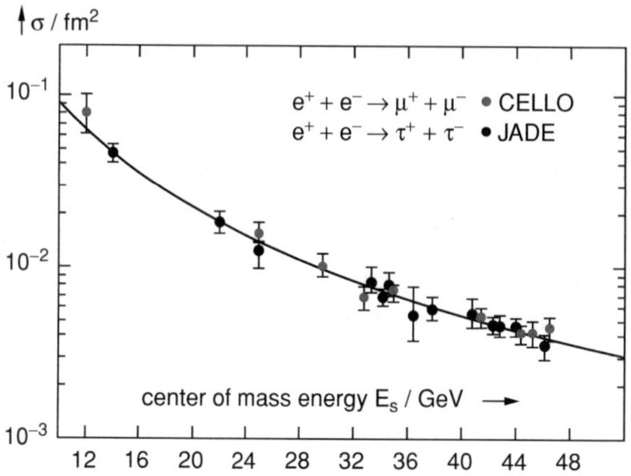

Fig. 7.5 Cross section of the reaction $e^+ + e^- \to \mu^+ + \mu^-$ (red points) and $e^+ + e^- \to \tau^+ + \tau^-$ (black points). (JADE and CELLO. Collaboration DESY [7])

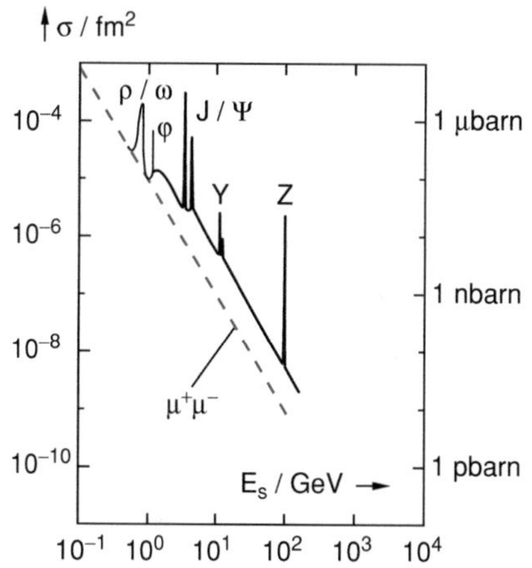

Fig. 7.6 Resonances in the cross section $\sigma(E_s)$ of the reaction $e^- + e^+ \to$ Hadronen as a function of the center of mass energy [8] For comparison the cross section for the direct muon production is marked

The ρ^0 meson decays into two π-meson

$$\varrho^0 \to \pi^+ + \pi^-. \tag{7.19}$$

At the energy $E_s = 1019$ MeV a relative narrow resonance occurs with a half width $\Delta E = 4.4$ MeV. It can be assigned to a hadron, the Φ-particle, which decays with a mean lifetime $\tau = 5 \cdot 10^{-21}$ s into two K-meson.

$$\phi \begin{matrix} \nearrow K^+ + K^- \\ \searrow K^0 + \overline{K}^0, \end{matrix} \tag{7.20}$$

With the masses $m(K^\pm) = 494$ MeV/c^2 and $m(K^0) = 498$ MeV/c^2.

The **K-mesons** (also called **Kaons**) are governed by the strong interaction (because they originate by the decay of the hadron ϕ) but they decay only via the weak interaction

$$K^+ \begin{array}{ll} \nearrow & \mu^+ + \bar{\nu}_\mu \quad (64\%) \\ \rightarrow & \pi^+ + \pi^0 \quad (21\%) \\ \searrow & \pi^+ + 2\pi_0. \end{array} \tag{7.21}$$

and have therefore a long lifetime $\tau = 5 \cdot 10^{-8}$ s. It was at first unclear, why only the weak interaction is responsible for the decay, although the Kaon can also decay into two π-mesons which underlies the strong interaction. The quark model (see Sect. 7.4) could explain this fact, because the decay of the Kaon causes a change of a strange quark into an up quark which is only possible by the weak interaction. Another hint is the fact that the parity changes at the Kaon decay, which excludes the strong interaction where the parity is conserved. A quantum-theoretical description can explain the experimental facts (see Sect. 7.6).

The Kaons, which are generated via the strong interaction but decay by the weak interaction, are called **strange particles** because this behavior seemed to be very strange before one could explain it.

An extremely sharp resonance at $E_S = 3097$ MeV was discovered 1974 independently by a group at the Stanford linear accelerator SLAC headed by *Burton Richter* and by the Brookhaven National Laboratory headed by *Samuel Ting*. Its resonance width was only $\Delta E = 87$ keV, corresponding to a lifetime of $\tau = 7.2 \cdot 10^{-21}$ s which is very long for hadronic unstable particles and is about 1000 times longer than expected.. The particle was named by Richter a ψ-particle and by Ting as J-particle. It is therefore generally cited as Jψ-particle. Richter and Ting received both the Nobel-prize in 1976. We will discuss this particle in Sect. 7.4.3 within the framework of the quark model.

After the discovery of the positron e^+ as anti-particle of the electron e^- by *C. Anderson* 1932, it took 32 more years until the anti-proton p^- was found by *O. Chamberlain, E. Segre*, C. *Wiegand* and *T. Ypsilontis* [9] in collisions of protons with $E = 6.2$ GeV and a copper target where the nucleus of the copper target acts as catalysator for the production of a proton-antiproton pair according to the reaction

$$p + p \rightarrow p + p + p + \bar{p} \tag{7.22a}$$

The experimental difficulty was the detection of this very rare process, which is superimposed by the much more frequent processes

$$p + p \rightarrow p + p + \pi^+ + \pi^-, \tag{7.22b}$$

$$p + n \rightarrow p + n + \pi^+ + \pi^-. \tag{7.22c}$$

For each generated anti-proton about $6 \cdot 10^4$ π-mesons are produced. It was therefore necessary to separate the few anti-protons from the large background of π^--mesons. The experimenters used momentum selection in magnetic fields with shielding blocks and small slits between the successive magnetic fields (Fig. 7.7). The anti-protons have, because of their larger mass, a slightly smaller velocity. Therefore a further separation against the π^- background can be achieved by time of flight measurements. The distance between D_1 and D_2 in Fig. 7.7 was 12 m.

A further discrimination against the π-background can be realized by a cunning anti-coincidence circuit. The Cerenkov counter C_1 in Fig. 7.7 (see Sect. 4.3.5) detects only particles with velocities $\beta = v/c > 0.79c$ while C_2 is constructed in such a way that it detects only particles with $0.75c < \beta < 0.78c$. The anti-protons have velocities after traversing the scintillation counter of $v = 0.765 \cdot c$, while $v(\pi^-) \approx 0.99c$. Therefore C_1 detects π^- mesons but not anti-protons, whereas C_2 detects anti-protons but not π^-mesons. The anti-coincidence circuit $\overline{C_1} \cdot C_2$ (not C_1 but C_2) only counts anti-protons, the circuit $C_1 \cdot \overline{C_2}$ (C_1, not C_2) only π^- mesons.

This illustrates that for such difficult experiments much phantasy and ingenuity is demanded in order to identify unambiguously such rare events.

Besides the anti-proton and several mesons a large variety of particles with spin 1/2ħ have been discovered. Because of their half-integer spin they are Fermions. Most of them have masses larger than the proton mass. They interact with other particles via strong forces and are therefore Hadrons. They are called **Hyperons,** which include the Δ-particle, the Λ-, Σ- and the Ω-particle (see Table 7.2). Most of them have a very short lifetime and they show up as resonances in the excitation cross section. They decay after 10^{-18}–10^{-24} s into lighter particles. Examples of reactions where Hyperons have been observed are

$$\pi^+ + p \rightarrow \Delta^{++} \rightarrow p + \pi^+,$$

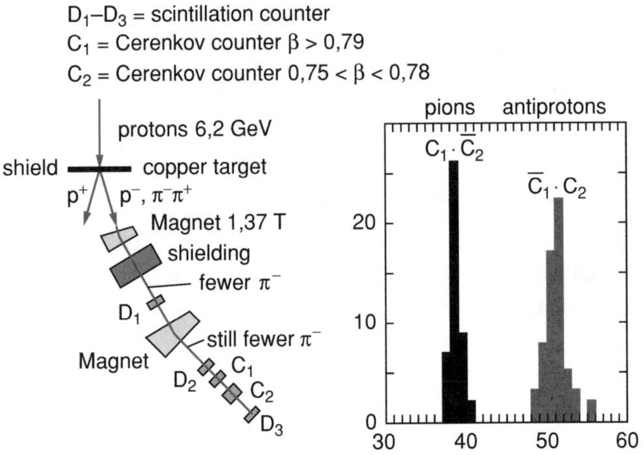

Fig. 7.7 Experimental setup and result of the first discovery of the anti-proton. (after O. Chamberlain [8])

where the doubly charged Δ^{++} particle decays again after $5 \cdot 10^{-14}$ s into the initial reaction partners.

The Λ-particle was found as resonance in the reaction

$$p + \pi^- \rightarrow \Lambda^0 \rightarrow p + \pi^-$$
$$\rightarrow n + \pi^0$$

In Table 7.2 the characteristic properties of the unstable elementary particles are summarized.

7.2.5 Classification of the Elementary Particles

The particles found up to now can be classified into two groups:

- **Leptons** (from the Greek word λεπτÓζ = weak). This includes all particles which show no strong interaction, but interact with other particles only via the weak or the electro-magnetic interaction. Examples are the electron e^- and its anti-particle, the positron e^+; the muon μ^- and its anti-particle μ^+, the τ-lepton τ^+ and its antiparticle τ^- and the three corresponding neutrinos ν_e, ν_μ, and ν_τ with their anti-neutrinos.
- **Hadrons** (from the Greek word αδρoζ = tough, heavy). All particles which are subjected to the strong interaction belong to this group. They are subdivided into **baryons** (heavy particles such as the proton and neutron, or the hyperons Λ, Σ, Ω) and **mesons** (medium heavy particles such as π^+, π^-, π^0. K^+, K^-, K^0, η, δ, ϕ ψ).

All hadrons, except the proton p^+ and the anti-proton $\bar{p}^-$ are unstable and decay either into p^+ or $\bar{p}^-$ or into mesons, which then decay further into leptons or photons. The decay due to the strong interaction is fast (lifetimes of 10^{-20}–10^{-24} s).

There are also hadrons which decays only by weak interaction. Their lifetime is relatively long (typically 10^{-20}–10^{+3} s). One example it the lifetime (τ = 887 s) of the neutron which decays due the weak interaction into a proton, an electron and an anti-neutrino (β-decay, see Sect. 3.4)

$$n \rightarrow p^+ + e^- + \bar{\nu}.$$

The question is now, whether it is possible to subsidies the many particles into a schematic order (besides the coarse division into Leptons and Hadrons). This will be discussed in the next section.

7.2.6 Quantum Numbers and Conservation Laws

In order to find systematic relations between the different elementary particles and regularities for their different reactions and decays, one must find the energy of specific states and assign quantum numbers to them. This is quite similar to the situation in the atomic electron shell or to excited states of atomic nuclei.

Here the energies correspond to the masses of the particles, which are generally given in units of MeV/c^2. The parity of the particles gives the behavior of their wave function under reflection of all coordinates at the origin.

All baryons are characterized by their baryon number $B = 1$. The anti-baryons (e.g. the anti-proton) get $B = -1$. For all other particle (non-baryons) is $B = 0$.

All experiments up to now have proved that for all reactions or decays of baryons the baryon number is conserved, i.e. $\Delta B = 0$.

Example For the β-decay of the neutron $n \rightarrow p + e^- + \bar{\nu}_e$ the baryon number is 1 → 1 + 0 + 0.

A baryon can decay in leptons or mesons only if simultaneously another baryon is created.

This selection rule justifies why the proton as the lightest baryon is stable and does not decay, because there is no lighter baryon which could be produced by the decay. There has been intense research for a possible decay into mesons or leptons which would violate the selection rule $\Delta B = 0$,

Up to now no proton decays have been observed, Experiments have proved that the lower limit for the lifetime of the proton is $\tau > 10^{34}$ years.

An essential property of particles is their angular momentum ***I*** with the amount $I = (I(I+1))^{1/2}$ (spin of the particle with the spin quantum number *I*).

Furthermore the isospin *T* and its component T_3 is an important means for the classification of particles (see Sect. 5.3). It describes the difference $1/2T_3 = (n_u - n_d)$ between the number n_u of up-quarks and n_d of down quarks in a particle (see Sect. 7.4).

The isospin component T_3 is related to the charge quantum number $Q = q/e$ and the baryon number B by

$$T_3 = Q - \frac{1}{2}B \tag{7.23}$$

The classification by the values of T_3 means a distinction according to the charge of a particle.

Examples For the proton is $Q = 1$ and $B = 1$. Therefore is the isospin component $T_3 = +1/2$. For the neutron is $Q = 0$ and $B = 1 \to T_3 = -1/2$. Proton and neutron are regarded as two components of the isospin-doublet, because they have the T_3. The isospin component $T_3 = +1/2$. for the proton and $T_3 = -1/2$ for the neutron.

The pion π^+ has $Q = 1$, $B = 0 \to T_3 = 1$ while for the π^- is $Q = -1$ $B = 0 \to T_3 = -1$.

A further quantum number for the characterization of particles is the **Strangeness quantum** number **S**. It was introduced after "strange" particles had been found (e.g. K-mesons and Λ-particles), which can be produced via the strong interaction but decay via the weak interaction and have therefore relatively long lifetimes (10^{-9}–10^{-11} s).

The K^0-mesons get the strangeness number $S = 1$ while the Λ-particles get $S = -1$, the nucleons and the pions get $S = 0$.

Later it turned out that all strange particles contain the strange quark.

For all processes due to the strong and the electro-magnetic interaction the strangeness quantum number is preserved ($\Delta S = 0$), whereas for the weak interaction is $\Delta S = \pm 1$,

Example Production of strange particles via the strong interaction.

$$p + \pi^- \to K^0 + \Lambda^0,$$

For the change ΔS of strangeness S we obtain

$$S = 0 + 0 \to 1 + (-1) \Rightarrow \Delta S = 0.$$

Decay of K^0 via the weak interaction

$$K^0 \to \pi^+ + \pi^- \to \mu^+ + \mu^- + \nu + \bar{\nu},$$

$$S = 1 \to 0 + 0 \Rightarrow \Delta S = -1.$$

For all particles observed up to 1974 the quantum numbers S for strangeness, the baryon number B and the isospin component T_3 could be related to the charge quantum number Q by

$$Q = \frac{1}{2}(B + S) + T_3. \tag{7.24}$$

Example

1. π^--meson

$$Q = -1, \quad B = 0, \quad S = 0 \Rightarrow T_3 = -1.$$

2. Δ^{++}-hyperon

$$Q = 2, \quad B = 1, \quad S = 0 \Rightarrow T_3 = \frac{3}{2}.$$

3. Ξ *-particle*:

$$Q = -1, \quad B = 1, \quad S = -2 \Rightarrow T_3 = -\frac{1}{2}.$$

7.3 Leptons

All particles with spin 1/2, which do not interact via the strong interaction are called Leptons. They are characterized by the lepton quantum number $L = 1$. Their antiparticles get $L = -1$. For all reactions involving leptons it was found that the lepton number does not change: $\Delta L = 0$.

For the lepton quantum number we therefore obtain:

$$\sum_i L_i = \text{const.} \Rightarrow \Delta L = 0. \tag{7.25}$$

Examples

1. Decay of the neutron.

$$n \to p + e + \bar{\nu} \quad L : 0 = 0 + 1 - 1$$

2. Neutrino production by $e^+ + e^-$ annihilation

$$e^+ + e^- \to \nu + \bar{\nu}$$
$$L : -1 + 1 \to 1 + (-1)$$

We have already discussed in Sect. 7.1 the muons with the mass $m_\mu = 105.7$ MeV/c^2, discovered by *Andersen* 1933 in the cosmic rays. The negative muon decays with a mean lifetime $\tau = 2.2$ μs in an electron and two neutrinos

$$\mu^- \to e^- + \nu_\mu + \bar{\nu}_e, \tag{7.26}$$

where the lepton number L is preserved if w assign the lepton number $L = +1$ to the μ^- muon. The following experiment, performed 1961 by *L. Ledermann, M. Schwarz and J. Steinberger* showed that the two neutrinos are of a different kind.

The proton beam of the Brookhaven Proton-Synchrotron is focused onto a Beryllium target and produces π^+- and π^- mesons, which form a collimated beam in the forward direction (Fig. 7.8). The pions decay on their way into muons.

$$\pi^+ \to \mu^+ + \nu_\mu; \quad \pi^- \to \mu^- + \overline{\nu}_\mu, \tag{7.27}$$

where neutrinos and antineutrinos are created. They fly through a big iron block which shields the pions and muons, while the neutrinos pass the block effectively unhindered and hit an aluminum target.

The neutrinos are detected by looking for reactions which are induced in the Aluminum target, such as

$$\nu_\mu + \mathrm{n} \to \mu^- + \mathrm{p}, \tag{7.28a}$$

$$\overline{\nu}_\mu + \mathrm{p} \to \mu^+ + \mathrm{n}. \tag{7.28b}$$

The produced muons are detected. If the neutrinos would be of the same kind as those created in the ß-decay of the neutron, the following reactions should be also observed:

$$\nu + \mathrm{n} \to \mathrm{e}^- + \mathrm{p}, \tag{7.29a}$$

$$\overline{\nu} + \mathrm{p} \to \mathrm{e}^+ + \mathrm{n} \tag{7.29b}$$

In spite of intense investigations these reactions could not be found. Therefore the muon-neutrinos ν_μ created in the muon-decay must be different from the electron-neutrinos ν_e produced in the ß-decay. They are distinguished by a proper lower index: ν_μ, ν_e.

In 1975 a new lepton was discovered as a product of $e^+ + e^-$ collisions:

It has a mass $m = 1777$ MeV and is therefore much heavier than the proton. It was named τ-lepton and can be positively or negatively charged. It is not stable but decays with a mean lifetime of $3 \cdot 10^{-13}$ s into the products shown in (7.18b, c).

An example of such a decay is shown in Fig. 7.9. The collision between e^+ and e^- occurs in the crossing point of the anti-collinear e^+- and e^- beams (indicated by the red point in Fig. 7.9 where the colliding beams come from above and below the drawing plane). The $\tau^\pm$-leptons are emitted with opposite momentum in the drawing plane perpendicular to the $e^\pm$-beams. Because of the short lifetime one cannot see their track in Fig. 7.9. For $T = 3 \cdot 10^{13}$ s their track is only 80 μm long. One can see, however, the track of the muons produced by the decaying τ-lepton:

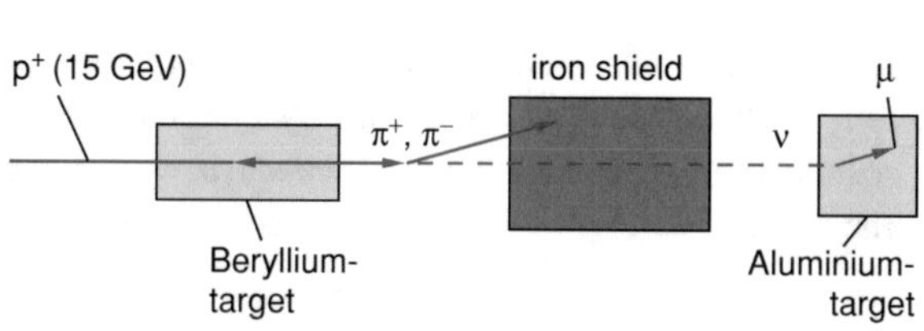

Fig. 7.8 Experimental setup for the verification of two different neutrino species ν_μ and ν_e

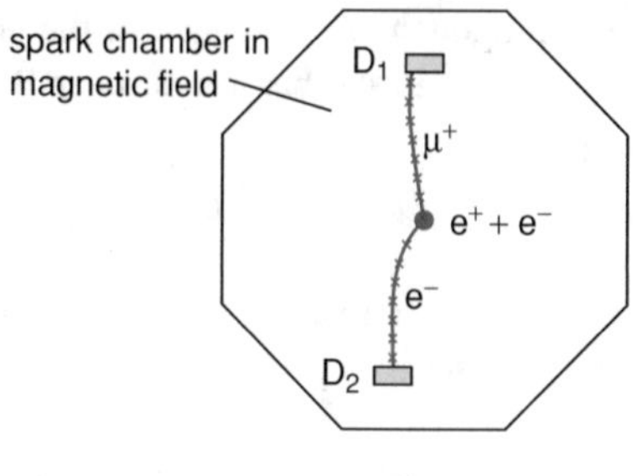

Fig. 7.9 Experimental detection of the production of a τ^- and τ^+ pair. (from G. J. Feldmann, M. I. Perl: physics reports 19C, 233 (1975))

$$\tau^+ \to \mu^+ + \nu_\mu + \overline{\nu}_\tau$$

and the track of the electron from the reaction

$$\tau^- \to e^- + \overline{\nu}_e + \nu_\tau.$$

The tracks of the charged particles are curved in the external magnetic field.

Electron and muon do not strictly fly anti-collinear, because a fraction of the total momentum is carried by the neutrinos. The sum of the kinetic energies $E(\mu^+) + E(e^-)$ is essentially smaller than the center of mass energy $E_{cm}(e^+) + E_{cm}(e^-)$ at the production of the $\tau^+ + \tau^-$) pair. The difference is due to the neutrino energies and can be used for their determination. Energy- and momentum conservation give together an unambiguous proof of the reactions (7.18a, b, c, d).

The interaction of the τ-neutrinos could not directly be observed up to date but could be only deduced indirectly.

For leptons the "weak isospin T" has been introduced with a component T_3 which is defined as:

$$T_3 = -Q - L/2$$

Where L is the lepton number.

Examples For the electron e^- is $T_3 = +1 - 1/2 = 1/2$. For the positron e^+ is $T_3 = -1 + 1/2 = -1/2$. For the τ^+ is $Q = +1$, $L = -1 \to T_3 = -1/2$, because the τ^+ is the antiparticles to the tau- lepton τ^- and has therefore $L = -1$.

If to each lepton the corresponding neutrino is ascribed one obtains a scheme of three "lepton families, which are compiled in Table 7.3. Each of these families has its own

Table 7.3 The three lepton families

	Lepton	Lepton number	Masse in MeV	Lebensdauer
1	e^-	$L_e = +1$	0.511	∞
	e +	$L_e = -1$	0.511	∞
	ν_e	$L_e = +1$	$< 10^{-6}$	∞
	$\overline{\nu}_e$	$L_e = -1$	$< 10^{-6}$	∞
2	μ^-	$L_\mu = +1$	105.7	$2.2 \cdot 10^{-6}$ s
	μ^+	$L_\mu = -1$	105.7	$2.2 \cdot 10^{-6}$ s
	ν_μ	$L_\mu = +1$	<0.25	∞?
	ν_μ	$L_\mu = -1$	<0.25	∞?
3	τ^-	$L_\tau = +1$	1777	$3 \cdot 10^{-13}$ s
	τ^+	$L_\tau = -1$	1777	$3 \cdot 10^{-13}$ s
	ν_τ	$L_\tau = +1$	<35	∞?
	$\overline{\nu}_\tau$	$L_\tau = -1$	<35	∞?

lepton number L_e, L_μ and L_τ. For each reactions observed up to now, the lepton number within the family is conserved, i.e.

$$\sum L_e = \text{const.} \quad \sum L_\mu = \text{const.}, \quad \sum L_\tau = \text{const.} \tag{7.30}$$

The total lepton number is the sum of all family lepton numbers

$$L = L_e + L_\mu + L_\tau, \tag{7.31}$$

which is, of course also conserved.

According to our present knowledge only three lepton families do exist, i.e. there are only the three leptons e^-, μ^- τ^- and their anti-particles e^+. $\mu^+$and τ^+ plus the corresponding neutrinos ν_e, ν_μ, ν_τ with their anti-neutrinos.

This can be deduced from the lifetime of the Z^0-particle, which decays via the strong interaction into a quark-antiquark pair q $\overline{q}$(meson) and via the weak interaction into lepton-antilepton pairs or into neutrinos (see Fig. 7.10). The lifetime T is related to the resonance width $\Gamma = 1/T$ of

hadronic	leptonisch visible	leptonisch unvisible
$Z^\circ \rightarrow q\overline{q}$	$Z^\circ \rightarrow e^+e^-$	$Z^\circ \rightarrow \nu\overline{\nu}$
	$Z^\circ \rightarrow \mu^+\mu^-$	
	$Z^\circ \rightarrow \tau^+\tau^-$	

Fig. 7.10 The possible decay channels of the Z-Boson

the excitation cross section σ(E) as the function of the excitation energy E. The total width Γ is the sum

$$\Gamma = N_L \cdot \Gamma_L + N_\nu \cdot \Gamma_\nu + N_h \cdot \Gamma_h$$

where N_B is the number of decays into baryons., N_L into leptons and N_ν into neutrinos. Since the rate of decays into the different leptons is equal we get

$$\Gamma_L = \Gamma_e = \Gamma_\mu = \Gamma_\tau$$

The number N_ν of decays into neutrinos cannot be observed directly, because the neutrinos are barely absorbed by the detectors. However, they can be deduced indirectly in the following way:

$$\Gamma_{\text{invisible}} = N_\nu \cdot \Gamma_\nu = \Gamma - N_h \cdot \Gamma_h - N_L \cdot \Gamma_L$$
$$\Rightarrow N_\nu = (1/\Gamma_\nu) \cdot (\Gamma - N_h \cdot \Gamma_h - N_L \cdot \Gamma_L)$$

The energy width Γ of the resonance cross section σ(E) for the production of Z^0 bosons has been precisely measured and can be also calculated. Plotting width and height of the resonance curve for different values of N_ν, one gets a coincidence with the measured curve only for the number $N_\nu = N_L = 3$ of lepton families (Fig. 7.11).

This proves that there exist only 3 lepton families.

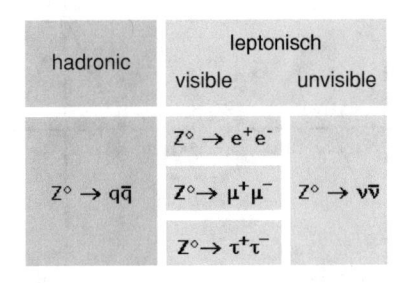

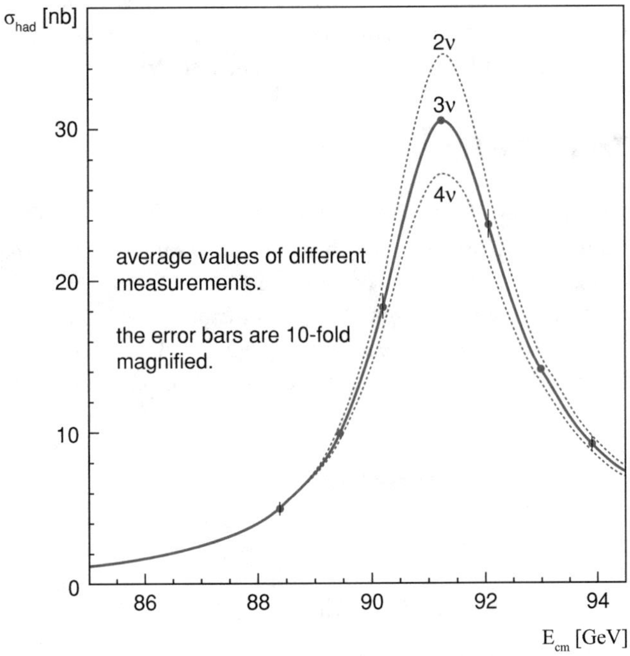

Fig. 7.11 Resonance of the production rate $N_Z(E)$ of Z^0- Bosons. The points give the average of the results from 4 different detectors. The error bars are 10fold enlarged. The curves were calculated for $N_{\text{lepton}} = 2; 3; 4;$ lepton families. The best fit to the experimental points arises for $N_{\text{lepton}} = 2.97$. (from a lecture of Eberhard, Karlsruhe Institute for Technology)

7.4 The Quark Model

The essential breakthrough for the achievement of an ordering scheme for all elementary particles was the postulation of the quark model, developed independently by *Murray Gell-Mann* and *George Zweig*. This quark model describes the particles regarded so far as elementary, now as systems with internal structure, composed of a few still more elementary particles called **Quarks.**

Plotting all known hadrons in a diagram with the mass as ordinate and the spin as abscissa (Fig. 7.12) it becomes evident, that they appear in groups with nearly equal masses, spins and parity in each group.

This suggests that all hadrons must be composed of a few elementary particles where all hadrons of one group have the same composition and the different groups differ by their composition of different quarks.

A convincing evidence for structured hadrons is the magnetic moment of the neutron. If it were a particles with no structure and no charge it should not have a magnetic moment. The quark model postulates that it is composed of sub-particles with different charges which compensate to a total charge of $Q = 0$.

Furthermore the scattering of high energy electrons at protons and neutrons revealed that both nucleons have many excited states while real elementary particles should have no internal structure and therefore also no different energy levels.

The hypothesis of the quark model becomes more and more convincing with increasing number of experimental facts, that can be well described by this model.

7.4.1 The Eightfold Way

Initially only three of such elementary building blocks of hadrons were assumed which were named by Gell-Mann as quarks (after the novel of James Joice: "*Finegin's Wake*" were the sentence occurs: "*three quarks for muster Mark*".

These three quarks are the up-quark (u), the down quark (d) and the strange quark (s) which differ from each other by the quantum numbers T of the isospin, its component T_3, the electric charge q and the strangeness quantum number S (Fig. 7.13 and Table 7.4).

To each quark with charge q an anti-quark exists with charge –q. Also the quantum numbers T_3, baryon number B and strangeness S have different signs for quarks and anti-quarks (see Table 7.4). This table also shows that all quarks have the spin 1/2.

All quarks have the spin s = 1/2. They are therefore Fermions.

Nearly all particles shown in Fig. 7.12 and in Table 7.2 (mesons and baryons) can be composed of these three quarks. In Table 7.5 some examples are given. For instance the proton is composed of 2 up quarks with charge $q = (2 \cdot 2/3) \cdot e$ and one down quark with charge $-(1/3) \cdot e$. The total charge is then $Q = +1 \cdot e$.

Examples

1. The total spin of the proton {2u, d} is $I = \Sigma Iq = (1/2)\hbar$ because the two u-quarks must have opposite spins, due to the Pauli- principle). The Baryon quantum number B is $B = 2 \cdot 1/3 + 1/3 = 1$ and the isospin component is $T_3 = 2 \cdot ½ - 1 \cdot ½ = ½$
2. The neutron n = {u, 2d} is composed of 1 up quark and two down quarks with the total charge $Q = (2/3 - 2 \cdot 1/3) \cdot e = 0$, the spin $I_n = ½$, $T_3 = -1/2$

All Baryons must be composed of three quarks, because their spin is half-integer. The mesons with spin $I = 0$ or $I = 1$ are composed of two quarks, namely one quark and one anti-quark.

Since each quark has the spin 1/2ħ, baryons which are composed of three quarks can have the spin $s = 1/2h$ or

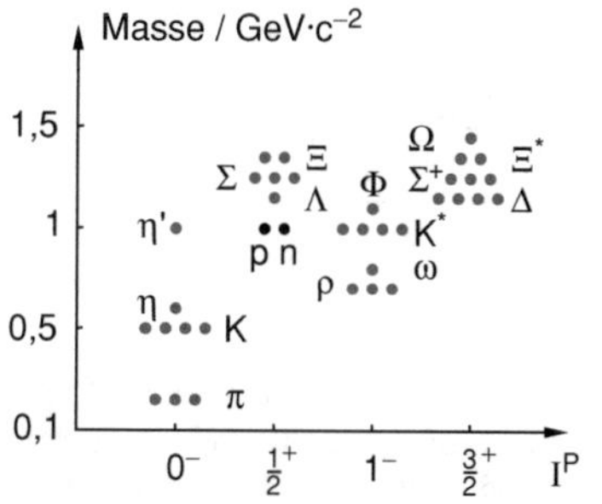

Fig. 7.12 Classification of the hadrons according to their masses, their spin and their parity

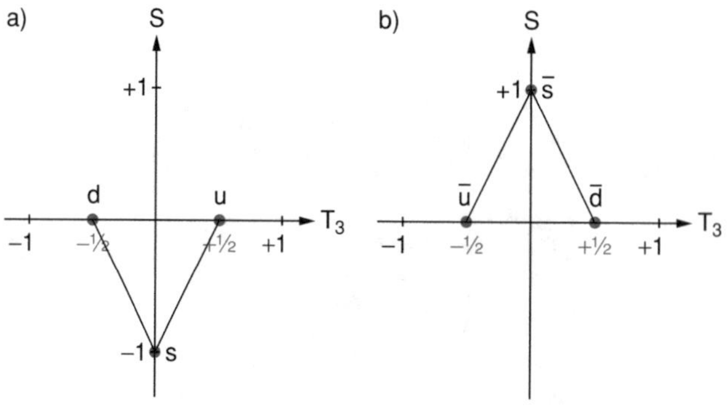

Fig. 7.13 Geometrical representation of the quarks in a T_3-S diagram, **a** u; d; s **b** the anti-quarks $\bar{u}$, $\bar{d}$, $\bar{s}$

Table 7.4 Quantum numbers of the first 4 quarks and their anti-quarks

Baryons quark composition								
Quantum numbers	u	d	s	$\bar{u}$	$\bar{d}$	$\bar{s}$	c	$\bar{c}$
Spin I	1/2	1/2	1/2	1/2	1/2	1/2	1/2	1/2
Isospin T	1/2	1/2	0	1/2	1/2	0	0	0
Isospin-component T_3	+ 1/2	−1/2	0	−1/2	+ 1/2	0	0	0
Charge Q	2/3	−1/3	−1/3	−2/3	+ 1/3	+ 1/3	+ 2/3	−2/3
Baryon number B	1/3	1/3	1/3	−1/3	−1/3	−1/3	1/3	−1/3
Strangeness S	0	0	−1	0	0	+ 1	0	0
Charm C	0	0	0	0	0	0	1	1

Table 7.5 Quark composition of some hadrons and mesons

Baryon		Meson	
Particle	Quark composition	Particle	Quark composition
Proton p	2 u + d	π^-	$d+\bar{u}$
Neutron n	u + 2d	π^-	$u+\bar{d}$
Σ^-	2d + s	K^-	$s+\bar{u}$
Σ^+	2 u + s	K^0	$d+\bar{s}$
Σ^0	u + d + s	K^+	$u+\bar{s}$
Ξ^-	d + 2 s	π^0	$u\bar{u}+d\bar{d}$
Ξ^0	$\bar{u}+2s$	η η'	$u\bar{u}+d\bar{d}+s\bar{s}$ $u\bar{u}+d\bar{d}+s\bar{s}$

$s = (3/2)\hbar$, depending on the vector addition of the three spins, whereas mesons must have the spin $s = 0$ or $s = 1 \cdot \hbar$. These results of the quark model are fully confirmed by experiments.

It turns out that the baryons with spin $(3/2)\hbar$ are excited states of the corresponding baryons with spin $1/2\hbar$. Likewise mesons with spin $I = 1 \cdot \hbar$ can be regarded as excited states of mesons with spin I = 0.

Particles which contain an s-quark get the strangeness quantum number $S = -1$. If they contain an $\bar{s}$-anti-quark, they have the strangeness quantum number $S = +1$. Within the quark model the strangeness of a particle is equal to the difference between the number of antiquarks $\bar{s}$and quarks s.

$$S = N(\bar{s}) - N(s)$$

This means that only those baryons have a strangeness quantum number $S \neq 0$ which contain only either 1 s-quark or 1 $\bar{s}$-antiquark.

In the quark model of Gell-Mann and Zweig all particles can be described by 8 quantum numbers. The access to the explanation of the substructure of hadrons was called "*the eightfold way*" where in addition to the 7 quantum numbers in Table 7.4 as 8th quantum number c of the color charge was introduced (see Sect. 7.4.6). This name was chosen following Buddha's eightfold way to reach enlightenment, where 8 virtues are required on the way to strive for perfection (*right understanding, right intention, right speech, right action, right livelihood, right effort, right mindfulness and right* concentration) [11].

The mathematical justification of this 8fold way is based on a special Lie-group called SU(3) group, which consists of 8 three-row matrices as group elements [12–13].

It is possible to classify all hadrons in families in such a way. That these families correspond to different representations of the SU(3) group. The representations are homomorphous images of the group [12, 13]. The possible representations of the SU(3) group have the dimensions 1, 3, 6, 8, 10 or even more. Therefore the families of hadrons consist of 1, 3, 6, 8, 10 ore more members.

Example The representation of the SU(3) group with 3 elements corresponds to the quark triplet of 3 quarks u, d, s. In an S-T_3 diagram they form the corners of a triangle (Fig. 7.13).

All hadrons with the same total spin I are now regarded as a family of hadrons.

7.4.2 Quarkmodel of Mesons

The families of mesons with spin $I = 0$ and $I = 1$ are represented in the diagram of Fig. 7.14a, b. These diagrams illustrate the symmetry properties of mesons, discussed later. In such a diagram these group elements of mesons form a hexagon.

We start with mesons which are composed of one quark and one antiquark.

Since altogether 9 possible combinations of quark-antiquark pairs can be formed of the 3 quarks u,d s the meson families with $I = 0$ and $I = 1$ have 9 members each. They can be classified, according to their isospin T into different isospin multiplets: a singlet ($T = 0$, $S = 0$) and an

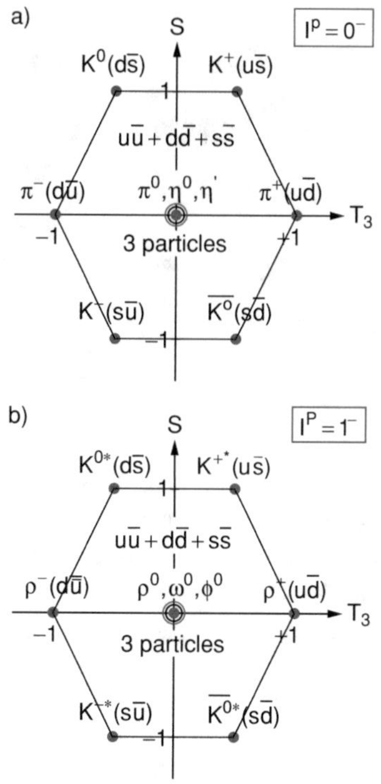

Fig. 7.14 Quark-model of the charm-free mesons **a** with spin $I = 0$ **b** with $I = 1$. All mesons have negative parity

octet ($T \neq 0$, $T_3 = 0, \pm 1$, $S = 0, \pm 1$). They can be arranged in a diagram with the strangeness S as ordinate and the isospin component T_3 as abscissa (Fig. 7.14a, b). Another way for their representation is a diagram with the charge Q as abscissa and the strangeness S as ordinate (Fig. 7.14x).

The mesons with I = 0 and negative parity are often called "*pseudo-scalars*". All mesons which contain no s-quark or the combination s + $\bar{s}$ have the strangeness quantum number S = 0. They are located in Fig. 7.14 on the central horizontal line S = 0. These are the three pions π^+ (u $\bar{d}$), π^0 ($u\bar{u}$ + $d\bar{d}$) and π^- (($d\bar{u}$), which form the isospin-triplet with T = 1 and T_3 = + 1 0, -1. The π^0-pion is its own anti-particle. It must therefore remain identical if the quarks d and u are changed into their antiparticles. The quark representation must be then

$$|\pi^0\rangle = 1/\sqrt{2}\{|u\bar{u}\rangle + |d\bar{d}\rangle\}. \tag{7.32}$$

There are two further mesons η and η′ with $T_3 = 0$ and $S = 0$, which contain s-quarks and their $\bar{s}$-anti-quarks. They have the charge $Q = 0$ and are identical to their anti-mesons. Their quark composition must be therefore

$$|\eta\rangle = 1/\sqrt{6}\{|u\bar{u}\rangle + |d\bar{d}\rangle - 2|s\bar{s}\rangle\},$$

$$|\eta\rangle = 1/\sqrt{6}\{|u\bar{u}\rangle + |d\bar{d}\rangle - 2|s\bar{s}\rangle\}, \tag{7.33}$$

where the pre-factors take care that the square of the integral over the wave function is normalized to 1, i.e.

$$|\eta\rangle = 1/\sqrt{6}\{|u\bar{u}\rangle + |d\bar{d}\rangle - 2|s\bar{s}\rangle\},$$

The strange particles with S = + 1 are the K^0-Meson ($d\bar{s}$) and the K^+-meson ($u\bar{s}$). Their charges are (see Table 7.4) q (K^0) = 0 and $q(K^+) = 1{\cdot}e$. Their anti-particles $\bar{K}^+$ ($\bar{u}s$) and $\bar{K}^0$ ($\bar{d}s$) have the strangeness quantum number $S = -1$.

In an analogue way the mesons with I = 1 (called *vector bosons*) can be arranged. Figure 7.14b shows that the quark composition of the vector bosons with $I = 1$ marked by the points is equivalent to that of the mesons with $I = 0$ in Fig. 7.14a at equivalent points. The mesons with $I = 1$ must be therefore excited states of the corresponding mesons with I = 0. This is completely equivalent to the excited states in the atomic electron shells, which also have the same composition of electrons, protons and neutrons as the ground state of the same atom. There is one big difference: The energies of the excited meson states are higher by many orders of magnitude than those of the excited states of the atomic electron shell. This can be quantitatively demonstrated by the example of the ψ-meson which will be treated in the next section.

7.4.3 Charm-Quark and Charmonium

In 1974 two research groups around *B.Richter* at SLAC and *S.Ting* at Brookhaven discovered nearly simultaneously and independent of each other a new particle with a mass m = 3097 MeV/c^2 [14–15]. The Stanford group observed an extremely sharp resonance at E = 3097 MeV when electrons and positrons collide (Fig. 7.15) at the process

$$e^+ + e^- \to \psi + c \text{ handrons};$$

$$\psi \to e^+ + e^-, \tag{7.34a}$$

They attributed a particle named ψ to this resonance, which decays into an electron–positron pair:

At Brookhaven National Laboratory the reaction

$$p + p \to J + X; \quad J \to e^+ + e^- \tag{7.34b}$$

was investigated and again a sharp resonance was found. The investigators named the particle J corresponding to the sharp resonance at 3097 MeV/c^2. It turned out that the two particles ψ and J were identical and the particle was therefore named J/ψ. More detailed investigations showed that the ψ/J-particle could not be composed of the quarks known up to now, because its mass were larger than that of a pair of known quarks. The width of the resonance shown in

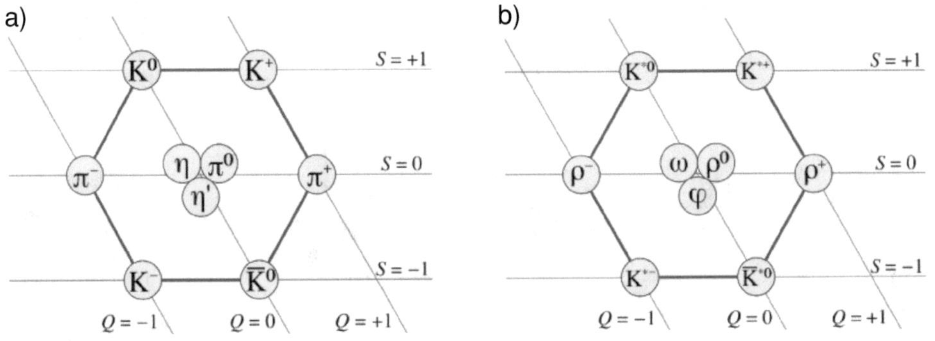

Fig. 7.14x Nonet of mesons a) with spin 0 and b) with spin 1. https://commons.wikimedia.org/w/indexphp?curid=3915252

Fig. 7.15 of about 2 MeV is mainly determined by the energy uncertainty of the electrons or protons. More accurate measurements showed later that the real width is only $\Gamma = 88$ keV corresponding to a lifetime of the particle of $\tau = 7{\cdot}10^{-21}$ s. This is about 3 orders of magnitude longer than typical decays caused by the strong interaction.

With the postulation of a new quark, called *charm quark* with the charge q = + 2/3e, the J/ψ-particle could be interpreted as $(c\bar{c})$ combination of the charm quark c and its anti-quark $\bar{c}$. The charm quark had been already theoretically postulated in 1970, long before the J/ψ-particle was experimentally discovered.

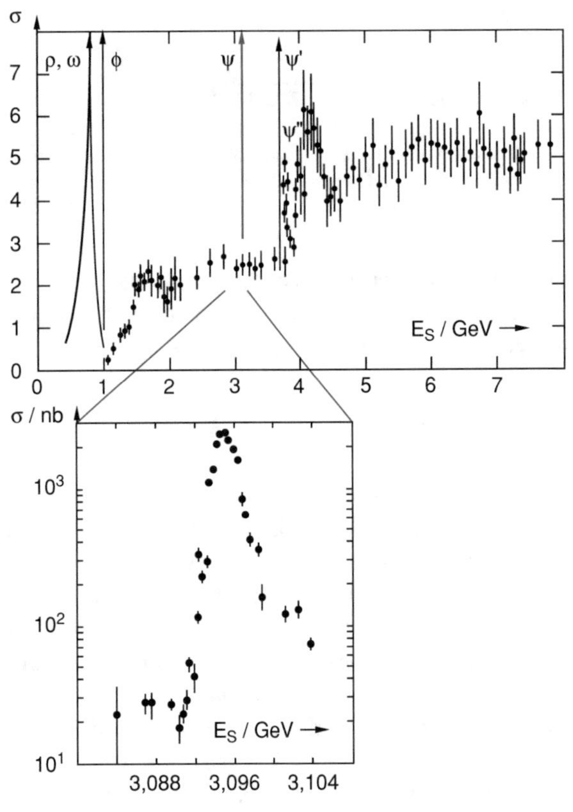

Fig. 7.15 Discovery of the ψ-meson, which appears as very sharp resonance in the reaction cross section $e^+ + e^- \rightarrow \psi$. (from G.J. Feldman M.L.Perl Phys. Reports 19C, 233 (1975))

The J/ψ-meson is a vector boson with a spin $I = 1{\cdot}\hbar$ and a negative parity as all other mesons.

Some days after the discovery of the ψ-meson another resonance was found at $E = 3.7$ GeV, The corresponding particle was named ψ', because it decays according to the scheme

$$\psi' \rightarrow \pi^+ + \pi^- + \psi \rightarrow \pi^+ + \pi^- + e^+ + e^-$$

which suggested, that ψ′ is an excited state of ψ. This was fully confirmed after several further excited states of the ψ-meson were found.

The J/ψ-meson as a two-quark system $(c\bar{c})$ is analogously built up as the positronium (e^+e^-) (Fig. 7.16). It was therefore named **charmonium**. In the following we will use the simpler notation ψ instead of J/ψ.

The excited states appear as resonances in the cross section of the reaction.

$$e^+ + e^- \rightarrow \psi^*$$

and were named as ψ, ψ′, χ_0-χ_1, χ_2 η_c-η_c' as indicated in Fig. 7.16.

Note

1. The different energy scale in the two systems e^+e^- and $(c\bar{c})$.
2. The nomenclature of the energy states differs in the two systems regarding the principal quantum number n which is defined for the positronium (according to the rules of atomic physics) as $n = N + 1 + \ell$ (N = number of nodes in the radial wave function and ℓ is the angular momentum quantum number) whereas for the charmonium $n = N + \ell$.

The accurate measurements give information about the interaction potential between the charm-quark and its anti-quark. The potential can be written in the form

$$E_{\text{pot}} = -A/r + B \cdot r$$

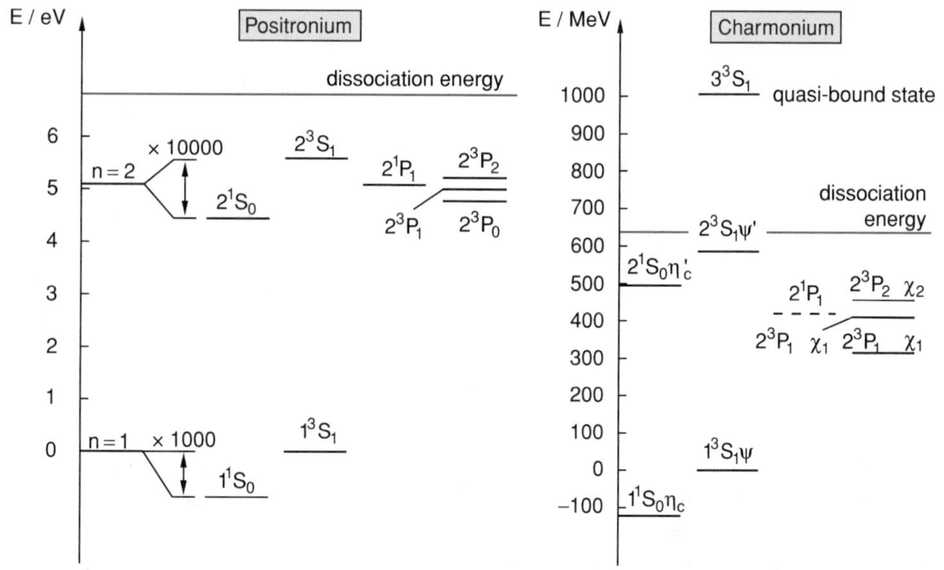

Fig. 7.16 Comparison of the energy term schemes of positronium and charmonium. Note the different energy scale!

It contains the attractive Coulomb potential between the charged quarks and a term that increases with r. This last term gives a barrier for the dissociation of the two quarks.

The excited states of the Charmonium can be determined by measuring with energy-resolving detectors the γ-quanta emitted when the excited charmonium decays into the ground state (Fig. 7.17).

The real ground state 1^1S_0 cannot be reached by $e^- + e^+$ collisions because of symmetry reasons. It can be only populated by γ-emission from higher excited states. Therefore the zero point of the energy scale in Fig. 7.16 was placed at the lowest state 1^3S_1 that can be reached by $e^+ + e^-$ collisions. The collisional formation of $(c\bar{c})$ proceeds via the production of a γ-quant

$$e^+ + e^- \to \gamma(1^-) \to c\bar{c}.$$

The γ-quant has the angular momentum $1 \cdot \hbar$ and negative parity, whereas the 1^1S_0 state has the angular momentum quantum number 0 and positive parity. The lowest excited state that fulfills the symmetry conditions is the 1^3S_1 state which therefore can be populated by collisions. It is this state where the corresponding resonance in Fig. 7.15 was first found.

The ψ-meson cannot decay into hadrons containing a charm quark, because its energy is not sufficient to produce a hadron pair, where one hadron contains a c-quark the other an anti-$\bar{c}$ quark. The excited $^3S_1$-state cannot decay directly into other hadrons because of symmetry reasons, but it can decay by an indirect route via 3 Gluons (see Sect. 7.5.1). This drastically diminishes the decay probability through the strong interaction and is mainly responsible for the long lifetime. Nevertheless the decay induced by the strong interaction amounts to 70% of all decay channels while the rest of 30% is caused by electromagnetic and weak interaction. The decay due to the weak interaction ends up in lepton pairs $e^+ + e^-$ or $\mu^+ + \mu^-$.

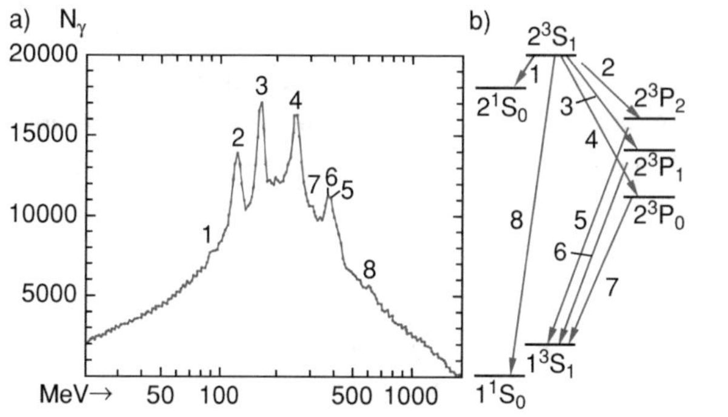

Fig. 7.17 **a** Experimental γ-spectrum of charmonium reflecting the different energy states. **b** Corresponding energy term scheme

7.4.4 Quark-Composition of Mesons

Taking into account the charm quark, all mesons with a given spin can be arranged in a **super-multiplet**. This is illustrated in Fig. 7.18a for mesons with spin $I = 1$ and in Fig. 7.18b for mesons with spin $I = 0$. The location of a meson in these diagrams is determined by the three quantum numbers isospin component T_3, strangeness S and charm quantum number c. The 16 mesons with spin $I = 0$ and also those with $I = 1$ occupy the 12 corners of the cubo-octaeder and its center where 4 mesons are located. The red central plane $C = 0$ contains the charm-free mesons with $c = 0$, which have been already shown in Fig. 7.14. and the mesons containing a $(c\bar{c})$-

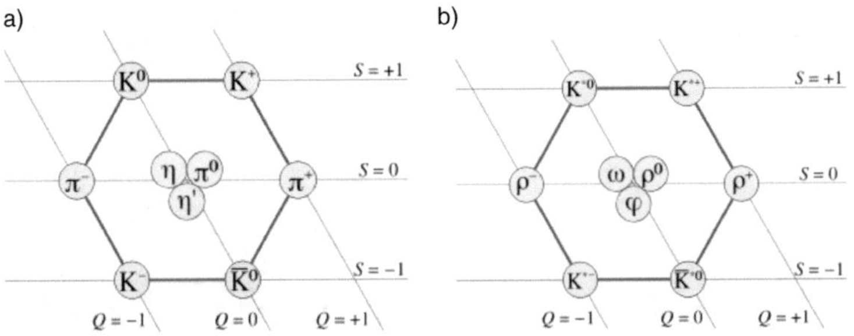

Fig. 7.17x Nonet of the quark composition of mesons with $c = 0$. **a**) with I = 0, **b**) with I =1

pair. It turns out, that the mesons with spin $I = 1$ can be regarded as excited states of the mesons with $I = 0$ at the corresponding locations. For example the $(s\bar{u})K^{*-}$ meson with $I = 1$ is the excited state of the $(\bar{s}u)K^-$ with $I = 0$ (Fig. 7.17x).

The question is now: Why do the pairs of quark and antiquark not immediately annihilate by emission of a γ-quantum with the energy $E = 2\ mc^2$ where m is the mass of the quark? In fact the lifetime of mesons that consist of a quark and its anti-quark, such as $(u\bar{u})$ or $(c\bar{c})$ is generally very short ($<10^{-23}$ s) while that of mesons consisting of a quark and another anti-quark (such as $u\bar{d}$) or $(d\bar{s})$ is much longer, because one of the two quarks has to change its flavor (see Sect. 7.4.6) in order to change the meson composition into a pair of quark and its own anti-quark. This can only happen by the weak interaction which causes a much longer lifetime of the mesons.

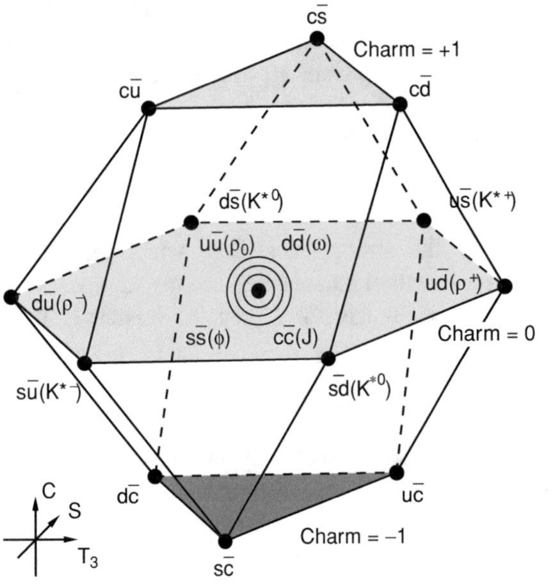

Fig. 7.18a Super-multiplet of mesons with spin $I = 1$ (From S.L. Glashow, Sci. Am. Oct. 1975)

7.4.5 Quark Composition of the Baryons

Since the spin of quarks is $I = 1/2$, all baryons that are composed of three quarks must have a half-integer spin quantum number $I = 1/2$ or $I = 3/2$.

> The spin of baryons is always a half-integer of ħ contrary to the mesons with a spin which is an integer multiple of $\hbar$.

Similar to the mesons the different baryons with $c = 0$ can be plotted in symmetric S-T_3 diagrams in the plane $c = 0$. The baryons with $I = 1/2$ form an octet (8 baryons) while the baryons with $I = 3/2$ form a decuplet (10 baryons) (Fig. 7.19).

This corresponds again to the possible dimensions of the representation of the $SU(3)$ group. As can be seen in Fig. 7.19a there are two baryons with $I = 1/2$ and $S = 0$. These are the two nucleons proton and neutron. There are 4 baryons with $I = 1/2$ and $S = -1$ which contain 1 s-quark (Σ-hyperons) and 2 baryons with 2 s-quarks (Σ-baryons). In Fig. 7.19a the charm-free baryons with spin $I = 1/2$ are listed

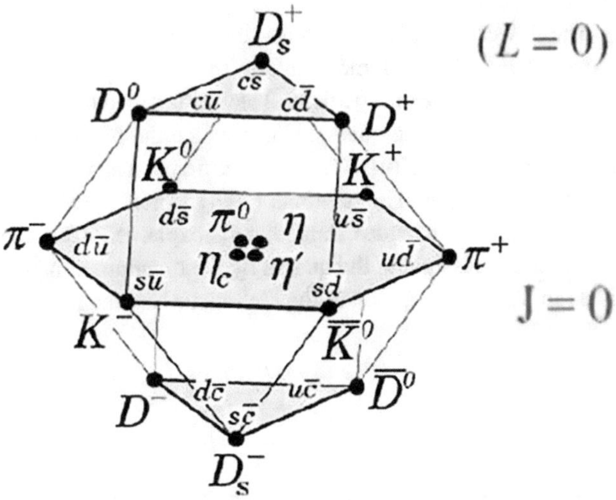

Fig. 7.18b Super-multiplet of mesons with spin $I = 0$ (Wikipedia)

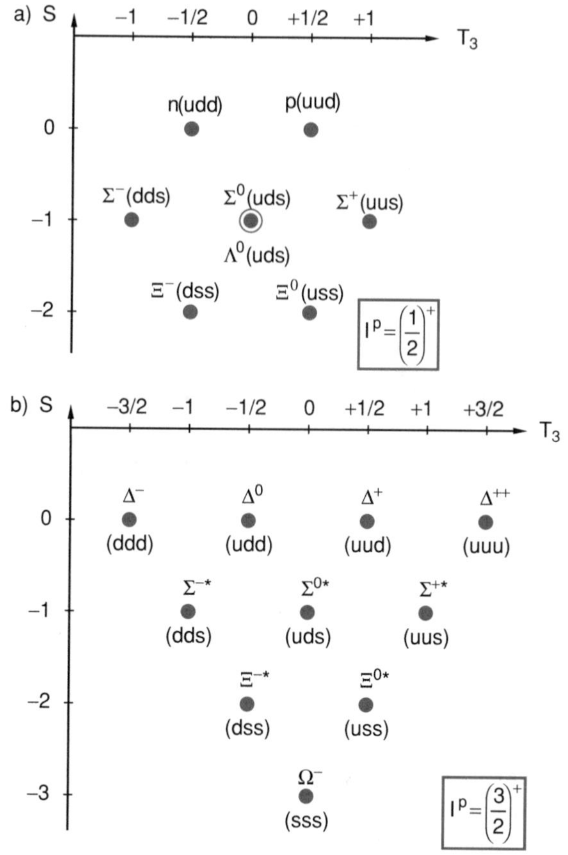

Fig. 7.19 **a** Baryon Octet of all charm-free baryons with I = 1/2. **b** Decuplett with all charm-free baryons with I = 3/2

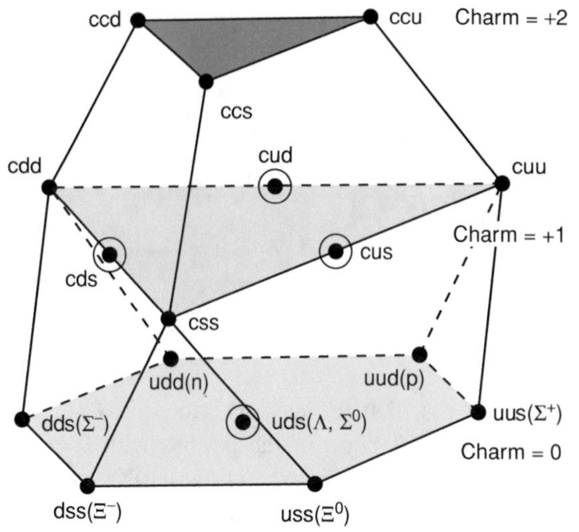

Fig. 7.20 Supermultiplet in a T_3- S- c-Space for all baryons with spin 1/2. The points surrounded bay circles are doubly occupied

and in Fig. 7.19b the charm-free baryons with I = 3/2 are listed. Besides the Ω^- the Δ^- and the Δ^{++}.baryons the other baryons in Fig. 7.19b are excited states of the baryons with spin 1/2., because they have the same quark composition.

Similar to the diagrams of the mesons also the baryons can be arranged in a three-dimensional scheme if the charm quark is included (Fig. 7.20).

The lower red plane contains the charm-free baryons shown in Fig. 7.19a. Some points in the diagram are occupied by more than one particle. They differ in the relative spin-orientation of the quarks. For instance the two compositions (u↑d↑s↓) and (u↑ds↓s↑) form different baryons with the same total spin. The relations (7.24) which were found empirically, can be now immediately explained with the quark model: Assigning the quarks by the quantum numbers listed in Table 7.4 we obtain the Gellmann-Nishina relation

$$Q = \frac{1}{2}(B + U - D + S + C)$$
$$= T_3 + \frac{1}{2}(B + S + C),$$

where B is the baryon number, S the strangeness quantum number, $U = n(\mathrm{u}) - n(\bar{\mathrm{u}})$, $D = n(\mathrm{d}) - n(\bar{\mathrm{d}})$ the difference of the u and $\bar{\mathrm{u}}$-quarks resp. the d and $\bar{\mathrm{d}}$ quarks, $T_3 = 1/2(U - D)$ and $C = n(\mathrm{c}) - n(\bar{\mathrm{c}})$.

Examples

1. The Σ^+ baryon (uus) has $B = 1$, $S = -1$, $C = 0$, $T_3 = 1$, and $D = 0$. Q is therefore $Q = 1/2(1 + 2 - 1 + 0 + 0) = 1$
2. The Ω^--particle (sss) has $T_3 = 0$, $B = 1$, S = −3, C = 0 ⇒ Q = −1.

The quark model was convincingly confirmed when the Ω^--particle was found, which had been already predicted as a possible quark combination of strange quarks. Also its mass and its possible decay channels had been correctly predicted.

It is produced by the reaction

$$\mathrm{K}^- + \mathrm{p} \to \Omega^- + \mathrm{K}^+ + \mathrm{K}^0$$

induced by the strong interaction, which preserves the strangeness quantum number S. Since the strangeness of K^- mesons is $S = -1$ and for K^+ and K^0-mesons is $S = +1$ (see Sect. 7.2.6) the strangeness S of the Ω^--particle must be

$$S = -1 + 0 - 1 - 1 = -3.$$

The Ω^- baryon cannot decay by the strong interaction (which preserves B, S and Q) because its mass is smaller than any possible baryon combination that fulfills the condition $B = 1$, $S = -3$ and $q = -$e. It can therefore only decay through the weak interaction, which results in a much longer lifetime. Its ionization track in a bubble chamber should be

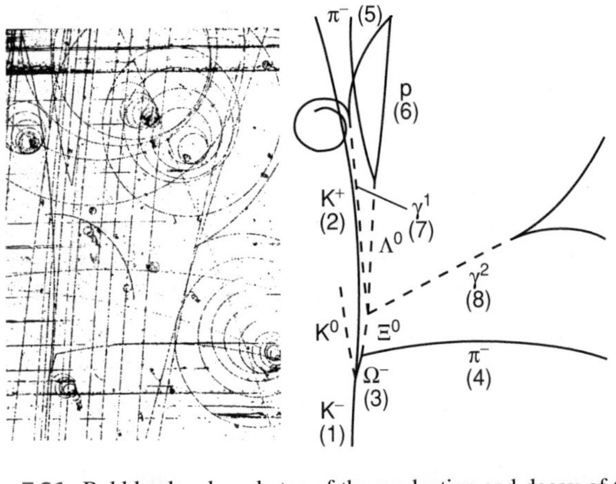

Fig. 7.21 Bubble-chamber photos of the production and decay of the Ω^--baryon. The dashed curves in the right drawing represent neutral particles which do not show any spur. The left photo illustrates that besides the wanted spur a wealth of other events from which the wanted event has to be extracted (V.E. Barnes et al. Phys. Rev. Lett.12, 204 (1964))

therefore observable, before it decays according to the scheme (Fig. 7.21)

$$\left.\begin{array}{ll}\Omega^- \to & \Xi^0 + \pi^- \\ & \quad \llcorner \Lambda^0 + \pi^0 \\ & \qquad \llcorner p + \pi^-\end{array}\right\} \Rightarrow \Omega^- \to 2\pi^- + \pi^0 + \mathrm{p} \qquad (7.35)$$

The bubble chamber track of the Ω^- particle illustrates that the analysis is by no means trivial, because many other tracks from various particles make the unambiguous reconstruction difficult and demands much intuition.

The Ω^- baryon as well as the Δ^{++}-baryon with spin 3/2 imposed a great challenge for the quark model discussed so far, because these particles are composed of 3 identical quarks with parallel spins. Since the quarks with spin 1/2 are Fermions, they have to obey the Pauli-principle, which excludes three identical particles in equal quantum states.

It has been extensively discussed whether the quark model should be abandoned which had proved very successful for the description of all other particles or whether the Pauli-Principle should be dropped which had not been doubted up to now.

The salving idea was proposed by *O.W. Greenberg*, who postulated to introduce a further quantum number which he called "**Color**". It has nothing to do with the usual meaning of color in daily life, but should only serve as a way to distinguish quarks that are identical in all other characteristic quantities (spin, parity, mass, charge, isospin and its component T_3). Greenberg postulated that all quarks could exist in three colors which he named red, green and blue. In this extended quark model the three quarks in the Ω^-- and the Δ^{++}-baryons are assumed to have a red, green and blue color and therefore differ in their color and are no longer identical particles. This saves the Pauli-principle even for the Ω^- and the Δ^{++}-baryons.

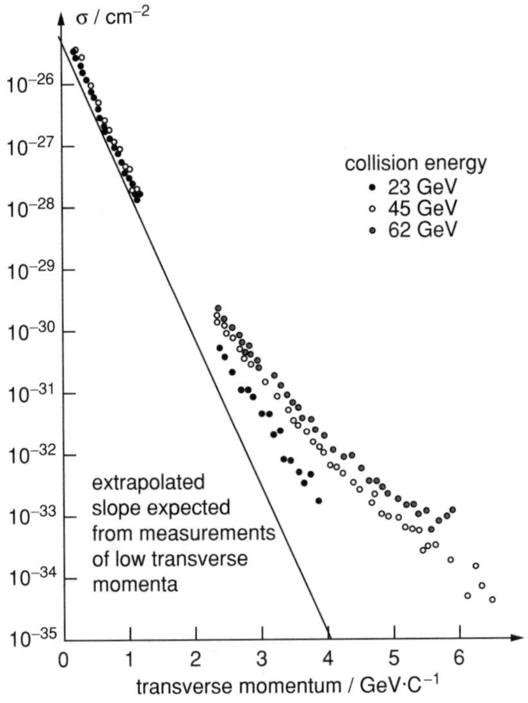

Fig. 7.22 Cross section of the reaction p + p → new particles as a function of the transverse momentum transfer during anti-collinear collisions. (From M. Jacob, P. Landshoff, Spektrum der Wissenschaft Mai 1980)

This idea initially met with great skepticism, because the ad-hoc introduction of a new property of quarks which had not been proved by any experimental evidence and with the only advantage that the Pauli principle could be saved, was not very convincing.

It turned out, however, that the introduction of this new property of color had many far reaching consequences which contributed to a much better understanding of the strong interaction. The property of quark colors is responsible for the strong interaction which holds the quarks within nuclei together against the repulsive electro-magnetic interaction between the positively charged quarks (for instance between the 2 u-quarks with charges (2/3)e). Analogous to the electro-magnetic interaction caused by electric charges the color property of the quarks responsible for the strong interaction is called **color charge**.

While the electro-magnetic interaction is caused by the exchange of photons, the strong interaction is effected by the exchange of **Gluons**, which interact with the color charges. There are now a wealth of experimental results which support the existence of color charges.

7.4.6 Color Charges

The color model triples the number of different quarks. One might naively assume that the number of possible combinations of two quarks in mesons now increases by a factor of 9 ($3^2 = 9$) and the number of baryons even by a factor of $3^3 = 27$. However, this is in fact not the case, because of certain combination rules which restrict the number of possible color combinations and which also explain, why up to now no free quarks have been observed.

These combination rules say that only those color combinations are allowed in an observable particle, that result in a white (colorless) combination.

- A possible color combination is for instance red–green–blue. All baryons which consist of 3 quarks must be composed of 1 red, 1 green and 1 blue quark. The anti-quarks carry the color anti-red, anti-green and anti-blue. Only those particles are observable which have a total color charge white. The combination of the color charge of a quark with the anti-color charge of another quark does not necessarily give a white color. Therefore a second combination rule must ensure that mesons, which consist of one quark and one anti-quark are always white, because they are observable. Some mesons are composed of a quark and an anti-quark with equal colors. For instance the π^+-meson can occur as
- $\pi^+ = (u_r, \bar{d}_r)$ or $(u_g, \bar{d}_g)$ or $(u_b, \bar{d}_b)$
- Since each of these combinations is equally probable, the pion should be described by the color combination
- $\pi^+ = \frac{1}{3}\{u_r\bar{d}_r + u_b\bar{d}_b + u_g\bar{d}_g\}$
- Which is white (colorless).
- The hypothesis of color charges appears at a first sight bizarre, but has been later strongly supported by a thorough theory called **Quantum Chromo-Dynamics** (see Sect. 7.5), which correctly explains all experimental findings up to now.

7.4.7 Experimental Hints to the Existence of Quarks

Although the quark-hypothesis has successfully brought a certain systematic order in the particle zoo of an increasing number of different particles, many physicists believed that the quarks were pure hypothetical particles without any real existence, because they had not been found as free single quarks in spite of intense search.

However, there are many experimental proofs of the real existence of bound quarks inside observable particles.

1. The fact that the neutron has a magnetic moment proves that it must consist of charged particles with a total charge zero.
2. If the proton were a real elementary particle with no substructure of sub-particles, its positive charge distribution for the model of a conductive sphere should be concentrated on the surface of the proton because of the repulsive force between the positive charges. For a dielectric sphere the charge should be uniformly distributed over the volume of the sphere. Both models differs from the charge distribution confirmed by scattering experiments, which proves that none of these models could be correct.
3. Convincing hints to the quark-composition of hadrons is given by inelastic high energy collisions between hadrons in colliders where new particles are produced, which could be readily explained by the quark model. If, for instance, the proton were a homogeneous charged particle, one would observe in the crossing points of a high energy collider scattered protons under a scattering angle that depends on the center of mass energy and the impact parameter. The scattering cross section should decrease with increasing scattering angle. i.e. increasing transverse momentum as shown by the black line in Fig. 7.22.

Measurements for large impact energies performed at CERN showed, however, essentially larger cross sections for larger scattering angles at higher energies. Furthermore not only elastic scattering of protons were observed but also the production of new particles. This cannot be explained if the proton were an unstructured homogeneous particle.

Measuring the vector sum of the momenta of all scattered particles, it should be zero, because the total momentum of the two colliding anti-collinear protons is zero. The total energy of all scattered particles must be equal to the total energy of the colliding particles before the collision. The Feynman diagram of one of such processes

$$p + p \rightarrow n + \pi^+ + n + \pi^+ + \pi^0$$

within the framework of the quark model is shown in Fig. 7.23. It illustrates that new quark pairs are produced (e.g. $d\bar{d}$ on the left side or $u\bar{u}$ and $d\bar{d}$ on the right side), which separate and recombine with other quarks to form new particles. The collisions of the two protons is in fact a collision between the quarks in the protons.

Also in anti-collinear electron-positron collisions at high energies new particles are formed, as for instance the ψ-meson discussed in Sect. 7.4.3. The production mechanism in the quark model proceeds via a virtual photon:

$$e^+ + e^- \rightarrow \gamma \rightarrow c + \bar{c}, \quad (7.36)$$

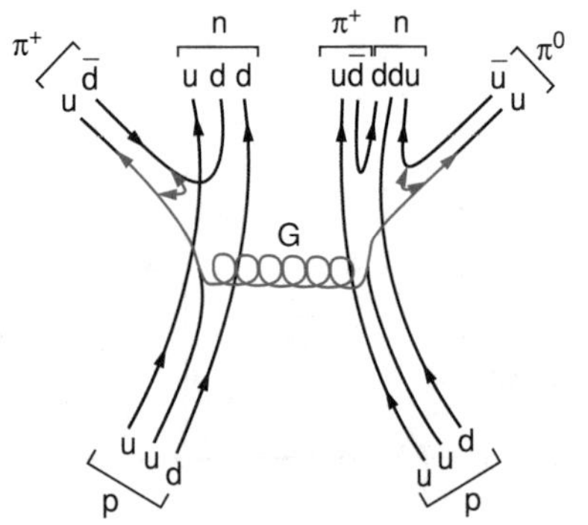

Fig. 7.23 Feynman Diagram of the reaction p + p → hadrons (from M.Jacob, P. Landshoff, Spektrum der Wissenschaft May 1980)

which serves as intermediary for the annihilation of the e^+e^- pair and the generation of the c+$\bar{c}$-charm quark pair forming the ψ-meson.

In general the γ-quantum produced in central collisions e^++e^- decays into a quark-anti-quark pair q+$\bar{q}$. Because of momentum conservation the two quarks must fly into opposite directions transverse to the directions of the colliding electron-positron particles. Before they have separated by more than a hadron radius the field energy of their strong interaction is converted into new particles (hadron jets see Sect. 7.5). Therefore one cannot see the free quarks but only the newly produced particles.

Also at collisions of high energy neutrinos with protons new particles can be produced, as shown in the bubble chamber photograph in Fig. 7.24. Here the track of the neutral neutrino, coming from the left, cannot be seen. The analysis of the produced hadron jets shows that the neutrino hits a proton and the quarks in the proton produce new hadrons.

All these and many more experimental results have meanwhile convinced the large majority of Physicists of the real existence of quarks. Whether the quarks themselves are really elementary particles, or whether they are composed of still smaller sub-particles is still a matter of controversy discussion. Also the question why no free quarks have been found was long discussed, until plausible models have been developed which can convincingly explain the "confinement" of quarks, which excludes free quarks (see Sect. 7.5.2).

7.4.8 Quark Families

During recent years two new quark-types have been discovered which have gotten the names "**bottom-quark**" (b-quark) and "**top-quark**" (t-quark). This increases the total number of quark-types to 6 (Fig. 7.24x) with additional 6 anti-quarks (not shown).

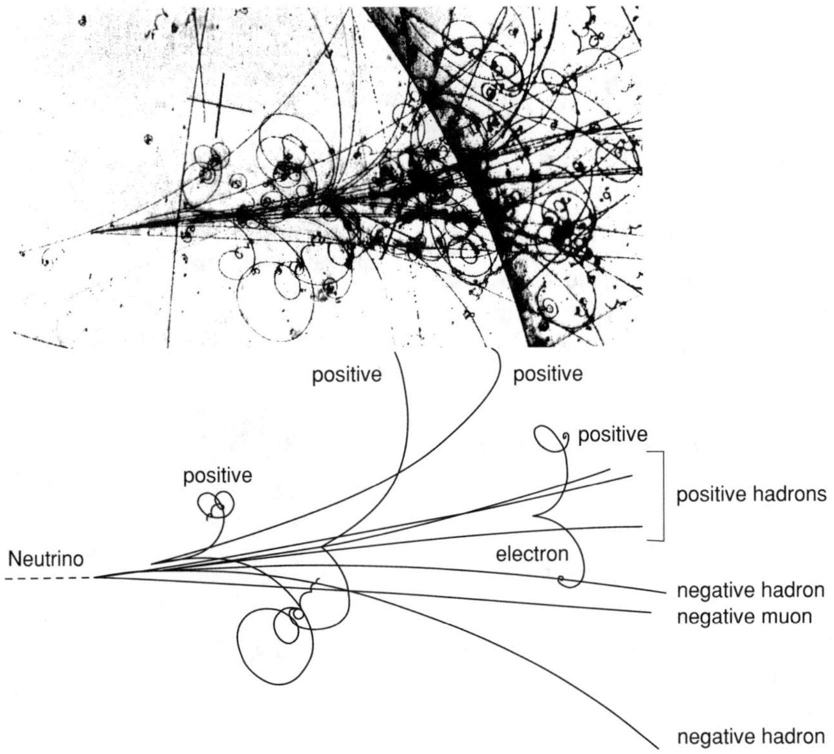

Fig. 7.24 Bubble-chamber photo of the reaction ν_e + p → hadron + muon. The lower drawing extracts the relevant spurs from the vast number of photographed spurs in the upper part (with kind permission of CERN)

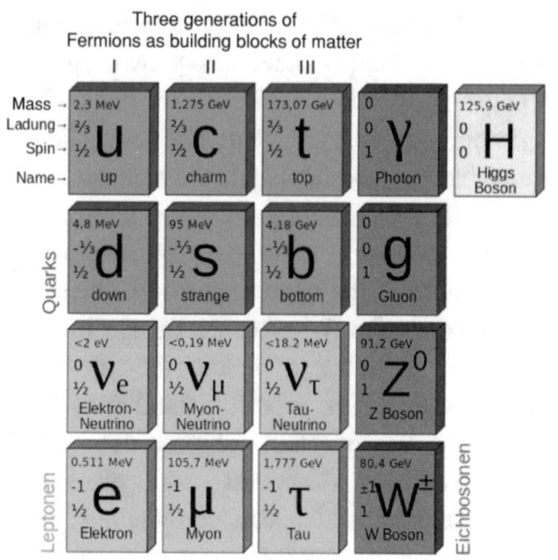

Fig. 7.24x The six quarks and 6 leptons as Fermions and the 4 bosons corresponding to the different interactions and the Higgs boson (Wikipedia)

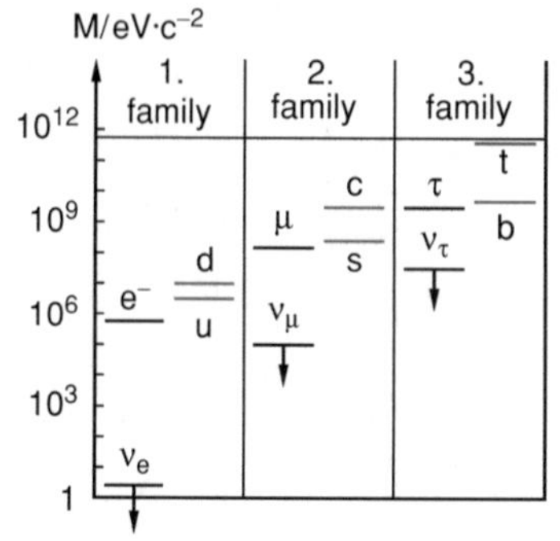

Fig. 7.25 Mass-scale of the Fermions i.e. 6 leptons and 6 quarks. (From H, Schopper: Materie und Antimaterie (Piper München 1989))

Each of these quarks can exist in 3 colors. The masses of these quarks is illustrated in Fig. 7.25 on a logarithmic scale and compared with the masses of the 6 leptons. For the distinction between the 6 quark types the name "*flavor*" was introduced. For example u and d-quarks have different flavors, i.e. they are different quark types, but each quark flavor can exists with three different color charges.

By its flavor and its color charge each quark is uniquely characterized. We will see later that the quark flavor can be changed only by the weak interaction while the color charge can be altered only by the strong interaction.

The 6 different quark flavors can be divided in 3 families with increasing mass (Table 7.6). Together with the anti-quarks each family contains 4 members. Taking into account the 3 color charges we get 12 members in each quark family.

Note that this division into families is quite similar to the 3 lepton families (see Table 7.3) apart from the color charge intermediated by the strong interaction, which does not exist for the leptons.

It seems that there are definite symmetries in nature, which are made apparent by the quark model. Some years ago it was experimentally proved that only 3 lepton families exist. Applying this symmetry principle to the quarks one might conclude that there are also only three quark families. This would imply that already all quark families have been found. This is, however, only a belief, but no proof, and it cannot be excluded that our present quark model has to be extended.

7.4.9 Valence Quarks and Sea-Quarks

The quarks that define the quantum numbers of the baryons, are called "valence quarks. Examples are the *u*- and the *d*-quark which define the quantum numbers of the nucleons proton and neutron. The name "valence quarks" has been borrowed from atomic physics where the valence electrons in the atomic shell determine the chemical properties of the atoms. Besides these "real" quarks there are also "virtual" quarks, which are formed through the interaction with the gluons for a short time as quark-anti-quark pairs and then decay again into gluons (see Sect. 7.5).

The major contribution to the masses of the quarks is given by the valence quarks, while the virtual quarks contribute only a small percentage to the nuclear mass. In this model (Fig. 7.26) which postulates that the baryon (e.g. the proton) consists of a sea of virtual quarks (which are therefore also named "sea-quarks" which surround the fixed valence quarks. The sum of the quantum numbers of these sea-quarks is always zero, therefore they do not influence the measurable properties of the baryons. Their influence

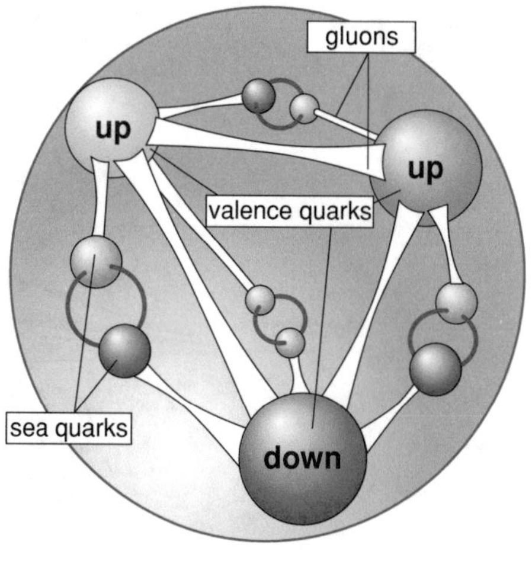

Fig. 7.26 Illustrative model of valence- and sea-quarks [22]

Table 7.6 The three quark families with two-quarks and two anti-quarks in each family

1. Familiy		2. Familiy		3. Familiy	
Quark	Charge	Quark	Charge	Quark	Charge
d	−1/3*e*	s	−1/3*e*	b	−1/3*e*
u	+ 2/3*e*	c	+ 2/3*e*	t	+ 2/3*e*
$\bar{d}$	+ 1/3*e*	$\bar{s}$	+ 1/3*e*	$\bar{b}$	+ 1/3*e*
$\bar{u}$	−2/3*e*	$\bar{c}$	−2/3*e*	$\bar{t}$	−2/3e

becomes only visible at inelastic high energy collisions, because the valence quarks take only about 50% of the momentum transferred at the collision. The rest is taken by the sea-quarks and the gluons. The sum of the masses of all valence quarks is therefore smaller than the baryon mass.

7.5 Quantum-Chromodynamics

The quark model discussed so far leaves a couple of open questions:

1. Are the quarks real or are they only fictive particles which serve as excellent working hypothesis, which is very useful for explaining the properties of all particles in the particle zoo?
2. Why were no free quarks found in spite of intense search?
3. What is the physical meaning of the color charges?
4. Why is it possible to produce quarks which are connected with the strong interaction by lepton collisions which are not related to strong interaction but obey only weak and electro-magnetic interaction?
5. Why can hadrons decay into leptons (preserving the hadron- and lepton numbers)

These questions are mainly answered by a comprehensive theory, called **Quantum-Chromodynamics, which** is structured analogue to quantum-electrodynamics. Since this theory is mathematically demanding, it will surpass the level of this textbook. We will therefore report its most important points in a more descriptive way.

While electrodynamics deals with the electromagnetic interaction which acts between electric charges, quantum-chromodynamics (from the Greek word χρωμα = color) describes the strong interaction between quarks caused by their color charges. We will call therefore this interaction a *color-force* or *color interaction.*

7.5.1 Gluons

The electro-magnetic interaction can be described by the exchange of photons with rest mass zero and spin 1 h (vector bosons). This results in the distance dependence for the Coulomb force $F_C \propto 1/r^2$. Analogue is the color force caused by the exchange of other vector bosons with rest mass zero and spin 1·ħ, which are called **Gluons** (because they act as glue to bind the quarks together). With Feynman diagrams (Fig. 7.27) one can depict the different interactions in a shorthand form (see Sect. 5.4) and illustrate their common features. There is, however, an essential difference between photons and gluons: While the photons do not carry charges and therefore there is no interactions between photons, the gluons do carry color charges, which enables them to interact with each other.

Analogue to the quarks the gluons can exist in three different colors. Each gluon carries a specific color or its anti-color. One would therefore expect 3 × 3 = 9 different gluons. In Fig. 7.28 eight of them are illustrated, where, because of typographical reasons, the colors are described by the corresponding letters *r*, *g* and *b* for red, green and blue and $\bar{r}$, $\bar{g}$ and $\bar{b}$ for the anti-colors. It should be emphasized again, that these color charges have nothing to do with the ordinary colors in daily life, but are solely used as

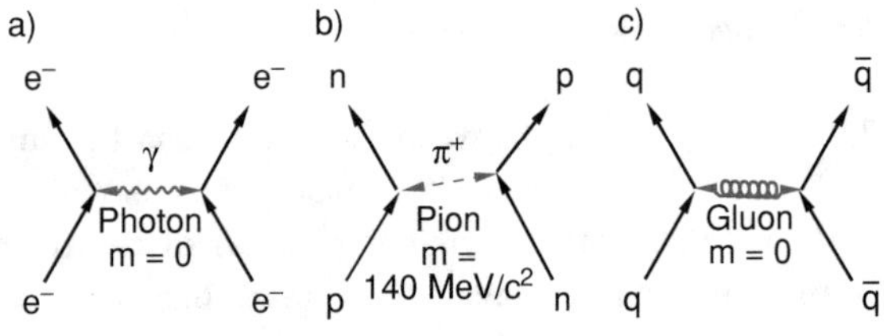

Fig. 7.27 Feyman diagrams of **a** electro-dynamic interaction **b** strong interaction **c** color interaction between quarks

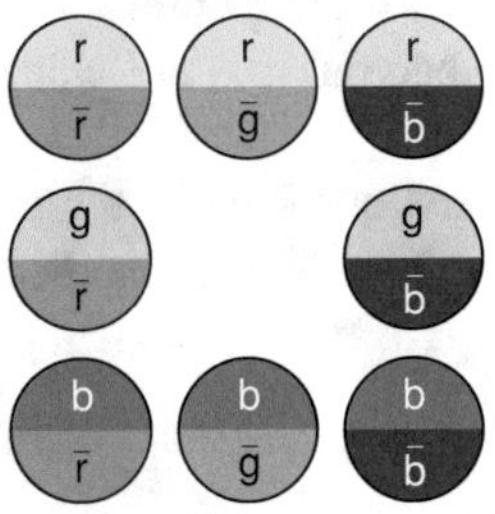

Fig. 7.28 The 8 possible gluons with their color charge which contribute to the strong interaction

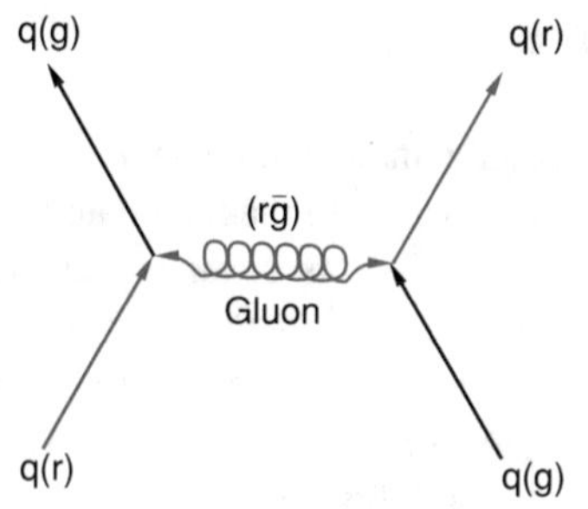

Fig. 7.29a Color-force between two quarks by exchange of gluons

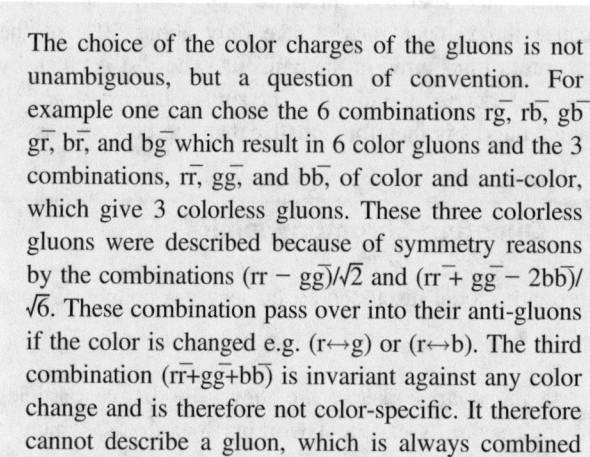

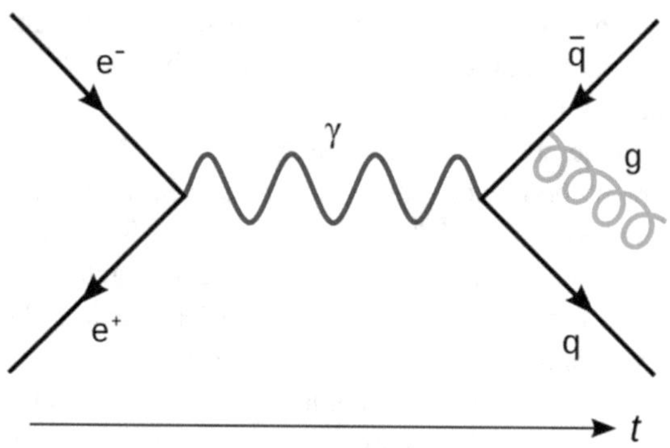

Fig. 7.29b Feynman diagram of the production of a quark-antiquark pair by $e^+ + e^-$ collisions

convenient and vivid way to distinguish the different gluons (Fig. 7.29b).

The color interaction between the quarks can be simplified as follows: Two quarks can exchange gluons with each other. This changes the color composition of the quarks. Every time a gluon is emitted by a quark the quark color changes. The emitted gluon travels with the speed of light until it is absorbed by another quark in the same baryon (Fig. 7.29a). The absorbing quark changes also its color in such a way, that the color changes of the emitting and that of the absorbing quark just compensate. This ensures that the color charge of the baryon does not change.

Example When a red quark emits a gluon ($r\bar{g}$) it becomes green (g). When the gluon ($r\bar{g}$) is absorbed by a green quark, the quark becomes red, because $g + \bar{g}$ gives a white color and therefore only red is left.

The quarks inside a hadron are hold together by a continuous exchange of gluons. Although they change their color at the emission and the absorption of gluons the hadron itself remains always white as postulated by the QCD-theory for observable particles.

While the quark color changes for every emission or absorption of gluons on the time average each color is equally present in a quark.

The hypothesis that quarks and gluons carry color charges can be proved by experiments. Calculating the interactions at collisions between baryons or between leptons and baryons and deriving the reaction cross sections, the calculations differ considerably from the experimental results if the quarks had no color charges. They agree excellently if the color charges are taken into account.

The choice of the color charges of the gluons is not unambiguous, but a question of convention. For example one can chose the 6 combinations $r\bar{g}$, $r\bar{b}$, $g\bar{b}$ $g\bar{r}$, $b\bar{r}$, and $b\bar{g}$ which result in 6 color gluons and the 3 combinations, $r\bar{r}$, $g\bar{g}$, and $b\bar{b}$, of color and anti-color, which give 3 colorless gluons. These three colorless gluons were described because of symmetry reasons by the combinations $(r\bar{r} - g\bar{g})/\sqrt{2}$ and $(r\bar{r} + g\bar{g} - 2b\bar{b})/\sqrt{6}$. These combination pass over into their anti-gluons if the color is changed e.g. (r↔g) or (r↔b). The third combination $(r\bar{r}+g\bar{g}+b\bar{b})$ is invariant against any color change and is therefore not color-specific. It therefore cannot describe a gluon, which is always combined with a specific color. Instead of the 9 possible color combinations there are only 8 gluons, which contribute to the interaction between the quarks.

7.5.2 Quark-Gluon-Model of Hadrons; Quark-Confinement

The question is now: Why are there no free quarks because up to now they could not be found?

Numerical calculations using the lattice gauge theory [16] and the experimental measurements of the energy states of the quarkonium have given definite hints that the attractive force between quarks does not decrease with increasing distance such as the electric force ($1/r^2$) but remains constant when increasing r. This means that the potential energy increases with r (Fig. 7.30). It can be written as

$$E_{pot} = -A/r + B \cdot r,$$

where the first term describes the attractive Coulomb potential between the electrically charged quarks which decreases as $1/r$ with increasing distance, while the second term represents a barrier for the separation of the quarks, because for $r \to \infty$ this term converges also $\to \infty$. The

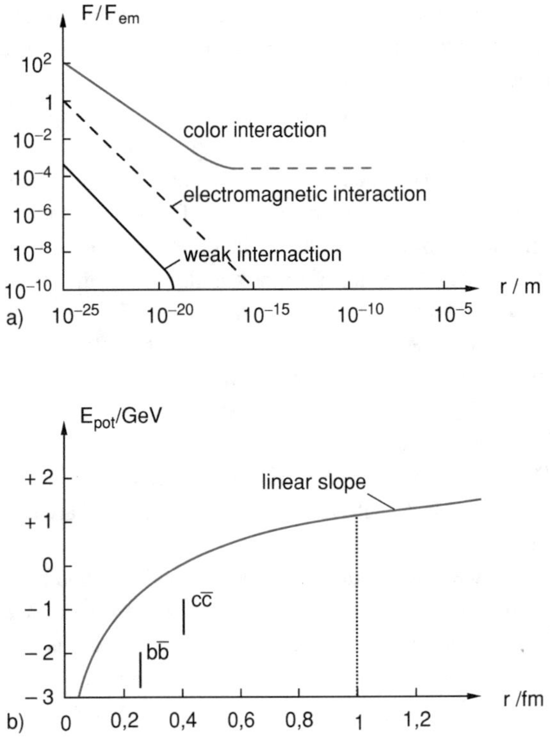

Fig. 7.30 **a** Comparison of the different interacting forces and their range-dependence **b** interaction energy between two quarks as a function of their mutual distance

Coulomb interaction is much weaker than the strong interaction. Therefore A << B.

In a vivid picture the quarks are bound together by a strong tear proof rubber-band that cannot break and prevents the quarks to separate (**Quark confinement**). The reason for this confinement is the fact that the gluons themselves carry color charges and can interact with each other. When the quark distance increases, the potential energy of the binding forces also increases, until new gluon-antigluon pairs are generated which forms a quark-antiquark pair (Fig. 7.30x).

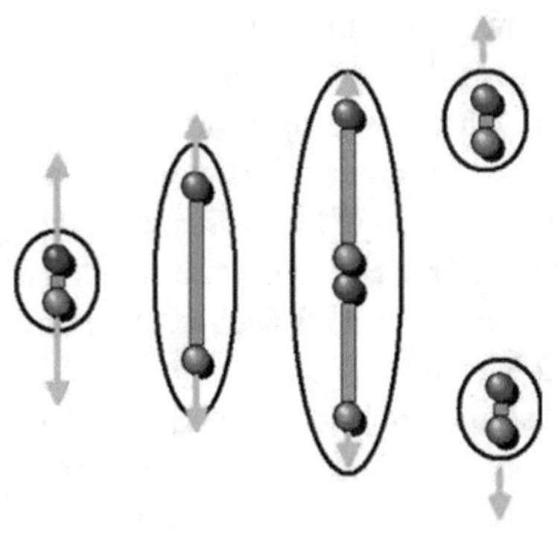

Fig. 7.30x Formation of new quark-antiquark pairs when two quarks are separated beyond a critical distance

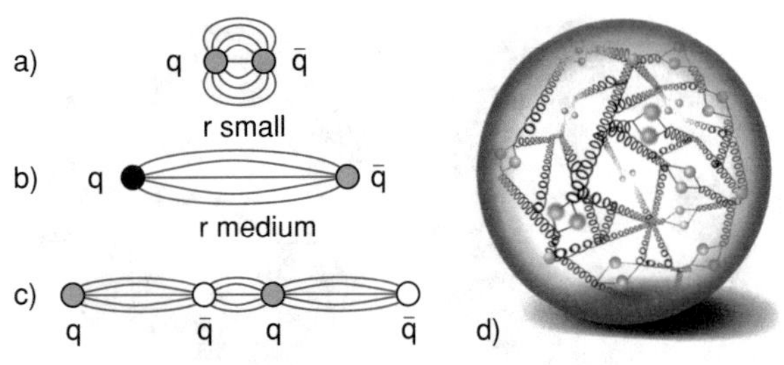

Fig. 7.31 Illustrative model of the quark confinement

Now we come to the essential point: When sufficient energy is put into a hadron (for instance by high energy collisions) to generate a quark-antiquark pair by interaction with the gluons, a meson is produced (Fig. 7.30a). The energy is not used to separate the quarks above a critical distance, but for the production of new colorless particles. This implies that the quarks remain enclosed inside colorless i.e .traceable particles. This production of new particles becomes possible through the interaction with the gluons (Fig. 7.31).

The gluons therefore play an important role for the production of new particles at high energy collisions.

The experiments show in deed that at high energy collisions of hadrons or of leptons a wealth of new particles is produced. One observes so called **jets of hadrons** which are produced at anti-collinear lepton-lepton collisions at very high energies. They are generated according to the reaction scheme

$$e^+ + e^- \rightarrow \gamma \rightarrow \overline{q} + q \rightarrow \text{Hadrons.}$$

The two quarks fly into opposite directions perpendicular to the directions of the leptons. However, before they are separated by more than the hadron radius they produce new hadrons. These hadrons form two jets which contain hadrons and leptons. They can be detected and analyzed by sophisticated systems of different detectors, which are arranged around the collision zone (Figs. 7.32, 4.87 and 4.88).

If the energy of the colliding leptons is sufficiently high, also gluons can be produced analogue to the production of photons (bremsstrahlung) at the deceleration of electrons in the Coulomb field of atomic nuclei. These gluons can be converted into a hadrons according to the process

$$\begin{aligned} e^+ + e^- \rightarrow q + \overline{q} + g &\rightarrow \text{Hadrons} \\ &|_{\rightarrow}\text{Hadrons} \\ &|_{\rightarrow}\text{Hadrons} \end{aligned}$$

Therefore three hadron jets are generated (Fig. 7.32b)

Scattering experiments at very high energies (called *deep inelastic scattering*) allow the determination of the momenta of the quarks inside the hadrons. The results of such measurements are astonishing: At p + p collisions the u- and

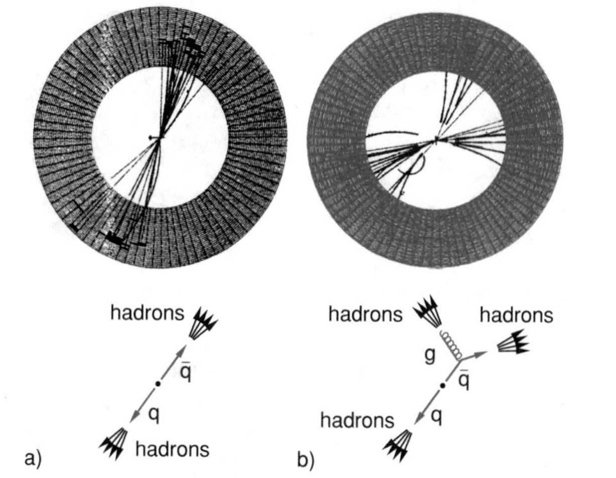

Fig. 7.32 Observation of hadron jets ar high energy collisions of electrons and positrons. **a** two-jets event of the process $e^+ + e^- \rightarrow q + \bar{q} \rightarrow$ hadrons **b** three-jet event of the process $e^+ + e^- \rightarrow q + \bar{q} \rightarrow$ hadrons

d-quarks get only less than half of the initial energy $p^2/2m$ of the two colliding protons.

It furthermore turned out that the masses of the u- and d-quarks are much smaller than expected from the mass of the proton. For example the mass of the u-quark is between 1.5 and 3 MeV/c^2, that of the d-quark between 4 and 8 MeV/c^2 while the mass of the proton is 938 MeV/c^2. Therefore the conclusion immediately suggests itself that there must be other particles besides the u- and the d-quark inside the proton. The suggested model is that quark-antiquark pairs and gluons continuously interchange, i.e quark-antiquark pairs convert into gluons which again convert into quark-antiquark pairs. Such virtual quark-antiquark pairs are called Sea-quarks, because the real u-and d-quarks move inside the proton like solid particles in a sea of virtual quarks and gluons. The energy $E = mc^2$ of this quark-gluon sea attributes an additional part to the mass of the proton. However, this additional contribution to the proton mass represents only a small part. The main contribution comes from the zero point energy

$$E = \delta p^2/2m = \bar{h}^2/(\delta r^2 2m)$$

Table 7.7 Masses and charges of the 6 valence quarks

	Quark	m · c²/MeV	Charge/e
1	u	1.5–5	+ 2/3
	d	17–25	−1/3
2	s	60–170	−1/3
	c	1100–1400	+ 2/3
3	b	4100–4400	−1/3
	t	173.800 ± 5200	+ 2/3

based on the uncertainty relation of the energy of the quarks inside the hadron, which are confined within the spatial volume $V = r^3$ (see Sect. 7.5.3).

The "real" quarks define the quantum numbers of the hadron such as spin, charge and isospin. They are therefore called "**valence quarks**". They determine the static behavior and the spectroscopic properties of the hadrons. The sea-quarks do not contribute to the quantum numbers of the hadron, because for the $q\bar{q}$-quark-antiquark pairs all quantum numbers are zero. They have, nevertheless, a certain influence on the observable effects at the deep-inelastic electron-proton or proton-proton scattering.

In Fig. 7.31d the "interior life" of a proton is illustrated in a simple but vivid way.

The masses and charges of the valence quarks are compiled in Table 7.7. With the general word "quark" always the valence quarks are meant.

Since also the gluons can interact with each other due their color charges, one expects, as for the quarks a confinement of the gluons. This implies that no gluons can be observed outside a hadron. This impedes their experimental detection which is possible only by indirect ways.

7.5.3 The Masses of Quarks

We have seen in the previous section that according to our present knowledge, there are no free quarks. It is therefore not possible to measure their mass directly, but only in an indirect way. The question is now, how the mass of a particle enclosed within a small volume can be measured? Due to the uncertainty relation $\Delta p \cdot \Delta r \geq \hbar$ which demands that the product of the momentum uncertainty Δp and the local uncertainty Δr cannot be smaller than ħ. The zero point kinetic energy $E_{kin} = p^2/2\ m$ of an confined particle is at least $\hbar^2/(2\ m{\cdot}\Delta r^2)$. An up quark, confined in a proton must then have a zero point energy of 300 MeV and similar values apply to the d-quark. The major part of the rest mass 938 MeV/c^2 of the proton can be therefore explained by the zero point energy of the confined quarks. When the "naked mass" of a quark (also called *stream mass*) is defined as the mass of a hypothetical free quark, one obtains values which have up to now large error bars. All authors agree, however, that this stream mass amounts only to a few percent of the mass of the hadron that is composed of the included quarks. For example the stream mass of the u-quark is between 1.8 MeV/c^2 and 3 MeV/c^2, that of the d-quark is about 4.8–5.2 MeV/c^2.

For the charmonium with the mass m 3100 MeV/c^2 and its potential energy obtained from the measured energy states one can get more accurate values for the stream mass m_c = 1240–1300 MeV/c^2 of the charm quark c., while the strange quark s has a stream mass of 100 MeV/c^2. The heaviest quarks are the bottom quark ($m \approx 4$ GeV/c^2) and the top quark t (≈ 174 GeV/c^2).

From the above discussion it becomes evident, that the distinction between mass and energy of the quarks is no longer meaningful. In this sense the quarks are no longer particles in the usual meaning, where to each particle a unique mass can be attributed. Therefore the question, whether the quarks might be further divisible becomes meaningless.

For very close distances the force between the quarks becomes small and they can move nearly free (**asymptotic freedom**). Their interaction energy becomes then smaller than their mass energy. On the other side the energy due to the smaller local uncertainty becomes larger.

The particle model with defined mass fails for short separations as well as for large separations ($r \to \infty$).

At very high temperatures (anti-collinear collisions of heavy nuclei at energies up to 5 TeV per nucleon achieved at the Large Hadron Collider at CERN) quarks and gluons become quasi-free and form a state of very hot matter called **quark-gluon plasma**. When the heavy nuclei (for instance gold nuclei) penetrate into each other the energy of the nucleons becomes so high, that quasi-free quarks and gluons are produced. The quarks and the gluons interact with each other and distribute their energy uniformly over all quarks and gluons, generating a thermodynamic equilibrium at a very high temperature.

Such a hot quark-gluon plasma was apparently formed within the first piccoseconds after the big bang when our universe started from a hot fireball.

7.6 Strong and Weak Interaction

The strong interaction was introduced in Sect. 5.4 as the nuclear force that holds the nucleons in a nucleus together despite the repulsive Coulomb-force between the charged protons. Before the quark model was introduced, the nuclear forces were explained by the exchange of π-mesons. In the quark model the quarks interact with each other (and are hold together) by the color forces described by the exchange of gluons. The forces between the nucleons can then be regarded as residual forces which appear when a quark-antiquark pairs forms a π^+-meson ($u\bar{d}$) which as a colorless particle can leave the nucleon and is absorbed again by the neighboring nucleon.

The quark model extends the Yukawa model to a more general and fundamental concept of the color forces and the exchange of gluons.

The analogue situation appears for the *van der Waals binding* between two neutral atoms (Vol. 3, Sect. 9.4.3). It is caused by the shift (polarization) of the atomic charge distribution in one atom caused by the charge distribution of the other atom. This results in an attractive electric potential because the superposition of the positive and negative charges do not exactly compensate at a point between the atoms. The van der Waals interaction is an interaction between two induced dipoles. It can be therefore regarded as a not completely compensated Coulomb-interaction between the positive and negative excess charge caused by the shift of the charge distribution of the two natural atoms.

> The nuclear forces are in the quark model the residue of the not completely compensated color forces.

This illustrates that the color forces between the quarks must be stronger than the rest-forces between the nucleons.

We have already pointed out that hadrons can decay via the weak interaction. One example is the beta decay of the neutron

$$n \to p + e^- + \bar{\nu}_e.$$

Here the neutron as a hadron converts into the slightly lighter proton (also a hadron) and an electron (lepton) plus neutrino (anti-lepton).

In the quark model this decay is described as

$$(udd) \to (uud) + e^- + \bar{\nu}_e \qquad (7.37)$$

> While in processes governed by the strong interaction the color charge of a quark changes, processes of the weak interaction change the quark type (flavor).

In the quark model the β-decay can be written as the flavor change of an u-quark into a d-quark.

$$d \xrightarrow{\text{weak interaction}} u + e^- + \bar{\nu}_e.$$

Examples

(1) We will illustrate the difference between strong and weak interaction by two more examples:
In the decay of a sigma baryon into a proton and a π^+. meson

$$\Sigma^+(uus) \to p(uud) + \pi^0(u\bar{u} + d\bar{d}) \qquad (7.38)$$

an s-quark is changed into a d-quark. The excess energy $\Delta E = (m_{\Sigma+} - m_p)c^2$ allows the simultaneous creation of a quark-anti-quark pair, which forms a π^0-meson. The conversion s $\rightarrow$ d of an s-quark into a d quark is only possible by the weak interaction. It therefore occurs relatively slow ($\tau = 8{\cdot}10^{-11}$ s).

(2) In contrast to this slow conversion the decay of the Δ-baryon caused by the strong interaction

$$\Delta^+(\text{uud}) \rightarrow \text{p}(\text{uud}) + \pi^0(\text{u}\bar{\text{u}} + \text{d}\bar{\text{d}}), \tag{7.39}$$

occurs in $6{\cdot}10^{-24}$ s which is faster than the reaction (7.37) by 13 orders of magnitude. Here the quark type (flavor) is not changed. The Δ^{++} baryon represents an excited state of the proton and has the same quark composition.

When the interaction between the quarks is quantified by the coupling constant α, the ratio of the lifetimes due to strong and weak interactions is inversely proportional to the square of the ratio of these coupling constants

$$\frac{\tau_{\text{strong}}}{\tau_{\text{weak}}} = \left(\frac{\alpha_{\text{weak}}}{\alpha_{\text{strong}}}\right)^2. \tag{7.40}$$

This is completely analogue to the conditions in the atomic electron shell, where the transition probabilities are proportional to the squares of the corresponding matrix elements.

Using a mean value of the many experimental results for the lifetimes due to weak and strong interaction one obtains the ratio of the coupling constants

$$\frac{\alpha_{\text{weak}}}{\alpha_{\text{strong}}} = 10^{-6} \tag{7.41}$$

Comparing this with the ratio

$$\frac{\alpha_{\text{weak}}}{\alpha_{\text{em}}} \approx 10^{-4}, \tag{7.42}$$

of the coupling constants of weak and electromagnetic interaction one gets the relative order of magnitudes

$$\alpha_{\text{strong}} : \alpha_{\text{em}} : \alpha_{\text{weak}} = 1 : 10^{-2} : 10^{-6} \tag{7.43}$$

The question is now whether it is possible to describe the weak interaction by the exchange of particles, as for the strong and the electromagnetic interaction. Which properties should such exchange particles have?

7.6.1 W- and Z-Bosons as Exchange Particles of the Weak Interaction

There are two different processes which are based on the weak interaction:

1. Processes which change the charge of the reactants. They are also called **Reactions of charged currents**. One example is the ß-decay of the neutron
   $n \rightarrow p^+ + e^- + \bar{\nu}_e$
2. Processes without any change of the charge (**neutral currents**). One example is the decay of the τ-lepton:

$$\begin{aligned}\tau^- &\rightarrow \mu^- + \bar{\nu}_\mu + \nu_\tau \\ &\hookrightarrow e^- + \nu_\mu + \bar{\nu}_e\end{aligned} \tag{7.44}$$

Another example is the elastic scattering of neutrinos by protons or by electrons

$$\nu_\mu + \text{p} \rightarrow \nu_\mu + \text{p} \tag{7.45}$$

$$\nu_e + e^- \rightarrow \nu_e + e^- \tag{7.46}$$

In all of these processes 4 fermions participate.

If these processes of the weak interaction should be described by the exchange of particles, as is the case for all other interactions, the exchange particle should be charged for all charged currents, but should be neutral for all processes of neutral currents. The charged exchange particle is called the **W-boson,** (it is a boson because it has an integer spin s = 1·ħ, as will be shown later). For the neutral currents the exchange particle is the neutral **Z^0-boson**. Since the interaction range of the weak force is very small (Fig. 7.30a), the mass of the exchange particles must be accordingly large. It can be estimated by two different ways:

Yukawa assumed for the range of the nuclear forces a value of $r = 1.5{\cdot}10^{-15}$ m (about the proton radius) and got the mass of the exchange particle (π-meson) m = 140 MeV/c^2.

With an estimation $r \approx 2{\cdot}10^{-18}$ m for the range of the weak interaction the mass of the exchange particle should be obtained from the uncertainty relation

$$\begin{aligned} &mc^2 \cdot \Delta t \geq \hbar \text{ with } \quad \Delta t = r/c \\ &\Rightarrow m \geq \frac{\hbar}{r \cdot c} \approx 1.5 \cdot 10^{-25}\text{kg} \\ &\Rightarrow mc^2 \geq 80\text{GeV}. \end{aligned} \tag{7.47}$$

The mass of the exchange particle is therefore much larger than the nucleon mass.

With the Fermi-coupling constant.
$\mathbf{G_F} = \sqrt{(2/8)}g^2/m_w{}^2c^2 = 1.166{\cdot}10^{-5}$ GeV^{-2}.
(g = coupling constant of the weak interaction).
derived by Fermi and used in the theory of the ß-decay, which is proportional to the probability W_{if} in Eq. (3.40), we can again estimate the mass m_w of the exchange particle and get again $m_w \approx 8{\cdot}10^{10}$eV/$c^2$ (Fig. 7.33).

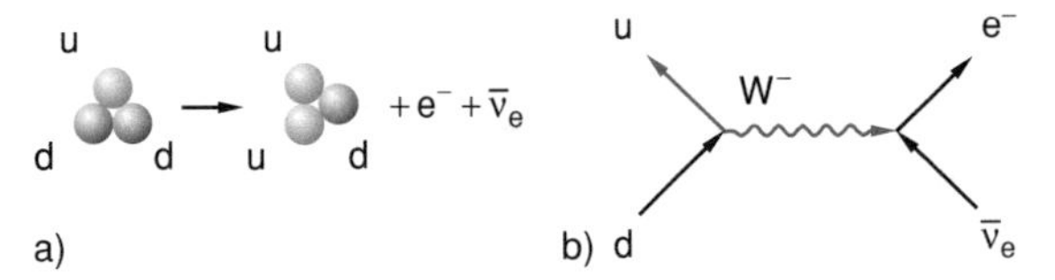

Fig. 7.33 Beta-decay of the neutron **a** Schematic illustration **b** Feynman diagram with exchange of a W^- boson

In this model the β-decay proceeds as follows:

The neutron converts into a proton by converting a d-quark into an u-quark, where a virtual W*Boson is emitted, which decays into an electron and an anti-neutrino

$$d \rightarrow u + W^{*-} \quad (7.49a)$$

$$W^{*-} \rightarrow e^- + \bar{\nu}_e, \quad (7.49b)$$

The star of the W^* boson should indicate that we talk about a virtual W-Boson).

The total process of the ß-decay is then

$$d \rightarrow u + e^- + \bar{\nu}_e \quad (7.49c)$$

Because the total charge is preserved in all allowed processes, the charge of the W^- boson must be $q = -e$·

One can illustrate the weak interaction and its analogy with the electro-magnetic interaction in the following way.

The electric charge is a measure of the coupling strength between the charged particle with the photon (as exchange particle of the el. magn. interaction). In analogy one introduces the "weak charge" which describes the coupling strength between a particle and the W-boson as exchange particle of the weak interaction. The reaction (7.46) of the scattering of neutrinos by electrons can be divided into two steps (Fig. 7.34a)

$$e^- \rightarrow \nu_e + W^{*-}, \quad (7.50a)$$

$$\nu_e \rightarrow e^- + W^{*+}, \quad (7.50b)$$

i.e. in the first step the electron converts into a neutrino by emitting a virtual W^- boson, and in the second step the neutrino converts into an electron by emitting a virtual W^+ boson.

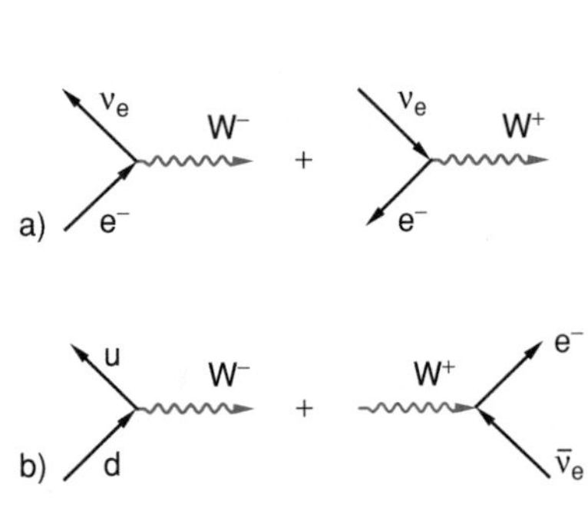

Fig. 7.34 Segmentation of the electron-neutrino-scattering (**a**) and the ß-decay (**b**)

In a similar way the β^-decay can be divided into the two steps

$$d \rightarrow u + W^{*-}, \quad (7.51a)$$

$$W^{*-} \rightarrow e^- + \bar{\nu}_e \quad (7.51b)$$

Or the β^+-decay in

$$u \rightarrow d + W^{*+}, \quad (7.52a)$$

$$W^{*+} \rightarrow e^+ + \nu_e. \quad (7.52b)$$

This division into two steps facilitates the description. One no longer has to consider the interaction between 4 fermions, but has to deal only with the interaction between two fermions and one boson. This makes the treatment of processes based on the weak interaction easier, analogue to those based on the electromagnetic interaction.

The coupling strength can be specified just by one number, namely the coupling constant α_{weak} in (7.41) and (7.42) analogue to the coupling constant (= fine structure constant α = 1/137) of the electromagnetic interaction.

Writing the leptons and the quarks as two-dimensional column vectors.

$$\begin{pmatrix} \nu_e \\ e^- \end{pmatrix}, \quad \begin{pmatrix} u \\ d \end{pmatrix}, \quad (7.52b)$$

the weak interaction (exchange of W-bosons) interchanges the upper component into the lower one. This is quite analogue to the isospin-formalism where a proton is changed into a neutron, i.e. an u-quark into a d-quark (see Sect. 5.3).

7.6.2 Real W- and Z-Bosons

The $W^{\pm}$ and the Z^0 bosons are not mere hypothetical particles which were introduced in order to get an analogue description of the weak interaction as the photons for the electro-magnetic interaction, but they are real particles which can be observed.

In order to generate $W^{\pm}$ or Z^0-bosons either a quark and an anti-quark or a lepton and an anti-lepton must react with each other. The minimum kinetic energy E_s necessary for the production of a neutral Z^0-boson in the process

$$e^+ + e^- \rightarrow Z^0$$

is in the center of mass system $E_s \geq M_Z{\cdot}c^2$.

The charged vector bosons W^+ and W^- were first discovered at CERN by *C. Rubbia* and coworkers in collisions

$$p + \bar{p} \rightarrow W + X \quad (7.53)$$

between protons and anti-protons, where X stands for further eventually generated particles. The W-boson is generated by the weak interaction between a quark in the proton and an anti-quark in the anti-proton:

$$\mathrm{u} + \bar{\mathrm{d}} \rightarrow \mathrm{W}^+, \tag{7.54a}$$

$$\bar{\mathrm{u}} + \mathrm{d} \rightarrow \mathrm{W}^-. \tag{7.54b}$$

Shortly after this discovery also the neutral Z^0-boson was found which is generated in the reaction

$$\mathrm{u} + \bar{\mathrm{u}} \rightarrow \mathrm{Z}^0 \tag{7.54c}$$

The W-bosons decay into leptons

$$\mathrm{W}^+ \rightarrow \mathrm{e}^+ + \nu_e, \tag{7.55a}$$

$$\mathrm{W}^- \rightarrow \mathrm{e}^- + \bar{\nu}_e. \tag{7.55b}$$

The Z^0 boson decays into a lepton-anti-lepton pair

$$\mathrm{Z}^0 \rightarrow \mathrm{e}^+ + \mathrm{e}^- \quad \text{or} \quad \mathrm{Z}^0 \rightarrow \mu^+ + \mu^-. \tag{7.55c}$$

The detection of the Z^0-boson is possible by observing the high energy $e^+ - e^-$ pair or the $\mu^+ - \mu^-$ pair, where lepton and anti-lepton fly into opposite directions.

One measures with calorimeter detectors surrounding the $p^+ + \bar{p}$ collision zone the transverse energy of electron and positron as a function of the polar angle ϑ and the azimuthal angle φ. In Fig. 7.35 the result of such measurements is shown for the reaction

$$\mathrm{p} + \bar{\mathrm{p}} \rightarrow \mathrm{Z}^0 \rightarrow \mathrm{e}^+ + \mathrm{e}^-.$$

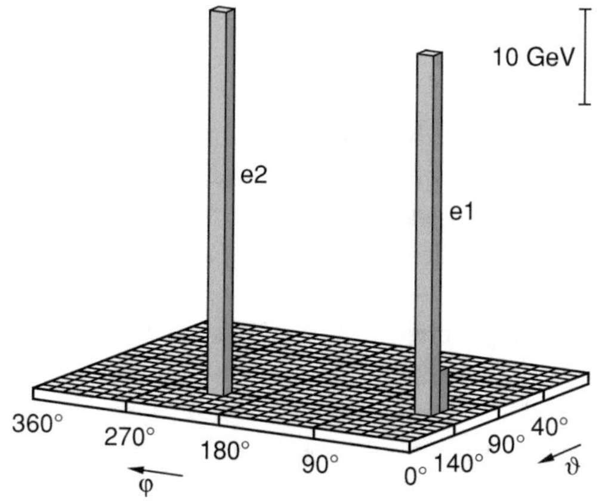

Fig. 7.35 Diagram of the first measured decays $Z^0 \rightarrow e^+ + e^-$ which was the first detection of the Z^0-boson. Here the transverse energy of anti-collinear collisions $e^+ + e^-$ as the function of the scattering angles ϑ and φ has been measured by a calorimeter (VA2-collobaration CERN, P. Bagnaia et.al. Phys. Lett. B 129, 110 (1983))

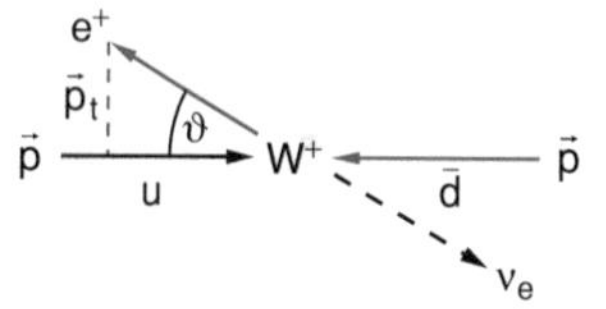

Fig. 7.36 Momentum diagram of the reaction $u + d \rightarrow W^+ \rightarrow e^+ + \nu_e$ (from Povh, Rith, Scholz, Zetsche: Teilchen and Kerne (Springer Heidelberg 1995)

At the decay of the charged bosons $W^\pm$ (Equ. 7.55a and 7.55b)only one charged particle is observable because the neutrino is not visible in the detector. It is therefore necessary to measure the sum of the transverse momenta and the energies of all detected particles. The sum of all measured momenta is not zero because the non-observable momentum of the neutrino has to be taken into account. Also the total energy of all detected particles is smaller than the primary energy of the collision partners p + $\bar{\mathrm{p}}$. Again the missing energy is attributed to the neutrino.

When we assume that the W^+.-boson generated in the anti-collinear collision $p + \bar{p}$ is at rest when it decays into $e^+ + \nu_e$ and that the e^+-positron is observed in the direction ϑ against the direction of $\bar{p}$ (Fig. 7.36) than the transverse momentum is

$$p_t(e^+) = \frac{M_W \cdot c}{2} \cdot \sin\vartheta. \tag{7.56}$$

because the neutrino takes 1/2 of the total momentum. The angular distribution of the reaction products depends on the cross section $\sigma(p_1)$ of the reaction $W^+ \rightarrow e^+ + \nu_e$.

The dependence of the cross section $\sigma(p_1, \vartheta)$ on the angle ϑ can be written as:

$$\frac{d\sigma}{dp_t} = \frac{d\sigma}{d\cos\vartheta} \cdot \frac{d\cos\vartheta}{dp_t}. \tag{7.57}$$

Inserting $\cos\vartheta = \sqrt{1 - \sin^2\vartheta}$ into (7.56) yields

$$\frac{d\sigma}{dp_t} = \frac{d\sigma}{d\cos\vartheta} \cdot \frac{2p_t}{M_W \cdot c \cdot \sqrt{(M_W \cdot c/2)^2 - p_t^2}}. \tag{7.58}$$

This shows that the *transveres momentum* p_1 is maximum for $p_t = M_w \cdot c/2$.

Generally the W-boson will not be generated at rest, because also other reaction products are generated at the collision $p + \bar{\mathrm{p}}$. Therefore the curve $\sigma(p_1)$ will be smeared out. The maximum of the curve $\sigma(p_1)$ allows the determination of the mass of the W^+-boson.

The width of the observed resonance in the cross section $\sigma(E_s)$ as a function of the center of mass energy E_s of the $p + \bar{\mathrm{p}}$ collision gives the lifetime of the W^+-boson. The results of such measurements are given in Table 7.8.

Table 7.8 Masses and Resonance widths of W- and Z-bosons

Boson	Mass in GeV/c^2	Resonance width Γ in GeV
$W^{\pm}$	80.2 ± 0.26	2.08 ± 0.07
Z^0	91.18 ± 0.004	2.497 ± 0.004

The necessary center of mass energy for the generation of W-bosons by $p + \bar{p}$ collisions is much higher than the mass energy $M_w c^2$. The reason is the fact that for the generation of the W-bosons in p + $\bar{p}$ collisions not the whole nuclei participate but in fact only the collision of two quarks contribute according to (7.54). In a fast proton about half of the momentum is taken by the gluons, (see Sect. 7.5). The rest is distributed onto the three quarks. The estimation shows that the quark, necessary for the reaction (7.54) gets only 12% of the proton momentum. Therefore the minimum energy of the proton-anti-proton beams is about 300 GeV in order to produce $W^{\pm}$ bosons in the reaction.

$p + \bar{p} \rightarrow W^{\pm}$

For the production of Z^0-bosons in electron-positron storage rings by the reaction

$$e^+ + e^- \rightarrow Z^0 \quad (7.59a)$$

on the other hand only the center of mass energy Es = 90 GeV of the colliding e^+ + e^- leptons is necessary. With the LEP accelerator at CERN this energy could be reached and meanwhile a large number of Z^0 bosons have been produced.

For the production of W^+ and W^- bosons in the reaction

$$e^+ + e^- \rightarrow W^+ + W^- \quad (7.59b)$$

twice the energy $2M_w \cdot c^2$ is needed, i.e.160 GeV, which means 90 GeV for each of the colliding p and $\bar{p}$ beams.

7.6.3 Parity Violation for the Weak Interaction

While for all processes based on the electro-magnetic or the strong interaction the parity remains preserved, in 1956 *Dao Lee* and *Chen Ning Yang* predicted that for the weak interaction the parity might not be preserved. This could be already 1957 experimentally confirmed by Mrs. *C.S.Wu* and coworkers. In this experiment cobalt nuclei were oriented in an external magnetic field B, due to their nuclear magnetic moment μ_{Co} = + 3.75 μ_N.

In order to reach a sufficiently large orientation the sample must be cooled down to temperatures below 1 K (Fig. 7.37). The nuclei decay by β^--emission

$$^{60}Co \rightarrow {}^{60}Ni + e^- + \bar{\nu}_e \quad (7.60)$$

with a mean lifetime of 5.26 years (Fig. 7.38). Measuring the angular distribution of the emitted electrons (Fig. 7.37a), it was found that more electrons are emitted into the opposite direction of B, i.e. opposite to the direction of the nuclear spin, than in the direction of B (Fig. 7.39).

If now an analogue experiment is performed for the β^+-. decay of ^{58}Co, it turned out that the positrons were preferentially emitted into the direction of B, i.e. into the direction of the nuclear spin, opposite to the direction of electrons..

In both cases the angular distribution is axially symmetric but not spherical symmetric.

Inverting the direction of B and therefore also the direction of the nuclear spin, the direction of maximum β-emission is

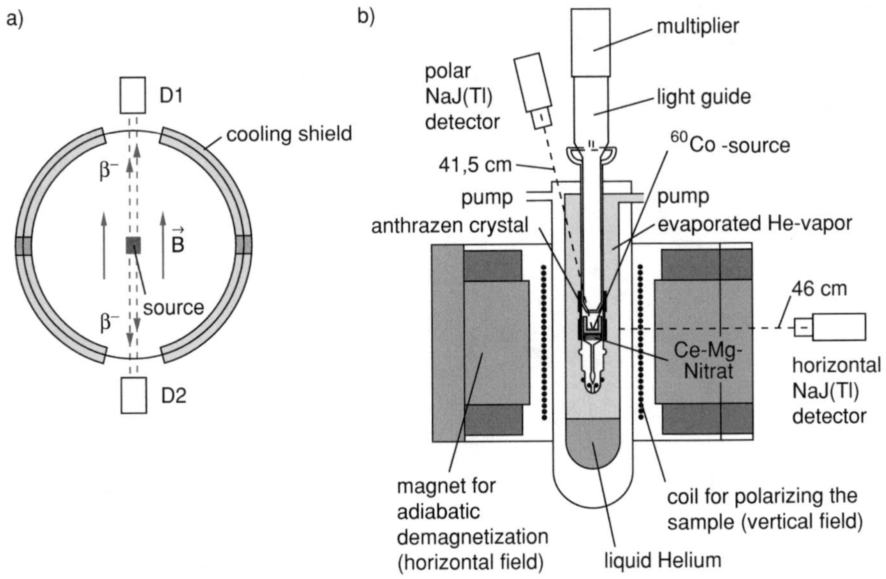

Fig. 7.37 Experimental setup for measuring the parity violation at the ß-decay. **a** Schematic drawing **b** Experimental arrangement of the Wu. group (C. S. Wu et al. The Experimental test of parity conservation in ß-decay. Phys. Rev. 107, 1413 (1957))

Fig. 7.37x Madame Chien Shiung Wu in her lab (Wikipedia)

also inverted. From Fig. 7.39 one can see, that this asymmetry also means a violation of parity conservation. Since the nuclear spin is an axial vector its helicity is inverted under a reflection at a plane. If the parity were conserved the angular distribution should be the same when inverting the direction of the nuclear spin.

This clearly proves that parity conservation is violated.

In order to analyze this asymmetry we introduce the helicity

$$H = \frac{\mathbf{s} \cdot \mathbf{v}}{|\mathbf{s}| \cdot |\mathbf{v}|} \tag{7.61}$$

of a particle with spin **s** which moves with the velocity **v**.

An electron with its spin direction parallel to the flight direction has a positive helicity. Transformation to a coordinate system that moves with the velocity $v^* > v$ into the direction of v the velocity **v** of the electron becomes **v' = v –** v***** anti-parallel to the direction of the electron spin **s**. This shows that the helicity of a particle with rest mass m_0 which always moves with a velocity $v < c$ depends on the choice of the coordinate system.

The situation is different for neutrinos with a rest mass that is below 0.1 eV/c^2 and probable zero.. They move with the speed of light independent of the chosen coordinate system (see Vol. 1, Sect. 3.4).

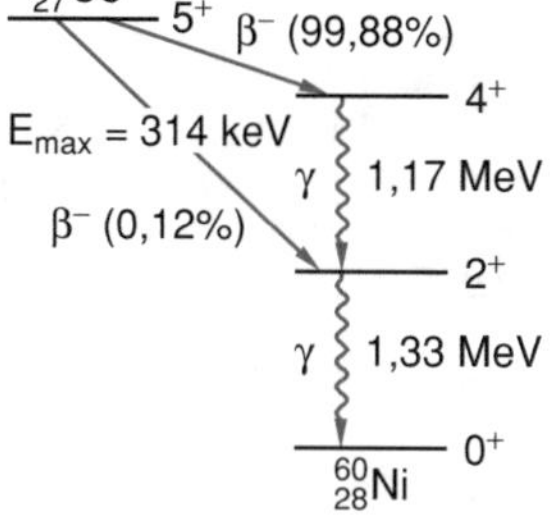

Fig. 7.38 Decay scheme of $^{60}_{27}$Co

The surprising experimental result is that the helicity of neutrinos is $H = -1$. This means that the spin of the neutrinos always points into the direction opposite to the flight direction.

Such particles are called **left-handed**. The anti-neutrinos on the other hand are **right-handed**, i.e. their spin points into the flight direction (Fig. 7.40).

It turns out that the asymmetry of the β-decay (7.60) is related to the right-handiness of the anti-neutrinos. This can be understood as follows.

In Fig. 7.41 the spin- and parity conditions at the ß-decay of ^{60}Co are illustrated. The ^{60}Co nucleus has the spin 5ħ and positive parity. The daughter nucleus ^{60}Ni has the nuclear spin 4ħ and also positive parity. Therefore the spins of the anti-neutrino $s_{an} = 1/2\hbar$ and $s_e = 1/2\hbar$ of the electron must be parallel. The anti-neutrino has the helicity $H = +1$, its spin must therefore point into the flight direction. At the decay of the Co-nucleus at rest the momenta of the emitted electron and anti-neutrino must point into opposite directions, i.e. $p_{\overline{\nu}} = -p_e$ because of the conservation of momentum. Therefore the electron spin must be anti-parallel to its momentum, i.e. the electron flies into a direction anti-parallel to the direction of the nuclear spin.

A further example of parity violation is the decay of the muon:

$$\mu^- \rightarrow e^- + \nu_\mu + \overline{\nu}_e, \tag{7.62a}$$

$$\mu^+ \rightarrow e^+ + \overline{\nu}_\mu + \nu_e \tag{7.62b}$$

When a muon at rest decays, the electron gets the maximum energy, if the momenta of the two neutrinos are parallel but anti-parallel to the momentum of the electron. Since in this case the spins of ν_μ and $\overline{\nu}_e$ are anti-parallel the spin of the electron must be parallel to that of the muon, because otherwise the angular momientum conservation would be violated (Fig. 7.42).

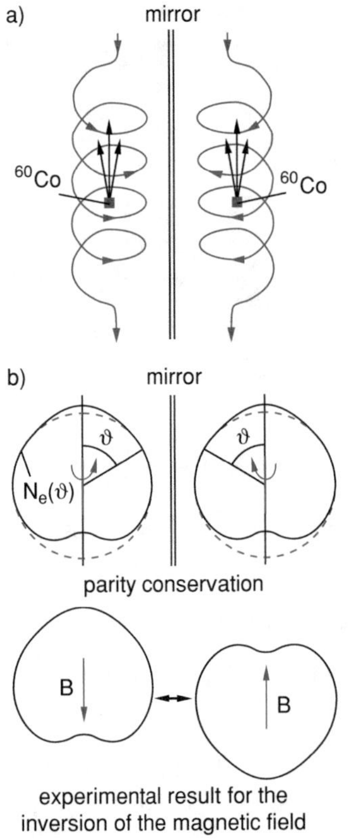

Fig. 7.39 Parity violation in the ß-decay **a** schematic drawing **b** angular distribution of electrons for parity conservation (upper drawing) and experimental results when the magnetic field direction is reversed (lower picture)

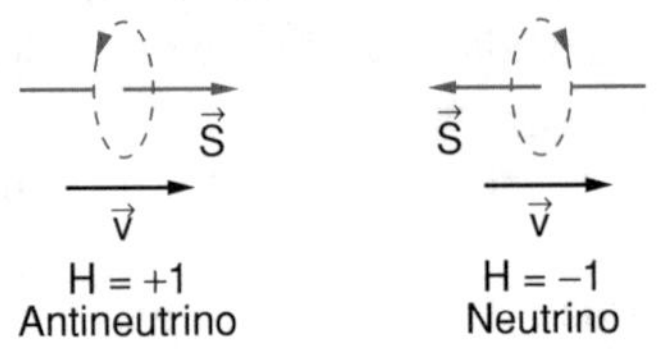

Fig. 7.40 Helicity of neutrino and anti-neutrino

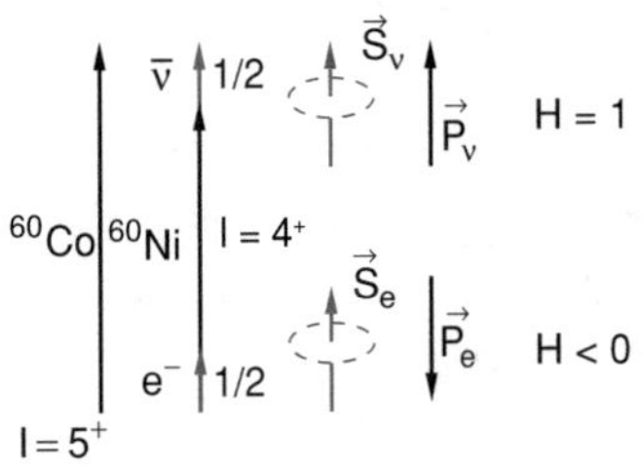

Fig. 7.41 Explanation of parity violation due to the helicity of the anti-neutrino

The experiment shows that at the decay of polarized muons with oriented spins, electrons are emitted preferentially in the direction anti-parallel to the muon spin direction.

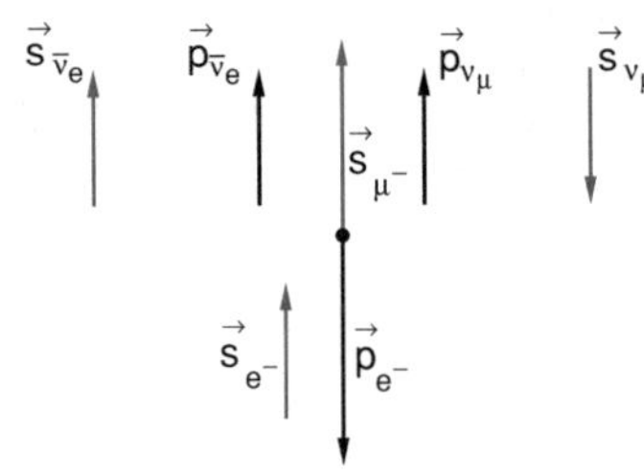

Fig. 7.42 Momentum and spin balance for the decay of the muon $\mu^- \rightarrow e^- + \bar{\nu}_e + V_\mu$

They are therefore preferentially left-handed. This left–right-asymmetry is a proof of the parity-violation.

7.6.4 The CPT-Symmetry

Three important symmetry operations in particle physics are.

- The particle-anti-particle conjugation $\widehat{C}$ (also called charge conjugation) where a particle transforms into its anti-particle.
- Parity operation $\widehat{P}$, which corresponds to a reflection $\boldsymbol{r} \rightarrow -\boldsymbol{r}$ at the origin of a coordinate system
- Time inversion $\widehat{T}$ where the passage of time is inverted $t \rightarrow -t$.

It turns out that for all processes, observed so far in nature the product C·P·T of the three operations remains invariant, i.e it is preserved.

This can be deduced from quite general symmetry principles. There is no consistent theory which allows the violation of the CPT-symmetry. This symmetry postulates that particle and its anti-particle have the same mass and the same lifetime, the same magnitude but opposite signs of charge and magnetic moment.

One can immediately see that due to the definition of the helicity of neutrinos, the C-symmetry (particle-antiparticle conjugation) alone is **not** preserved, because changing the neutrino into the anti-neutrino left-handed neutrinos transfer into right-handed anti-neutrinos i.e. the helicity changes. Left-handed neutrinos have not been found in nature.

Quite general it was found that all processes based on the weak interaction violate the C-symmetry.

On the other side the combined C·P-symmetry of particle-anti-particle conjugation and spatial inversion transforms a left-handed neutrino into a right-handed anti-neutrino.

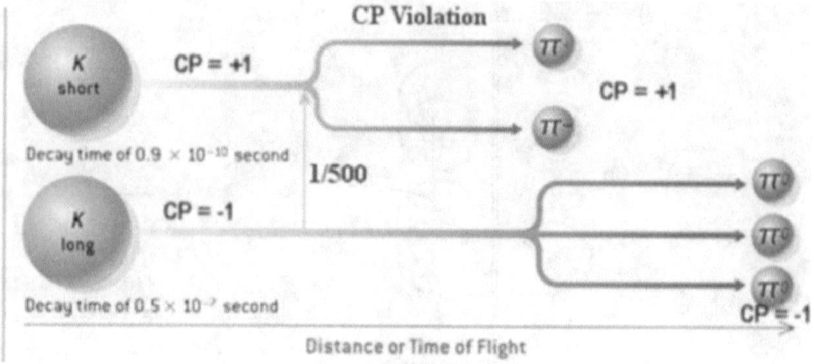

Fig. 7.42x Decay of the K^0 meson into two and into three pions

This implies that the CP-symmetry operation is a possible process which agrees with the experimental results.

There is, however, an observed process where the CP-symmetry is violated. This is the decay of the neutral K^0-meson (Kaon), which can decay into two as well as into three pions. It was found that a short-lived K^0-meson exists, which decays into two pions according to

$$\begin{aligned} K_S^0 &\rightarrow \pi^+ + \pi \rightarrow \\ &\rightarrow \pi^0 + \pi^0 \end{aligned} \tag{7.63a}$$

and a long-lived Kaon $K_L{}^0$ which decays into three pions

$$\begin{aligned} K_L^0 &\rightarrow \pi^+ + \pi^- + \pi^0 \\ &\rightarrow \pi^0 + \pi^0 + \pi^0 \end{aligned} \tag{7.63b}$$

The question is now, why there are two neutral K-mesons with the same mass but different lifetimes and different decay products.

The short-lived Kaon has positive parity because it decays into two pions (see Sect. 7.2.3), while the long-lived Kaon with three decay products must have negative parity. Careful measurements have shown, that there is no difference between the two K^0 particles. They therefore should have the same parity. The fact that both decay process (7.62) and (7.63) occur in nature is therefore a further proof of parity violation (Fig. 7.42x).

In the quark model, which could solve this puzzle, the quark composition of the kaons is $\overline{K} = (d\bar{s})$ and $\overline{K}^0 = (\bar{d}s)$. The strangeness quantum number of K^0 is $S = +1$ and of $\overline{K}^0$ is $S = -1$. Caused by the weak interaction K^0 and $\overline{K}^0$ can blend into each other, which changes the quark flavor ($s \leftrightarrow \bar{s}$, d $\bar{d}$) and is equivalent to the exchange of two W-bosons (Fig. 7.42y). Such a quark-mixture is possible, if the Kaon lives sufficiently long to facilitate the exchange of the W-bosons.

The result of this quark-mixture are two states which correspond to a linear combination of K^0 and $\overline{K}^0$ (Fig. 7.42z). They are

$$\begin{aligned} |K_1^0\rangle &= \frac{1}{\sqrt{2}}\left\{|K^0\rangle + |\overline{K}^0\rangle\right\} \\ &\text{with } \hat{C}\cdot\hat{P}|K_1^0\rangle = +1|K_1^0\rangle, \end{aligned} \tag{7.64a}$$

$$\begin{aligned} |K_2^0\rangle &= \frac{1}{\sqrt{2}}\left\{|K^0\rangle - |\overline{K}^0\rangle\right\} \\ &\text{with } \hat{C}\cdot\hat{P}|K_2^0\rangle = -|K_2^0\rangle. \end{aligned} \tag{7.64b}$$

If the CP-symmetry should be preserved, the K_1^0 must decay into two pions and the K_2^0 into three pions (Fig. 7.42z).

If an experiment is designed where the Kaons fly over a larger distance with a flight time long compared to the lifetime of the short lived kaons, the detector should only see long-lived kaons and therefore only the kaon decay (7.63b) into three pions should be observed. It turns out that the long lived kaon decays predominantly into three pions but a small fraction (0.3%) decays into two pions.

This would be a first hint to a violation of the CP-symmetry [16].

In addition one has observed a semi-lepton decay (Fig. 7.42XX) with an asymmetry between the two decay channels

$$K_1^0 \rightarrow \pi^+ + \mu^- + \bar{\nu}_\mu \tag{7.65a}$$

$$\rightarrow \pi^- + \mu^+ + \nu_\mu \tag{7.65b}$$

The decay (7.65b) occurs more often (by a factor 1.0033), than the decay channel (7.65a). This also indicates a small violation of the CP-symmetry.

If the CP-symmetry were not violated we could not tell inhabitants of other universes whether our matter consists of particles or anti-particles, since all reactions would occur the same way if we interchange particles and anti-particles.

However, due to the CP-symmetry violation we can give the following statement: We define as antiparticles π^- and μ^+ those particles which occur at the $K_1{}^0$-decay with a higher probability than the particles π^+ and μ^-. In a similar way we

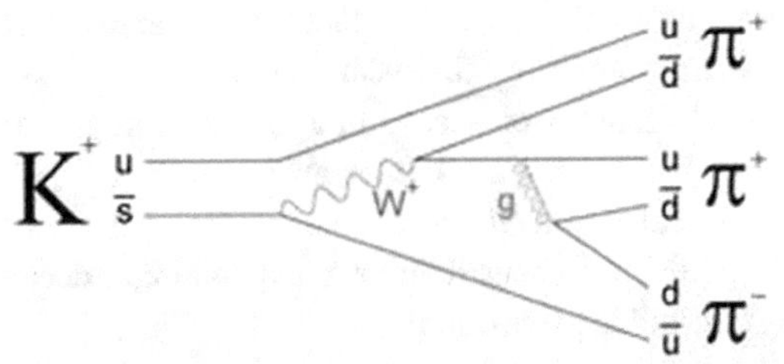

Fig. 7.42y Decay of the K^0-meson into three pions by converting the *s*-quark into an u-quark through the interaction with a W^+-boson and a gluon

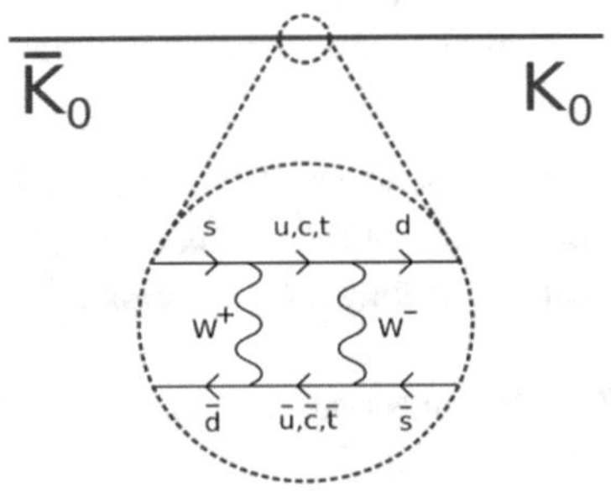

Fig. 7.42z Quark model of the mixing of the K^0 with its anti-particle $\overline{K}^0$

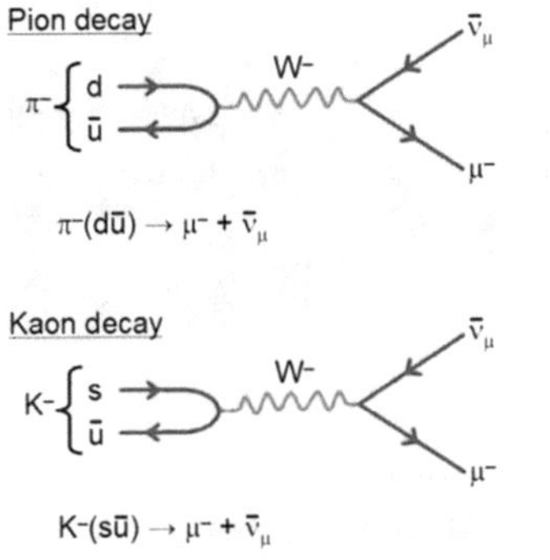

Fig. 7.42XX Decay of Kaons into leptons ν and μ

can distinguish between right and left, as illustrated in Fig. 7.43. The neutrino created at the π^-decay has a helicity $H = -1$ corresponding to a left-handed screw., while in the μ^- decay the anti-neutrino with $H = +1$ is generated which corresponds to a right-handed screw.

If the product $\widehat{C}\cdot\widehat{P}\cdot\widehat{T}$ of the three symmetry operations is always preserved, processes which violate the $C{\cdot}P$-symmetry must also violate the time reversal symmetry T. This implies that for such processes the probability of the process $P(t)$ differs from that for the reverse process $P(-t)$.

Up to now, no direct experimental proof for the T-violation has been observed. Such a proof would be, for

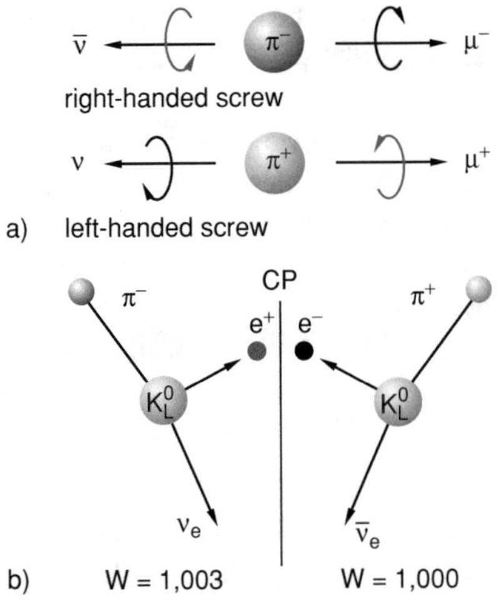

Fig. 7.43 Possibility **a** of distinguishing between right and left **b** between particle and anti-particle due to the CP violation at the K^0-decay

instance, the discovery of an electric dipole moment of the neutron. For a neutron at rest, isolated from any external influence, the only preferential direction is given by its spin *s*. This would be also the direction of a possible electric dipole moment d_n. The interaction energy with an external electric field *E* is then

$$W = -d_n E = sE$$

A parity transformation $\boldsymbol{r} \rightarrow -\boldsymbol{r}$ would correspond to the transformation $(\boldsymbol{s}, \boldsymbol{E}) \rightarrow (\boldsymbol{s}, -\boldsymbol{E})$, a time reversal would correspond to $(\boldsymbol{s}, \boldsymbol{E}) \rightarrow (-\boldsymbol{s}, \boldsymbol{E})$ (Table 7.9). Inversing the electric field or the spin direction should change the energy by $\Delta W = 2\ W$.

Very careful measurements have, however, up to now not found any evidence of such a change of the energy. The experimental uncertainty gives an upper limit for a possible electric dipole moment of the neutron

$$d_n < 2.3 \cdot 10^{-46}\mathrm{C} \cdot \mathrm{m}$$

In Table 7.9 the symmetry properties of some important physical values under CPT-symmetry operations are compiled.

Table 7.9 Transformation properties of some quantities in Physics such as mirror-imaging, charge conjugation $\widehat{C}$ time reversal $\widehat{T}$

Quantity	$\hat{P}$	$\hat{C}$	$\hat{T}$
Location vector r	$-r$	r	r
Time t	t	t	$-t$
Momentum p	$-p$	p	$-p$
Angular momentum L	L	L	$-L$
Spin s	s	s	$-s$
Electric field E	$-E$	$-E$	E
Magnetic field B	B	$-B$	$-B$

7.6.5 Conservation Laws, Symmetries and Stability of Particles

In the foregoing sections we have seen that for the classification of elementary particles the knowledge of their characteristic properties, such as mass, energy, momentum, angular momentum, parity P, charge quantum number C and certain further quantum numbers as baryon number B or lepton number L is essential. For the understanding of nuclear reactions the conservation or the change of these properties are of fundamental importance.

We can formulate a general principle of nuclear reactions:

> Each particle decays in other particles with lower mass, unless this decay is forbidden by conservation laws.

Examples

1. The photon is stable during the time between emission and absorption by matter. The reason for its stability is its rest mass zero. Therefore there are no lighter particles into which it could decay.
2. The electron neutrino is stable because there are no lighter leptons.
3. The electron e^- and its anti-particle the positron e^+ are stable, because there are no lighter charged leptons and the conservation of charge is a rigorous general law, which forbids the decay into neutrinos.
4. The proton is probably stable (the lower limit of its lifetime is $T > 10^{34}$ years.

The conservation of the baryon number B prohibits the decay in lighter leptons.

Measurements of lifetimes of instable particles give important information about the character of the interaction of the particle with other particles. For example the fast decay of the Δ-baryon

$$\Delta^{++} \rightarrow p^+ + \pi^+$$

is induced by the strong interaction, while the much slower decay

$$\pi^0 \rightarrow \gamma + \gamma$$

of the π^0 –meson occurs via the electro-magnetic interaction because the reaction products are two photons.

An example of a decay due to the weak interaction is the decay of the baryon sigma Σ^-.

$$\Sigma^- \rightarrow \mu^0 + e^- + \bar{\nu}_e,$$

because one of the decay products is the anti-neutrino $\bar{\nu}_e$ which interacts only by the weak interaction.

Typical lifetimes of decays due to the strong interaction are about 10^{-24} s, for weak interactions about 10^{-21}–10^{-16}s and for the electromagnetic interaction 10^{-16}–10^{+3} s where the long lifetimes occur if the excited state can decay only through forbidden transitions.

It is interesting that the conservation of the lepton number L and the baryon number B cannot be derived from symmetry principles but is only an empirical rule. But up to now no exception from this rule has been found.

In several laboratories intense efforts are undertaken to detect a possible decay of the proton, but without success, although the Georgi-Glashow model predicts a proton decay.

$p \rightarrow e^+ + \pi^0; \quad \pi^0 \rightarrow 2\gamma.$

With a lifetime between 10^{31} and 10^{36} years. The experimental lower limit of the proton lifetime is $T_p > 10^{34}$ years. Note, that this is longer by a factor 10^{24} than the age of the universe.

In Table 7.10 the conservation laws valid for the different interactions are listed (Fig. 7.43x).

> Emmy Noether postulated 1917 a fundamental theorem that every symmetry found in nature corresponds to a conservation law. For example the homogeneity of space causes the invariance of natural laws under translation. This determines the conservation of momentum. The invariance under transformations in time causes the conservation of energy. The isotropy of space causes the invariance under rotations which determines the conservation of the angular momentum.

7.7 The Standard Model of Particle Physics

The Standard model of Particle Physics summarizes all experimental results discussed in the previous sections. It also makes predictions about particles not found by now. It is based on a **gauge theory** which contains parameters that can be renormalized in order to compare the theoretical predictions with experimental results. Such parameters are for instance, reaction cross sections, the mass of exchange particles for the different interactions or the values of decay probabilities of instable particles [16].

The gauge theory contains characteristic quantities, such as the electro-dynamic potentials $\phi(\boldsymbol{r}, t)$ and $\boldsymbol{A}(\boldsymbol{r}, t)$ which remain constant under gauge-transformations. Therefore it is possible to fix their values by a gauge, e.g. the Coulomb gauge (see Vol. 2, Sect. 5).

Table 7.10 Conserved quantities for the different interactions in particle physics energy

Quantity	El. magn. Interaction W.W	Weak Interaction W.W	Strong Interaction W.W
Energy	Yes	Yes	Yes
Momentum	Yes	Yes	Yes
Angular momentum	Yes	Yes	Yes
Baryon number B	Yes	Yes	Yes
Lepton number L	Yes	Yes	Yes
Parity P	Yes	No	Yes
Charge quantum number C	Yes	No	Yes
Product $C{\cdot}P$	Yes	No	Yes
Time reversal invariant T	Yes	No	Yes
Product $C{\cdot}P{\cdot}T$	Yes	Yes	Yes

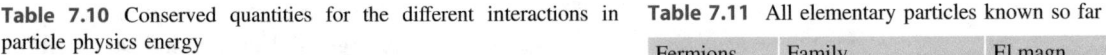

Fig. 7.43x Emmy Noether

Table 7.11 All elementary particles known so far

Fermions	Family			El.magn.	Color	Spin
	1	2	3			
Lepton	ν_e	ν_μ	ν_τ	0	–	1/2
Quarks	e^-	μ^-	τ^-	–1	–	1/2
	u	c	t	+ 2/3	r, g, b	1/2
	d	s	b	–1/3	r, g, b	1/2

Table 7.12 The 4 interactions and their exchange particle

Interaction	Couples to	Exchange particle	Mass (GeV/c^2)	I^P
Strong	Color charge	8 gluons	0	1^+
Electro-magnetic	Electrical charge	1 Photon	0	1^+
Weak	Weak charge	3 $W^\pm$, Z^0	80, 90	1
Gravitation	Mass	Graviton	0	2

Table 7.13 All elementary components of our world (12 Fermions and their anti-particles with masses, and 5 types of bosons, which are responsible for the interaction between the particles

	Fermions			Bosonen
Quarks	u d	c s	t b	g γ
Lepton Families	ν_e	ν_μ	ν_τ	Z
	e	μ	τ	$W^\pm$
	1	2	3	H

g = gluon,gamma = photon, Z = neutral Z-boson, W upper index + - = charged W-boson, H = Higgs boson

According to our present knowledge the total matter and all interactions between its components can be reduced to three groups of particles:

Leptons, Quarks and Exchange Particles

They are compiled in the Tables 7.11, 7.12 and 7.13. Leptons and quarks are Fermions, the exchange particles are bosons. The six leptons (plus their anti-leptons) are classified according to the charge Q, their electron-lepton number L_e, their muon lepton number L_μ and their tau-lepton number L_τ.

The six quarks (plus their anti-quarks) differ by their flavor, and each quark type can have one of the three color charges. The leptons and the quarks are divided into families.

For each kind of interaction there are exchange particles with integer spin, which impart this interaction:

1. The **photon** for the electro-magnetic interaction
2. **Three bosons** (W^+, W^-, W^0) for the weak interaction
3. **Eight gluons** for the strong interaction
4. The **graviton** for the gravitational interaction (which is not included in the Standard Model.

In Fig. 7.44 the four interactions and their dependence on the distance r are shown on a logarithmic scale.

According to the Standard Model there are 12 leptons; 12·3 = 36 Quarks and 12 exchange particles (1 Photon, 2 W-bosons, 1 Z^0-boson and 8 Gluons). In addition there is a new particle, the Higgs boson, which has been predicted already 1965 but only found experimentally in 2012 by collision experiments in the large hadron collider LHC at

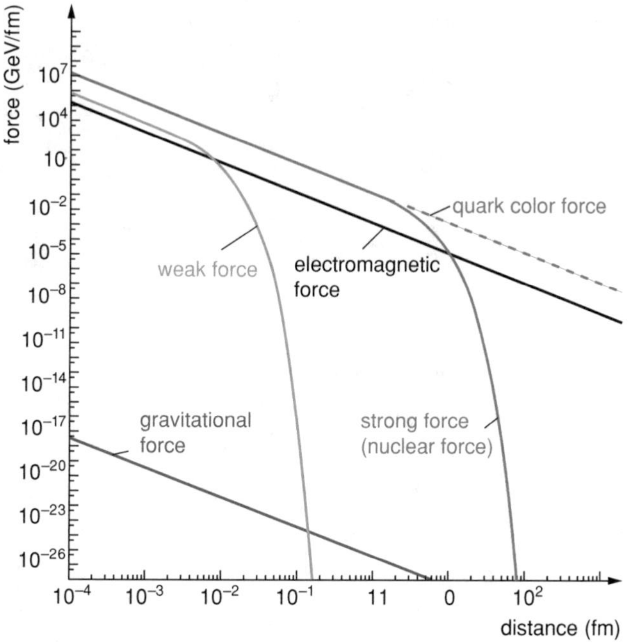

Fig. 7.44 The fundamental forces in physics and their ranges

CERN. It is also postulated by the Standard Model. It is responsible for the masses of the particles.

There are therefore altogether 61 particles in the Standard Model.

Remark The gravitational interaction is not treated in the Standard Model, The Graviton is therefore not included in the 61 particles.

The physical basis of the Standard Model is the quantum field theory. It explains how a general field and particles can be described by the quantum theory and the special relativity theory. It is based on the following fundamental principles:

1. The gauge symmetry which demands that the field equations must be independent of location and time.
2. The spontaneous symmetry breaking, which occurs if the ground state of the system has no longer the full symmetry of energetically excited states.

Such a spontaneous symmetry breaking appears in many areas of physics, if a system is energetically degenerate because of symmetry reasons. One example is a triatomic molecule which has at the equilateral triangle symmetry (D_{3h}) two degenerate states. It transfers spontaneously into a configuration of lower symmetry (isosceles triangle with apex angle $\neq 60°$) where the two degenerate states are split.

The great goal of physicists is to reduce all 4 interaction to a common principle (grand Unification Theory GUT). A first step on this way was the unification of the electrical with the magnetic interaction by the Maxwell equations in the 19th century (see Vol. 2. Chap. 4). In the twentieth century the weak and the strong interaction were discovered. This increased the number of interactions again instead of reducing it. The task to unify them is more difficult and there is no solution up to date.

However, S. L. Glashow St. Weinberg and A. Salam succeeded to reduce the electro-magnetic interaction and the weak interaction to a common "electro-weak" force. The GWS-theory postulates 4 massless exchange particles, where 3 of them (W^+, W^- and Z^0) get their mass through **spontaneous symmetry breaking**.

In High energy physics the particle energies are determined by their mass $m = E/c^2$. They can then transform by spontaneous symmetry breaking into lower energy states. This transformation splits the 4 particles into a massless photon and 3 W^+, W^- and Z^0-bosons.with masses. The GWS theory could predict the masses of these exchange particles. These predictions were convincingly confirmed when the W^- and the Z-bosons were experimentally discovered. Another prediction of the theory was the existence of the Higgs-boson, which was discovered 2012 at CERN as a particle with a mass $m = 125$ GeV/c^2.

The Higgs boson has no charge and the spin 0. It couples to other particles with a coupling strength that is proportional to the mass of the particle. Therefore the Higgs particle is responsible for the masses of the other particles. One says: It gives these particles their mass.

This Higgs particle is the last one of all particles postulated by the Standard Theory of particle physics.

The search for new particles would surpass the Standard Theory. If they were found, one has to upgrade the Standard Model or even correct it.

Presently an intense search for new particles is undertaken which are called "WIMPS" (weakly **i**nteracting massive particles) which might be the particles to explain the "dark matter" [12, 25].

7.8 New Theories, Up to Now Not Confirmed Experimentally

A theory, which tries to refer the strong interaction to a common cause with the electro-weak interaction is the "**Supersymmetry-Theory**" It assumes that each lepton and each quark has a boson as partner. This results in a symmetry between Fermions and Bosons. These supersymmetric pairs are called "*Squarks*" and "*Sleptons*". Their masses are estimated to be 0.3–1TeV (300–1000 GeV). Even the largest colliders before the operation of the LHC did not reach sufficient energies for the production of such pairs. Future experiments at the LHC will prove whether such super-symmetric pairs do exist or not.

One of the most promising models for the realization of the "*Grand Unification Theory* " is the **super-string theory**. It assumes that all elementary particles can be regarded as different energy states of a single particle, the **string**. They can be illustrated as very thin vibrating strings which oscillate in a higher dimensional space [23]. The different vibrational frequencies correspond to the different elementary particles (e.g. electron, quark). The strings must be very small, even smaller than the Planck-length ($\sim 10^{-35}$ m), their vibrational energies are very high, more than 10 TeV. They can therefore not be detected experimentally, even not with the LHC. Because the experimental proof of this theory is not possible up to now, there are still hot discussions about the validity of the superstring theory [21].

Summary

- In high energy collisions of stable particles (electrons, positrons, protons) a large number of new particles can be generated. They are, however, instable and decay through several reaction chains into stable particles as final products of the collision.
- Each particle can be characterized by its mass m, its charge Q, its spin I, its isospin T and the component $T_3 = T_z$, its parity P and its lifetime τ.
- All known particles can be divided into two classes: **Leptons** (e^-, μ^- τ) the corresponding neutrinos (ν_e, ν_μ, ν_τ) and their anti-particles and **Hadrons** (Mesons and Baryons).
- Leptons interact by the weak interaction. If they have a charge $Q \neq 0$, also by the electro-magnetic interaction). They are characterized by the lepton number L
- The hadrons interact by the strong interaction. If they are charged in addition the electro-magnetic interaction applies. They are characterized by the baryon number B. For Mesons is $B = 0$ for baryons is $B = 1$.
- For all reactions found up to now the lepton number L and the baryon number B are conserved.
- The quark model postulates 6 different quark types (flavor) and their anti-quarks. Each quark can exist in three different color charges. There are therefore 36 (6·2·3) different quarks.
- The quarks have half-integer spin and the charge q = ± 1/3 e or ± 2/3 e. They are therefore Fermions.
- All quarks and leptons can be sorted into 3 families. Each family contains 2 quarks, 2 leptons and the corresponding exchange particles and the antiparticles.
- Besides their mass, their spin and isospin the quarks have an additional property, called the color charge. It is responsible for the strong interaction.
- In the quark-model mesons consist of two quarks (1 quark and 1 anti-quark), baryons are composed of three quarks.
- The strong interaction is caused by the exchange of gluons. Gluons are massless vector bosons with spin $1 \cdot \hbar$. They also carry color charges but only in the combination: color-anti-color which makes them colorless. Due to their color charge gluons can interact with each other. There are 8 different allowed color combination and therefore also 8 different gluons.
- The color force between quarks does not decrease with increasing distance. Therefore the potential energy increases with the distance between two quarks. Applying sufficient energy the quarks are not separated but new quark-anti-quark pairs are generated. Therefore there exists no free quarks (quark confinement). It is not possible to produce free quarks.
- The color composition of quarks in observable particles is always color-neutral. This means that all observable particles are colorless.
- The nuclear forces are residuals of the color forces. This is analogue to the van-der-Waals binding of neural molecules, where the electro-static interaction is not completely cancelled because of the shift of the electron distribution in the atomic shells due to the mutual interaction between the charges. The compensation becomes better with increasing distance r. Therefore the nuclear forces decrease exponential with r while the van-der Waals interaction (induced dipole-induced dipole interaction) decreases with $1/r^6$.
- The very short range of the weak interaction is caused by the exchange of the vector bosons W and Z. They have a large mass ($M_W = 80$ GeV/c^2, $M_Z = 90$ GeV/c^2, which implies the short range $r \approx \hbar/(M \cdot c)$.
- In processes induced by the weak interaction different quark flavors transform into each other. The strong interaction changes the color charge but the quark flavor is preserved.
- The quark model has been confirmed by many experimental results, although no free quarks have been ever observed. The confinement hypothesis explains why no free quarks exist.

- Many heavy hadrons can be regarded as excited states of lighter hadrons with the same quark composition.
- According to our present knowledge all existing particles can be reduced to 6 quarks, 6 leptons and their corresponding anti-particles. The quarks can carry 3 different color charges
- The interaction between particles can be described by the exchange of bosons:
  The massless photon (electro-magnetic interaction), three massive vector bosons (W^+, W^- and Z^0) (weak interaction) and 8 gluons (strong interaction).

Problems

7.1 Show that the cross section for the reaction $\pi + p \rightarrow \pi + d$ be written as

$$\sigma = A \cdot \frac{(2I_\pi + 1) \cdot (2I_d + 1) \cdot p_\pi^2}{v_{pp} - v_{\pi d}},$$

where A is a constant, p_π the momentum of the pion, I the spin quantum number, v_{pp} the relative velocity of the two protons and $v_{d\pi}$ that between d and π.

7.2 What are the minimum and maximum momentum of the electron, if a muon μ^- at rest decays according to $\mu^- \rightarrow e^- + \nu_\mu + \bar{\nu}_e$. ?

7.3a The coupling constant of the weak interaction at the energy of 1 GeV is $\alpha_w \approx 10^{-6}$. Estimate the absorption cross section of a neutrino with $E = 1$ GeV, when it passes through matter.

b What is the mean free path length of this neutrino on its way through the earth?

7.4 The parity of the nucleon is defined as $P = +1$. Which parity have the $\bar{u}$- and the d-quark and the deuteron?

7.5 Why is the transition of the charmonium $\psi' \rightarrow \psi + \gamma$ in. Fig. 7.16 forbidden?

7.6 Estimate the range of the strong and the weak interaction from the mass of the corresponding exchange particles π-meson and W-boson, if you assume that the exchange particles cannot move faster than the speed of light.

7.7a Two particles with equal masses m (e.g. particle and anti-particle) collide with equal but opposite velocities $v = (3/5)c$ in a central collision. Which energy $E = Mc^2$ has the unified system? (Compare M with 2 m!)

b A particle with mass M decays into two particles with equal masses m. Is this always possible? With which velocity fly the two particles apart ?

7.8 The kinetic energy of 14 MeV neutrons shall be measured with a time-of flight apparatus. How long must be the flight length in order to reach an energy resolution of 0.5 MeV if the time resolution is $\Delta t = 10^{-9}$ s?

References

1. J. G. Wilson: Cosmic Rays. (Taylor and Francis London 1976)
2. O. C. Alkofer: Introduction to Cosmic Radiation. (Thieme München 1975)
3. S. Neddenmeyer, C.D.Anderson: Note on the Nature of Cosmic Ray Praticles. Phys. Rev. **51,** 884 (1937)
4. C. Lattes, H. Muirhead, G. Occhialini, C. F. Powel: Observation on the tracks of slow mesons in photographic emulsions. Nature **160**, 453 (1947)
5. J. W. Rohlf: Modern Physics from α to Z^0 (Wiley, New York 1994)
6. O. Chamberlain, R.F.Mosley, J. Steinberger,C.Wiegand:A Measurement of the positive π-μ-decay lifetime: Phys. Rev. **79,** 394 (1950)
7. DESY Jahrbuch 2000, JADE- and CELLO. Kollaboration.
8. P. Große-Wiesmann: CERN.Courier **31,** 15 (April 1991)
9. O. Chamberlain, E.Segrè, C.Wiegand, T.Ypsilantis: Observation of Anti-Protons, Phys. Rev. **100**, 947 (1955)
10. L. M. Ledermann, Observations in Particle Physics from two neutrinos to the Standard Model: Rev. Mod.Pphys. **61**, 547 (1989)
11. M. Gell-Mann: the Eightfold Way (Addison Wesley new York 1998)
12. Ashish Kumar Sharma: Introduction of Algebra (Oct. 2019)
13. J. Q. Chen: Group Representation Theory for Physicists, 3rd ed. (World Scientific Singapore 1089)
14. J. J. Aubert et al: Experimental Observation of a Heavy Particle: Phys. Rev Lett. 33, 1404 (1974)
15. J. E. Augustin et al: Discovery of a Narrow Resonance in e^+e^- Annihilation. phys. Rev Lett. 33, 1406 (1974)
16. K. Kleinknecht: CP-Violation (Springer Tract in Modern Physics 2010)
17. István Montvay und Gernot Münster: *Quantum Fields on a Lattice* Cambridge University Press, Cambridge 1994
18. https://en.wikipedia.org/wiki/Proton
19. Joy Moody: Modern Particle Physics (ML Books International 2015)
20. R. Alkofer, J.Greensite, Quark Confinement; The hard Problem in Hadron Physics. J. Phys. G 34, S.3 (2007)
21. https://en.wikipedia.org/wiki/Physics_beyond_the_Standard_Model
22. htpp://www.solstice.de/grundl
23. P. Davies, J.R.Brown, Superstrings 3.ed. (dtc München 1996)
24. H. Snyder:Quantized Space time. Phys. Rev. **67**, 38 (1947)
25. S. Maijid (ed) on Space and Time. Cambridge Univ. Press 2008
26. I. Nicolson: Dark Side of the Universe (John Hopkins Univ. press Baltimore 2007

8 Applications of Nuclear- and High Energy Physics

The scientific results and the technical developments of nuclear and high energy physics have meanwhile found multifarious applications in Biology, Medicine, Environmental Sciences, Geology, Archeology and Energy Technique. In this chapter we will shortly present some of these applications in order to illustrate that basic science, which is primarily not concerned with applications, can have an important influence on our daily life.

8.1 Radio-Nuclide Applications

Many natural radio-active substances, but in particular artificially produced radio-nuclides have found numerous applications which have brought an undeniable benefit for mankind. Their possible dangers are controversially discussed in public, because their longtime effects on human tissue and the resulting damages of cells are diversely estimated. In order to obtain quantitative statements about the radiation exposure we will define at first some terminology and units used in radiation metrology. In particular the comparison of radiation from natural sources with that of artificial sources will give a more realistic estimation of real and felt dangers.

8.1.1 Radiation Doses, Units and Measuring Techniques

As has been discussed in Sect. 4.2 energetic particles (e^-, e^+, p^+, α^{++}, n, γ) passing through matter, will transfer part of their energy onto the atoms and molecules of the substance. This results in excitation, ionization or dissociation of atoms or molecules, where the ionization is the most important energy transfer process. Therefore, the incident radiation is also called "**ionizing radiation**". This radiation can come from natural radio-active sources such as soil or materials grubbed from it, from the cosmic radiation and its reaction products in the earth atmosphere, or from man-made sources such as artificial radioactive nuclides, X-ray tubes and the bremsstrahlung of particle accelerators.

This radiation is characterized by the following measurable quantities:

- The **particle flux density**

$$\Phi = \frac{d^2N}{dA \cdot dt}, \quad [\Phi] = 1\,m^{-2}s^{-1} \tag{8.1}$$

  gives the number of particles passing per s through the unit area of a surface surrounding the source.
- The **activity A** of a radio-active substance is the number of decaying nuclei per second. Its unit is

$$[A] = 1\,Bq\,(Becquerel) = 1\,decay/s = 1\,s^{-1}. \tag{8.2}$$

  In the older literature one can find the unit

$$1\text{ Curie} = 1\text{ Ci} = 3.7 \cdot 10^{10}\,Bq.$$

- For the radiation exposure of a sample the **energy dose D** is the essential quantity. It gives the total radiation energy absorbed per kg in the body. Its unit is

$$[D] = 1\,Gy\,(Gray) = 1\,J/kg. \tag{8.3}$$

  The energy dose of x Gy corresponds to the absorbed radiation energy of x Joule per kg in the substance exposed to ionizing radiation.
  In former times the unit 1 Rad (= rad = absorbed radiation dose) was used and can be found in the older literature. The conversion factor is

$$1\,Gy = 100\,Rad \Leftrightarrow 1\,Rad = 0.01\,Gy. \tag{8.4}$$

- The actual ***dose rate* d*D*/d*t*** is the energy dose absorbed per sec with the unit [1 Gy/s].

W. Demtröder, *Nuclear and Particle Physics*, Undergraduate Lecture Notes in Physics,
https://doi.org/10.1007/978-3-030-58313-2_8

- The different kinds of ionizing radiation cause different impairments of the human body. For example the energy dose of 1 Gy α-radiation causes a much larger damage than 1 Gy of X-rays. In order to quantify the damage of tissue by the different kinds of ionizing radiation the dimensionless quality factors Q are introduced. The **equivalent dose** H is defined as the product

$$H = D \cdot Q \tag{8.5}$$

of energy dose D and quality factor Q. Its unit is

$$1\,\mathrm{Sv\,(Sievert)} = 1\ \mathrm{J/kg}. \tag{8.6}$$

The quality factors Q of the different kinds of ionizing radiation are listed in Table 8.1.
In former times the unit rem was used.

$$1\ \mathrm{rem} = 0.01\ \mathrm{Sv} = 10\ \mathrm{mSv}$$
$$1\ \mathrm{mrem} = 10^{-5}\ \mathrm{Sv} = 10^{-2}\ \mathrm{mSv}$$

Example: A person stands 2 m apart from a neutron source which emits 10^{10} neutrons/s with the energy 1 MeV uniformly into all directions. The particle flux density through his body is then

$$\Phi = \frac{10^{10}}{4 \cdot 4\pi}\mathrm{m^{-2}s^{-1}} \approx 2 \cdot 10^8\,\mathrm{m^{-2}\,s^{-1}}.$$

If 20% of the neutron energy (1 MeV = $1.6 \cdot 10^{-13}$ J) are absorbed in the body the actual dose rate is for a body cross section of 0.6 m^2 and a body weight of 75 kg

$$\begin{aligned}\frac{\mathrm{d}D}{\mathrm{d}t} &= 0.2 \cdot 2 \cdot 10^8\,\mathrm{m^{-2}\,s^{-1}} \cdot 1.6 \cdot 10^{-13}\mathrm{J} \\ &\quad \cdot 0.6\,\mathrm{m^2}/75\,\mathrm{kg} \\ &\approx 5 \cdot 10^{-8}\,\mathrm{J/(kg \cdot s)}.\end{aligned}$$

The equivalent dose rate dH/dt is with the quality factor of $Q = 10$ for fast neutrons

$$\frac{\mathrm{d}H}{\mathrm{d}t} = 10\frac{\mathrm{d}D}{\mathrm{d}t} = 5 \cdot 10^{-7}\,\mathrm{Sv/s}.$$

If the person stays for 10 min at this radiation exposed location he receives a total equivalent dose H = 0.3 mSv.

Table 8.1 Quality factors for radio-active radiation

Q	Kind of radiation
1	X-rays, γ-radiation, electrons
2.3	Thermal neutrons, protons and charged pions
10	Fast neutrons, protons and light charged particles
20	α-particles and heavy ions, 10 MeV neutrons

The radiation exposure caused by all natural sources (cosmic radiation, radon and radio-active materials in houses and in the soil) amounts to 4.3 mSv /year. This implies that the person close to the neutron source gets the same radiation exposure within 1 h as that from natural sources within 1 year. More information about the different definitions and their units can be found in [1–5].

For the measurements of such radiation exposure dose special dosimeters are used, which measure the momentary dose-rate (gas-ionization—or Scintillation detectors) or dose-meters which integrate over the dose within the exposure time Δt [6, 7].

As dosimeters for X-rays, γ-radiation or fast electrons, ionization chambers are used that operate in the proportional range (see Sect. 4.3.1). They have for γ-radiation in the range 10–1000 keV a relatively constant sensitivity. Often Geiger-Müller counters are used, which, however, do not directly measure dose rates but only the flux density of ionizing particles independent of their energy. In order to directly measure the equivalent dose, tissue-specific detectors have been developed, where the filling gas and the wall material of the detector have the same absorption characteristics as the human soft tissue.

Neutron detectors use the recoil protons produced at elastic collisions of neutrons with hydrogen atoms. A tissue equivalent detector sensitivity can be realized by proportional counters with an ethylene gas filling and walls covered with polyethylene. Dose meters, which measure the integrated dose, are often built in the form of fountain pens as shown in Fig. 8.1. These are small air-filled ionization chambers with an isolated inner electrode, which is charged. The ionizing radiation, absorbed within the gas volume, discharges the electrode and the amount of the discharge is proportional to the absorbed energy dose. It is monitored by a small electrometer (fused silicon wire). The disadvantage of this small detector is its sensitivity against vibrations.

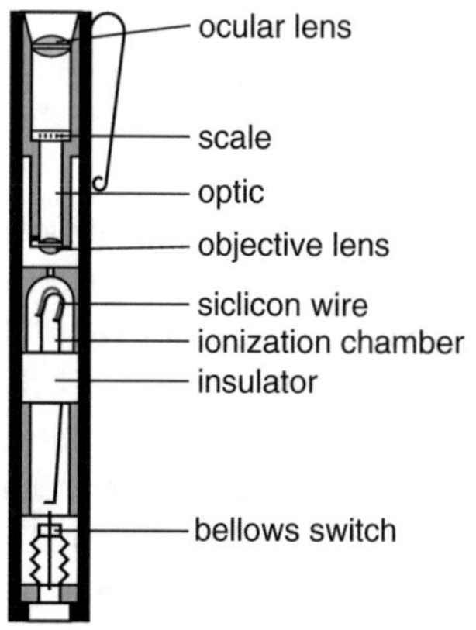

Fig. 8.1 Fountain pen dosimeter (after Kiefer, Koelzer: Strahlen und Strahlungsschutz Springer Heidelberg 1987)

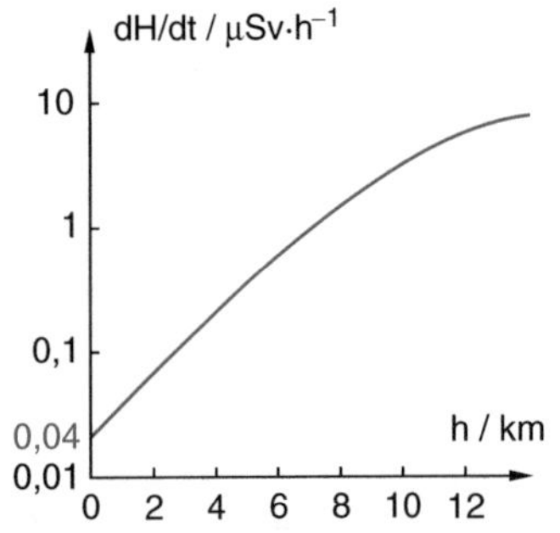

Fig. 8.2 Equivalence dose of the cosmic radiation at a mean geographical latitude as a function of the heights h above ground

In order to compare the radiation exposure by natural sources with that caused by radio nuclides in medical treatment and diagnostics, we will shortly discuss the different natural sources of ionizing radiation.

An important source is the cosmic radiation. Its equivalent dose is shown in Fig. 8.2 as a function of the heights above ground, given in µSv/h.

A further source is the manifold of radio-nuclides in the earth crust. In Table 8.2 the mean activity concentrations are averaged over the different regions in Germany. The units for solid radioactive substances are given in Bq/kg, for the gaseous Radon in Bq/(m^3 air).

The natural radiation exposure of a person is due to two effects: The external exposure (about 25% due to external radioactive sources) and the internal exposure (75%). The external exposure comes from the cosmic radiation (50%), 25% from radioactive Potassium (^{40}K) and 25% from nuclides in the Uranium and Thorium decay chain.

The internal exposure is due to inhalation of radon in the room air (68%) which is emitted by bricks in the room wall. The less the rooms are ventilated, the higher is the Radon concentration. Furthermore Thoron ($^{220}_{86}$Rn) and its decay products are present in the room air. Another source of internal exposure are radioactive isotopes of potassium and sodium in the food.

The effective equivalent dose caused by natural radiation exposure is in Germany between 1.5 and −4 mSv/year, depending on the specific location. The average value is about 2.2 mSv/year = 220 mrem/year/see Table 8.3 and [8–10]). In addition a dose of about 0.5–1.6 mSv/year is caused by medical investigations.

Table 8.2 Natural radio-activity outside and inside von houses

Outdoor area		Load in the house	
Rock	*A*/Bq/kg	Source	Activity
Granite	1000	Radon	≈ 50 Bq/m^3 air
Slate	700	Tap water	1–30 Bq/dm^3
Sandstone	350	Potassium in the body	4500 Bq
Basalt	250		
Garden soil	400		

Table 8.3 Mean natural and man-made radiation exposure per capita in Germany, given as effective equivalent dose in units of milli-Sievert/year

Source	Type of radiation	$(dH/dt)_{eff}$ (mSv/year)
External exposure	γ, p, n	0.7
Cosmic radiation		0.3
Ground, building		0.4
Inner exposure	$^{3}_{1}H, ^{14}_{6}C$ β	0.01
(Inhalation, Ingestion)	$^{22}_{11}Na, ^{40}_{19}K$ $\beta\gamma$ Uranium Radon α, β, γ	0.17 0.7–1.4
Total	Thorium Radon α, β, γ	0.4 1.9–2.4
Medical radiation exposure	γ, β	0.5–1.6
Emissions from coal and Nuclear power plants	γ, β	<0.003
Professional		
Radiation exposure		<0.3
Total burden		2.7–4.3

8.1.2 Technical Applications

The absorption and scattering of radioactive radiation form the basis for numerous radiometric measuring techniques [3]. For instance radiometric devices for thickness measurements allow a contact-free inspection of material sheets during the continuous production process. Examples are thickness measurement of paper, plastic foils, glass or metal sheets (Fig. 8.3). Radioactive sources for such measurements are radio-nuclides with a sufficiently long lifetime, which emit β- or γ-radiation (Table 8.4).

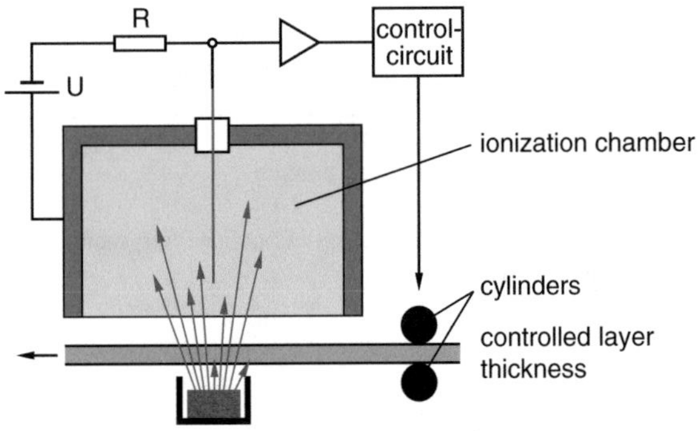

Fig. 8.3 Principle of thickness measurements by detecting the transmitted intensity

Table 8.4 Radio-nuclides used for measurement of material thicknesses

Radionuclide	Type of radiation used and energy (MeV)	Half-life	Additional radiation
^{226}Ra	α, 4.59	1620 a	β, γ
^{238}Pu	α, 5.50	88 a	
$^{3}_{1}H$	$β^-$, 0.018	12.3 a	–
^{22}Na	$β^+$, 0.55	2.58 a	γ
^{32}P	$β^-$, 1.71	14.3 d	–
^{85}K	$β^-$, 0.67	10.7 a	γ
^{90}Sr	$β^-$, 0.55	28 a	–
^{60}Co	γ, 1.33; 1.17	5.26 a	$β^-$
^{137}Cs	γ, 0.662	30 a	$β^-$
^{99}Tc	γ	6 h	
^{132}I	γ	2.3 h	$β^-$

Generally ionization chambers or semi-conductor counters are used as detectors. The exit signal of the detector is fed into a feedback circuit which controls the production facility in such a way that always a given absorption and with it a given thickness is kept constant.

With the optimum choice of the radio nuclides in Table 8.4 a mass thickness between 10^{-2}–10^{+3} kg/m^2 can be measured (Fig. 8.4). For aluminum sheets this corresponds to a thickness range between 4 μm and 0.4 m.

Instead of the attenuation of the transmitted radiation also the back scattering of γ-radiation can be used, which increases with increasing thickness up to a maximum limit which is set by multiple scattering (Fig. 8.5).

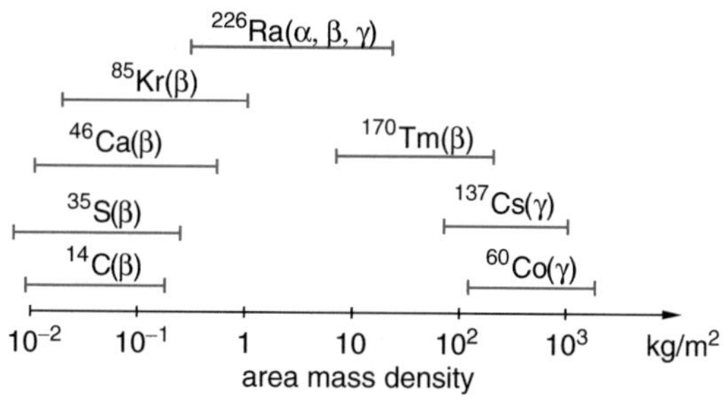

Fig. 8.4 Application of different radio-nuclides for materials with different masses per unit area

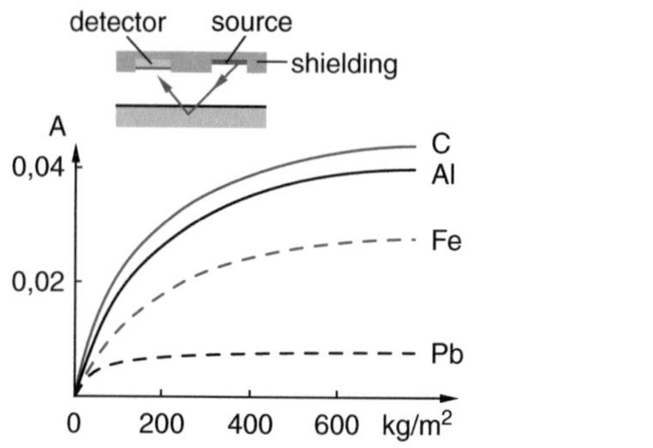

Fig. 8.5 Allbedo = ratio of backscattered to incident radiation flux of a backscattering layer for γ-radiation from ^{60}Co as a function of the mass per m^2

With $\dot{N}_0$ particles (β- or γ-radiation) incident per s onto the scattering sheet and $\dot{N}_R$ particles scattered back into the half space above the sheet the ratio

$$A = \frac{\dot{N}_R}{\dot{N}_0} \tag{8.7}$$

Is called the **albedo** of the scattering sheet. The albedo A approaches with increasing thickness a limiting value depending on the material of the sheet (Fig. 8.5).

The attenuation of the transmitted γ-radiation can be used for liquid level inspection in containers which cannot be opened (Fig. 8.6). It can also be applied to bulk cargo such as grain or sand. Here generally the radio nuclides $^{60}_{27}Co$, $^{137}_{55}Cs/^{137}_{56}Ba$ are used. From the difference of the signals monitored by the detectors D_i in Fig. 8.6 the liquid level can be unambiguously determined. There are no measurements inside the container necessary. This is particular advantageous if there is no access to the inner volume such as for blast furnaces, high pressure tanks or containers with chemical aggressive materials.

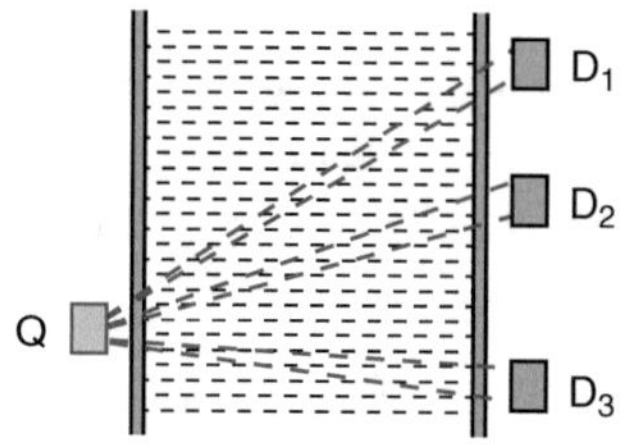

Fig. 8.6 Inspection of filling level in a container using a radio-active source and several detectors

The radiation of radio isotopes can be used not only for the diagnostics of materials but also for the purposeful change of material characteristics. One example is the controlled change of the electrical conductivity of semiconductors by directed irradiation. Also the tear resistance of polyethylene foils can be improved by irradiation. More infomation can be found in [11, 12].

8.1.3 Applications in Biology and Food Inspection

Ionizing radiation is used on a large scale to tackle micro-organisms, such as bacteria, microbes and to kill pathogenic germs. Examples are the disinfection of chirurgic instruments and of wound dressings. Also the storability of food can be often improved by exposure to ionizing radiation. For example for potatoes, carrots and onions the storage and transport times can be prolonged by irradiation with β rays without spoiling the food quality. Of cause such methods require a careful preceding inspection of possible health implications.

Radio nuclides are often applied in combination with biological pest control. Since the gonads are very sensitive against ionizing radiation, male insects can be sterilized without losing their biological activities. Such sterilized insects can be abandon in areas which are infested by dangerous vermin. This results in a strong reduction of the vermin production rate up to their total extinction.

8.1.4 Application in Medicine

The different applications of radio-active nuclides in medicine have increased considerably during recent years.

There is now an extra medical section called **nuclear medicine** which is specialized on radiation treatment of patients. Here we can only present some examples. For a more detailed presentation we refer to the extensive literature [13–17]. For the nuclear-medical diagnostics radio-active compounds (radio-medication) are feed to the patient, either by injection or orally. They spread through the body and they act the same way under transport processes and metabolism reactions as the corresponding inactive isotopes. Following this spread out by detecting the radiation with local resolution it is possible to study these transport processes in organisms without penetrating into the body.

The radio-drugs have to fulfill the following conditions:

- The radiation must be detectable outside the body. Therefore only γ-radiation or positron-emitter can be used where for the positrons the annihilation reaction $e^+ + e^- \rightarrow 2\gamma$ with detection of the γ-quanta is utilized.
- The retention time of the drug inside the body should not be much longer than the inspection time in order to reduce the radiation exposure to a minimum value. As a measure of the optimum retention time the effective half-lifetime

$$T_{\text{eff}} = \frac{T_{1/2} \cdot T_{\text{B}}}{T_{1/2} + T_{\text{B}}} \tag{8.8}$$

is used which is defined by the physical half lifetime $T_{1/2}$ of the radio-nuclide and the biological lifetime T_{B}, where T_{B} gives the time after which the body has exuded half of the applied drug. Equation (8.8) can be derived as follows:
The decay of the radio-active nuclide is given by the sum

$$\frac{\mathrm{d}N}{\mathrm{d}t} = -A \cdot N - B \cdot N \tag{8.8a}$$

of radio-active decay—$A \cdot N$ and the biological decomposition—$B \cdot N$. Integration gives

$$N = N_0 \cdot e^{-(A+B) \cdot t} = N_0 \cdot e^{-t/T_{\text{eff}}} \tag{8.8b}$$

with $T_{\text{eff}} = 1/(A+B) = 1/\left(\frac{1}{T_{1/2}} + \frac{1}{T_{\text{B}}}\right)$.

As important example the radio-nuclide test for the diagnostics of thyroid disease will be discussed. This disease becomes perceivable by the characteristic change of the iodine metabolism.

The patient has to take orally the substance Na ^{131}I with the radioactive isotope ^{131}I. It is then measured how fast the thyroid absorbs the applied iodine and how fast it is released again (Fig. 8.7).

Instead of iodine nowadays the radio-active isotope $^{99}_{43}$Tc of technetium is used because it is stored in the thyroid the same way as iodine but has a much shorter lifetime of $T_{1/2} = 6$ h and therefore the exposure of the body to radiation is much smaller, because a smaller quantity of the radio-active substance can be used for the same radiation dose rate.

For the local diagnostics of the thyroid the γ-radiation of the excited radio nuclide $^{99}_{43}$Tc is measured by a detector foil

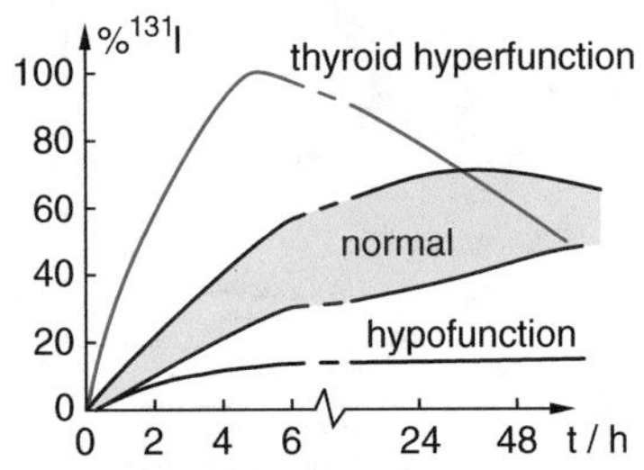

Fig. 8.7 Intake of radio-active iodine by the thyroid gland for normal operation, hypo-function and hyper function

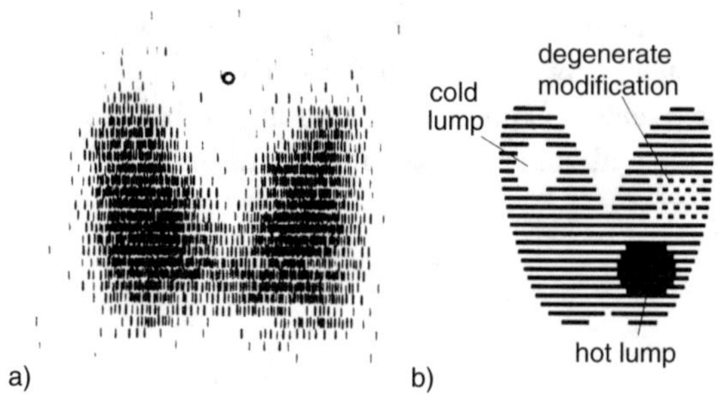

Fig. 8.8 **a** T hyroid scintillation diagram **b** Schematic illustration of cold and hot lumps

placed around the neck of the patient. With such an arrangement the scintigram of the thyroid can be obtained (Fig. 8.8). In the right part of the figure the detection of hot and cold lumps in the thyroid are shown. The hot = active lumps (carcinomas) store more technetium, the cold = inactive lumps less [18].

An advanced modern technique for obtaining images with a high contrast of patho-physiological processes inside the body is the **positron-.emission tomography** (PET) [19, 20]. Here positron emitting nuclides are used in special substances, which are drunk by the patient, or are injected. They are transported by the body to special internal organs. The following nuclides can be used: ${}^{11}_{6}C$; ${}^{13}_{7}N$; ${}^{15}_{8}O$ and ${}^{18}_{9}F$. The positrons collide with electrons of body-molecules and annihilate in the process

$$e^+ + e^- \rightarrow 2\gamma.$$

The γ-quanta fly into opposite directions and are detected in coincidence by two confronting detectors (Fig. 8.9). In modern devices detector array are used which surround the patient. Similar to the X-rays tomography two-dimensional sectional images can be obtained. The advantage compared to X-ray tomography is the fact that the β^+-emitter is located in the sectional plane and therefore the superposition with other planes is avoided. This makes the images free of background noise and they show a greater contrast than X-ray images.

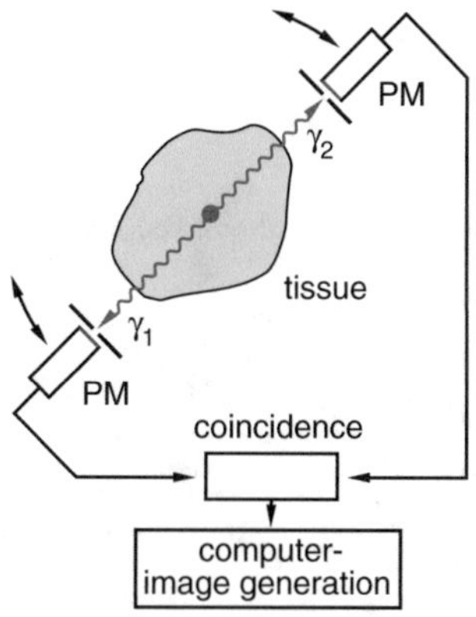

Fig. 8.9 Positron-emission tomography

Examples of applications are the investigation of blood circulation disturbances in the brain or in the heart. Also malfunctions of the glucose metabolism in specific organs can be inspected.

An important application of radio-nuclides is the treatment of localized cancer cells. If they are detected early enough that no metastases have been spread, the limited cancer volume can be very successfully treated by a small glass tube filled with the radio-active substance, which is inserted into the body to the cancer location. With this technique the damage of healthy cells is minimized. One example is the cure of cervix cancer which has been treated by this technique since long. For more examples of mecical radioactive treatment see [20–26].

8.1.5 Detection of Tiny Atom Concentrations by Activation of Radio-Active Nuclei

If a sample is irradiated by neutrons, protons or γ-quanta the atoms in the sample can be partly converted into radio-. active nuclides (see Sect. 6.4). The radiation of these nuclides can be detected with very high sensitivity [21]. This allows the quantitative determination of even very small atomic concentrations (Fig. 8.10).

One example is the activation reaction

$$\begin{aligned} n + {}^{16}_{8}O &\rightarrow {}^{16}_{7}N^* + p \\ &\underset{7{,}2\,s}{\hookrightarrow} {}^{16}_{8}O + \beta^-(10{,}4\,\text{MeV})\,, \end{aligned} \tag{8.9}$$

where the β^--radiation of the ${}^{16}_{7}N_7{}^{16}N$-nuclids is detected in order to determine the concentration of O-atoms (for instance in metals).

The concentration of nitrogen atoms N can be measured, for instance, by the activation reaction

$$\begin{aligned} n + {}^{14}_{7}N &\rightarrow {}^{14}_{6}C + p \\ &\underset{5730\,a}{\hookrightarrow} {}^{14}_{7}N + e^-(0{,}156\,\text{MeV})\,. \end{aligned} \tag{8.10}$$

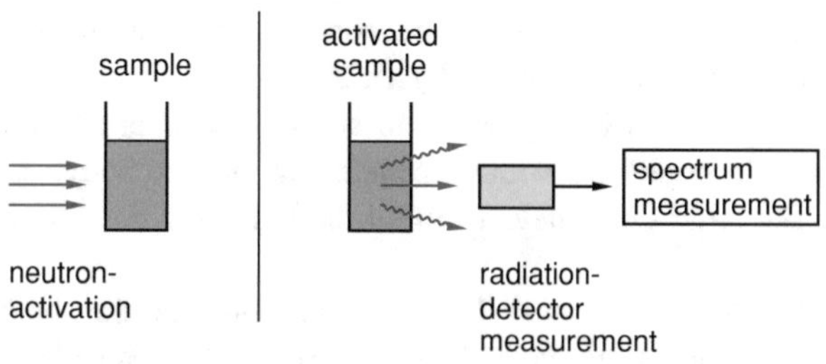

Fig. 8.10 Sensitive detection of small atom concentrations based on neutron-activation of radio-active isotopes

The quantitative activation analysis is based on the proportionality between the activity of the produced radioactive nuclides and the number of the parent nuclei of the wanted substance. Since it is difficult to determine the fraction of the activated nuclei, one uses the activation of a known number of atoms of the same element under the same conditions to get a reference. The ratio of the measured activities of reference and unknown sample gives the ratio of the numbers of atoms in the reference and the inspected sample number.

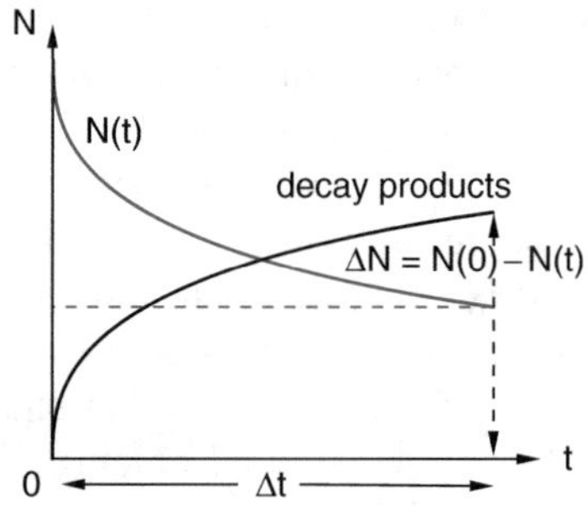

Fig. 8.11 Age determination of archeological discoveries based on the decay curves of radio-active nuclides

8.1.6 Radio-Active Age Determination

For the age determination of geological or archeological objects often the time dependent activity of natural radioactive substances is used.

If the number $N(t_0)$ of atoms in a radioactive sample is known at time t_0 the number $N(t)$ at a later time t is related to $N(t_0)$ by the exponential law

$$\begin{aligned} N(t) &= N(t_0) \cdot e^{-\lambda(t-t_0)} \quad \text{with} \quad \lambda = \ln 2/T_{1/2} \\ &= N(t_0) \cdot 2^{-(t-t_0)/T_{1/2}} \end{aligned} \tag{8.11}$$

where $\lambda = 1/\tau$ is the decay constant. The time difference

$$\Delta t = t - t_0 = \frac{1}{\lambda} \ln \frac{N(t_0)}{N(t)} \tag{8.12}$$

can be therefore determined by measuring $N(t)$.

If $N(t_0)$ is not known, often the number $\Delta N = N(t_0) - N(t)$ of particles, that have decayed during the time difference Δt can be measured. This number is equal to the number of stable nuclides into which the radioactive nuclides have decayed (Fig. 8.11). The time difference can be then obtained from

$$\Delta t = \frac{1}{\lambda} \ln \left(\frac{\Delta N}{N(t)} + 1 \right). \tag{8.13}$$

For a sufficiently accurate determination of Δt the lifetime τ of the radioactive nuclide should not deviate from Δt by more than a factor of 10.

This method of age determination will be illustrated by some examples:

Best known is the age determination with the ^{14}C nuclide, which is based on the following arguments [22–24]:

Nitrogen- and Oxygen nucleons in the higher atmosphere of the earth are hit by protons emitted by the sun and by high energy particles in the cosmic radiation. These collisions break the nuclides of N and O into nucleons. The resulting free protons and neutrons penetrate further into the atmosphere. While the protons are slowed down by collisions already in the higher atmosphere, the neutrons can reach the lower atmosphere where they collide with atoms and molecules. In the reaction

$$^{14}_{7}N + n \rightarrow ^{14}_{6}C + p$$

The radioactive nucleus of the carbon isotope ^{14}C is produced which reacts with oxygen to form the radioactive carbon dioxide $^{14}CO_2$.The mean production rate of carbon atoms per m^2 of the earth surface is about $2.5 \cdot 10^4/m^2 \cdot s$.

The present ratio $^{14}C/^{12}C$ which equals the ratio $^{14}CO_2/^{12}CO_2$ is $1.2 \cdot 10^{-12}$. Due to the natural gas mixing in our atmosphere the CO_2 reaches the earth surface and is incorporated by plants during their assimilation. It reaches by food intake the human and animal bodies.

In Fig. 8.12 the different sources and sinks for the production, resp. the reduction of ^{14}C are represented. The concentration of ^{14}C in the atmosphere is determined by the production rate and the exchange rate between atmosphere biosphere, surface of earth and oceans.

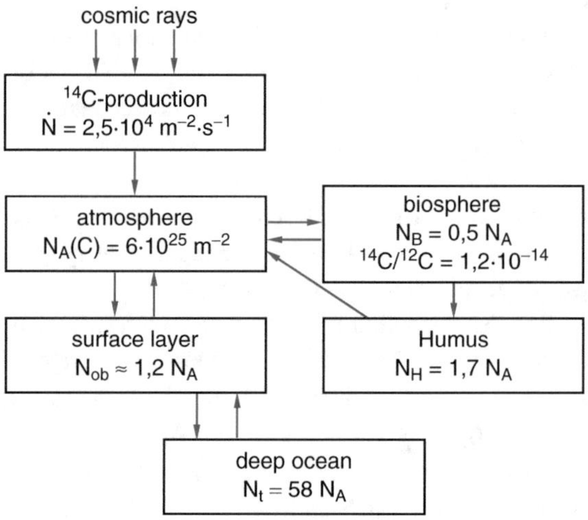

Fig. 8.12 Carbon reservoir and its exchange with the environment (N_i gives the number of C-atoms per m^2 of the earth surface) (after H. Wilkomm; Altersbestimmjung im Quartär. Thiemig, München 1976)

While the mixing of CO_2 within the different layers of the atmosphere takes only a few months, it takes several tens of years to penetrate into the deeper layers of the oceans. The numbers N_A in Fig. 8.12 give the carbon concentrations in the atmosphere per m^2 of the earth surface. Furthermore the relative values N/N_A are listed for the biosphere, the humus layer of the earth surface and the deep ocean layers.

The ratio $^{14}CO_2/^{12}CO_2$ in the whole biosphere equals that in the atmosphere. At the death of a biological creature the intake of CO_2 stops. While the concentration of ^{12}C remains constant, that of ^{14} C decreases with the half lifetime $T_{1/2}$ = 5730 years. The ^{14}C-concentration can be measured by its β^- activity (E_{max} = 0.155 MeV).

With this ^{14}C-method the age determination of archeological finds (bones, fossils excavated residuals of wooden houses) can be performed for times between 1000 and 7500 years. For a reliable age determination the following facts have to be taken into account:

- Due to burning of coal which has been mined out of deep layers of the earth, the concentration of ^{14}C is lower than at the earth surface. Therefore the ratio $^{14}CO_2/^{12}CO_2$ in our atmosphere has slightly decreased over the last 100 years.
- Nuclear weapons tests in the atmosphere around 1960 have increased the ^{14}C-concentration.by about 4%. The neutrons released at the explosion of a hydrogen bomb induce the reaction

$$n + {}^{14}_{7}N \rightarrow p + {}^{14}_{6}C$$

  and produce additional ^{14}C similar to the neutrons in the secondary cosmic rays.
- In addition long term natural fluctuations of the ^{14}C concentration occur, which are caused by changes of the cosmic ray intensity, the sun activity and changes of the earth magnetic field.

The evaluation of ${}^{14}_{6}$C-concentration measurements for the age determination therefore demand a very careful analysis of all interference effects [25].

The age determination of rocks and minerals, which are much older require radioactive nuclides with longer lifetimes. The natural radioactive nuclei such as ${}^{238}_{92}$U ($T_{1/2} = 4.5 \cdot 10^9$ a), ${}^{235}_{92}$U ($T_{1/2} = 7 \cdot 10^8$ a), ${}^{232}_{92}$Th ($T_{1/2} = 1.4 \cdot 10^{10}$ a) decay by α-emission and generate the decay chains shown in Fig. 3.12 which end in stable lead isotopes (see Sect. 3.2.3). The longest half life time in the chain determines the ratio of the concentrations of the unstable members in the chain to that of the stable final product. From the thorium decay chain, for example, the number of stable lead nuclei is according to Eqs. (8.11) and (8.13):

$$\begin{aligned} N_{Th}(t) &= N_{Th}(0)e^{-\lambda t} \\ N_{Pb}(t) &= N_{Th}(0) - N_{Th}(t) = N_{Th}(t)\left(e^{+\lambda t} - 1\right) \\ \Rightarrow t &= \frac{1}{\lambda_{Th}}\ln\left(\frac{N_{Pb}(t)}{N_{Th}(t)} + 1\right), \end{aligned} \tag{8.14}$$

where we have assumed that at $t = 0$ no lead was present in the investigated mineral. Measuring the number ratio of lead and thorium (for instance with a mass spectrometer) the age τ of the mineral can be determined, provided the life time of thorium is known. If there was already lead in the mineral at $t = 0$ the number of lead atoms at the time t is

$$N_{Pb}(t) = N_{Pb}(0) + N_{Th}(t)\left(e^{\lambda t} - 1\right)$$

Measuring the ratio

$$N_{Pb}(t)/N_{Th}(t) = (N_{Pb}(0)/N_{Th}(0))e^{\lambda t} + e^{\lambda t} - 1,$$

one can deduce the original ratio at $t = 0$. The age determined this way assuming $N_{Pb}(0) \neq 0$ is shorter than that obtained from (8.14) with the assumption of $N_{Pb}(0) = 0$ (Fig. 8.13).

If the samples also contain the uranium isotopes ${}^{235}_{92}$U and ${}^{238}_{92}$U one has to take into account that the measured lead concentration originates from all radioactive elements contained in the sample.

Many minerals contain potassium which includes 0.01% of the radioactive isotope ${}^{40}_{19}$K ($T_{1/2} = 1.28 \cdot 10^9$ a). It decays by electron capture into ${}^{40}_{18}$Ar. If the ratio $N({}^{40}_{18}Ar)/N({}^{40}_{19}K)$ is measured, the age of the mineral can be determined in the same way as in (8.14) for lead.

The potassium/argon method anticipates, however, that no argon has escaped the sample during the time t. The combination of the lead-and the potassium-argon techniques gives the age $t = 4 \cdot 10^9$ years for the oldest minerals on earth, which gives a lower limit for the age of the earth.

A particular sensitive and accurate method of age determination based on radio-active nuclides is the accelerator mass spectrometry (Fig. 8.13X, [26]). The radio-active atoms contained in the sample are evaporated and ionized

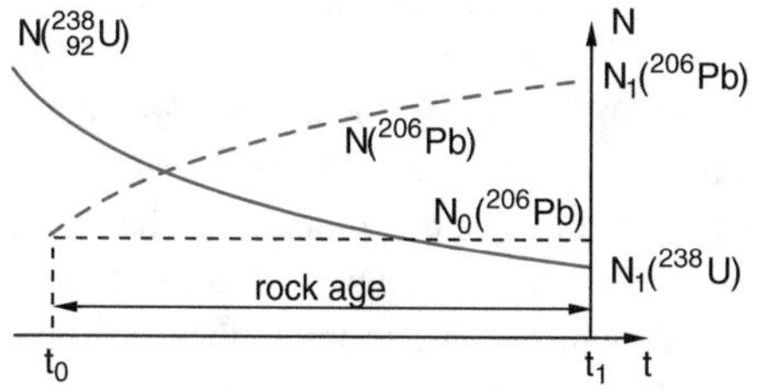

Fig. 8.13 Radio-active age determination of rocks which contained already at the time of their formation stable decay products of radio-active isotopes

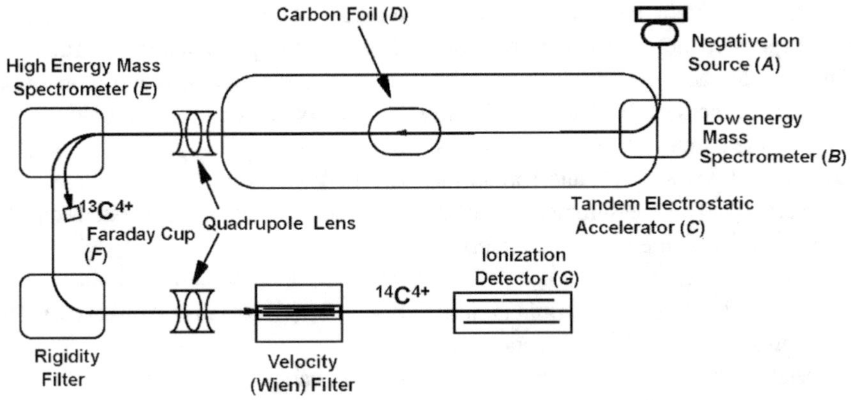

Fig. 8.13X Accelerator mass spectrometry [Hah; BioMed Central Ltd., CC BY 2.0, https://commons.wikimedia]

in the ion source of a mass spectrometer. Often electron-attachment techniques are used to produce negative ions (e.g. C^-). The ions are accelerated and sent through an electric field in a 90° cylinder capacitor and a 90° magnetic sector field (see Vol. 3, Sect. 2.7). In a tandem accelerator (see Sect. 4.1.3) the negative ions are recharged into positive ions and further accelerated.

Then a further mass selection is provided by an electric and magnetic sector field. With this double arrangement the mass selection is increased and it becomes possible to detect tiny concentration of a wanted isotope in the presence of much more abundant neighboring isotopes.

With these techniques many of the ^{14}C isotope concentrations have been investigated. With the half life time of ^{14}C ($T_{1/2} = 5730\ a$) only a small fraction of the ^{14}C-atoms decay during the measuring time of a few hours. Therefore one needs a sufficiently large sample with enough ^{14}C-atoms. The advantage of the accelerator technique is that it does not measure the decayed nuclei but the number of presently existing ^{14}C-atoms. Therefore a 1000 times smaller sample

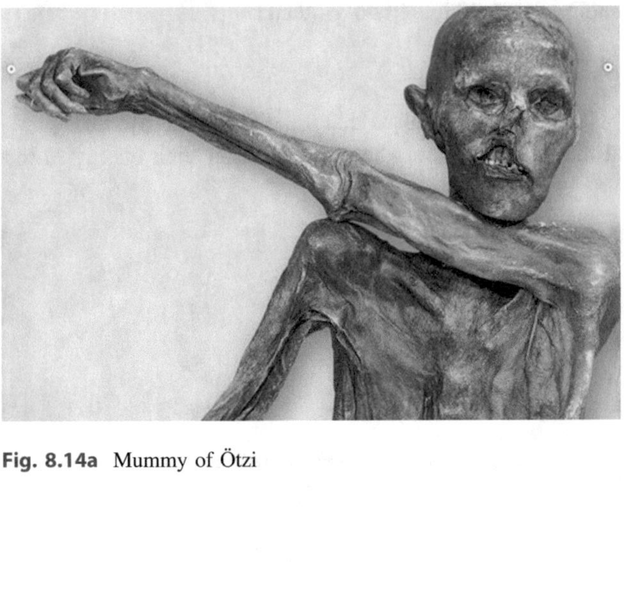

Fig. 8.14a Mummy of Ötzi

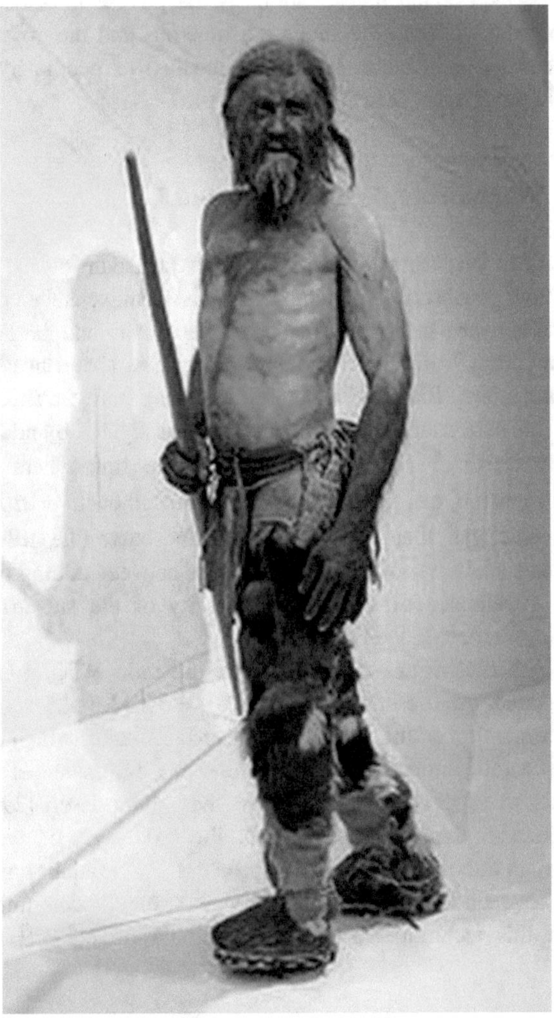

Fig. 8.14b Reconstruction of Ötzi

down to 1 mg is sufficient to obtain accurate results. Its drawback is the large expensive apparatus and the high experimental expenditure for the accelerator.

One example for the effectiveness of this technique is the determination of the age of the shroud of Turin, which was reputedly used for the burial of Jesus Christ and should be therefore at least 2000 years old. The measurements proved that it was only 800 years old and therefore a counterfeit made in the twelfth century [27, 28].

In particular interesting was the accurate determination of the age of the "Ötzi", a corpse found 1991 by hikers in the ice of a glacier in the "Oetztal Alpes" on the border between Austria and Italy. His body and his clothes were completely preserved, because he had been covered by ice for many years. His age was determined as 5250 ± 100 years. The importance of the investigations is the precise information about the clothing, the shoes the weapons and the possible illnesses (for instance intestinal parasites) of people living 5000 years ago [28] (Fig. 8.14a, 8.14b).

8.1.7 Hydrological Applications

It is often interesting and necessary to investigate the course and the flow of subsurface running water. One example is the upper Danube river where part of its water seeps away into the limestone in the "white Jura" and appears again as abundant water source in the "Blautopf" about 100 km away from the sink.

This subsurface water flow can be investigated by adding some ${}^3_1H_2O$ or HTO isotopomers to the sink where the hydrogen ${}^1_1H$ has been replaced by the radioactive tritium ${}^3_1H$. One can then measure at which source the tritium appears again. From the time difference between adding time and appearance time the flow velocity of the subsurface water can be deduced (Fig. 8.15).

With the tritium method, where the ratio HTO/H_2O is measured, the lifetime of tritium ($T_{1/2}$ = 12.43 a) allows the determination of the age of the subsurface water, which can be several hundred years old. There are, for instance, old sweet water reservoirs deep below the Sahara desert [29].

Another important subject is the exchange of water between atmosphere and earth surface. If its circulation time lies between some years and 100 years it can be determined with this technique. The concentration ratio c(T)/c(H) is measured either by the acceleration mass spectrometry or by detecting the β^- radiation of tritium with a proportionality counter after enrichment of tritium by electrolytic techniques.

The exploration of oil fields is possible by inserting radioactive krypton ${}_{36}{}^{85}Kr$ into a borehole and measuring the exit of krypton from another borehole. This gives information about the contiguous extension of the oil field.

8.2 Applications of Accelerators

Besides radio-active nuclides for the therapy of cancer also accelerators are used. In many clinics Betatrons are very useful instruments for the irradiation of localized cancer cells. Either the collimated high energy electron beam (E = 1–10 MeV) is used which can be guided by electron-optical elements and focused onto the cancer region, or the electrons hit a copper or tungsten anode where they produce high energy γ-radiation. The cross section of this collimated beam of γ-radiation is narrowed by tungsten apertures in order to reach the optimum adaption to the cancer area. The patient lies on a special turntable and is continuously turned during the irradiation treatment in order to reach the maximum irradiation in the cancer region but the minimum irradiation in the surrounding healthy tissue (Fig. 8.16).

In special clinics which are connected to large accelerators, tissue irradiation with π-mesons is applied. For a proper energy of the pions they penetrate into the tissue exactly until the cancer region. Here they decay according to

$$\pi^+ \rightarrow \mu^+ + \nu$$

where the muon decays further into

$$\mu^+ \rightarrow e^+ + \nu_e + \bar{\nu}_\mu$$

The positron collides with electrons of the tissue and annihilates producing two γ-quanta:

$$e^+ + e^- \rightarrow 2\gamma$$

The essential point is that the different decays take place at the same location where they transfer their energy to the

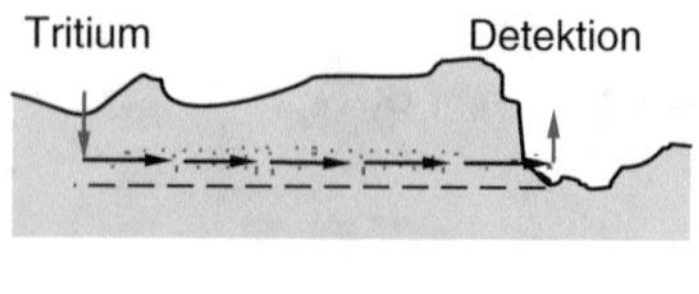

Fig. 8.15 Principle of radio-active mark of underground water flow

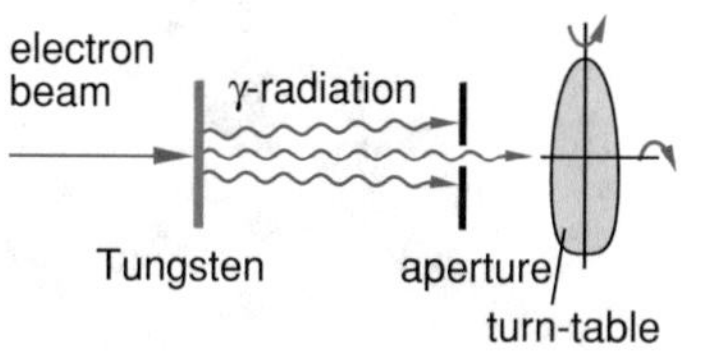

Fig. 8.16 Application of Betatrons for radiation therapy in medicine

cancer cells and kill them. This technique is therefore the optimum cancer therapy with a minimum damage of the healthy cells. It is, of course, a very expensive treatment and can be therefore only applied for special cases. One example is the irradiation of eye cancer where pointedly special small areas are irradiated by the focused π-beam which ensures that the eye is protected as much as possible [30].

Such irradiations are for instance performed at the Paul-Scherrer-Institute in Villingen, Switzerland.

8.3 Nuclear Reactors

Although nuclear reactors which convert energy from the fission of heavy nuclei, such as Uranium, into electrical energy, are presently under critical discussion, they are in many cases still necessary for the provision of CO_2-free energy. As we will see later, their main problem is the safe storage of the nuclear waste. In order to competently judge their merits and drawbacks it is worthwhile to be informed about their construction, the principal basis of their operation and necessary safety measures.

In order to compare them with other CO_2 –free renewable energy sources such as wind energy one should just consider the numerical facts:

An on-shore wind-converter with a maximum power output of 5 MW delivers per year the average power of about 1 MW, depending on the wind conditions. A medium size nuclear power station delivers about 1 GW and runs continuously over the year (apart from about 3 days when the fuel rods are replaced. One needs therefore about 1000 wind converters to replace 1 nuclear power plant.

It is therefore still useful to study the physical basis of nuclear reactors [31, 32].

In Sect. 6.5 is was shown that the fission of an uranium nucleus releases the energy of about 200 MeV while the burning of coal for each reaction $C + O_2 \rightarrow CO_2$ releases the energy of only 13.5 eV, which is smaller by a factor of $1.5 \cdot 10^5$. This means that the fission of 1 kg uranium provides the same energy as burning 750 tons of coal. In order to compare the operation costs for both types of reactors one has also to take into account the costs of mining and processing the fuel, which is more expensive for uranium than for coal.

We will now discuss the physical basis of nuclear reactors.

8.3.1 Chain Reactions

The neutrons emitted at the nuclear fission can induce further fissions. They can also be absorbed before they can start new fission processes. In order to realize a controlled chain reaction where the neutrons of a fission generates exactly one new fission, the number of neutrons produced at a fission and the number of absorbed neutrons must have the correct relationship.

In Sect. 6.5 it was shown, that the uranium isotope ${}^{238}_{92}U$ can be split only by fast neutrons ($E_{kin} > 1$ MeV) whereas the isotope ${}^{235}_{92}U$ can be split also by thermal neutrons. The important fast is now that the cross section σ for neutron induced nuclear fission increases with decreasing neutron energy as $1/\sqrt{E}$ (Fig. 6.17).

Example: $\sigma(E = 0.01\ \text{eV}) \approx 10^4 \cdot \sigma(1\ \text{MeV})$.

Figure 8.18 shows that the isotope ${}^{235}_{92}U$ is much more efficient for nuclear fission than ${}^{238}_{92}U$. In order to reach a sufficiently high probability for nuclear fission, one has to enlarge the concentration of ${}^{235}_{92}U$, which is naturally only 1%, to about 3.5%.

Since the fission cross section increases with decreasing neutron energy one must decelerate the fast neutrons generated at the fission process. This is achieved with moderators (water, graphite) (Fig. 8.17). The demands for a good moderator material are

- A high efficiency in slowing down the neutrons by elastic collisions. Therefore one has to use moderators with light nuclei (e.g. H_2O, D_2O, graphite)
- The absorption of neutrons should be small (Fig. 8.17)

  If the mean deceleration is described by the logarithmic energy decrement

$$\xi = \overline{\ln(E_1/E_2)}, \tag{8.15}$$

where E_1 is the energy of the neutron before the collision and E_2 after the collision. One can derive an approximate formula

$$\xi = \frac{6}{2 + 3M + 1/M^2}, \tag{8.16}$$

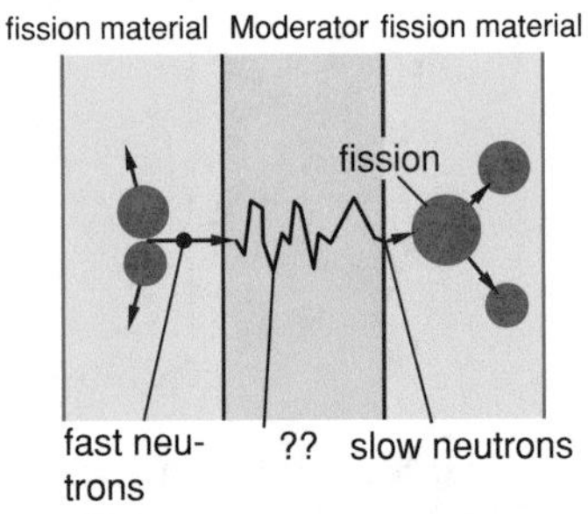

Fig. 8.17 Schematic illustration of the slowing down of fast fission neutrons and the new fission generation induced by slow neutrons

M is the mass of the decelerating nucleus in atomic units (e.g. $M = 1$ for H-nuclei in H_2O).

Examples For hydrogen is $M = 1 \rightarrow \xi = 1$ which gives $\overline{E_1/E_2} = e \approx 2.72$. The average relative energy loss per collision is then $(E_1 - E_2)/E_1 = \Delta E/E_1 = (1 - 1/e) = 0.63$.

For carbon nuclei is $M = 12 \rightarrow \xi = 0.158 \rightarrow \Delta E/E = 0.146$.

The average number q of collisions which is necessary to slow the neutrons down from the initial energy E_0 to thermal energy E_{th} is deduced from (8.15) as

$$C = \frac{1}{\xi} \ln E_0/E_{th}. \tag{8.17}$$

Example $E_0 = 2\,\text{MeV}, E_{th} = 0.052\,\text{eV} \Rightarrow C = 18.2/\xi$. For hydrogen nuclei is $\xi = 1 \rightarrow q = 18.2$. There are 18 collisions necessary to slow the neutrons down to thermal energies. For a graphite moderator is $\xi = 0.3 \rightarrow q = 60.7$. The graphite moderator is much less efficient to decelerate the neutrons.

There are several reasons, while, nevertheless, in several nuclear reactors (most of the Russian reactors) graphite is used as moderator material. One of these reasons is the low absorption coefficient of neutrons in graphite, which is lower than that in water.

Not only in the moderator but also in the fission material neutrons are absorbed before they can induce a fission process. In Fig. 8.18 the absorption coefficients for the reaction

$$n + U \rightarrow U^* \rightarrow U + \gamma$$

are illustrated as a function of the neutron energy. At certain energies sharp absorption resonances occur which are drawn in Fig. 8.19 and schematically in Fig. 8.18. The envelop in Fig. 8.18 gives the limits for the maximum and minimum absorption coefficients. In order to illustrate the manifold of such sharp resonances for illustration in Fig. 8.19 the absorption cross section $\sigma(n,\gamma)$ for part of these resonances are shown in more detail within the limited energy interval from 600 to 900 eV. For comparison the fission cross section $\sigma(n,f)$ is shown in red.

We will study in Fig. 8.20 the life cycle of N_n thermal neutrons during one cycle from the generation by fission over the thermalization until the next fission process. Before their deceleration they can split $^{238}_{92}U$ as well as $^{235}_{92}U$, after their deceleration they cannot split $^{238}_{92}U$ but only $^{235}_{92}U$. However, they can be absorbed by both isotopes without inducing a fission. We assume that the fraction N of the N_n neutrons induces a fission process in ^{235}U, which produce $N \cdot \eta\,(\eta > 1)$ new fission neutrons. These neutrons have an energy distribution shown in Fig. 6.20. The fast neutrons can also split ^{238}U. Altogether there are after the nth fission

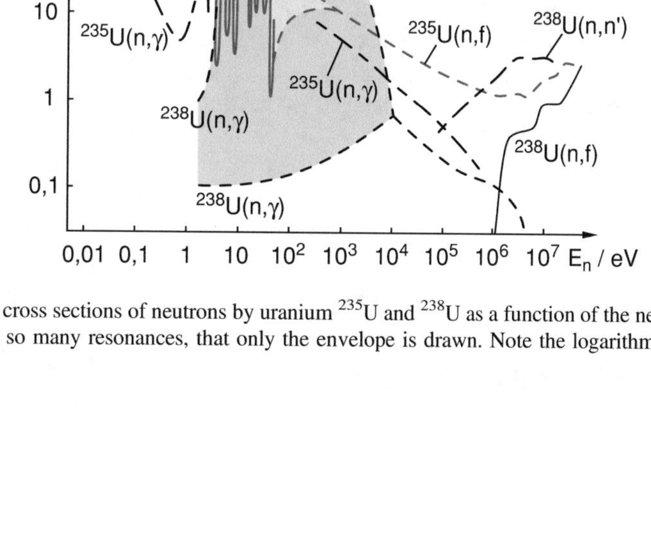

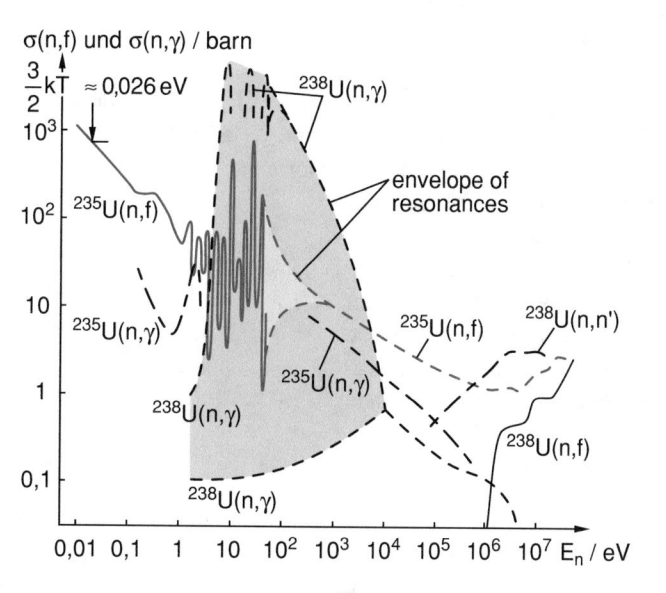

Fig. 8.18 Absorption- and fission cross sections of neutrons by uranium ^{235}U and ^{238}U as a function of the neutron kinetic energy. Because in the energy range 10–800 eV there are so many resonances, that only the envelope is drawn. Note the logarithmic ordinate scale

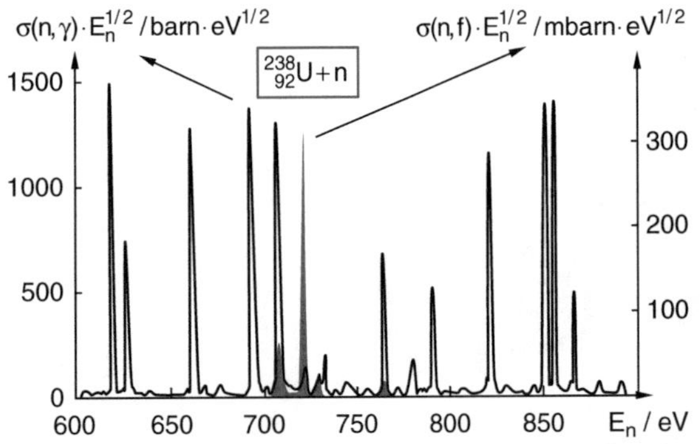

Fig. 8.19 Absorption cross section σ(n,γ) (black) and fission cross section σ(n,f) (red) of neutrons in the energy range 600–900 eV. Note, that the ordinate scale for fission (given in mbarns) is about 4000 times smaller than that for absorption (given in barns). (From Musiol: Kern-und Elementarteilchen-Physik, VEB-Verlag Berlin 1988)

generation $N_n \cdot \eta \cdot \varepsilon\,(\varepsilon > 1)$ neutrons. The product $\eta \cdot \varepsilon \cdot N_n$ gives the mean number of neutrons produced by the fission of ^{238}U by fast neutrons and by the fission of ^{235}U by all neutrons. This is the number of neutrons which is available after the nth fission generation.

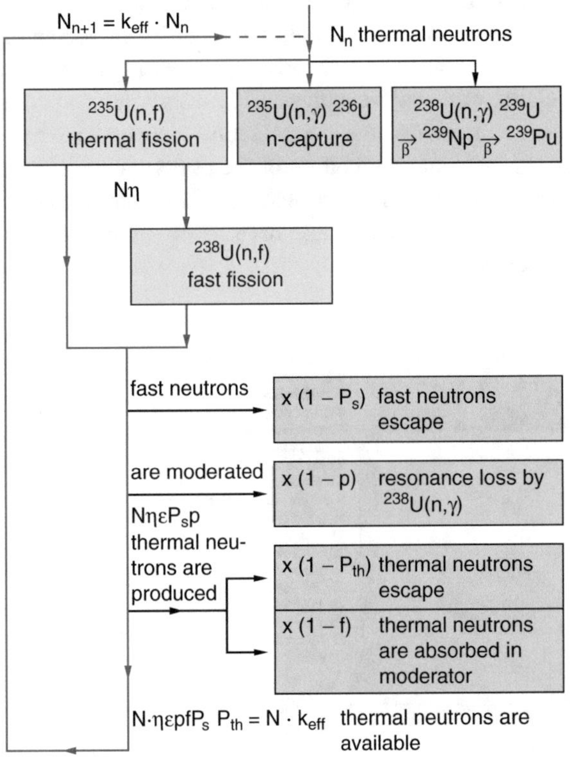

Fig. 8.20 Life-cycle of N_n thermal neutrons during the interval between two successive fission generations. (From Mayer-Kuckuk: Kernphysik, Teubner Stuttgart 1992)

The fast neutrons can partly escape out of the reactor core and are therefore lost for the next fission generation. If $(1–P_s)$ is the probability for escape of a fast neutron, p the probability that a neutron is slowed down without being absorbed, $(1–P_{th})$ the escape probability of a thermal neutron and $(1–f)$ the probability that a neutron is absorbed in the moderator, only the number

$$N_{n+1} = N_n \cdot \eta \cdot \varepsilon \cdot p \cdot f \cdot P_s \cdot P_{th} = k_{eff} \cdot N_n \qquad (8.18)$$

of thermal neutrons are available for the fission of the $(n + 1)$th generation. With the mean cycle time T between two successive generations the increase dN of the neutron number during the time interval dt becomes

$$dN = \frac{k_{eff} - 1}{T} \cdot N \cdot dt. \qquad (8.19)$$

In order to reach a stationary operation of the reactor the product

$$k_{eff} = \eta \cdot \varepsilon \cdot p \cdot f \cdot P_s \cdot P_{th} \qquad (8.20)$$

must be equal to 1. For $k_{eff} < 1$ the chain reaction diminishes for $k_{eff} > 1$ the number of neutrons increases exponentially. From (8.19) it follows:

$$N = N_0 \cdot e^{(k_{eff}-1)\cdot t/T}. \qquad (8.21)$$

For the standard dimensions of some meters for the core of nuclear reactors the escape probability for fast and thermal neutrons is very small. This means $P_s \approx 1$ and $P_{th} \approx 1$. For the limiting case of a reactor with infinite large dimensions we get the four-factor formula

$$k_\infty = \eta \cdot \varepsilon \cdot p \cdot f \qquad (8.22)$$

for a stationary operation of a reactor.

The quantity

$$\varrho = \frac{k_{eff} - 1}{k_{eff}} \qquad (8.23)$$

is the ***reactivity*** of the reactor. For the stationary operation it must always be $\varrho = 0$. This can be achieved by placing absorbing materials in the moderator and install in addition controllable absorber rods which can be shifted under controlled conditions into the reactor core.

If the reactor output power should increase the absorber rods are pulled a little bit out of the core, if the output power should decrease the rods must be shifted further into the core. This power regulation and its safety problems will be discussed in more detail in the Sects. 8.3.3 and 8.3.5.

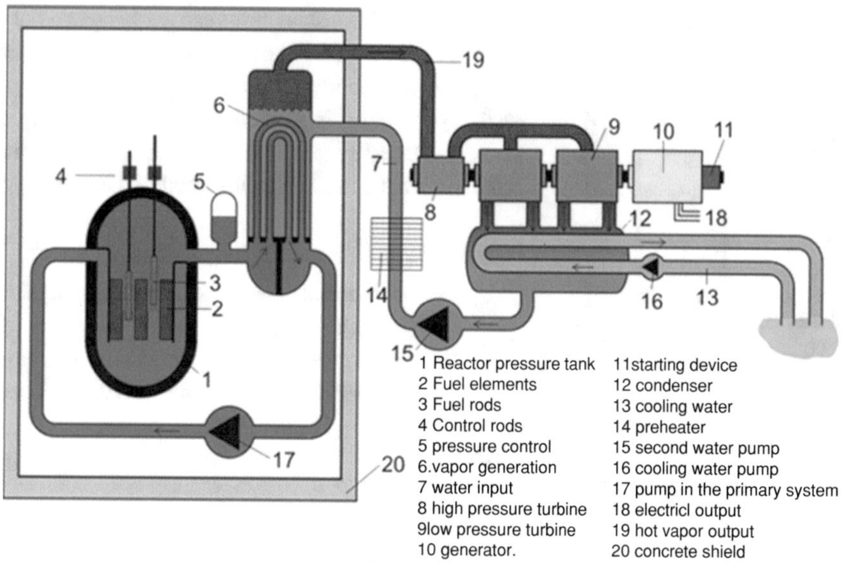

Fig. 8.20X High pressure nuclear reactor [Wikipedia]

8.3.2 Classification of Nuclear Reactors

The central region of a nuclear reactor is the reactor core. There are reactors with a ***homogeneous core*** where the fission material and the moderator are uniformly intermixed, or ***heterogeneous reactors***, which have a spatial separation of uranium and moderator. The main number of reactors operating today are of the heterogeneous type. There are reactors with graphite moderator (most Russian reactors, as for instance Chernobyl) and water moderated reactors (all western reactors). Here we distinguish between high pressure water reactors and boiling water reactors. The high pressure reactors have a primary circulation of the cooling water at high pressure (up to 160 bars) and a secondary circulation which is heated up by heat exchange with the first circulation (Fig. 8.20X). The operation temperature in the high pressure tank is just below the boiling temperature at this pressure. Because of the lower pressure in the secondary circuit the water boils and the steam drives a turbine where it cools down below the condensation temperature and is pumped back into the bottom of the primary circuit. It then rises up around the nuclear heating rods (= fuel rods).

In the boiling water reactor there is only one circuit (Fig. 8.20Y).

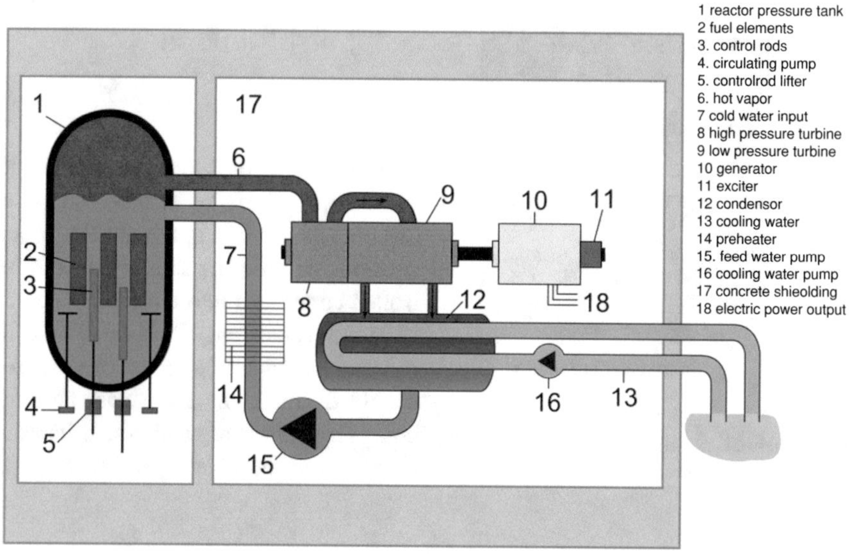

Fig. 8.20Y Boiling water nuclear reactor [Wikipedia]

Fig. 8.20z Cooling towers for power plants. Left are two forced draft wet circulation cooling towers (34 m high) at right is a natural draft wet cooling tower (122 m high) [Wikipedia]

The hot steam above the boiling water surface is directly guided to the turbine, where it cools down below the condensation temperature and the water is pumped back into the reactor core.

The advantage of the high pressure reactor is the separation of the radioactive primary circuit and the substantially clean secondary circuit. It prevents radioactive substances, for example tritium, to penetrate into the cooling water and to contaminate the secondary circuit. A further advantage is the higher temperature which allows a higher efficiency and higher energy output for a given amount of nuclear fuel. Its drawback is the necessary thick walls of the reactor pressure container and possible leaks due to the high pressure and high temperatures. It therefore needs more frequent inspections.

The drawback of the boiling water reactor is its lower energy efficiency (due to the lower temperature) and the spread of the radioactive cooling water over the whole system.

The cooling system uses for both reactor types either water from a river or cooling towers, which are cooled by atmospheric air (Fig. 8.20z).

8.3.3 Structure of a Nuclear Reactor

In Fig. 8.21 the different components of a heterogeneous high pressure nuclear reactor are shown: The fissionable material (generally uranium oxide UO_2) is pressed into tablets. About 200 of such tablets are put into gas-tight welded nuclear fuel rods.The inner part of these rods must have sufficinet free space in order to keep the gas pressure of the gaseaous fission products (Kr, Xe) sufficiently low, even at temperatures of 500–600 °C. About 20–25 of these rods are then arranged in the quadratic fuel assembly as shown in Fig. 8.21. They form the reactor core. Several control rods are placed between the quadratic fuel rods. They contain materials, such as Boron or Cadmium which have large

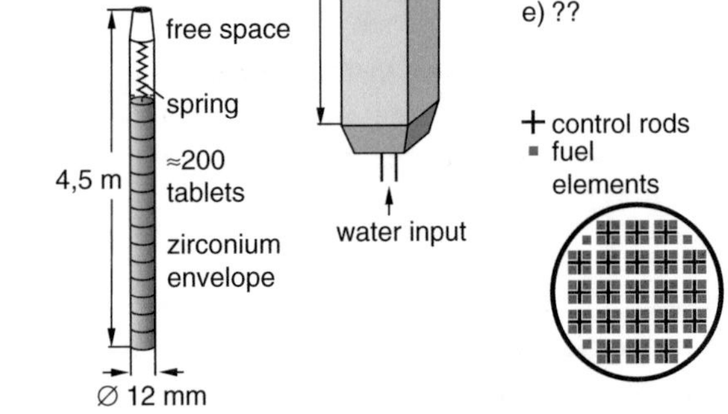

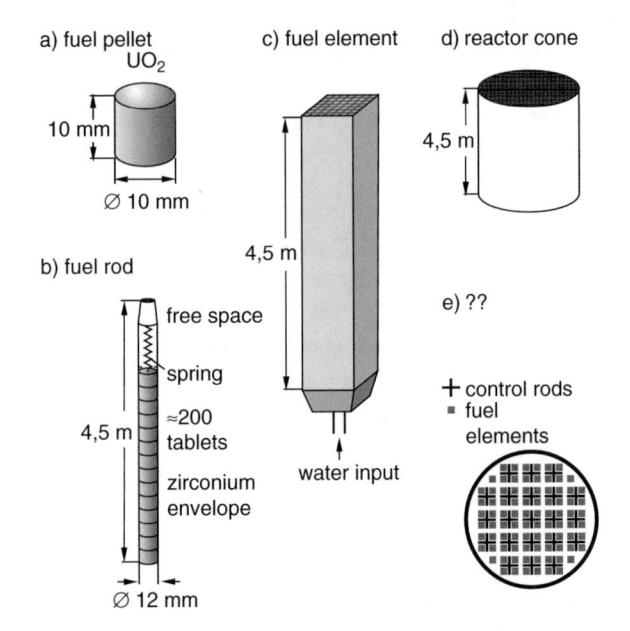

Fig. 8.21 The components of the reactor core of a water-cooled heterogeneous reactor

absorption cross sections of neutrons. They can be drawn more or less out of the reactor core in order to controll the chain reaction. The water which is pumped through the space between the fuel rods has to transport the energy produced in the rods to the output of the reactor chamber. It has the additional task to moderate the neutron flux and slow down their velocity. A schematic drawing of the total reactor system is shown in Fig. 8.22.

A reactor with an electric output power of 2000 MW contains about 45,000 fuel rods and 100 tons of enriched uranium. The energy conversion efficiency of nuclear reactors is about 35%. With a thermal power generation of 3 GW the electric output power is about 1.05 GW.

About 70,000 t cooling water are pumped per hour through the reactor core which is heated up to 326 °C and transfers its heat energy through a heat exchanger to the secondary circuit. Here the pressure is only 66 bar and the water boils already at 300 °C. The steam drives a turbine which is connected to an electric generator.

Typical electric output powers are, for the example of the nuclear power plant Brockdorf in Germany, 1400 MW at the voltage of 27 kV. About 70 MV of this output are used for the operation of the power plant which reduces the effective output power to 1330 MW.

The thermal power production inside the reactor core is about 40 kW per kg uranium. This corresponds to a power density of 100 MW/m^3.

For a thermal power output of 3 GW about 90 t enriched uranium are necessary. The achievable mean burn-up of the fissionable uranium at an electric power output of 10^4 MW-days are reached per ton of enriched uranium. This means that in a nuclear power plant that delivers the electric power output of 1 GW, each year about 90 tons of uranium have to be replaced.

The technical data of a high pressure reactor are compiled in Table 8.5 [33, 34].

Table 8.5 Technical data of high pressure reactors given for the example of the reactor in Brokdorf, Germany. (Informationskreis Kernenergie Bonn 1993)

Nuclear Fuel	UO_2
Enrichmnent of U^{235}	1.9%; 2.5%; 3.5%
Quantity of fuel	103 t
Number of fuel elements	193
Number of fuel rods per fuel element	236
Number of control rods	61
Absorber material	InAgCd
Cooling substance and moderator	desalinated H_2O
Thermal output power	3765 MW
Electric gross output power	1395 MW
Electric net output power	1326 MW
Net efficiency	35.5%
Mean power density in the reactor core	92.3 kW/dm^3
Produced energy per ton uranium	53.000 MWd

8.3.4 Operation and Control of a Nuclear Reactor

In Sect. 8.3.1 it was shown that the temporal progression of the density of fissionable neutrons and with it the density of fissions follows the exponetial function

$$N = N_0 \cdot e^{\rho \cdot k_{\text{eff}} t/T} \tag{8.24}$$

In order to reach a stationary operation one has to guarantee that the reactivity $\rho = (k_{\text{eff}} - 1)/k_{\text{eff}}$ is reliably kept at $\rho = 0$. For the typicasl reactor period (the time between two successive fission generations) $T = 1$ μs this would demand a fast controll system in order to react fast enough on short time fluctuation of ρ.

Fortunately the delayed neutrons, emitted by the fission products (see Sect. 6.5.4) are helpful to improve the stable operation of the reactor. About 0.75% of all neutrons availabe for the fission, are emitted by the fission products with a time delay between 0.10 s and 80 s. One example is the decay of the fission product bromium:

$${}^{87}_{35}\text{Br} \xrightarrow{\beta^-} {}^{87}_{36}\text{Kr}^* \xrightarrow{n} {}^{86}_{36}\text{Kr},$$

where the neutron are emitted with a time delay of 76.4 s against the "prompt neutrons" emitted during the fission of ${}^{235}_{92}$U. These delayed neutrons prolong the reactor period markedly.

If the reactor is operated in such a way, that ρ without the delayed neutrons the mechanical control has sufficient time to compensate the fluctuations of the reactivity by inserting or to draw out the control rods (neutron absorber) and therefore keeping the reactivity exactly at $\rho = 0$.

During the start up of the reactor the reactivity must be of course $\rho > 0$. The start is initiated by an artificial neutron source in the reactor core. Initially the reactivity is kept at $\rho \approx 10^{-3}$, allowing a slow increase of the neutron number. Then the reactivity is gradually increased until the produced power is equal to the demanded one, depending on the wanted electrical power output.

The arrangement of the fuel elements in the reactor core (Fig. 8.23) are arranged in such a way that the elements with the most enriched uranium are placed in the outer range of

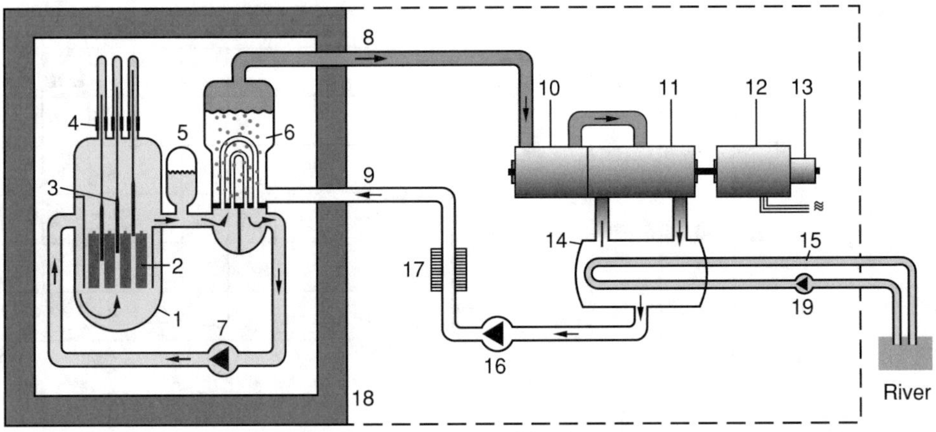

Fig. 8.22 Schematic drawing of a water-cooled high pressure reactor (Informationskreis Kernenergie Bonn 1993)

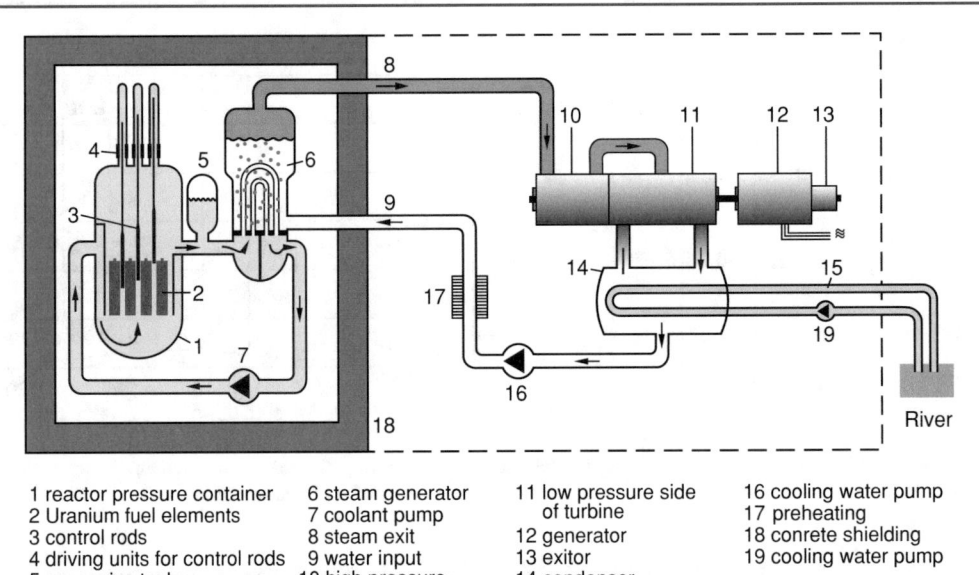

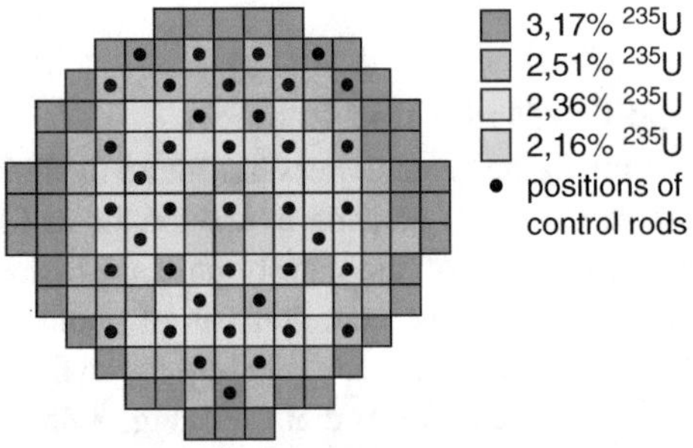

Fig. 8.23 Spatial arrangement of the nuclear fuel rods with different enrichment percentages of ^{235}U in the reactor core

the reactor core, the elements with the lowest enriched uranium in the inner part of the core., because the loss of neutrons by diffusion out of the core is lowest in the inner part (Fig. 8.23).

After about 2 years of full operation the reactor is shut down for a few days in order to rearrange and replace the fuel rods. About 1/3 of all elements namely those from the inner part of the core are replaced because they have the largest burn-up due to the higher neutron flux. The new elements are placed in the outer range and the old ones are now placed in the inner range of the core. By this way the reactivity remains approximately constant over the whole space of the core. This rearranging is repeated twice. After about 6 years of full operation all fuel elements have to be exchanged. They are stored in special water basins on the area of the reactor until the radiation of the fuel elements has decreased down to an acceptable value and the cooled reactor rods are later transported in radiation safe containers to a reprocessing plant where the uranium thorium and plutonium are separated and used for the production of new fuel elements.

The fission products are partly neutron absorbers. This decreases the reactivity of the reactor. Generally the neutron capture cross sections are small except for Xenon and Samarium. During the reactor operation Xenon is produced according to the scheme

$$\begin{aligned} &U \xrightarrow{\text{fission}} {}^{135}\text{Te} \xrightarrow[2\,\text{min}]{\beta^-} {}^{135}\text{I} \\ &\Rightarrow \xrightarrow[6.7\,\text{h}]{\beta^-} {}^{135}\text{Xe} \end{aligned} \tag{8.25}$$

Xenon decays again

- by β^- emission into the stable isotope ^{135}Ba

$$^{135}\text{Xe} \xrightarrow[9.2\,\text{h}]{\beta^-} {}^{135}\text{Cs} \xrightarrow[2\cdot 10\text{a}]{\beta^-} {}^{135}\text{Ba}, \tag{8.26}$$

- by (n,γ) transformation ($\sigma_a = 2.9 \cdot 10^6$ barn)

$$^{135}\text{Xe} + \text{n} \rightarrow {}^{136}\text{Xe}^* \xrightarrow{\gamma} {}^{136}\text{Xe} \tag{8.27}$$

into the stable isotope ^{136}Xe which does not absorb neutrons.

During the stationary reactor operation a stable equilibrium will adjust itself with a stable concentration of ^{135}Xe where the production rate is equal to the decay rate (***xenon contamination***).

$$V = \sum_{\text{gap}} \sigma_{\text{section}} / \sum_{\text{burn}} \sigma_{\text{section}} \tag{8.28}$$

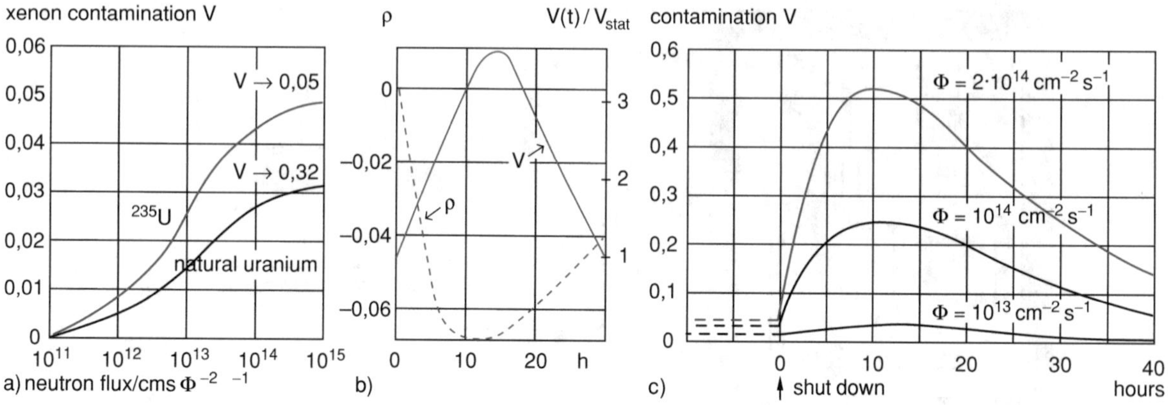

Fig. 8.24 Contamination $V(t)$ by fission products **a** during the stationary operation for ^{235}U and natural uranium. **b** Reactivity $\varrho(t)$ and relative contamination $V(t)$ after the shut-down of the reactor at $t = 0$. **c** Contamination for different neutron fluxes

After the shutdown of the reactor the production of ^{135}Xe by the fission products still goes on for a while but its loss by the (n, γ)-reaction (8.27) stops because the neutrons from the fission of U are missing. Therefore the concentration of Xenon increases (Fig. 8.24) until the iodine ^{135}I which produces xenon in the reaction (8.25) has decayed.

Generally the reactor with the new fuel elements has sufficient reactivity to reach $\varrho > 0$. In order not to rely solely on the control rods one adds neutron-absorbing boron to the cooling water in order to decrease the reactivity of the reactor with new fuel elements. During the operation time, when the concentration of enriched uranium decreases because of the burn-up, the boron concentration is gradually reduced.

8.3.5 Nuclear Reactor Types

Besides the high pressure water-moderated reactor described above which is the most frequently built reactor type, there are several other reactor types which will be shortly discussed.

8.3.5.1 Graphite Moderated Reactors

Particular attention in the general press has found the boiling water pressure reactor with graphite as moderator in Chernobyl (Ukraine), because of the catastrophic explosion in 1986. In this reactor type the core consists of about 1700 tons of graphite bricks which are piled-up as cylindrical block with 7 m height and 12 m diameter (Fig. 8.25). The

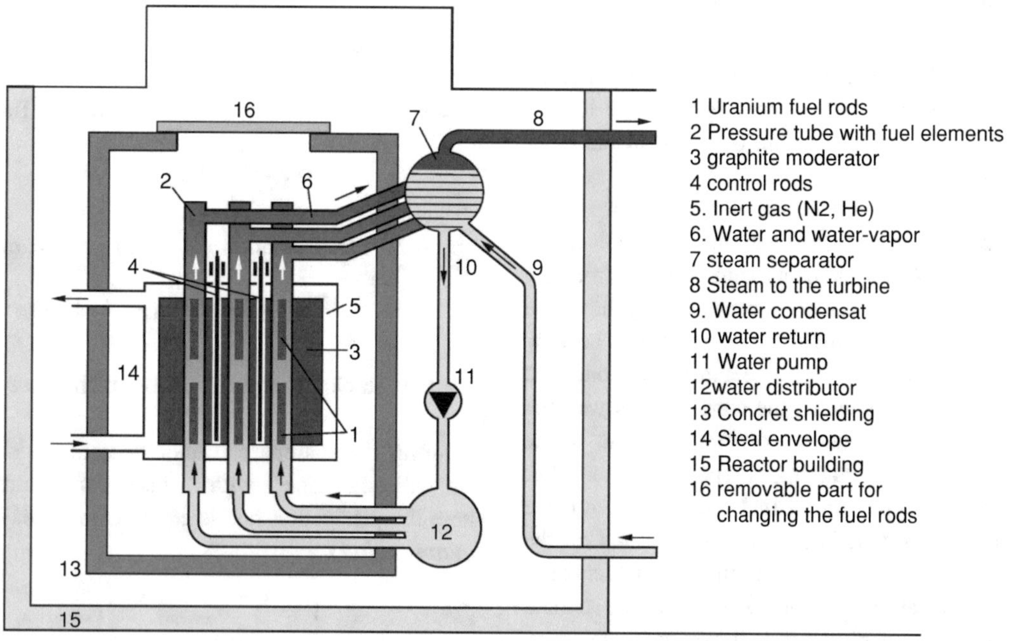

Fig. 8.25 Schematic representation of the graphite-moderated pressure-tube reactor in Tschernobyl

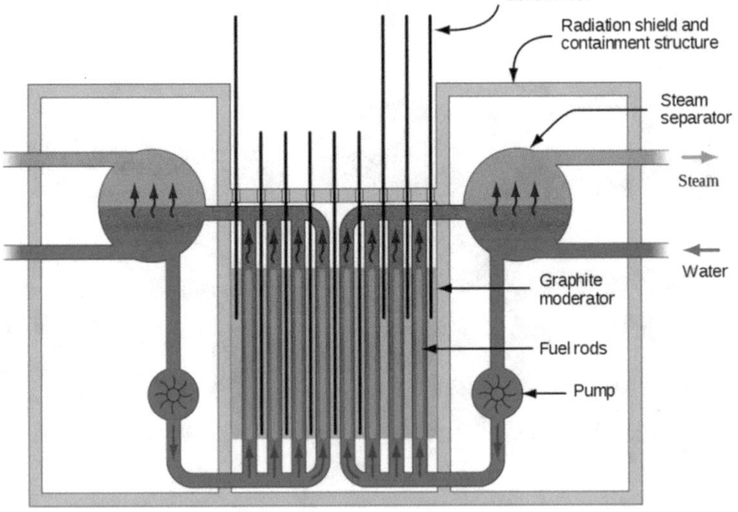

Fig. 8.25X Graphite-moderated reactor (Chernobyl type) (https://commons.wikimedia.org/w/index.php?curid=3579057)

fuel elements are 3.65 m long and contain about 115 kg enriched uranium. They are suspended in vertical pressure tubes inside the graphite block through which the cooling water under high pressure flows (Fig. 8.25X).

In the Chernobyl reactor there are 1661 of such pressure tubes. There are furthermore 211 boreholes for the control and absorber rods (Fig. 8.25Y).

The whole graphite block is embedded into a steel vessel which is filled with inert gases such as He of N_2 in order to prevent burning of the graphite.

The volume of the reactor core is about 10 times as large as that of the high pressure reactors in the western world.

The heat produced during the fission is absorbed by the cooling water which partly evaporates. The steam-water mixture is separated by a steam separator and the steam drives a turbine.

The economic advantages of this reactor type, which have convinced the Russian technicians to choose this type, are the following:

- a loos of cooling water due to leaks in the pressure tubes would be relevant only for some of the cooling tubes but not for the whole reactor which excludes a total cooling loss.
- The planning of larger reactors does not require a completely new design but only the addition of equal elements.
- A change of the fuel elements is possible without shut down of the reactor, because the roof of the reactor could be removed and the rods could be pulled out and again inserted into the core. This allows a continuous operation of the reactor. Furthermore the removal time of the fuel elements can be chosen such that the optimum concentration of plutonium, suitable for atomic bombs is reached in the fuel elements. This last point might be a major reason for selecting this reactor design.

These advantages are, however, confronted with serious drawbacks:

- The reactivity ρ of graphite-moderated reactors has a positive temperature coefficient ($d\rho/dT > 0$), because with increasing temperature more cooling water evaporates, thus decreasing the absorption of neutrons. **Note** that here the moderation occurs in the graphite block and therefore the loss of water does not essentially decrease the moderation capability of the reactor. This means that the absorption of neutrons decreases but the efficiency of slowing down the neutron velocity and therefore the fission yield does not essentially decrease. The reactor therefore tends to an unstable operation and demands a very careful control.

Because of the large core volume the control of the chain reaction is more difficult, because local fluctuations of the neutron flux can happen.

- There are no reactor pressure containers and extra safety enclosure, because the reactor is accessible during its operation.

The reason for the accident were experiments of the service team who wanted to find out, at which minimum reactivity the reactor could be started again and what would

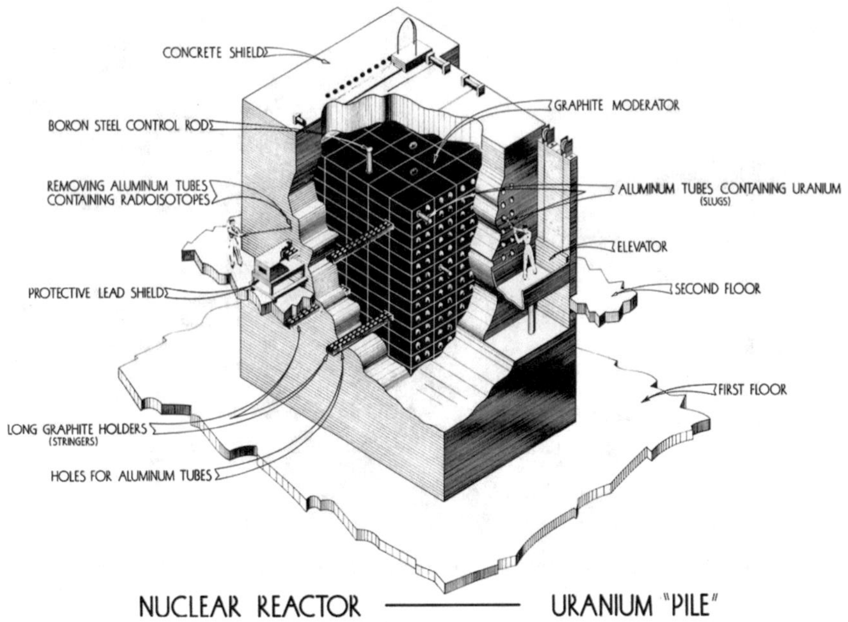

Fig. 8.25Y Graphite-moderated nuclear reactor [By doe-oakridge - Nuclear Reactor Uranium Pile, Public Domain]

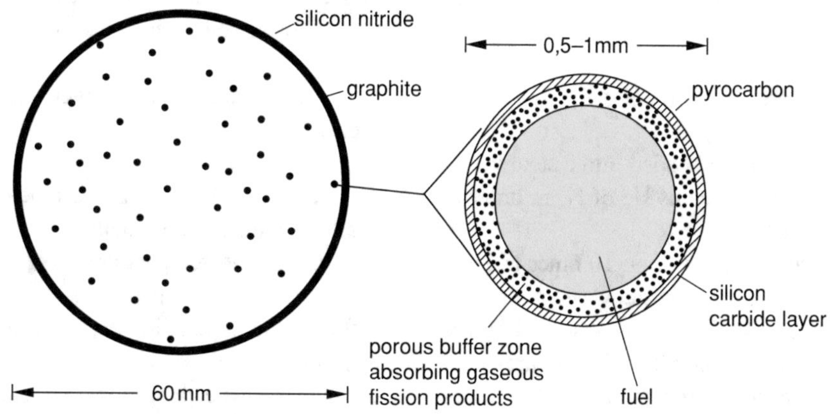

Fig. 8.26 Fuel pellets in a hollow graphite sphere in a High temperature reactor

be the minimum time for reaching full operation. During these experiments the absorber rods were partly drawn out of the core but could not be drawn in fast enough when the reactivity suddenly increased which caused an exponential growth of the fission rate until the explosion of the whole reactor core started. Since the upper containment was only very weak it was destroyed and the exploding material was tossed into the atmosphere and spread out over several European countries.

Note: The high pressure water-moderated nuclear reactors have a negative temperature coefficient ($d\rho/dT < 0$), because here the evaporation of water decreases the moderation efficiency and therefore decreases the fission rate. Furthermore the temperature increases broadens the resonance lines of neutron absorption in uranium, thus increasing the total neutron absorption and therefore decreasing the reactivity.

8.3.5.2 The High Temperature Pebble Bed Reactor

A very promising and particular safe reactor is the helium-cooled high temperature thorium reactor (Fig. 8.26). The fuel elements are graphite hollow pellets with a diameter of about 6 cm. They are covered with a protective coating of silicon nitrite (Fig. 8.27) which is very hard and protects the graphite balls from abrasion. They contain per pellet about 1 g ^{235}U as fuel and 10 g ^{232}Th as breeding material. Both

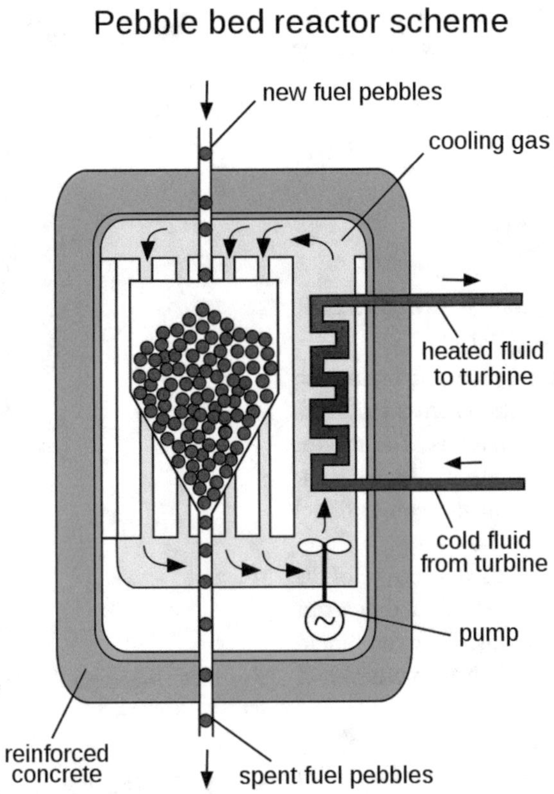

Fig. 8.27 High temperature reactor with graphite balls containing fuel pellets. [Pebble bed reactor scheme (italiano) svg.svg:, CC0, https://commons.wikimedia.org/w/index.php?curid=24456605]

materials are formed as small pellets with a diameter of about 0.5–1 mm. One graphite hollow sphere contains about 35.000 of these small pellets. About 360,000 graphite hollow spheres are filled into a large container of graphite which also contains about 280,000 pure graphite balls without fuel which act as moderator. All pellets are uniformly mixed. The reactor is therefore of the homogeneous type, where the fuel and the moderator are uniformly distributed over the reactor core [35].

The fission neutrons produced by the fission of ^{235}U can convert the thorium isotope ^{232}Th into the isotope $^{233}_{90}$Th, which decays into Palladium according to the reaction

$$^{232}_{90}\mathrm{Th} + \mathrm{n} \rightarrow ^{233}_{90}\mathrm{Th} \rightarrow ^{233}_{91}\mathrm{Pa} + \mathrm{e}^-$$

The palladium can be split by thermal neutrons similar to ^{235}U. The graphite serves as moderator. Different from the Chernobyl reactor here the cooling is not done by graphite but by helium. This allows the increase of the operation temperature above 800 °C which enhances the thermodynamic efficiency up to 70%.

The helium gas flows from above with 250 °C into the reactor, is heated up by the hot graphite balls and leaves the reactor at the bottom with about 800 °C. Through a heat exchanger it converts its heat energy to a water vapor circuit where a turbine is driven.

The advantage of the Helium is that it is not activated by the radio-active radiation in the reactor core and the cooling system outside the core is not contaminated. Since graphite melts only at temperatures above 3650 °C the graphite balls do not melt if the reactor has to be shut down because of a failure of the cooling system and the reactor would be heated up due to the residual heat delivered by the decay of the fission products. At the higher temperature the heat energy is radiated ($\propto T^4$!) to the outside and the reactor will not be damaged. The helium-cooled high temperature reactor has a negative temperature coefficient ($d\rho/dT < 0$) different from the Chernobyl reactor where $d\rho/dT > 0$ (see above.).

Although this high temperature reactor was successfully operated in Hamm-Uetrop in Germany and represents the best and safest reactor type build up to now, it was shut down after the Chernobyl catastrophe because of political reasons.

8.3.5.3 Breeding Reactors

In order to increase the exploitation of uranium, breeding reactors have been developed. They use the fission of ^{238}U and ^{239}Pu by fast neutrons for energy production and for the generation of more fission neutrons which can induce the reactions:

$$^{238}_{92}\text{U} + \text{n} \rightarrow ^{239}_{92}\text{U} \xrightarrow[23.5\,\text{min}]{\beta^-} {}^{239}\text{Np}, \tag{8.29a}$$

$$^{239}_{93}\text{Np} \xrightarrow[2.36\,\text{d}]{\beta^-} {}^{239}_{94}\text{Pu} \xrightarrow[2.4\cdot 10^4\,\text{a}]{\alpha} {}^{235}_{92}\text{U}. \tag{8.29b}$$

In these processes the uranium isotope $^{235}_{92}$U is "breeded" from $^{238}_{92}$U via the plutonium isotope $^{239}_{94}$Pu.

Since this reactor type uses fast neutrons for fission, no moderator is necessary. However, the fission cross section is smaller for fast than that for thermal neutrons. The reactor core has therefore to be compact and needs a higher concentration of fissionable material than in the conventional high pressure reactors.

As cooling agent water cannot be used because of its absorption of neutrons and its capability to slow down fast neutrons resulting in a loss of fast neutrons. Therefore liquid sodium is used for the cooling, which has, moreover, a higher thermal conductivity than water.

The central fission zone where ^{238}U and ^{239}Pu are split by fast neutrons is surrounded by the breeding zone (Fig. 8.28) where ^{238}U is converted by neutron capture into ^{235}U and ^{239}Pu according to the reaction (8.29).

The reactor can be optimized such that more ^{239}Pu is breeded than consumed by fission. The breeding reactor can then deliver fissionable material to other nuclear reactors.

The technological risks of such a fast breeding reactor are:

- the safe control of hot liquid sodium is by no means trivial. It is chemical aggressive and can corrode metal tubes and walls
- If it is exposed to air it will burn
- The sodium flowing through the reactor core becomes radio-active under neutron bombardment according to the reaction

$$^{23}_{11}\text{Na} + \text{n} \rightarrow ^{24}_{11}\text{Na} \xrightarrow[15\,\text{h}]{\beta^-} {}^{24}_{12}\text{Mg}. \tag{8.30}$$

- The fast breeder has a positive temperature coefficient ($d\varrho/dT < 0$) and must be therefore carefully controlled

In order to keep the radio-active sodium inside the safety zone, three cooling circuits are used (Fig. 8.29). In the primary circuit sodium enters the core from below at the temperature T = 395 °C and leaves the core at the top with T = 545 °C. The boiling point of sodium is T_B = 883 °C which ensures that also at lower pressures the sodium always remains liquid. It transfers its heat energy through 4 cooling coils to the secondary circuit which also uses sodium as cooling agent. However, this sodium is not radioactive because it never enters the radio-active zone. Through a second heat exchanger the energy is finally transferred into the water–vapor circuit where the turbine is driven by the water vapor [36].

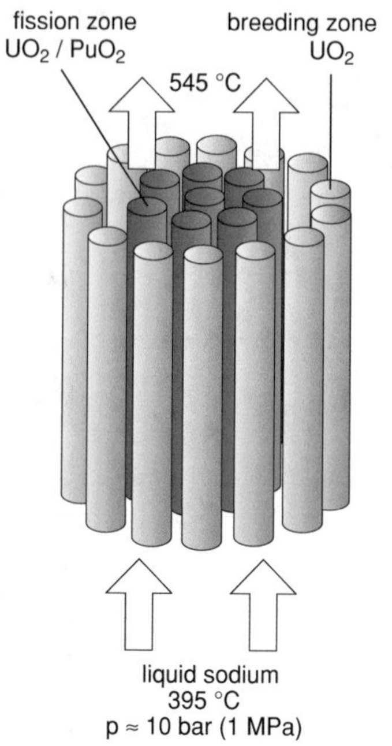

Fig. 8.28 Schematic drawing of the reactor core of a fast breeder reactor. (after M. Volkmer kernenergie Basiswissen Bonn 1993)

The largest breeder reactor is the *Superphenix Reactor* in Creys-Malville in France. It is operating since 1986.

8.3.6 Safety of Nuclear Reactors

The safety of nuclear reactors involves several aspects:

- It must be guaranteed, that the chain reaction is always under control. There should be no increasing number of neutrons in the core.
- The heat removal must be fail-safe, in order to prevent any overheating, which could result in a fracture or even melting of parts in the reactor core. In Fukushima this in fact happened because the cooling pumps were destroyed by the Tsumani, which aused a 10 m high water wave overflowing the safety walls with 7 m height.
- Any ionizing radiation must be shielded in order to prevent any thread of people.

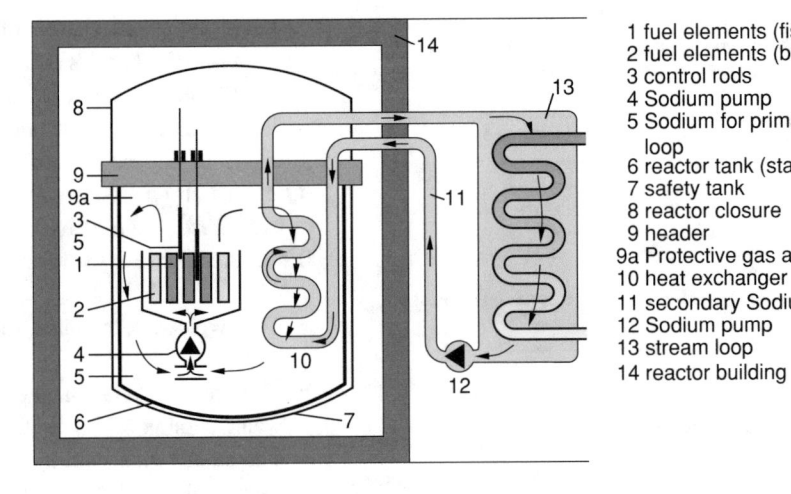

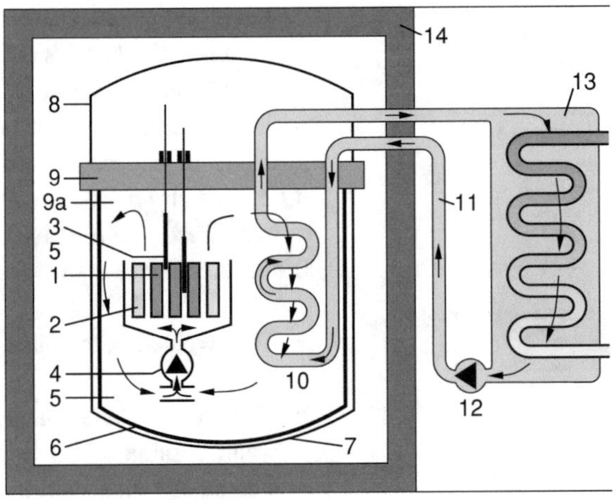

Fig. 8.29 Design principle of a nuclear power plant with a fast breeder reactor. According to M. Volkmer: Basic knowledge of nuclear energy (information group on nuclear energy, Bonn 1993)

- Any radio-active materials must be safely enclosed in order to exclude radio-active waste in the biosphere, neither during the transport nor at the storage and final disposal.

We will discuss these points in more detail:

8.3.6.1 Control of the Chain-Reaction

The reactivity ρ of a reactor is determined by the balance between neutron production by fission and neutron loss by neutron absorption and escape from the core.

The reactivity depends on the temperature in the reactor core. There are several reasons for this dependence: With increasing temperature the spectral width of the Doppler-broadened absorption lines for neutron absorption by uranium increases. (Fig. 8.19) This increases the chance for neutron absorption during the slowing down of the neutrons. This leads to a negative contribution to the term $\mathrm{d}\rho/\mathrm{d}T$. A larger effect is, however, caused by the cooling water. In reactors where the water serves as cooling agent and moderator the evaporating water and the formation of steam bubbles in the water decrease the moderator efficiency and therefore decrease the rate of slow neutrons and of the fission (*steam bubbles effect*). This leads therefore to a negative temperature coefficient $\mathrm{d}\rho/\mathrm{d}T < 0$. The evaporation of water decreases also the neutron absorption which results in a positive temperature coefficient $\mathrm{d}\rho/\mathrm{d}T > 0$. The first effect is larger than the second which therefore results in a total negative coefficient $\mathrm{d}\rho/\mathrm{d}T < 0$.

In the graphite-moderated reactors of the Chernobyl type the neutrons are slowed down in the graphite and therefore the effect of diminished neutron deceleration in water disappears. The sum of all effects results here in a positive temperature coefficient.

Note: Also reactors with positive temperature coefficient can be safely operated if all instructions are strictly obeyed. The Chernobyl catastrophe was caused by several severe and careless mistakes of the operation crew, who wanted to make experiments during the shutdown phase of the reactor. They neglected the alarm signals and just switched them off. The report of the atomic energy commission about the details of this accident reads like a horror story.

Reactors with a negative temperature coefficient are not as critical against mistakes of the operation crew, because the reactivity decreases automatically with increasing temperature.

8.3.6.2 Heat Removal

The second safety aspect with respect to the failure of the cooling system (e.g. when the main cooling pipe breaks) is based on the following physical effect: Even if the reactor is shut down by inserting the absorber rods farther into the core, there will be still heat produced by the radioactive decay of the fission products. This heat energy amounts directly after the shut down to about 5–8% of the thermal energy production before the shutdown. For a thermal output power of 3 GW this amounts to about 240 MW which has to be dissipated in order to prevent the melting of the reactor core. Therefore there must be several independent emergency cooling systems (Fig. 8.29X).

The residual heat energy decreases in time with the decay times of the fission products. The emergency cooling therefore has to operate at least for some hours after the shutdown of the reactor. In the case of the Fukushima reactor in Japan the cooling pumps had been destroyed by the Tsunami and also the emergency cooling systems were not functioning. Therefore the core melting could not be prevented.

When the reactor core melts the local density of the fissionable material suddenly increases and exceeds the critical value for starting the chain reaction. Since the chain reaction could no longer be controlled (the control rods were melted and mixed with the other materials) the fission rate increased

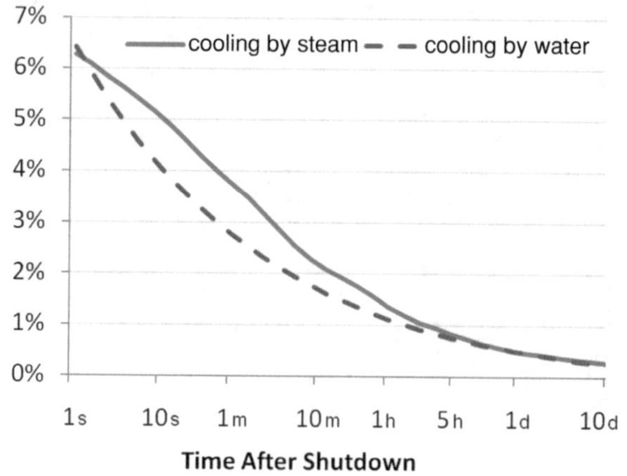

Fig. 8.29X Residual Heat production after the shutdown of a nuclear Reactor. The ordinate gives the fraction of the heat production before the shutdown

exponential and resulted in an explosion of the whole reactor. The main reasons for this Fukushima catastrophe were the Tsunami and the careless attitude of the responsible people who did not follow the urgent advice to install the cooling systems on a safe location.

8.3.6.3 Radiation Protection

The reactor core represents a source of intense ionizing radiation with the following components:

- Fission neutrons which might escape the reactor core and penetrate through the shielding. This is, however, only a small fraction of all generated neutrons.
- The fission products are artificial radio-active nuclides which emit β- as well as γ-radiation [37].
- Also stable non-radioactive nuclides can be converted into radio-active elements by the neutron bombardment, Examples are the reactor walls and the pipes of the cooling system-
- Also the fission material itself can be converted by neutron capture into radio-active isotopes which decay by β-emission. Examples are the reactions

$$\begin{aligned}&{}^{238}_{92}\mathrm{U}+\mathrm{n}\rightarrow{}^{239}_{92}\mathrm{U}\xrightarrow[23.5\,\mathrm{min}]{\beta}{}^{239}_{93}\mathrm{Np}\\&\xrightarrow[2.0135\,\mathrm{d}]{\beta}{}^{239}_{94}\mathrm{Pu}\xrightarrow[2.4\cdot10^{4}\,\mathrm{a}]{\alpha}{}^{235}_{92}\mathrm{U},\end{aligned}\tag{8.31a}$$

$${}^{239}_{94}\mathrm{Pu}+\mathrm{n}\rightarrow{}^{240}_{94}\mathrm{Pu}^{*}\xrightarrow{\gamma}{}^{240}_{94}\mathrm{Pu},\tag{8.31b}$$

$$\begin{aligned}&{}^{240}_{94}\mathrm{Pu}+\mathrm{n}\rightarrow{}^{241}_{94}\mathrm{Pu}^{*}\xrightarrow{\gamma}{}^{241}_{94}\mathrm{Pu}\xrightarrow[14.4\mathrm{a}]{\alpha}{}^{237}_{92}\mathrm{U}\\&\Rightarrow\xrightarrow[6.7\,\mathrm{d}]{\beta}{}^{237}_{93}\mathrm{Np}\xrightarrow[2.2\,10^{6}\,\mathrm{a}]{\alpha}{}^{233}_{91}\mathrm{Pa}.\end{aligned}\tag{8.31c}$$

Besides the α-radiation of the uranium nuclei and the γ-radiation of other fissionable material the fission products represent the by far most abundant additional source of ionizing radiation. There are more than 35 different fission products in the reactor core which exist in more than 200 different radio-active isotopes.

All these radio-active radiation sources are enclosed inside the reactor by several protection barriers:

- The welded zirconium rods which contains the UO_2 tablets
- The pressure tank which encloses the reactor core and the water cooling tubing of the primary cooling system
- The safety shell with a tightness skin
- The slight under-pressure inside the reactor container which prevents that gaseous contaminants (such as radio-active Xenon) can escape into the outer atmosphere.
- Retention of liquid and gaseous materials.

The reactor pressure container consists of a cylindrical stainless steel vessel with 17 cm thick walls, which can withstand an internal pressure $p \geq 150$ bar.

This steel container is enclosed by a concrete double wall where between the two walls always an under-pressure is maintained, which again prevents gaseous contamination from escaping into the outer atmosphere.

A more detailed representation of all safety measure of a nuclear reactor can be found in [38, 39].

8.3.7 Radio-Active Waste Disposal and Management

The supply and waste disposal of the radio-active fuel of a nuclear power plant is schematically illustrated in Fig. 8.30. Uranium containing minerals are extracted either by surface- or underground- mining. By mechanical or chemical fragmentation uranium is separated and by a chemical reaction with Fluor, gaseous UF_6 is obtained. This natural uranium contains about 0.8% ^{235}U, which is not enough to serve as fuel in nuclear reactors. Therefore the isotope ^{235}U is enriched by diffusion techniques, in fast ultra-centrifuges or by the separation nozzle technique [40]. This enriched (2–4% gaseous UF_6 is converted into solid uranium oxide powder UO_2 which is compressed in form of tablets which are filled into the fuel rods. The annual consumption of enriched uranium in Germany had been before the shut down of nuclear reactors about 1000 tons. The burned out rods are stored for a while in water basins on the area of the reactor until the fast decaying fission products are diminished and are later on brought to a reprocessing plant, where new fuel

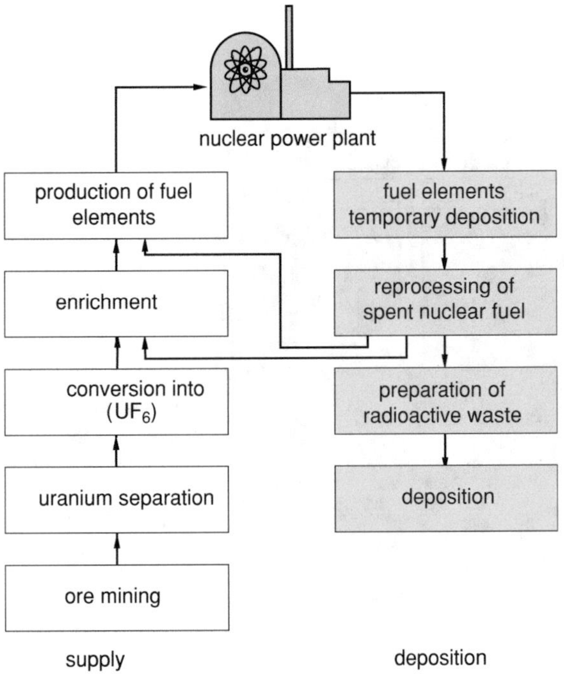

Fig. 8.30 Fuel cycle of a water-cooled reactor with enriched ^{235}U. (after M.Volmer Kernenergie Basiswissen Bonn 1993)

rods are manufactured. During this fabrication process radio-active waste is produced which has to be safely stored either in glass blocks or in stainless steel containers, which are stored in abandoned salt mines. Fuel elements which cannot reconditioned are packed into large stainless steel containers and brought to their final storage place. What is the best place for storage over thousands of years is still not clear and to find the optimum location is the task for a special investigating commission [41].

8.3.8 New Concepts

Recently an interesting proposal was published by scientists in Los Alamos for an intrinsic safe nuclear fission reactor and how the nuclear waste can be greatly reduced and used for energy production [42, 43]. This proposal can be realized as follows:

Protons are accelerated in a linear accelerator up to 800–1000 MeV and shot onto a target of molten lead–bismuth eutectic mixture, where a lot of fast neutrons are produced. This method of neutron production, where neutrons are evaporated from heavy nuclei is called ***Spallation***. These fast neutrons are slowed down by a graphite moderator which surrounds the spallation source in a cylindrical form, in order to enhance their absorption by the fission products in the nuclear reactor (Fig. 8.31).

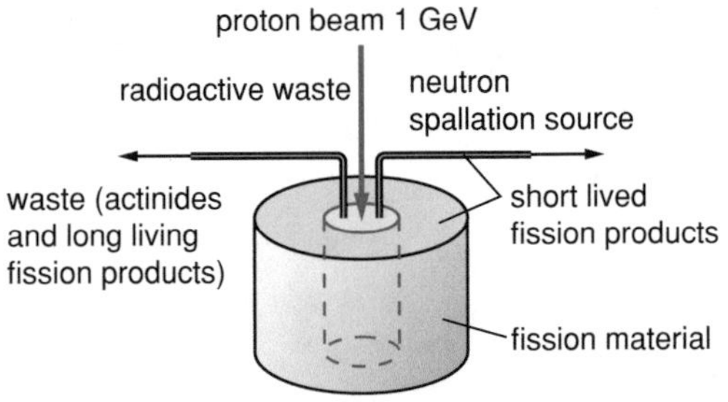

Fig. 8.31 Principal scheme of the transmutation of radio-active fission products int stable elements by irradiation with high energy protons

The waste products of the nuclear reactor are brought into a liquid phase which flows through the moderator. These waste products consist of two classes:

1. The actinides (Plutonium, Neptunium, Americium, and Curium) which are generated by neutron attachment and which have a very long lifetime, and
2. the fission products (Iodine, Cesium Krypton, Xenon etc.).

By attachment of slow neutrons in the moderator the actinides can be converted into short lived isotopes or into fissionable nuclei. The fission products are either converted into short lived isotopes which decay in the reactor and contribute to the energy production of the reactor, or into stable nuclei, which are not radio-active and can be therefore safely extracted from the reactor core. The long-lived isotopes are again send into the spallation volume and are irradiated by neutrons.

Estimations and first experiments have shown, that with this technique about 99.9% of all actinides can be fissured and about 99% of the fission products can be converted into non-radioactive isotopes.

If the spallation source is placed inside the reactor core, the reactor can be operated under-critical, because the additional neutrons bring the chain reactor above threshold. This would increase the stability of the reactor, because any shutdown of the accelerator would make the reactor under-critical. A simplified scheme of such a spalation-added reactor is shown in Fig. 8.32.

If this proposal turns out to be realizable (from the technical as well as the economical aspect) the largest problems of nuclear reactors, namely the nuclear waste problems, could be solved.

Of course any new development has to fight with difficulties. One of them is the instability of the proton beam which causes fluctuations of the produced neutron rate and therefore also of the reactivity.

A technical problem is the window where the proton beam enters the reactor core. It separates the vacuum of the accelerator and the high pressure in the reactor core. The

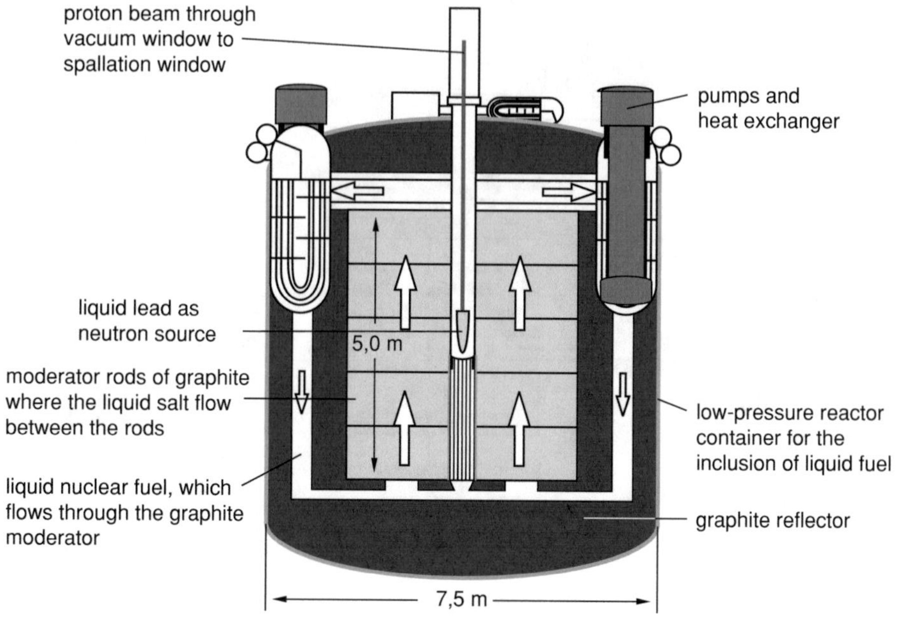

Fig. 8.32 Schematic plan for the realization of a trans-mutation nuclear reactor

transmitted high energy proton beam must be transmitted through this window. This brittles the window material and makes it mechanical unstable. Possible solutions are "plasma windows" which have no material window but just a hole. The gas close to the hole is ionized and an electric field prevents the ions from passing through the hole into the accelerator.

Further concepts for improving the existing reactors are mainly concerned with the safety conditions, which should exclude catastrophic events even in case of a failure of the cooling system. Any improvement of the thermo-dynamic efficiency would yield less waste at equal output power. There are several proposals in this direction [44, 45].

8.3.9 Advantages and Drawbacks of Nuclear Fission Energy

The advantages are:

- No CO_2-emission
- Uranium fuels are longer available then fossil fuels
- The emission of radio-active gaseous substances during the normal operation of a nuclear reactor is smaller than that in unfiltered fossil power plants at a comparable output power. This is due to the concentration of radio-active isotopes in coal and oil such as Potassium, Krypton, Iodine and Cesium. The filtering is more difficult and more expensive in coal power plants because of the larger gas turnover in the furnace.

The drawbacks are:

- The operation of a nuclear reactor, the transport of the radioactive fuel and in particular the final storage of the nuclear waste has to be performed with a large safety effort. The waste management must take care that the dangerous plutonium cannot leak into the biosphere.

In order to give a realistic estimation of advantages and drawbacks one should not rely on emotional judgements but should perform an unemotional consideration of all possible hazards.

1. How large is the risk that a worst case scenario happens? The safety concept of modern nuclear reactors implies that such a scenario should be controllable. This means that even in such a case no radio-active substances above the allowed concentration should leave the reactor encasement.
2. How large is the risk that radio-active substances stored in the final nuclear waste repository could penetrate into the ground water and into the biosphere?
3. How large is the risk that the CO_2- emission of fossil power plants results in a dangerous temperature increase and possible instability of our climate?

Many experts estimate the risk of point 3 by far higher than that of 1 and 2.

Remark

The reactor accident in Fukushima was due to the following circumstances:

The earthquake with the strength 9.0 on the Richter scale did not damage the nuclear reactor, which was automatically shut down. *Note* that the Richter scale is a logarithmic measure for the amplitude A of the deflection of an earthquake seismograph.

$$M_L = \log_{10}(A/A_0)$$

where A_0 is a normalization factor which depends on the distance between epicenter of the earthquake and seismograph. The intensity of the earthquake which is a measure of the amount of destruction, is proportional to A^2. The intensity ratio between $M_L = n$ and $M_L = n+1$ is $10^2 = 100$. Earthquakes with $M_L \geq 8$ cause disastrous devastation.

The subsequent Tsunami wave with an amplitude of about 20 m was higher than ever expected and surpassed the protection wall on the sea side of the reactor and destroyed the cooling pumps that were located just behind this wall, against an earlies advice to place them at a higher lying location. Without the cooling pumps the residual heat could not be removed. This leads to the melting of the reactor core with a sudden increase of the reactivity and a following explosion. The order of magnitude of the subsequent events was caused by the bad management and incompetence of the responsible people. It could have been avoided.

8.4 Controlled Nuclear Fusion

It is very tempting to realize the nuclear fusion, which occurs in the center of our sun, as controlled fusion in nuclear fusion power plants on earth. The energy released per kg of fusion material is much higher than that obtained for nuclear fission. Furthermore the nuclear waste is much less than that in fission reactors and it has furthermore a much shorter lifetime. It is therefore not surprising that during the last decades much effort has been spent to realize nuclear fusion reactors that can be safely controlled. Since the fuel of such reactors is hydrogen and deuterium which are abandon on earth there is therefore no lack of supply. Fusion reactors are also in this aspect superior to fission reactors [46, 47].

Although nuclear fusion could be realized for a short burning time in test reactors after many setbacks, it will take still many years before a reliable continuously working fusion reactor will be available. The immense technical problem is the maintenance of very high temperature (about 10^8 K) plasma for a sufficiently long time (at least some seconds). The high temperature is needed for the collision partners to gain enough kinetic energy for overcoming the Coulomb barrier of the nuclear potential. The most promising nuclear fusion reaction is the deuterium-tritium reaction

$$^2_1\mathrm{H} + {}^3_1\mathrm{H} \rightarrow {}^4_2\mathrm{He}(3.5\ \mathrm{MeV}) + \mathrm{n}(14.1\ \mathrm{MeV}), \qquad (8.32)$$

where the kinetic energy of 17.1 MeV is released per fusion process which is divided between the α-particle and the neutron according to their mass ratio $(m_n/m_\alpha) = 1:4$.

There are 3 different ways of realizing controlled nuclear fusion on earth:

- The entrapment of the deuterium and tritium nuclei in strong suitably formed magnetic fields and their heating by electric currents, by irradiation with high frequency power or by collisions with fast neutral particles which are injected into the hot plasma of deuterium and tritium nuclei.
- The heating of small solid deuterium–tritium balls by irradiation with high intensity laser radiation (laser-induced nuclear fusion) or by high energy particle beams. The hot plasma produced in this way is further heated and compressed by the recoil of the evaporating particles from the surface of the balls (inertial confinement).
- Nuclear fusion catalyzed by muons, which are generated by proton-proton collisions in fast anti-collinear proton beams $\mathrm{p} + \mathrm{p} \rightarrow \pi^+ + \mathrm{D}$ and $\pi^+ \rightarrow \mu^+ + \bar{\nu}$ (see Sect. 7.1). This reaction was successfully tested in the lab but will probably have no chance for nuclear fusion on a large scale.

Some years ago news about "cold fusion" have found much attention. However, it turned out by reexamination in other labs, that this was a flop without physical reality.

8.4.1 General Requirements

In order to overcome the Coulomb barrier the collision partners d and t in the reaction (8.32) must have sufficient kinetic energy. In Fig. 6.5 the fusion cross section is plotted as a function of the relative energy of the collision partners. This plot illustrates that the kinetic energy must be at least 10 keV in order to reach sufficiently large cross sections. This corresponds for a thermal energy distribution to a temperature of 10^8 K (1 eV = 11.605 K). At such a high temperature all light atoms are completely ionized and it exists a plasma which consists of free electrons, protons, deuterium nuclei and tritium nuclei.

Due to the tunnel-effect the minimum kinetic energy of the reaction partners required for penetrating the Coulomb barrier is slightly diminished. Therefore nuclear fusion (8.32) becomes already possible (although with a small probability) at $T = 10^7$ K.

The plasma is always quasi-neutral, which means that the density of the total ion charge must be equal to that of the electron charge. Since p, d and t have the positive charge +e, this implies

$$n_d + n_t = n_e = n. \tag{8.33}$$

There are small local deviations from the condition (8.33) possible within the Debye-length, but the resultant local electric field drives the electrons back into the equilibrium condition.

The Debye-length is defined as

$$D_B = \left(\varepsilon_0 k_B \cdot T / n_e e^2\right)^{1/2}.$$

Typical values for a hot fusion plasma are $D_B = 10^{-4}$ m.

The number Z_f of fusion processes per sec and volume (the fusion density rate) is given by the product

$$Z_f = n_d \cdot n_T \cdot \overline{\sigma_f(E) \cdot v} = \frac{n^2}{4} \cdot \overline{\sigma_f \cdot v} \tag{8.34}$$

of deuterium- and tritium density and the mean value of fusion cross section and relative velocity of the reaction partners. This number becomes only sufficiently large, if the time which the particles spend in the hot plasma must be longer than the time between two fusion processes.

In Fig. 8.33 the cross sections for the relevant fusion processes are plotted as a function of the deuterium kinetic energy. The red dotted curve gives the density $n(E)$ of deuterium ions at the temperature of 15 Million K which follows a Maxwell-Boltzmann distribution..

Note that the probability for collisions between two reaction partners is proportional to n^2.

The fusion rate Z_d (8.34) is plotted as the solid red curve. It shows that Z has a maximum at around 10 keV = 10^8 K. The diagram Fig. 8.33 illustrates that the by far most favorable fusion cross section is realized for the D + T reaction, according to

$$^2H + {}^3H \rightarrow {}^4He + n + 17.6\,MeV$$
$$\rightarrow {}^3He + 2n$$

Note the logarithmic ordinate scale! For igniting the fusion reaction, tritium has to be used. During the further progression of the reactor operation tritium can be breeded in the lithium mantle surrounding the hot plasma, following the reactions

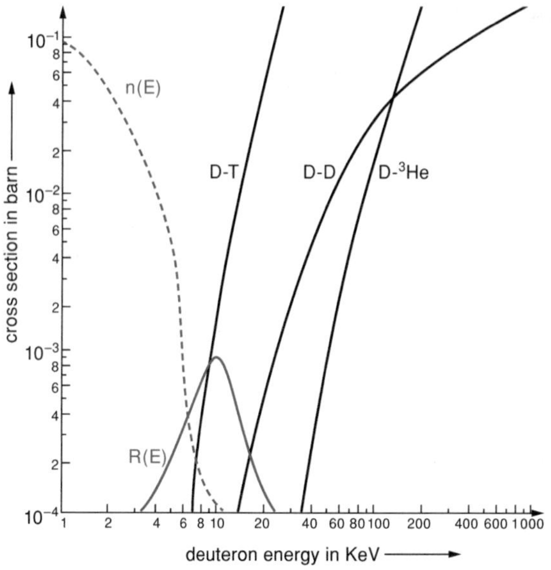

Fig. 8.33 Cross sections (in barn) for the fusion reactions D + T; D + D; D + ^{3}He (black curves), thermal energy distribution of reaction products at $T = 10^7$ K (dashed red) and reaction rate $R(E)$ in relative units

$$^6Li + n \rightarrow {}^4He + {}^3H + 4.8\,MeV$$
$$^7Li + n \rightarrow {}^4He + {}^3H - 2.5\,MeV$$

By these reactions more tritium is produced than used for the fusion reaction (Fig. 8.33X).

The important fusion parameter is the product

$$F = n \cdot \tau_e \cdot T \tag{8.35}$$

of particle density n, confinement time τ_c and temperature T.

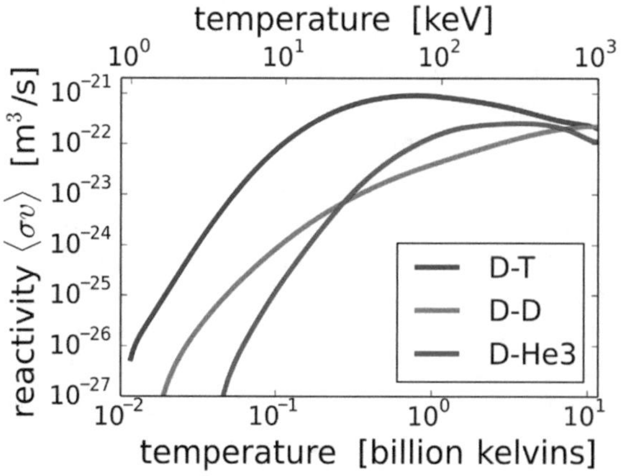

Fig. 8.33X Reactivity of fusion processes [By Dstrozzi—Own work, CC BY 2.5, https://commons.wikimedia

With $p = n \cdot k \cdot T$ the fusion parameter $F^* = k \times F$ can be also written as the product

$$F^* = p \cdot \tau_c \tag{8.36}$$

of plasma pressure p and confinement time τ_c.

In Fig. 8.34 these fusion parameters are plotted which could be realized during recent years.

If a fusion reactor should serve as energy source the output energy must be larger than the input energy.

Since neutrons cannot be entrapped by magnetic fields they escape the hot plasma and the heating of the plasma can be only caused by the kinetic energy $E_{kin} = 3.5$ MeV of the α-particles ($^4He^{++}$) which collide with the d- and t-nuclei. Besides the radiation of the hot plasma the kinetic energy of the neutrons represents the power delivered to the outside. The neutrons are slowed down in the mantel of lithium or other materials which surrounds the hot plasma. Here the radiation energy as well as the neutron energy is converted into heat, which drives as in conventional power plants a turbine to generate electric power.

The fusion power, available for heating the plasma is, according to (8.34)

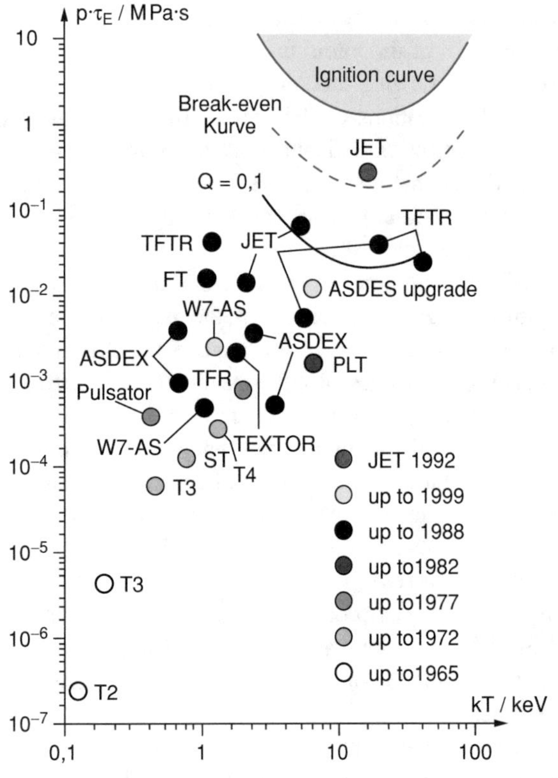

Fig. 8.34 Up to now realized ion temperature for the optimized product of particle pressure po, and confinement time τ_E and ignition curves. (after K Pinkau and U. Schumacher bH.G.Wolf: Phys. Blätter 45, 41 1989)

$$P_A = \frac{1}{4} n^2 \overline{\sigma_F \cdot v} \cdot E_\alpha \tag{8.37}$$

while the loss power density for a plasma temperature T and a confinement time τ_c is

$$P_V = 3nk \cdot T / \tau_E. \tag{8.38}$$

where we have assumed that the electron temperature T_e is equal to the ion temperature $T_{ion} = T$.

The condition $P_A \geq P_V$ for a fusion power plant yields the **Lawson Criterion** for a fusion reactor delivering more output power than it receives

$$n \cdot \tau_E > \frac{12kT}{E_\alpha \cdot \overline{\sigma_F \cdot v}}. \tag{8.39}$$

For the stationary operation the condition $P_A = P_V$ holds. The fusion cross section is within the operation range of the reactor proportional to T^2 and the relative velocity $\overline{v}$ increases proportional to $\sqrt{T}$. Therefore the minimum required product $n \cdot \tau_E$ decreases with increasing temperature. Inserting numerical values the Lawson Criterion (8.39) yields for the fusion parameter F in (8.35) the condition

$$F = n \cdot \tau \cdot T > 6 \cdot 10^{28} \mathrm{s} \cdot \mathrm{K} \cdot \mathrm{m}^{-3}.$$

Introducing the ignition parameter

$$ZP = n \cdot k \cdot T \cdot \tau_E = p \cdot \tau_E = F^* \tag{8.40}$$

as the product of plasma pressure p and confinement time τ_E, which is equal to the fusion parameter F^* in (8.36) we can see, that the Lawson Criterion can be written as $ZP > 1$. If this condition is fulfilled, the fusion power, remaining in the plasma becomes larger than the power losses caused by radiation, heat conduction and diffusion of particles to the wall of the container.

In order to ignite the fusion plasma, i.e to obtain so many fusion processes that the condition (8.39) is fulfilled, the product ZP must be at least

$$ZP = n \cdot k \cdot T \cdot \tau_E \geq 10^{21} \mathrm{keV} \cdot \mathrm{s} \cdot \mathrm{m}^{-3}$$

Remark

For laser-induced nuclear fusion the plasma density n must be extremely high ($>10^{22}$ cm^{-3}) in order to realize a sufficient large number of collisions during the short expansion time of 10^{-8} s.

For the magnetic confinement on the other side, one can only achieve substantially smaller plasma densities. Therefore the plasma has to be confined for much longer times before it reaches the walls, where the ions are adsorbed and contribute to the vaporization which causes the deterioration of the wall.

8.4.2 Magnetic Confinement

To obtain fusion in a hot plasma confined in a magnetic field the following experimental steps are necessary [48].

- A plasma with sufficient density must be generated
- The plasma has to be heated up to temperatures $T > 10^8$ K.
- The energy produced by nuclear fusion must be converted into heat which has to be transferred to turbines.
- the helium, produced by the fusion processes has to be fast removed from the plasma since it does no longer contribute to the fusion process but causes losses. It can be regarded as the "ash" like carbon dioxide in carbon burners.

The experimental problems arising when the above points shall be realized, are demanding and not all of them have been solved up to now. We will discuss them in more detail:

Up to now there are two different advanced solutions for the magnetic confinement:

(1) The **Tokamak** (Fig. 8.35) which was developed 1952 in Russia by *Igor Tamm* and *Andreij Sacharov*. Its name is an acronym for the Russian designation "toroidal'naya kamera s magnitnymi katushkami" = toridial chamber in a coil magnetic field. It has been operated for many years for experimental tests and is still a possible candidate for fusion energy production, but did up to now not reach the threshold of positive energy balance.

(2) The Stellerator, developed 1951 by Lyman Spitzer in Princeton, USA and further improved and operated by the Max Planck Institute for Plasma Physics in *Garching*, Germany [49].

The principal of the magnetic confinement in a Tokamak is illustrated in Fig. 8.35. It is based on the arrangement of three magnetic fields (Fig. 8.35x): A time varying ac-current through the transformer coils around the toroidal vacuum tube containing the fusion gas induces a current in the tube which acts as secondary winding of the transformer [50].

This current heats up the gas and ionizes it, thus producing a hot plasma. The toroid field coils generate a ring-shaped magnetic field along the toroid. The charged particles in the plasma travel on helices around the magnetic field lines. The plasma current itself also generates a magnetic field which surrounds the current filaments. Its field

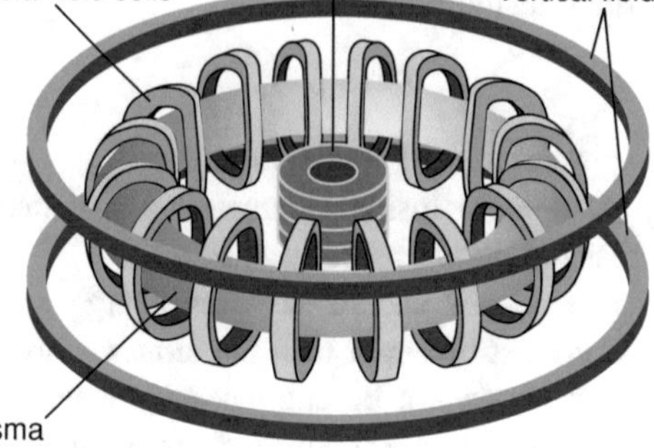

Fig. 8.35 Principle of the Tokamak

lines are helices around the plasma current. The resulting field lines of the total magnetic field are helical lines around the equilibrium orbit of the charged particles.

Finally a large horizontal pair of coils produces a vertical magnetic field which prevents the particles from escaping in radial direction.

All these three magnetic fields must be optimized in order to compensate the plasma pressure

$$p = n \cdot k \cdot T,$$

which is built up in the hot plasma with a particle density n. It keeps the plasma together and prevents it from reaching the cold walls of the toroid tube.

Now the following problems arise:

If small deviations of the plasma from the equilibrium conditions occur (e.g. fluctuations of current density and current directions) the pressure forces and the magnetic forces (Lorentz-force) do no longer exactly compensate. The residual forces may drive the plasma either back to its equilibrium conditions or away from them, resulting in instabilities before corrections of the magnetic fields can oppose these instabilities. If, for instance, the direction of the plasma current changes, also the magnetic field generated by this current changes. This results in a further change of the spatial density distribution of the charged particles.

Such instabilities had been out of control for many years and have avert to reach the conditions for igniting fusion.

There are several different instabilities, which can be divided into macroscopic and microscopic instabilities. The former result in macroscopic changes of the nominal values of pressure, temperature and electric current distribution, the latter show up as small local density and temperature fluctuations.

If, for instance, the direction of the electric current changes also the magnetic field generated by this current changes accordingly. This in turn modifies the local variation of the plasma density. Such instabilities can diminish

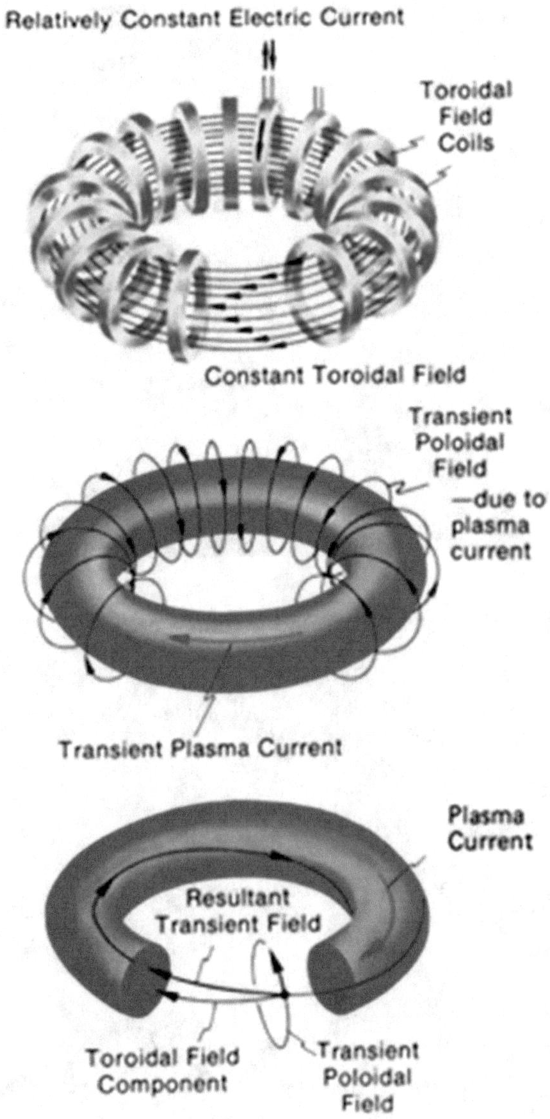

Fig. 8.35X The three magnetic fields in a Tokamak

the stabilizing effects of the magnetic fields. Meanwhile such instabilities can be minimized by a special shaping of the magnetic field coils and by fast control of the instabilities. This has resulted in substantially longer confinement times of the hot plasma.

Besides these instabilities another serious problem is caused by hot ions which impinge onto the walls of the fusion chamber. They sputter the wall material (e.g. iron or carbon). The evaporating atoms have a much higher nuclear charge number Z. If they reach the hot plasma they are completely ionized by electron impact. The resulting nuclei with high Z lead to large bremsstrahlung losses of the electrons in the plasma. They can become so large that they are no longer compensated by the plasma heating. This implies that the plasma temperature decreases.

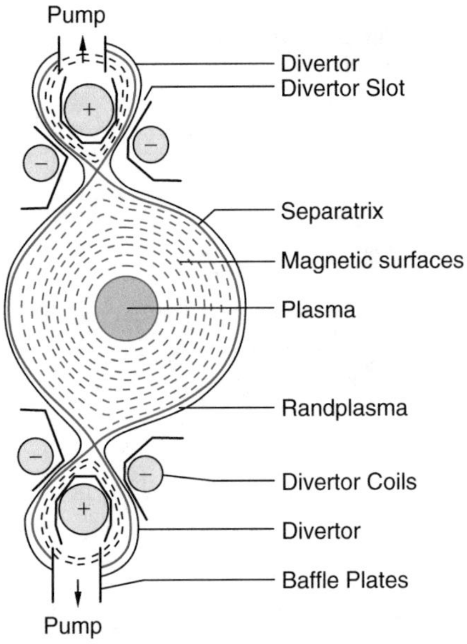

Fig. 8.36 The divertor principle

The installation of a *divertor* (Fig. 8.36) can improve this situation. It distracts the evaporated particles into a sidearm of the fusion chamber where they can be pumped out.

The considerations above illustrate that it is advantageous to build the fusion chamber as large as possible, because then the hot plasma is farther apart from the walls. Of course this demands larger magnetic field coils, In order to reduce the energy consumption for maintaining the magnetic field, supra- conducting coils are used. Since the magnetic forces are very large, stable support construction is necessary to avoid explosion of the whole system.

The largest fusion reactor (tokamak) realized up to now is the Joint European Torus *(JET)* (Fig. 8.37), Here for the first time 1991 a deuterium-tritium mixture was used as fuel for the reactor. The operation data were: $n_d = 10^{20}\,m^{-3}$, $T > 15$ keV, $\tau_E > 1$ s. With a tritium quantity of 0.2 g an ignition parameter $Z_{eff} > 1.5 \cdot 10^{21}$ was reached, which means the nuclear fusion started. About $1.5 \cdot 10^{18}$ fusion processes were detected. The maximum output power during the fusion process was 1.7 MW. [51]. However, this output power was still smaller than the input power necessary to maintain the hot plasma.

Meanwhile an international test reactor (ITER) is under construction in Cadarache, southern France, which should overcome the break-even limit, i.e. it should produce more output power than necessary to maintain its operation (Fig. 8.37x). It is planned to start its operation in 2025 where at first the filling gas will be stable Helium $_2^4He$, in order to study the plasma characteristics and possible instabilities [52–54] (Fig. 8.37x).

It is expected that ITER will deliver 500 MW with an input power of 50 MW [10] (Fig. 8.38x).

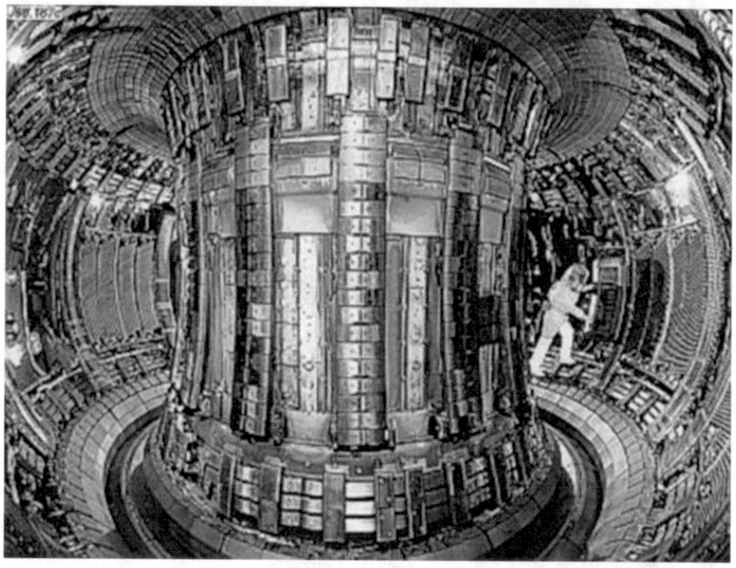

Fig. 8.37 Interior of the fusion reactor JET. The stainless steel walls are covered by carbon plates

Fig. 8.37x The central part of the fusion reactor ITER under construction

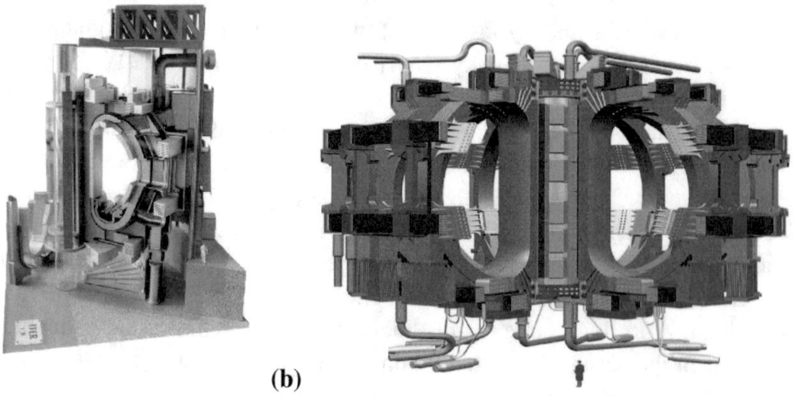

(a) (b)

Fig. 8.38x **a** Cut through the circular structure of the Tokamak ITER **b** full circular structure. [https://commons.wikimedia.org/w/index.php?curid=944817]

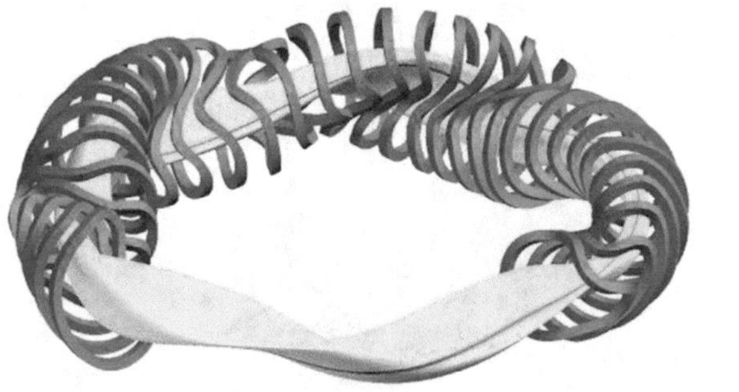

Fig. 8.38y Twisted field coils for the Stellerator *Wendelstein*

The alternative to the Tokamak is the Stellerator, which was developed in the Institute of Plasmaphysics in Garching near Munich and in an improved version, called *Wendelstein*, (a mountain in the German Alpes) operated in Greifswald Northern Germany. Here all experimental experiences about instabilities have been used for the calculation of the optimum coil configuration, which needs only one coil with a complicated form (Fig. 8.38y) instead of the three coil pairs in the Tokamak: This coil leads to a twisted plasma distribution (indicated by the yellow band in Fig. 8.38y)

This configuration is presently tested in Greifswald, Germany (Fig. 8.38z).

The plasma is ignited by injection of fast neutral particles or by irradiation by an external high frequency wave.

The big advantage of the Stellerator is the possibility of continuous operation while the Tokamak can be only run in a pulsed operation.

If test experiments for investigating the elimination of instabilities will be successful, the successor of ITER will be a Stellerator. The goal is to confine a stable plasma for 0.5 h at $T = 10^8$ K.

8.4.3 Plasma Heating

The electric resistance R of the plasma decreases with increasing temperature. Therefore the heating power $P = I^2 \cdot R$ decreases for a given allowed current I. The direct heating by the plasma current therefore becomes less effective and one has to look for other heating mechanisms.

Besides the current heating the plasma can be heated by injection of high energy neutral particles. For instance deuterium molecules D_2. are ionized and accelerated in an electric field. After having reached the necessary energy they are sent through a charge-exchange chamber where an alkali vapor is present (Fig. 8.39). When the fast D_2-molecular ions collide with an alkali atom they acquire an electron (because the binding energy of the electron in the D_2-molecular ion is larger than the ionization energy of the alkali atom). This charge exchange has a large cross section at streaking collisions. The fast neutralized D_2 particles are injected tangential

Fig. 8.38z Stellerator Wendelstein, Greifswald [Dmm2va7, CC BY 3.0, https://commons.wikimedia]

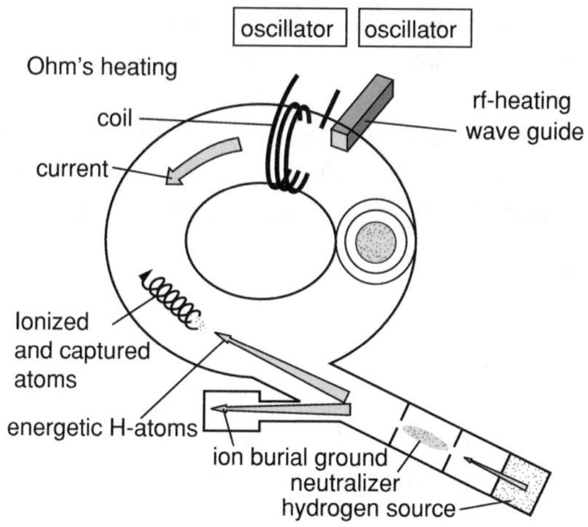

Fig. 8.39 Schematic drawing of the neutral pellet injection for the plasma heating and the coils for rf- heating. (Phys. in uns. Zeit 22, 119 (1991)

into the magnetic field and heat the plasma by collisions with the plasma ions, since their kinetic energy is higher than the mean thermal energy of the plasma ions [51].

If ions would be injected, they were deflected at the entrance into the magnetic field and would miss the central part of the plasma. On the other side, the neutral particles are not affected by the magnetic field. Only after penetrating into the magnetic field and reach the central part of the plasma where the density is maximum, they are ionized by collisions with the plasma ions.

A third method of plasma heating is the irradiation with high frequency (about 30 MHz) or microwave (140 GHz) power. While the hf is injected by antennas at the plasma edge, the microwave can be send by mirrors and waveguides into the plasma. The high frequency wave can directly heat the ions whereas the microwaves heats the electrons which transfer their energy by collisions to the ions.

Besides the heating processes also the loss mechanisms has to be considered. Particles which diffuse out of the hot plasma to the walls of the container have to be replaced and heated up again. A possible technique to refill the plasma uses small frozen Deuterium pellets. If they get to the center of the plasma they evaporate and deliver new deuterium atoms to the plasma. This increases the density at the center of the plasma compared to the density at the edges and diminishes the particle loss by diffusion to the walls [55].

8.4.4 Laser-Induced Nuclear Fusion

If the beam of a high intensity pulsed laser is focused onto a solid surface, sufficient energy is deposited within a small volume, rising the temperature T up to values of 10^5–10^8 K.

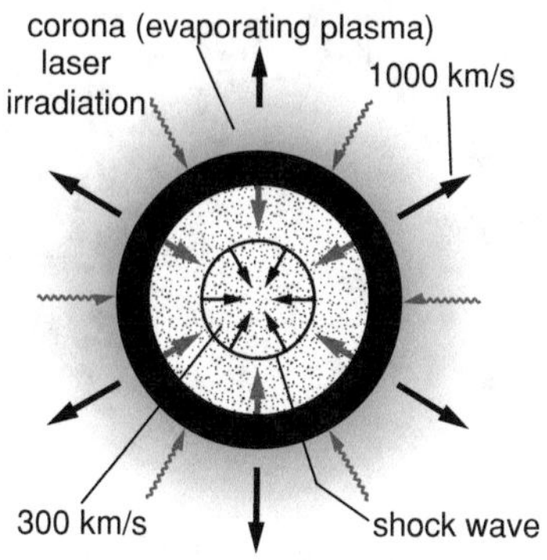

Fig. 8.40 Experimental arrangement of laser-induced nuclear fusion

The material in this volume evaporates and a small micro-plasm is generated which has a high density and temperature [56–58].

The recoil of the evaporating material produces a compression wave through the solid which rises the density up to 1000 times of the normal density.

If the solid material consists of small frozen deuterium–tritium pellets the high density and temperature of the compressed deuterium–tritium mixture can lead to nuclear fusion processes. Since the density is very high the time between subsequent fusion events in the chain reaction becomes very small and can result in an uncontrolled fusion of part of the pellet. This represents a micro nuclear fusion bomb As soon as the rapid expansion of the evaporating material has decreased the particle density below a critical value, the ignition parameter $ZP = n \cdot k \cdot T \cdot \tau_E$ falls below the ignition condition and the fusion processes stop. This time is about 10^{-8}–10^{-7} s.

In order to achieve spherical symmetry of the recoil forces the pellets are irradiated by 8–12-different laser beams from the same laser with directions, distributed symmetrically over all directions (Fig. 8.40). The high power output beam of the laser is divided into different partial beams which are all focused onto the target. The high power density achieved by the simultaneous irradiation of the pellet leads to a rapid temperature rise and the sudden increase of the density due to the recoil of the evaporating material increases the important product $n \cdot T$ above the limit for fusion ignition. The goal is to achieve fusion for a major part of the pellet. The frozen pellet is released through a pipe (Fig. 8.40X) at a time synchronized with the time of the laser pulse.

In Fig. 8.41 the up to now largest laser facility for laser fusion, the Shiva Nova Laser in Livermore, California is shown (named after the Indian God Shiva with many arms). It consists of a Neodymium Glass laser which radiates at $\lambda = 1.05$ μm. The output beam is amplified and then split into 8 partial beams. Each of these beams is further amplified

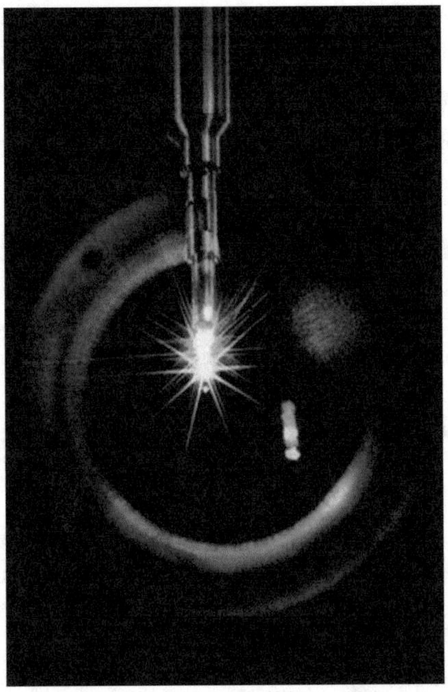

Fig. 8.40X Irradiated Deuterium/Tritium pellet (Lawrence-Livermore Facility)

in large Nd-glass rods pumped by flash lamps. In order to avoid too high power densities, which would destroy the amplifier rods, all beams are enlarged. Before they are focused into the target all beams are sent through nonlinear crystals where the frequency is quadrupled and the resulting ultra-violet beams are then focused onto the deuterium–tritium pellets, because the absorption of the radiation in the pellets is much higher for UV, than for infrared radiation. In 2003 the scientists at the Lawrence Livermore Facility succeeded for the first time to produce more fusion energy than the irradiation input energy. With a laser pulse energy of 45 kJ and a pulse length of 3 ns a peak power of $1.5 \cdot 10^{15}$ W could be realized. Focusing these laser pulses onto the deuterium pellet a yield of 10^{13} neutrons was measured, which indicated the ignition of nuclear fusion.

With an improved and larger laser version a pulse energy of 300 MJ could be reached with peak powers in the Petawatt range (10^{15} W). More information about nuclear fusion and modern reactor types can be found in the references [59, 60].

Fig. 8.41 Laser facility *Shiva Nova* in Livermore California for reaching the threshold for nuclear fusion

Summary

- For the quantitative determination of the impact of ionizing radiation on biological tissue a number of characteristic quantities are defined:
    1. The activity A of a radio-active substance gives the number of decays per s (Unit: 1 Bequerel)
    2. The energy doses D gives the total in the body per kg absorbed radiation energy (Unit: 1 Gy = 1 J/kg)
    3. The equivalent doses $H = Q \cdot D$, where the quality factor Q takes into account the specific tissue damage dependent on the kind of radiation. (Unit: 1 Sv = $Q \cdot 1$ Gray)
- The radiation power is measured by doses meters (ionization chambers). Devices which integrate the radiation power over the exposure time τ measure the exposure of the person to the total radiation energy during the time τ.
- The specific transmission or reflection of β- or γ-radiation depends on the material and can be used for precise thickness measurements.
- The damage of tissue by γ-radiation is used for the sterilization, vermin extinction and prolongation of the storage life of food.
- Radio-isotopes are used in medical diagnostics for marking thyroid changes or mutations, for investigations of metabolic processes in the body and for the positron-emission tomography PET.
- For radiation therapy (cancer treatment) γ-emitting radio-isotopes (e.g. ^{60}Co) are used as well as high energy electrons, protons and π-mesons from accelerators and the γ-radiation generated by them.
- Radio-active nuclides with known half-lifetimes are utilized for the age determination of archeological objects, of minerals or of meteorites.
- For the energy production by nuclear fission a controlled chain reaction has to be maintained. This can be achieved, for instance, by capture of neutrons by enriched Uranium (^{238}U + 3–5% ^{235}U). The fast neutrons emitted during the fission are slowed down by specific moderators, because slow neutrons have a larger fission cross section of ^{235}U than fast neutrons.
- The increase of the neutron number between two successive fission generation is described by

$$N(T) = N(0) \cdot e^{(k_{\text{eff}}-1) \cdot t/T}$$

where the reactor period T is the mean time between two fission generations. The multiplication factor $k_{eff} = \eta \cdot \varepsilon \cdot p \cdot f$ is the product of the mean number η of fission neutrons per ^{235}U nucleus, ε per ^{238}U nucleus, the slow-down probability p for a neutron and the escape probability f for a neutron which leaves the reactor core.

For a stationary operation ween the former points of the reactor the condition $k_{\text{eff}} = 1$. must be fulfilled.

- The safe control of a reactor is supported by the delayed neutrons which are emitted by the fission products.
- Energy gain through controlled nuclear fusion is possible by
    1. Confinement of the hot plasm in magnetic fields
    2. Generation and compression of a hot micro-plasma by irradiation of a deuterium–tritium pellet with very strong focused laser beams or by high energy particles (inertial confinement).
- Magnetic confinement of a plasma in a gas-filled torus is possible with specially formed magnetic coils (Tokomak or Stellerator)
- Heating of the plasma can be achieved with Ohm's heating, with injection of fast neutral deuterium molecules, or by irradiation with a hf-field or a microwave.

Problems

8.1 At an X-ray irradiation of the whole body with 50 keV X-ray quanta a patient (75 kg) receives the equivalent doses of 0.2 mS
How many X-ray quanta are absorbed in his body?
How large was the flux density of the incident X-rays for the irradiation time of 1 s if the irradiated area was 0.1 m^2 and 50% of the X-rays are absorbed in the body?

8.2 α-Particles with the energy $E_{\text{kin}} = 6$ MeV are slowed down to $E_{\text{kin}} = 0$ in an aluminum layer with the mass density $8 \cdot 10^{-3}\,\text{g/cm}^2$. What is their energy after having passed an aluminum foil of 20 μm thickness?

8.3 The mean range of β-particles with the energy $E_{\text{kin}} =$ 2 MeV in iron is about 1.5 mm. Which percentage of the incident particles penetrates iron sheets with thickness 1 mm; 1.5 mm and 2 mm?

8.4 The isotope ^{14}C is β-radioactive with a half-life time of $\tau = (5739 \pm 30)$ a. The specific ^{14}C-activity in living tissue containing the natural concentration of carbon is 0.255 Bq/g. The discovery of clay jugs 1947 in a cave close to Qumran containing the "Dead Sea Scrolls" was dated by Archeologists into the ninth century before Christ. For the book of the prophet Esaias (700 b.C.) Measurements 1952 of a sample containing 2 g carbon yielded the activity of 0.404 Bq.
1. Calculate the time of die back of the organic material and its error bars.
2. Calculate the number of the ^{14}C-atoms in the sample at the time of the measurement and at the time of die back of the organic material.

3. Estimate the isotope ratio $^{14}C/^{12}C$ in living tissue. Is there any difference of these ratios for plants close to the freeway and in the mid of a forest?

8.5 In an uranium mineral which initially did not contain any lead, Pb is generated by the radio-active decay of the isotopes ^{235}U (present abundance 0.72%, half-lifetime $7,038 \cdot 10^8$ a) and ^{238}U (present abundance 99.28% half-lifetime $4,468 \cdot 10^9$ a). With an age of the mineral of 600 Million years, what is the present weight ratio Pb/U in the mineral and what is the abundance ratio $^{207}Pb/^{206}Pb$?
Estimate the age of the earth assuming that in the beginning ^{205}U and ^{238}U had the same abundance.

8.6 The block 4 of the nuclear power plant in Chernobyl had been continuously operated before the accident at April 26th, 1986 with a thermal output power of 1000 MW. At 2% of all fission processes the iodine isotope ^{131}I is produced as fission product. It has a half-lifetime of 8.04 d. After which operation time an equilibrium is reached between production and decay of this isotope. What was the amount of ^{131}I at the time of the accident?
Assume the fission energy of 190 MeV per fission process

References

1. J. Magil, J. Galy: Radioactivity, Radionuclides. Radiation (Springer Heidelberg 2005)
2. M. S. Mustapha, A. S. Bukhari, H. E. S. Mohamed: Environmental Radioactivity (LAP Lambert Academic Publications 2011)
3. https://www.britannica.com/science/radioactivity
4. https://de.wikipedia.org/wiki/Radioaktivit%C3%A4t
5. https://www.sciencedirect.com/topics/earth-and-planetary-sciences/natural-radioactivity
6. https://de.wikipedia.org/wiki/Dosimeter
7. https://www.vedantu.com/physics/dosimeter
8. N. J. Carron, An Introduction to the Passage of Energetic Particles through Matter, 2007, Taylor and Francis Group
9. Glenn F. Knoll, Radiation Detection and Measurement, fourth edition, 2010, John Wiley and Sons, Inc.
10. https://en.wikipedia.org/wiki/Radiation_exposure
11. R.T.Overmann (ed): The technical Applications of Radioactivity (Taylor and Francis 2017)
12. E. Broda, Th.Schonfeld: Technical Applications of Radioavtivity (Pergamon Presss 1966)
13. Radioactive Sources: Applications and Alternative Technologies (National Academic Press 2021)
14. https://en.wikipedia.org/wiki/Radiation_therapy; https://www.epa.gov/radiation/radiation-sources-and-doses
15. Mayles, P; Rosenwald, JC; Nahum, A (2007). *Handbook of Radiation therapy Physics: Theory and Practice*. Taylor & Francis.
16. F. M. Khan: The Physics of Radiation Therapy (Lippincott Williams &Wilki 2009)Handbook of Radiation Therapy Zheory and Practice Institute of Physics Pub. 2007
17. https://en.wikipedia.org/wiki/Radiometric_datin
18. https://www.hindawi.com/journals/cmmm/2021
19. E. E. Kim u. a. (Hrsg.): Clinical Pet: Principles and Applications. Springer 2004
20. https://de.wikipedia.org/wiki/Positronen-Emissions-Tomographie
21. https://en.wikipedia.org/wiki/Neutron_activatio
22. R. E. Taylor Radio Carbon after four decades (Springer Heidelberg 1992). https://www.epa.gov/radiation/radiation-sources-and-doses
23. Lloyd A. Curie: The Remarkable Metrological History of Radiocarbon Dating [II]. J Res Natl Inst Stand Technol. 2004 Mar-Apr; 109(2): 185–217
24. *Rogers, Raymond N. (2005). "Studies on the radiocarbon sample from the shroud of turin". Thermochimica Acta. **425** (1–2): 189.* https://en.wikipedia.org/wiki/Radiocarbon_dating_of_the_Shroud_of_Turin
25. Radio-Carbon Calibration https://en.wikipedia.org/wiki/Radiocarbon_calibration
26. C.Tuniz: Accelerator Mass Spectrometry. Ultra-Sensitive Analysis for Global Science. (CRC Press New York 1998).
27. https://de.wikipedia.org/wiki/%C3%96tzi
28. https://www.wissenschaft.de/rubriken/dossiers/der-mann-aus-dem-eis/
29. https://www.radioactivity.eu.com/site/pages/Various_Applications.htm
30. P. Heydari: Principles of Modern Radiation Therapy Systems (2017). P. N. McDermont, C. G. Orton: The Physics and Technology of Radiation Therapy (Medical Physics Publ. 2018. **ISBN-13:** 978–1930524989. McGarry, M (2002). *Radiation therapy in Treatment*. AUSG Book
31. S. Marguet. The Physics of Nuclear Reactors (Springer, Heidelberg 2018)
32. Trenton Hensley: Nuclear Reactor Physics: NY Research Press 2019
33. John C. Lee: Nuclear Reactor physics and Engineering, Wiley 2020
34. https://www.iaea.org/topics/water-cooled-reactors, https://en.wikipedia.org/wiki/Pressurized_water, https://en.wikipedia.org/wiki/Breeder_reactorreactor
35. https://en.wikipedia.org/wiki/Pebble-bed_reactor
36. https://en.wikipedia.org/wiki/Breeder_reactor
37. https://en.wikipedia.org/wiki/Nuclear_fission_product
38. US NRC: Backgrounder on Nuclear Reactor Risk (2018)
39. https://en.wikipedia.org/wiki/Nuclear_reactor_safety_system
40. G. Janes et al. "Two-Photon Laser Isotope Separation of Atomic Uranium: Spectroscopic Studies, Excited-State Lifetimes, and Photoionization Cross Sections," IEEE J. Quantum Elect. 12, 111 (1976)
41. Jan Crossland: nuclear Fuel Cycle Science and Engineering Woodhead Publishing 2012. J. Zhang: Nuclear Fuel: Reprocessing and Waste Management (World Scientific, Singapore 2018). https://en.wikipedia.org/wiki/Radioactive_waste
42. H, Nifenecker: Accelerator Driven Subcritical Reactors. CRC-Press 2003
43. https://en.wikipedia.org/wiki/Accelerator-driven_subcritical_reactor
44. https://www.popularmechanics.com/science/energy/a35131133/advanced-nuclear-reactor-designs/
45. https://www.world-nuclear.org/information-library/current-and-future-generation/plans-for-new-reactors-worldwide.aspx
46. Mitsuru Kikuchi, Masafumi Azumi: Frontiers in Fusion Research (Springer,. Heidelberg 2015

47. J-P. Freiberg: Plasma Physics and fusion energy.(Cambridge Univ. Press 2007)
48. W. M. Stacey: Fusion: An Introduction to the Physics and Technology of Magnetic Confinement Fusion (Wiley VCH, Weinheim 2010)
49. https://en.wikipedia.org/wiki/Wendelstein_7-X
50. M. Akijama: Design Technology of Fusion Reactors (World Scientific Singapore 1990)
51. https://en.wikipedia.org/wiki/Joint_European_Torus
52. Richard Dinan: The Fusion Age: Modern Nuclear Fusion Reactors (Applied Fusion Systems 2017)
53. M. Classens: ITER, The Giant Fusion Reactor (Copernicus 2020) https://de.wikipedia.org/wiki/ITER
54. https://de.wikipedia.org/wiki/ITER
55. https://www.cnbc.com/2021/08/17/lawrence-livermore-lab-makes-significant-achievement-in-fusion.html40
56. S. Atzen, Meyer-ter-Ven: The Physics of Inertial Fusion. (Oxford Univ. Press 2004)
57. T. Ditmire et al.: Nuclear Fusion from Explosion of Femtosecond laser –Heated Deuterium Cluster. Nature 398, 489 (1999)
58. S. H. Glenzer et.al.: Symmetric Inertial Confinement Fusion. Imposions at ultra-high Laser Energies. (2010dx.doi.org./https://doi.org/10.1126/science.1185634)
59. https://en.wikipedia.org/wiki/Laser_Inertial_Fusion_Energy
60. https://www.cnbc.com/2021/08/17/lawrence-livermore-lab-makes-significant-achievement-in-fusion.html

9 Solutions to the Problems

Chapter 2

2.1 (a) For central collisions is

$$\frac{m}{2}v_0^2 = \frac{Z_1 Z_2 e^2}{4\pi\varepsilon_0\delta_0}.$$

with $Z_1 = 79$, $Z_2 = 2$, $e = 1{,}6 \cdot 10^{-19}$C, $m/2 \cdot v_0^2 = 5 \cdot 10^7 . 1{,}6 \cdot 10^{-19}\,\mathrm{J} \Rightarrow \delta_0 = 4{,}5 \cdot 10^{-15}\,\mathrm{m} = 4{,}5\,\mathrm{fm}$

(b) According to Eq. (2.6) is

$$\delta = \frac{1}{2}\delta_0\left[1 + \frac{1}{\sin\vartheta/2}\right].$$

This gives for $\vartheta = 60°$ with the result from (a)

$$\delta = \frac{3}{2}\delta_0 = 6{,}75\mathrm{fm}.$$

(c) It is $\delta < \delta_{min} = 6.5$ fm, only if

$$\sin\vartheta/2 > \frac{\delta_0}{2\delta_{\min} - \delta_0} = \frac{4{,}5}{13 - 4{,}5} = 0{,}53$$

$$\Rightarrow \vartheta > 64^{\circ}.$$

2.2 If the particles should be deflected by the angle $\vartheta \geq 90°$, they must have a minimum distance δ from the scattering center

$$\delta \leq \frac{1}{2}\delta_0\left[1 + \frac{1}{\sin\vartheta/2}\right] = \frac{1}{2}\delta_0(1 + \sqrt{2})$$

$$\delta_0 = 45\ \mathrm{fm} \Rightarrow \delta \leq 54{,}3 \cdot 10^{-15}\,\mathrm{m}.$$

The number of gold atoms per cm^2 in the foil is

$$n = \frac{M}{m} = \frac{\varrho \cdot V}{m} = \frac{\varrho \cdot 10^{-5}}{m} = \frac{19{,}32 \cdot 10^{-5}\mathrm{g/cm^2}}{197 \cdot 1{,}66 \cdot 10^{-24}\mathrm{g}} = 5{,}9 \cdot 10^{17}\,\mathrm{cm^{-2}}.$$

The fraction of particles deflected by $\delta > 90°$ is

$$\frac{\Delta N(\vartheta \geq 90°)}{N} = \pi \cdot \delta^2 \cdot 5{,}9 \cdot 10^{17} = 5{,}5 \cdot 10^{-5}.$$

2.3 (a) The mean square of the radius is

$$\langle r^2\rangle = \frac{\int_0^{R_0} r^4 \cdot \varrho(r)\mathrm{d}r}{\int^{R_0} r^2\varrho(r)\mathrm{d}r} = \frac{\int^{R_0} r^4(1 - ar^2)\mathrm{d}r}{\int r^2(1 - ar^2)\mathrm{d}r} = \frac{3}{7}R_0^2 \quad \text{with} \quad R_0 = 1/\sqrt{a}.$$

R_0 is the radius of a hard sphere, where $\varrho(r) = 0$

(b) Here is

$$\langle r^2\rangle = \frac{\int_0^\infty r^4 \mathrm{e}^{-r/a}\mathrm{d}r}{\int_0^\infty r^2 \mathrm{e}^{-r/a}\mathrm{d}r} = \frac{4! \cdot a^5}{2! \cdot a^3} = 12a^2$$

$$\Rightarrow \sqrt{\langle r^2\rangle} = 3{,}46a.$$

2.4 The deBroglie wavelength of the neutron is

$$\lambda = \frac{h}{\sqrt{2m_n \cdot E_{\mathrm{kin}}}} = 7{,}40 \cdot 10^{-15}\,\mathrm{m}$$

For $E_{kin} = 15$ MeV $= 2.4 \cdot 10^{-12}$ J is $E_{kin} << m_n c^2$, We can therefore use the nonrelativistic calculation. With a nuclear radius R_K(Ph) = $10.5 \cdot 10^{-15}$ m the first diffraction minimum appears at

$$\sin\vartheta_{\min} = 1{,}2 \cdot \frac{\lambda}{2R_{\mathrm{K}}} = \frac{8{,}88 \cdot 10^{-15}}{21{,}0 \cdot 10^{-15}} = 0{,}422$$

$$\Rightarrow \vartheta_{\min} = 25°$$

W. Demtröder, *Nuclear and Particle Physics*, Undergraduate Lecture Notes in Physics,
https://doi.org/10.1007/978-3-030-58313-2_9

2.5. The Fermi energy in the three-dimensional spherical potential is (see books on solid state physics)

$$E_F = \frac{\hbar^2}{2m_e}\left(3\pi^2 \cdot n_e\right)^{2/3} \quad \text{with} \quad n_e = \frac{A-Z}{\frac{4}{3}\pi R^3}.$$

$$E_{\text{total}} = \int_0^{E_F} 2E \cdot D(E)\mathrm{d}E \approx \frac{3}{5}E_F \cdot (A-Z).$$

Inserting the numerical values gives

(a) α-Particle: $A = 4$, $Z = 2$, $R = 1.8 \cdot 10^{-15}$ m

(b) $n_e = \dfrac{2 \cdot 10^{45}}{\frac{4}{3}\pi \cdot 1{,}8^3}\,\text{m}^{-3} = 8{,}2 \cdot 10^{43}\,\text{m}^{-3}$

$$\begin{aligned}\Rightarrow E_F &= \frac{1{,}05^2 \cdot 10^{-68}}{2 \cdot 9{,}1 \cdot 10^{-31}}\left(3\pi^2 \cdot 8{,}2 \cdot 10^{43}\right)^{2/3}\,\text{J}\\ &= 6 \cdot 10^{-39} \cdot \left(2{,}4 \cdot 10^{45}\right)^{2/3}\,\text{J}\\ &= 1{,}1 \cdot 10^{-8}\,\text{J} = 6{,}87 \cdot 10^{10}\,\text{eV}\end{aligned}$$

$$\Rightarrow E_{\text{total}} = \frac{3}{5} \cdot 6{,}87 \cdot 10^{10} \cdot 2\,\text{eV} = 8{,}2 \cdot 10^{10}\,\text{eV}.$$

(b) $A = 200$, $Z = 80R = 6.5 \cdot 10^{-15}$ m.

$$n_e = \frac{120 \cdot 10^{45}}{\frac{4}{3}\pi \cdot 6{,}5^3}\,\text{m}^{-3} = 1{,}0 \cdot 10^{44}\,\text{m}^{-3}$$

$$\begin{aligned}\Rightarrow E_F &= 6 \cdot 10^{-39} \cdot \left(3\pi^2 \cdot 10^{44}\right)^{2/3}\,\text{J}\\ &= 1{,}3 \cdot 10^{-8}\,\text{J} \approx 8{,}1 \cdot 10^{10}\,\text{eV}\end{aligned}$$

$$\Rightarrow E_{\text{total}} = 5{,}8 \cdot 10^{12}\,\text{eV} = 5{,}8\,\text{TeV}.$$

The electro-static energy of the Z electrons in the field of the Z protons can be estimated as follows.
We assume that the neutral nucleus has the energy zero.
If we remove one electron from the nucleus a charged sphere with the radius R and charge $+e$ remain. It has the potential energy

$$E_{\text{pot}} = \frac{e^2}{4\pi\varepsilon_0 R}$$

If we remove the remaining $(Z-1)$ electrons we have to spend the work

$$\begin{aligned}W &= E_{\text{pot}}(1+2+3+\cdots+Z) = (1+Z)\frac{Z}{2}E_{\text{pot}}\\ &= \frac{Z(Z+1)e^2}{8\pi\varepsilon_0 R}\end{aligned}$$

For $Z = 2$ this gives

$$\begin{aligned}W(Z=2) &= \frac{3}{4} \cdot \frac{1{,}6^2 \cdot 10^{-38}}{\pi \cdot 8{,}85 \cdot 10^{-12} \cdot 1{,}8 \cdot 10^{-15}}\,\text{J}\\ &= 3{,}8 \cdot 10^{-13}\,\text{J} = 2{,}4 \cdot 10^6\,\text{eV},\end{aligned}$$

For $Z = 80$ we obtain

$$\begin{aligned}W(Z=80) &= \frac{80 \cdot 81 \cdot 1{,}6^2 \cdot 10^{-38}}{8\pi \cdot 8{,}85 \cdot 10^{-12} \cdot 6{,}5 \cdot 10^{-15}}\\ &= 1{,}2 \cdot 10^{-10}\,\text{J} = 7{,}2 \cdot 10^8\,\text{eV}.\end{aligned}$$

In both cases is $E_{\text{pot}} << E_{\text{kin}}$.
Therefore the electrons cannot be kept inside the nucleus.

2.6 The total kinetic energy of the fragments is

$$\begin{aligned}E_{\text{kin}} &= h \cdot \nu - |E_B| = (2{,}5 - 2{,}2)\,\text{MeV}\\ &= 0{,}3\,\text{MeV}.\end{aligned}$$

it is evenly shared by proton and neutron, because $m_p \approx m_n$.

$$\begin{aligned}&\Rightarrow E_{\text{kin}}(\text{p}) = \frac{1}{2}E_{\text{kin}}^{\text{total}} = 0{,}15\,\text{MeV}\\ &\Rightarrow \frac{m}{2}v^2 = 0{,}15\,\text{MeV} = 2{,}4 \cdot 10^{-14}\,\text{J}\\ &\Rightarrow v = \sqrt{2E/m} = \left(\frac{4{,}8 \cdot 10^{-14}\,\text{J}}{1{,}67 \cdot 10^{-27}\,\text{kg}}\right)^{1/2}\\ &= 5{,}36 \cdot 10^6\,\text{m/s} = 0{,}018c.\end{aligned}$$

For the deflection in the magnetic field B is

$$\begin{aligned}&\frac{m \cdot v^2}{R} = e \cdot v \cdot B\\ &\Rightarrow B = \frac{m \cdot v}{e \cdot R}\\ &= \frac{1{,}6 \cdot 10^{-27} \cdot 5{,}36 \cdot 10^6}{1{,}6 \cdot 10^{-19} \cdot 0{,}1}\,\text{kg}/\left(\text{A} \cdot \text{s}^2\right)\\ &= 0{,}56\ \text{Tesla}.\end{aligned}$$

The focal length of a magnetic sector field with the sector angle 2φ is

$$f_0 = \frac{R}{\sin\varphi}$$

$$f_0 = \frac{R}{\sin 30^\circ} = \frac{0{,}1\,\text{m}}{0{,}5} = 0{,}2\,\text{m}.$$

The distance between source and focal point is then

$$d = 2f \cdot \cos\varphi/2 = 0{,}4\frac{1}{2} \cdot \sqrt{3}\,\text{m} = 0{,}35\,\text{m}.$$

2.7. For the magnetic moment of the deuteron we obtain

$$\left.\begin{aligned}\mu_p &= +2{,}79\mu_K\\ \mu_n &= -1{,}91\mu_K\end{aligned}\right\} \Rightarrow \mu_D = +0{,}88\mu_K,$$

Since the deuteron spin is $S = 1$ the two spins of the nucleons must be parallel.
The hyperfine-structure-splitting is $\Delta\nu_{HFS} \propto \mu_{nucleus} \cdot \mu_B \cdot |\psi(0)|^2$, where the last factor gives the electron density at the nucleus, which is equal for both isotopes. We therefore get

$$\frac{\Delta\nu_{HFS}(H)}{\Delta\nu_{HFS}(D)} = \frac{2{,}79}{0{,}88} = 3{,}17.$$

The HFS-splitting in the $^2S_{1/2}$ state of the isotope ${}^2_1D$ is therefore smaller by a factor 3.17 than for the ${}^1_1H$, where it is

$$\Delta\nu_{HFS}(H) = 1{,}42 \cdot 10^9\,s^{-1} = 1{,}42\,GHz$$

We therefore get for the isotope ${}^2_1D$

$$\Delta\nu_{HFS}(D) = 448\,MHz.$$

2.8 (a) The frequency of the Lyman α-line $1S \leftrightarrow 2P$ is according to volume 3 Eq. (3.72)

$$\nu = \frac{Ry^*}{h} \cdot Z^2\left(\frac{1}{1} - \frac{1}{2^2}\right) = \frac{3}{4}Ry^*/h.$$

The Rydberg constant is

$$Ry^* = \frac{\mu \cdot e^4}{8\varepsilon_0^2 h^2} \quad \text{with} \quad \mu = \frac{m_K \cdot m_e}{m_K + m_e}.$$

For the ${}^1_1H$-atom is $m_K = m_p = 1836\ m_e$. For the ${}^2_1H$ isotope is $m_K = m_p + m_n - E_B = 3672\ m_e$. Introducing the Rydberg constant

$$Ry_\infty^+ = \frac{m_e e^4}{8\varepsilon_0^2 h^2}$$

and using the approximation for the reduced mass $\mu \approx m_e$ we get

$$Ry^*({}^2_1H) - Ry^*({}^1_1H) = Ry^*_\infty \cdot 2{,}72 \cdot 10^{-4}$$

$$\begin{aligned}\Rightarrow \Delta\nu &= \nu({}^2_1H) - \nu({}^1_1H)\\ &= \nu(Ry^*_\infty) \cdot 2{,}72 \cdot 10^{-4}\\ &= 3{,}29 \cdot 10^{15} \cdot 2{,}72 \cdot 10^{-4}\,s^{-1}\\ &= 8{,}946 \cdot 10^{11}\,s^{-1} = 895\ GHz.\end{aligned}$$

(b) The energy shift of a level with quantum number $F = J + I$ and $J = L + S$ due to the electric quadrupole moment $QM = eQ$ for a nucleus with $I \geq 1$ is (see for instance H. G. Kuhn: Atomic Spectra Longman London 1964 or H. Haken, H. CH. Wolf: Atomic and Quantum Physics (Springe Heidelberg 1997)

$$\Delta E_Q = e \cdot Q \cdot \frac{\partial^2 V}{\partial z^2} \cdot \frac{\frac{3}{4}C(C+1) - I(I+1) \cdot J(J+1)}{2I(2I-1) \cdot (2J+1)}$$

with

$$C = F(F+1) - I(I+1) - J(J+1),$$

where $\partial^2 V/\partial z^2$ is the gradient of the electric field of the electron shell at the nucleus. For the spherical symmetric electron density in the 1s-state is $\frac{\partial^2 V}{\partial \partial^2} = 0 \Rightarrow \Delta E_Q = 0$.
This implies that the deuteron has no quadrupole shift in spite of the spin $I = 1$…

2.9 The quadrupole moment is

$$\begin{aligned}QM &= \int (3z^2 - r^2)\varrho(r)dV,\\ dV &= 2\pi R dR dz,\\ \varrho &= \varrho_0 = \text{const.},\\ r^2 &= x^2 + y^2 + z^2 = R^2 + z^2\\ \Rightarrow 3z^2 - r^2 &= 2z^2 - R^2.\end{aligned}$$

With the ellipsoid equation (Fig. 9.1)

$$\frac{z^2}{a^2} + \frac{x^2 + y^2}{b^2} = \frac{z^2}{a^2} + \frac{R^2}{b^2} = 1$$

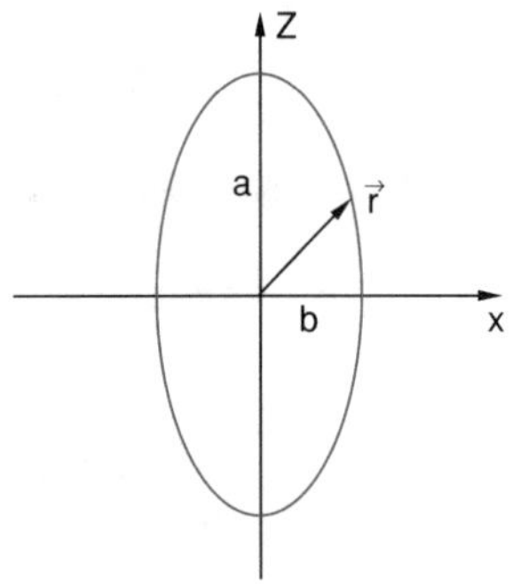

Fig. 9.1 To Solution 2.9

we get the quadrupole moment

$$\Rightarrow QM = 2\pi\varrho_0 \int_{z=-a}^{+a} \int_{R=0}^{b/a\sqrt{a^2-z^2}} (2z^2 - R^2) R\,dR\,dz$$

$$= 2\pi\varrho_0 \int_{z=-a}^{+a} \left| z^2R^2 - \frac{1}{4}R^4 \right|_0^{b/a\sqrt{a^2-z^2}} dz$$

$$= 2\pi\varrho_0 \int_{z=-a}^{+a} \left[\left(b^2 + \frac{1}{2}\frac{b^4}{a^2}\right) z^2 - \left(\frac{b^2}{a^2} + \frac{1}{4}\frac{b^4}{a^4}\right) z^4 - \frac{1}{4}b^4 \right] dz$$

$$= \frac{8\pi}{15}\varrho_0 \cdot a \cdot b^2 (a^2 - b^2).$$

The volume of a rotational ellipsoid is $(4/3)\pi ab^2$. If it has the charge $Z \cdot e$ this gives

$$Z \cdot e = \frac{4}{3}\pi ab^2 \cdot \varrho_0$$

$$\Rightarrow QM = \frac{2}{5} Z \cdot e (a^2 - b^2).$$

Defining $a = \overline{R} + \frac{1}{2}\Delta R$ and $b = \overline{R} - \frac{1}{2}\Delta R$. we get

$$\Rightarrow QM = \frac{4}{5} Z \cdot e \cdot \overline{R} \cdot \Delta R = \frac{4}{5} Z \cdot e \cdot \overline{R}^2 \cdot \delta$$

with $\delta = \Delta R/\overline{R}$. The quadrupole moment is therefore proportional to the deformation parameter δ.

2.10 According to (2.42) is the total Coulomb repulsion energy

$$E_C = \frac{3}{5} \frac{Z^2 e^2}{4\pi\varepsilon_0 R}.$$

With $Z = 80$ and $R = 7 \cdot 10^{-15}$ m this gives

$$E_C = \frac{3}{5} \frac{80^2 \cdot 1{,}6^2 \cdot 10^{-38}}{4\pi \cdot 8{,}85 \cdot 10^{-12} \cdot 7 \cdot 10^{-15}} \text{ J}$$
$$= 1{,}26 \cdot 10^{-10}\,\text{J} = 7{,}87 \cdot 10^8\,\text{eV}.$$

The mean Coulomb energy per proton is then

$$\overline{E}_C(\text{Proton}) = E_C^{\text{total}}/80 = 1{,}575 \cdot 10^{-12}\,\text{J}$$
$$= 9{,}8\,\text{MeV/Proton}.$$

Each energy level can be occupied by two protons, they lie in the proton potential box of Fig. 2.28 higher by ΔE = 19.6 eV than the corresponding levels in the neutron potential box.

Chapter 3

3.1 It is

$$N_A(t) = N_A(0) \cdot e^{-\lambda_1 t}$$

$$N_B(t) = N_{A_0} \cdot \frac{\lambda_1}{\lambda_2 - \lambda_1} \left(e^{-\lambda_1 t} - e^{-\lambda_2 t}\right)$$

$$\lambda = \frac{1}{\tau} = \frac{\ln 2}{T_{1/2}}$$

$$\Rightarrow \lambda_1 = \frac{\ln 2}{10\text{d}} = 0{,}069/\text{d}$$

$$\lambda_2 = \frac{\ln 2}{5\text{d}} = 0{,}139/\text{d}$$

$$\Rightarrow N_A(1\text{d}) = N_{A_0} \cdot e^{-0{,}069} = 0{,}933 N_{A_0}$$
$$N_A(10\text{d}) = N_{A_0} \cdot e^{-0{,}69} = 0{,}50 N_{A_0}$$
$$N_A(100\text{d}) = N_{A_0} \cdot e^{-6{,}9} = 1{,}0 \cdot 10^{-3} N_{A_0}$$

$$N_B(1\text{d}) = \frac{0{,}069 N_{A_0}}{0{,}069} \cdot \left(e^{-0{,}069} - e^{-0{,}139}\right)$$
$$= 0{,}063 N_{A_0}$$
$$N_B(10\text{d}) = N_{A_0} \cdot \left(e^{-0{,}69} - e^{-1{,}38}\right) = 0{,}25 N_{A_0}$$
$$N_B(100\text{d}) = N_{A_0} \cdot \left(e^{-6{,}9} - e^{-13{,}8}\right)$$
$$= 1{,}0 \cdot 10^{-3} N_{A_0}.$$

(b) Tritium ${}^3_1$H has a half lifetime $T_{½}$ = 12.3 a =>

$\lambda = \ln 2/T_{½} = 0.056/\text{a}$

$$\frac{N(t)}{N(0)} = 10^{-3} = e^{-\lambda t}$$
$$\Rightarrow t = \frac{1}{\lambda} \cdot \ln 10^3 = 123{,}4\text{a}.$$

After 123 years only 1g tritium is left from the original 1 kg.

3.2 (a) The transmission of a particle with mass m and energy E_1 through a rectangular barrier of heights h_0 and width a:

$$T \approx \frac{16E_1}{E_0^2}(E_0 - E_1) \cdot e^{-2a\cdot\alpha} \tag{9.1}$$

with

$$\alpha = \frac{1}{\hbar}\sqrt{2m(E_0 - E_1)},$$

if $a{\cdot}\alpha >> 1$

For our example with $m = 6.68 \cdot 10^{-27}$ kg; $E_0 = 11$ MeV; $E_1 = 6$ MeV; $a = 4 \cdot 10^{-14}$ m is
$a{\cdot}\alpha = 39.24 >> 1$.
This tells us, that we can use Eq. (9.1). Inserting the numerical values gives

$$T = \frac{16 \cdot 30}{11^2} \cdot e^{-78{,}48} = 3{,}3 \cdot 10^{-34}.$$

When the α-particle in the box potential with a depth $-E_2 = -15$ MeV has the kinetic energy $E_{kin} = (15 + 6)$ MeV = 21 MeV, its velocity is

$$\upsilon = \sqrt{2E_{kin}/m},$$

For a box width b it hits the outer potential wall

$$W_1 = \frac{v}{2b} = \frac{1}{2b}\sqrt{2E_{kin}/m}$$

times per sec. Inserting the numerical values $b = 6 \cdot 10^{-15}$ m, $E_{kin} = 21$ MeV $= 3.32 \cdot 10^{-12}$ J, gives $W_1 = 2.64 \cdot 10^{21}$ s^{-1}.
The decay-constant λ is then

$$\lambda = W_1 \cdot T = 8{,}7 \cdot 10^{-13}\,\mathrm{s}^{-1}$$

and the half-lifetime

$$T_{1/2} = \ln 2/\lambda = 8{,}0 \cdot 10^{11}\,\mathrm{s} \approx 2{,}53 \cdot 10^4 \mathrm{a}.$$

(c) The Coulomb potential corresponds to the potential energy

$$E_{pot}(r) = \frac{1}{4\pi\varepsilon_0}\frac{2Z \cdot e^2}{r} \quad \text{for} \quad r \geq b.$$

The tunnel probability is now (Fig. 9.2)

$$T = e^{-G}$$

with

$$G = \frac{2\sqrt{2m}}{\hbar}\int_{r_1}^{r_2}\sqrt{\frac{2Z \cdot e^2}{4\pi\varepsilon_0 r} - E_1 \mathrm{d}r}$$
$$= \frac{2}{\hbar}\sqrt{2mE_1}\int_{r_1}^{r_2}\sqrt{\frac{r_2}{r} - 1\mathrm{d}r}.$$

The Integral can be solved by the substitution $x = 1/r$ and gives

$$G = \frac{2}{\hbar}\sqrt{2mE_1} \cdot \left[-r_1\sqrt{\frac{r_2}{r_1} - 1} + r_2 \cdot \arctan\sqrt{\frac{r_2}{r_1} - 1}\right].$$

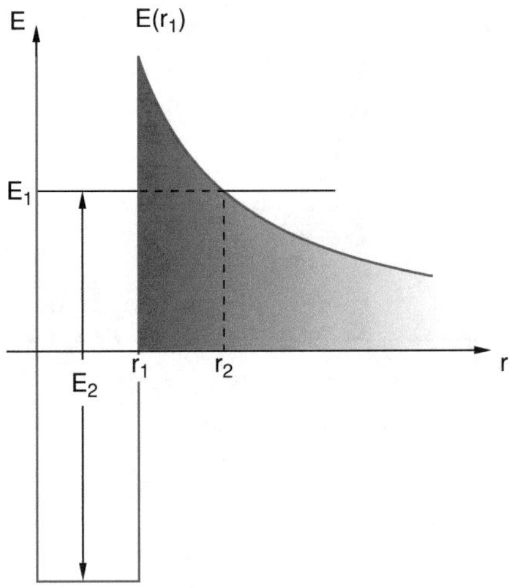

Fig. 9.2 To Solution 3.2

Inserting the numerical values:
$r_1 = b = 6{,}0 \cdot 10^{-15}\,\mathrm{m}, \;\; m = 6{,}5 \cdot 10^{-27}\,\mathrm{kg}, E_1 = 6\,\mathrm{MeV},$
$r_2 = 2Ze^2/(4\pi\varepsilon_0 E_1) = 4{,}3 \cdot 10^{-14}\,\mathrm{m}.$

yields the Gamov factor

$$G = 21{,}2 \cdot 10^{14} \cdot [-14{,}9 + 43 \arctan 2{,}48] \cdot 10^{-15}$$
$$= 77{,}25 \Rightarrow T = e^{-77{,}25} \approx 2{,}8 \cdot 10^{-34}.$$

The decay constant is then,

$$\lambda = W_1 \cdot T = 2{,}64 \cdot 10^{21} \cdot 2{,}8 \cdot 10^{-34}$$
$$\approx 7{,}5 \cdot 10^{-13}\,\mathrm{s}^{-1}.$$

The decay constant is therefore only slightly smaller than for the box potential in part a in spite of the fact that here the potential height with $E_{pot}(r_1) = r_2/r_1 \cdot E_{pot}(r_2) = (43/6.5) \cdot 6$ MeV = 39.7 MeV markedly higher is. The tunnel probability depends on the area

$$\int_{r_1}^{r_2}\sqrt{E_{kin}}\mathrm{d}r,$$

above the tunnel path, which is nearly equal in both cases.

3.3 An α-particle, approaching on a central path an $^{208}_{28}$Pb nucleus with the kinetic energy $E_{kin} = T_\alpha = 8.78$ MeV. In the center of mass system the relative energy is

$$T = \frac{T_\alpha}{1 + m_\alpha/m_{Pb}} = 8{,}61\ \mathrm{MeV}.$$

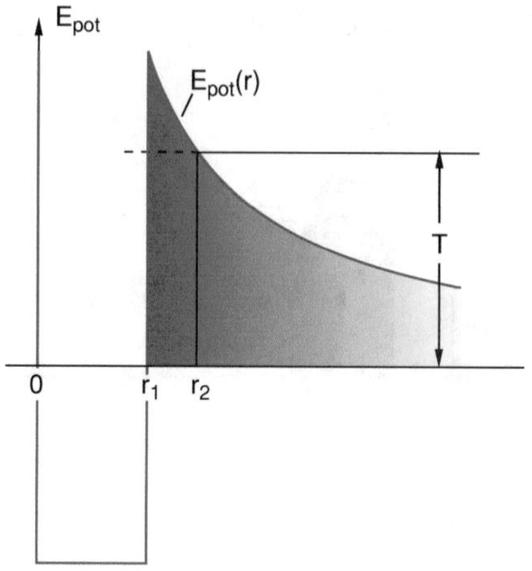

Fig. 9.3 To Solution 3.3

The reversal point is at a distance r_2 where $E_{kin}(r_2) = E_{pot}(r_2)$

$$T = \frac{Z \cdot e^2}{4\pi\varepsilon_0 \cdot r_2}$$

The gives for the distance r_2 (Fig. 9.3)

$$r_2 = \frac{2Ze^2}{4\pi\varepsilon_0 T} = \frac{2 \cdot 82 \cdot 1{,}6^2 \cdot 10^{-38}\,\text{m}}{4\pi\varepsilon_0 \cdot 8{,}61 \cdot 1{,}6 \cdot 10^{-13}} = 27{,}4\,\text{fm}.$$

With the nuclear radius

$$r_1 = b = 1{,}4\,\text{fm} \cdot \left(A_{Pb}^{1/3} + A_\alpha^{1/3}\right) = 10{,}52\,\text{fm}$$

we get, as has been shown in problem 3.2b with $E_1 = T$ the Gamov factor.

$$G = \frac{2}{\hbar}\sqrt{m_\alpha E_1}\left[-r_1\sqrt{\frac{r_2}{r_1} - 1} + r_2 \arctan\sqrt{\frac{r_2}{r_1} - 1}\right] = 29{,}05.$$

the transmission through the Coulomb barrier is then

$$T = e^{-G} = 2{,}42 \cdot 10^{-13}$$

This is also the probability that an α-particle, hitting the barrier from outside, will penetrate into the nucleus. The decay constant is, according to (3.26)

$$\lambda = W_0 \cdot W_1 \cdot T = \ln 2/T_{1/2} = \frac{0{,}693}{3 \cdot 10^{-7}}\,\text{s}^{-1} = 2{,}3 \cdot 10^6\,\text{s}^{-1}.$$

With a potential depth E_0 = 35 MeV and $b = r_1$ = 10.52 fm is

$$W_1 = \frac{v}{2b} = \frac{1}{2b} \cdot \sqrt{2(E_1 + E_0)/m} = 2{,}2 \cdot 10^{21}\,\text{s}^{-1}.$$

The probability W_0, that an α-particle is formed inside the nucleus then becomes

$$W_0 = \frac{\lambda}{W_1 \cdot T} \frac{2{,}3 \cdot 10^6}{2{,}2 \cdot 10^{21} \cdot 2{,}42 \cdot 10^{-13}} = 4{,}3 \cdot 10^{-3}.$$

3.4 The positron-decay in the reaction

$$^{63}_{30}\text{Zn} \rightarrow {}^{62}_{29}\text{Cu} + \beta^+ + \nu_e + E_{kin}(\beta^+) + E_{kin}(\nu_e).$$

has a maximum positron energy if $E_{kin}(\nu_e) = 0$.
In the case that the decaying nucleus $^{62}_{30}$Zn is at rest, the momentum of the daughter nucleus is

$$\boldsymbol{p}\left({}^{62}_{29}\text{Cu}\right) = -\boldsymbol{p}\left(\overline{\beta}^+\right).$$

because the total momentum must be zero.
With the relation $E_{kin} = \sqrt{(cp)^2 - (m_0c^2)^2} - m_0c^2$ we get for the momentum

$$p = \frac{1}{c}\sqrt{E_{kin}^2 + 2E_{kin}m_0c^2}.$$

For the maximum β^+-energy this becomes

$$p = \frac{1}{c}\sqrt{0{,}66^2 + 2 \cdot 0{,}66 \cdot 0{,}51\,\text{MeV}} = 5{,}6 \cdot 10^{-22}\text{N} \cdot \text{s}.$$

The decay energy of the nucleus $^{62}_{30}$Zn is then

$$Q = \frac{p^2}{2M(\text{Cu})} + E_{kin}^{max}(\beta^+) + m_e c^2 = 9{,}5 \cdot 10^{-6}\,\text{MeV} + 0{,}66\,\text{MeV} + 0{,}51\,\text{MeV}.$$

the recoil energy of the nucleus is then only 9.5 eV and therefore negligible. The maximum energy of the neutrino $E_{kin}(\nu_e)$ = 0.66 MeV is equal to the maximum β^+-energy.
For the electron capture is

$$e^- + {}^{62}_{30}\text{Zn} \rightarrow {}^{62}_{29}\text{Cu} + \nu_e$$

$$\Rightarrow E(\nu_e) = Q + m_e c^2 = E_{kin}(\beta^+) + 2m_e c^2 = 1{,}68\,\text{MeV}.$$

3.5 $\Delta M \cdot c^2 = m_e c^2 + E_{kin}(e^-) + E_{kin}(\overline{\nu}) + E_{kin}(M_2)$

(a) Neglecting the recoil is $E_{kin}(M_2) = 0$. For the maximum β^- energy is $E_{kin}(\nu) = 0$.

$$\Rightarrow E_{kin}^{max}(e^-) = \Delta M c^2 - m_e c^2 = (3 - 0{,}5)\ \text{MeV} = 2{,}5\ \text{MeV}.$$

(b) Taking into account the recoil we get for the case with maximum β-energy $E_{kin}(\overline{\nu}) = 0$.

$$p_{e^-} = -p_{M_2}$$
$$\Rightarrow E_{kin}(M_2) = \frac{p_e^2}{2M_2}.$$

Since $E_{kin}(e^-) >> m_e c^2$ one has to use the relativistic energy relation

$$E_{kin}^{el} = \sqrt{p^2c^2 + m_e^2c^4} - m_e c^2$$

$$\Rightarrow p_e^2 = \frac{1}{c^2}\left[(E_{kin} + m_e c^2)^2 - m_e^2 c^4\right] = \frac{1}{c^2}\left(E_{kin}^2 + 2E_{kin} m_e c^2\right)$$

$$\Rightarrow E_{kin}(M_2) = \frac{1}{2M_2c^2}\left(E_{kin}^2 + 2E_{kin} m_e c^2\right) = \frac{1}{70 \cdot 1836} \cdot \left(2{,}5^2 + 5 \cdot 0{,}5\right)\ \text{MeV} = 68\ \text{eV}.$$

The recoil energy is therefore very small.
The maximum energy of the neutrino becomes for $E_{kin}^{max}(e^-) = 0$

$$E_{kin}^{max}(\overline{\nu}) = 2{,}5\ \text{MeV}$$

The recoil of the nucleus is for this case (assuming $m_\nu = 0$)

$$p_{M_2} = -p(\overline{\nu}) = \frac{E(\overline{\nu})}{c},$$

which gives

$$E_{kin}(M_2) = \frac{p^2}{2M_2} = \frac{E^{max}(\overline{\nu})^2}{2M_2c^2} = \frac{2{,}5^2}{70 \cdot 1836}\ \text{MeV} = 48{,}6\ \text{eV}.$$

3.6 The mass of the nucleus changes by

$$\Delta m = m_e - E_{kin}(\nu_e)/c^2,$$

The atomic mass changes by

$$\Delta M = -(E_{kin}(\nu_e) + h \cdot \nu_K)/c^2,$$

where $h \cdot \nu_K = E_B(K) \approx 50$ eV.
The maximum energy of the neutrino is

$$E_{kin}(\nu_e) \le \left(m_e + m_p - m_n\right)c^2 = 0{,}783\ \text{MeV}.$$

Therefore is

$$\Delta m = [0{,}511 - 0{,}783]\ \text{MeV}/c^2 = -4{,}85 \cdot 10^{-31}\ \text{kg}$$

$$\Delta M = -[0{,}783 + 0{,}05]\ \text{MeV}/c^2 = -0{,}833\ \text{MeV}/c^2 = -1{,}48 \cdot 10^{-30}\ \text{kg}.$$

3.7 The radius R of the circle can be obtained from

$$m_e v^2/R = e \cdot v \cdot B \Rightarrow m_e \cdot v = R \cdot e \cdot B.$$

The momentum of the electron which equals the recoil energy of the nucleus is

$$p_e = m_e \cdot v = \text{R}e \cdot B.$$

The recoil energy is then

$$E_{kin}(M) = \frac{p^2}{2M} = \frac{R^2e^2B^2}{2M}.$$

and the electron energy

$$E_{kin}(e^-) = \frac{p^2}{m_e} = \frac{R^2e^2B^2}{2m_e}.$$

This allows to determine the excitation energy of the nucleus as

$$E_a = E_{kin}(M) + E_{kin}(e^-) + E_B(e^-).$$

With the numerical values $R = 0.1$ m $B = 0.05$ T, and $M(^{137}_{55}\text{Cs}) = 137 \cdot 1.66 \cdot 10^{-27}$ kg one obtains

$$E_{kin}(e^-) = 3{,}5 \cdot 10^{-13}\ \text{J} = 2{,}2\ \text{MeV},$$

$$E_{kin}(\text{Cs}) = 8{,}7\ \text{eV}.$$

$$E_B = \frac{4}{3} \cdot h\nu_{ik}(K_\alpha).$$

The binding energy of the K_α-electron is $h \cdot \nu(K\alpha) = 5.7$ keV (see Chemical Rubber Handbook on Chemistry and Physics)

$$\Rightarrow E_B(S, K\text{-Schale}) = 7{,}6\ \text{keV}$$

$$\Rightarrow E_a = (2{,}2 + 0{,}0076)\,\text{MeV} = 2{,}2076\,\text{MeV}.$$

This shows that the recoil energy of the nucleus is negligible.

3.8 (Author: Sauerland). β^- decay of the neutral tritium-atom:

$$^3_1\text{H} \rightarrow ^3_2\text{He}^+ + e^- + \bar{\nu}_e$$

Nuclei: $m(Z,A) = Z \cdot m_p + (A - Z)m_n - E_B/c^2$.
Atoms: $M(Z,A) = Z \cdot m_p + Z \cdot m_e(A - Z)m_n - E_B/c^2$.
For the β^- decay we obtain (with $m_\nu = 0$)

$$M(^3_1\text{H})c^2 = M(^3_2\text{He})c^2 + E_{kin}(e^-) + E_{kin}(\bar{\nu}_e)$$

$$\begin{aligned}\Rightarrow E_0 &= E_{kin}(e^-) + E_{kin}(\bar{\nu}_e)\\ &= \left[M(^3_1\text{H}) - M(^3_2\text{He})\right]c^2\\ &= (m_n - m_p - m_e)c^2\\ &\quad - \left(E_B(^3_1\text{H}) - E_B(^3_2\text{He})\right)\\ &= (0{,}7824 - 8{,}4819 + 7{,}7181)\,\text{MeV}\\ &= 0{,}0186\,\text{MeV}\\ &= 18{,}6\,\text{keV}.\end{aligned}$$

3.9 (Author; Sauerland): $^{12}_5\text{B} \rightarrow ^{12}_6\text{C} + e^- + \bar{\nu}_e\bar{\nu}_e$

For the nuclear masses we obtain:

$$m_B c^2 = m_C c^2 + m_e c^2 + E_{kin}(e^-) + E_{kin}(\bar{\nu}_e) + E_{kin}(C),$$

where the recoil energy E_{kin}/C) is very small, and can be neglected.
For the atomic masses is (Fig. 9.4)

$$M_B \approx m_B + Z_B \cdot m_e; M_C \approx m_C + Z_C \cdot m_e$$

$$\Rightarrow \underbrace{(m_B + Z_B \cdot m_e)}_{M_B} c^2 = \underbrace{[m_C + (Z_B + 1)m_e]}_{M_C} c^2 + E_{kin}(e^-) + E_{kin}(\bar{\nu}_e)$$

$$(M_B - M_C)c^2 = E_{kin}^{max}(e^-) = 13{,}369\,\text{MeV}.$$

For the positron decay is

$$^{12}_7\text{N} \rightarrow ^{12}_6\text{C} + e^+ + \nu_e,$$
$$m(e^+) = m(e^-) = m_e.$$

The energy balance of the nuclei is

$$m_N c^2 = m_C c^2 + m_e c^2 + E_{kin}(e^+) + E_{kin}\nu_e$$

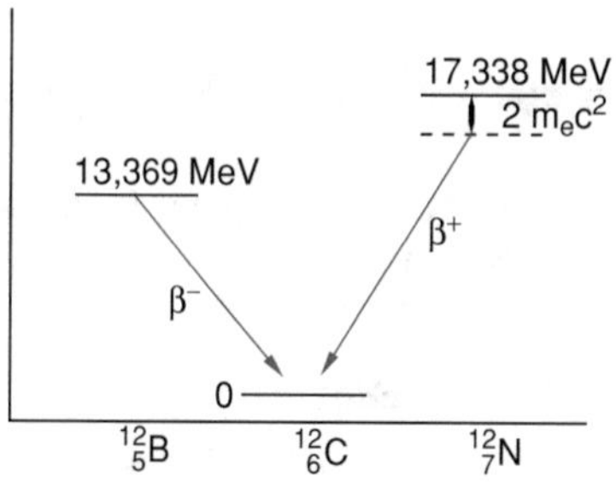

Fig. 9.4 To Solution 3.9

Adding the electrons gives

$$\underbrace{[m_N + (Z_C + 1)m_e]}_{M_N} c^2 = \underbrace{[m_C + (Z_C + 2)m_e]}_{M_C + 2m_e} c^2 + E_{kin}(e^+) + E_{kin}(\bar{\nu}_e),$$

because the β^+-decay produces a negatively charged daughter ion with 7 electrons in the atomic shell of the $^{12}_6$C-ion.
For the atomic masses we therefore obtain:

$$\begin{aligned}\Rightarrow (M_N - M_C)c^2 &= 2m_e c^2 + E_{kin}^{max}(e^+)\\ &= 1{,}022\,\text{MeV} + 16{,}3161\,\text{MeV}\\ &= 17{,}338\,\text{MeV}.\end{aligned}$$

3.10 (Author: Sauerland)

$$N_1 = N(^{226}\text{Ra}), \quad N_2 = N(^{222}\text{Rn})$$

$$\lambda_1 = \ln 2/T_{1/2}(^{226}\text{Ra}) = 1{,}187 \cdot 10^{-6}\text{d}^{-1}$$

$$\lambda_2 = \ln 2/T_{1/2}(^{222}\text{Rn}) = 0{,}181\text{d}^{-1}.$$

From (3.13) we get

$$N_2(t) = \frac{N_1(0) \cdot \lambda_1}{\lambda_2 - \lambda_1}\left(e^{-\lambda_1 t} - e^{-\lambda_2 t}\right).$$

The maximum of $N_2(t)$ is reached for $dN_2/dt = 0$

$$\Rightarrow t_m = \frac{1}{\lambda_2 - \lambda_1}\ln(\lambda_2/\lambda_1) = 65{,}87\text{d}$$

$$\Rightarrow N_2(t_m) = \frac{\lambda_1 N_1(0)}{\lambda_2 - \lambda_1}\left(e^{-\lambda_1 t_m} - e^{-\lambda_2 t_m}\right).$$

With

$$N_1(0) = \frac{10^{-2}\,\text{kg}}{226 \cdot 1{,}66 \cdot 10^{-27}\,\text{kg}} = 2{,}67 \cdot 10^{22}$$

we obtain

$$N_2(t_m) = 1{,}746 \cdot 10^{17}.$$

Applying the gas law

$$p \cdot V = N_2 \cdot k \cdot T$$

to the gaseous Radon we get with $N_2 = N_2(t_m)$ and $T = 293$ K for the volume V

$$V\left({}^{226}\mathrm{Ra}\right) = \frac{10\,\mathrm{g}}{5{,}5\mathrm{g/cm}^3} = 1{,}82\,\mathrm{cm}^3$$

$$\Rightarrow V\left({}^{222}\mathrm{Rn}\right) = 5 - 1{,}82 = 3{,}18\,\mathrm{cm}^3$$

$$\Rightarrow p = \frac{N_2 kT}{V} = \frac{1{,}75 \cdot 10^{17} \cdot 1{,}38 \cdot 10^{-23} \cdot 293}{3{,}18 \cdot 10^{-6}}$$
$$= 222\,\mathrm{N/m}^2 = 222\,\mathrm{Pa}.$$

At a total pressure of 10^5 Pa the pressure in the tube rises only by 0.222%. After 66 days the partial pressure decreases again because then more radon decays than is generated by the decay of radium. For long time delays one obtains

$$N_2(t) \approx \frac{\lambda_1}{\lambda_2} N_1(0) \mathrm{e}^{-\lambda_1 t},$$

This implies that the long time behavior is determined by the exponential radium decay with $T_{1/2} = 1600$ years.

Chapter 4

4.1 The kinetic energy is

$$E_{\mathrm{kin}} = mc^2 - m_0c^2 = 1\,\mathrm{GeV}$$
$$\Rightarrow mc^2 = E = 1\,\mathrm{GeV} + m_0c^2$$
$$E(\mathrm{e}^-) = 1\,\mathrm{GeV} + 0{,}5\,\mathrm{MeV} = 1{,}0005\,\mathrm{GeV}$$
$$E(\mathrm{p}^-) = 1\,\mathrm{GeV} + 938\,\mathrm{MeV} = 1{,}938\,\mathrm{GeV}$$
$$E(\alpha^{++}) = 1\,\mathrm{GeV} + 3{,}73\,\mathrm{GeV} = 4{,}73\,\mathrm{GeV}$$

$$E = mc^2 = \frac{m_0c^2}{\sqrt{1-\beta^2}}$$

$$\Rightarrow \beta = \sqrt{1 - \left(\frac{m_0c^2}{mc^2}\right)^2}$$

$$\Rightarrow \upsilon = c \cdot \sqrt{1 - \left(\frac{m_0c^2}{mc^2}\right)^2}$$

$$\upsilon(\mathrm{e}^-) = c \cdot \sqrt{1 - \left(\frac{5 \cdot 10^{-4}}{1{,}005}\right)^2}$$
$$\approx c\left(1 - 1{,}25 \cdot 10^{-7}\right) \approx 0{,}999999875\,c$$

$$\upsilon(\mathrm{p}^+) = c \cdot \sqrt{1 - \left(\frac{0{,}938}{1{,}938}\right)^2} \approx 0{,}718c$$

$$\upsilon(\alpha) = c \cdot \sqrt{1 - \left(\frac{3{,}73}{4{,}73}\right)^2} \approx 0{,}615\,c.$$

4.2 For the final energy $E_{\mathrm{kin}} = 1\,\mathrm{MeV} \Rightarrow E = E_{\mathrm{kin}} + m_0c^2 = 1{,}5\,\mathrm{MeV}$. With a radius $r_0 = 1$ m the magnetic flux is, according to (4.26)

$$\Phi = \frac{2\pi r_0}{e \cdot c} \sqrt{E^2 - (m_0c^2)^2}.$$

Inserting the numerical values gives

$$\Phi = 2{,}96 \cdot 10^{-2}\,\mathrm{Vs}.$$

The magnetic field at the electron ring path is

$$B(r_0) = \frac{p}{er_0} = \frac{\sqrt{E^2/c^2 - m_0^2c^2}}{er_0}$$
$$= 4{,}7\,\mathrm{mTesla}.$$

4.3 For $E = 400$ GeV is $\alpha = E_{\mathrm{kin}}/(m_0c^2) \approx 426$

According to (4.27) is then the magnetic field

$$B = \frac{m_0c^2}{c \cdot e \cdot r} \sqrt{2\alpha + \alpha^2};$$
$$m_0c^2 = 938 \cdot 1{,}6 \cdot 10^{-13}\,\mathrm{J}.$$
$$\Rightarrow B = 1{,}33\,\mathrm{Tesla}.$$

4.4 The velocity of the protons can be obtained from

$$E_{\mathrm{kin}} = (m - m_0)c^2 = m_0c^2\left(\frac{1}{\sqrt{1-\beta^2}} - 1\right)$$

$$\Rightarrow \beta = \frac{\upsilon}{c} = \sqrt{1 - \left(\frac{1}{1 + E_{\mathrm{kin}}/(m_0c^2)}\right)^2},$$

$$m_0c^2 = 938{,}27\,\mathrm{MeV}, \quad E_{\mathrm{kin}} = 100\,\mathrm{MeV},$$
$$\Rightarrow \upsilon = 0{,}428\,c.$$

The focal length of a longitudinal magnetic field is

$$f = \frac{2\pi m}{e \cdot B} \upsilon \Rightarrow B = \frac{2\pi m \cdot \upsilon}{e \cdot f}.$$

With $m = m_0 \cdot (1 - \upsilon/c)^{-1/2} = 1{,}32\, m_0$.
For $\alpha = 1 - m(v)/m_0$ follows

$$\alpha = 0.572$$

This gives

$$\Rightarrow B = \frac{2\pi \cdot 1{,}32 \cdot 1{,}67 \cdot 10^{-27} \cdot 0{,}428 \cdot 10^8}{1{,}6 \cdot 10^{-19} \cdot 10} \text{ Tesla}$$
$$= 1{,}11 \text{ Tesla}.$$

4.5 According to (4.54b) is the energy loss of an electron per roundtrip

$$\Delta E = \frac{e^2 \gamma^4}{3 \varepsilon_0 r_0}$$

with

$$\gamma = \frac{mc^2}{m_0 c^2} = \frac{5 \cdot 10^4\,\text{MeV}}{0{,}5\,\text{MeV}} = 10^5$$
$$\Rightarrow \Delta E = \frac{1{,}6^2 \cdot 10^{-38} \cdot 10^{20}}{3 \cdot 8{,}85 \cdot 10^{-12} \cdot 4 \cdot 10^3} \text{ J} = 2{,}4 \cdot 10^{-11} \text{ J}.$$

The circulating electron with velocity $v \approx c$ represents a circular current

$$I = 1{,}6 \cdot 10^{-19} \cdot \frac{c}{2\pi R} \text{A} = 1{,}9 \cdot 10^{-15} \text{A}$$

For a current of 0.1 A then

$$N = \frac{0{,}1}{1{,}9 \cdot 10^{-15}} = 5{,}3 \cdot 10^{13}$$

electrons are circulating in the ring. Their energy loss per roundtrip is then

$$\Delta E = 1{,}26 \cdot 10^3\,\text{J/Umlauf}.$$

The roundtrip time is $T = 2\pi R/c = 84$ µs.
The electrical power P is then

$$\Rightarrow \mathrm{d}E/\mathrm{d}t = P = 1{,}5 \cdot 10^7\,\text{W} = 15\,\text{MW}.$$

4.6 It is

$$E = mc^2 = \frac{m_0 c^2}{\sqrt{1 - \upsilon^2/c^2}}; \quad \upsilon = \upsilon(x).$$

$$\frac{\mathrm{d}E}{\mathrm{d}x} = \frac{m_0 c^2 \cdot \mathrm{d}\upsilon/\mathrm{d}x}{(1 - \upsilon^2/c^2)^{3/2}} \cdot \frac{\upsilon}{c^2}$$
$$a = \frac{\mathrm{d}\upsilon}{\mathrm{d}t} = \frac{\mathrm{d}\upsilon}{\mathrm{d}x}\frac{\mathrm{d}x}{\mathrm{d}t} = \frac{\mathrm{d}\upsilon}{\mathrm{d}x} \cdot \upsilon$$
$$= \frac{\mathrm{d}E}{\mathrm{d}x} \frac{(1 - \upsilon^2/c^2)^{3/2}}{m_0 \upsilon}$$
$$\propto \frac{\mathrm{d}E}{\mathrm{d}x}.$$

At the Stanford accelerator SLAC is

$$\frac{\mathrm{d}E}{\mathrm{d}x} = \frac{50\,\text{GeV}}{3200\,\text{m}} = 15{,}6\,\text{MeV/m}.$$

The accelerated electron with an energy gain dE/dx radiates on a straight line the power

$$P_\text{S}^\text{lin} = \frac{e^2 c}{6\pi\varepsilon_0 (m_0 c^2)^2} \cdot \left(\frac{\mathrm{d}E}{\mathrm{d}x}\right)^2.$$

On a circular path with radius R the additional circular acceleration results in a high radiation power. If the electron travels on the circle with constant velocity it radiates according to (4.54a) the power

$$P_\text{S}^\text{kreis} = \frac{e^2 c}{6\pi\varepsilon_0} \frac{E^4}{(m_0 c^2)^4 r_0^2}$$

The ratio

$$\frac{P_\text{S}^\text{lin}}{P_\text{S}^\text{kreis}} = \frac{(\mathrm{d}E/\mathrm{d}x)^2 \cdot (m_0 c^2)^2 r_0^2}{E^4}.$$

strongly depends on the energy E of the circulating electron. Inserting the numerical values dE/dx = 15.6 MeV/m, r_0 = 4000 m, E = 50 GeV, $m_0 c^2$ = 0.5 MeV gives

$$P_\text{S}^\text{lin}/P_\text{S}^\text{kreis} = 1{,}56 \cdot 10^{-10}!$$

This ratio becomes 1, if the acceleration on the linear straight path becomes

$$\frac{\mathrm{d}E}{\mathrm{d}x} = 8 \cdot 10^4 \cdot 15{,}6\,\text{MeV/m} = 1250\,\text{GeV/m}$$

Such high acceleration values are technical impossible, even with laser-driven accelerators.

4.7 (Author; Sauerland).

For all collision processes the square of the Four-momentum R must be preserved

$$|\mathcal{R}|^2 = E^2/c^2 - \boldsymbol{p}^2 \text{ with } E = \sum E_i, \boldsymbol{p} = \sum \boldsymbol{p}_i$$

It is then also conserved under transformation from the lab-system to the center of mass system.
If we denote the kinetic energy by the letter T we have in the lab-system

$$E_{1\mathrm{L}} = T_{\mathrm{L}} + m_1c^2, \quad \boldsymbol{p}_{1\mathrm{L}} \neq 0, \quad \boldsymbol{p}_{2\mathrm{L}} = 0,$$
$$E_{2\mathrm{L}} = m_2c^2,$$
$$c^2|\mathcal{R}|_{\mathrm{L}}^2 = \left[T + (m_1 + m_2)c^2\right]^2 - p_1^2c^2.$$

In the center-of-mass system we get $\boldsymbol{p}_1 + \boldsymbol{p}_2 = 0$

$$\begin{aligned}\Rightarrow c^2|\mathcal{R}|_{\mathrm{S}}^2 &= (E_1 + E_2)^2 = E^2\\ &= \left(E_{1\mathrm{L}} + m_2c^2\right)^2 - (c\boldsymbol{p}_{1\mathrm{L}})^2\\ &= c^2|\mathcal{R}|_{\mathrm{L}}^2 = E_{1\mathrm{L}}^2 - (cp_{1\mathrm{L}})^2\\ &\quad + 2E_{1\mathrm{L}}m_2c^2 + \left(m_2c^2\right)^2\\ &= 2E_{1\mathrm{L}}m_2c^2 + \left(m_1^2 + m_2^2\right)c^4,\end{aligned}$$

because $E_{1\mathrm{L}}^2 = m_1^2c^4 + (cp_{1\mathrm{L}})^2$.

(a) For the reaction in the lab-system is

$$\mathrm{p} + \mathrm{p} \to \mathrm{p} + \mathrm{p} + \overline{\mathrm{p}} + \mathrm{p}$$

(the baryon number must be preserved). We obtain with $m(\overline{\mathrm{p}}) = m(\mathrm{p})$

$$\begin{aligned}E_{\min^2} &= \left(4m_{\mathrm{p}}c^2\right)^2 \leq 2\left(m_{\mathrm{p}}c^2\right)^2 + 2E_{1\mathrm{L}} \cdot m_{\mathrm{p}}c^2\\ &= 2\left(m_{\mathrm{p}}c^2\right)^2 + 2m_{\mathrm{p}}c^2\left(T_1 + m_{\mathrm{p}}c^2\right)\\ &= 4\left(m_{\mathrm{p}}c^2\right)^2 + 2T_1m_{\mathrm{p}}c^2\\ \Rightarrow 2T_1^{\min} &= 16m_{\mathrm{p}}c^2 - 4m_{\mathrm{p}}c^2 = 12m_{\mathrm{p}}c^2\\ T_1^{\min} &= 6m_{\mathrm{p}}c^2 = 5{,}63\,\mathrm{GeV}\\ E_1^{\min} &= 7m_{\mathrm{p}}c^2 = 6{,}568\,\mathrm{GeV}.\end{aligned}$$

(b) If the two protons collide head-on. The total kinetic energy can be transferred.

$$\begin{aligned}&T_1 + m_{\mathrm{p}}c^2 + T_2 + m_{\mathrm{p}}c^2 \geq 4m_{\mathrm{p}}c^2\\ &\Rightarrow 2T \geq 2m_{\mathrm{p}}c^2,\\ &T \geq m_{\mathrm{p}}c^2.\end{aligned}$$

Each of the two colliding protons must have the minimum energy

$$E_{\min} = T_{\min} + m_{\mathrm{p}}c^2 = 2m_{\mathrm{p}}c^2 = 1{,}877\,\mathrm{GeV},$$
$$T_{\min} = m_{\mathrm{p}}c^2 = 0{,}938\,\mathrm{GeV}$$

in order to produce the proton-anti-proton pair.

4.8 (Author: Sauerland):

(a) According to (4.52) the reaction rate is

$$\dot{R} = L \cdot \sigma.$$

On the other side is

$$\dot{R} = \Phi_1 \cdot N_2 \cdot \sigma,$$

where $\Phi = n_1 \cdot A_1 \cdot v_1$ is the particle flux of the incident beam with particle density n_1, cross section area A_1 and velocity v_1.
The area density N_2 of the target with thickness d, area $A_2 > A_1$ and particle density n_2

$$N_2 = n_2 \cdot d.$$

The thickness d can be obtained from the mass area density $m/A_2 = 64\ \mu\mathrm{g/cm^2}$

$$d = \frac{m}{A_2 \cdot \varrho(^{12}\mathrm{C})} = \frac{64 \cdot 10^{-6}\,\mathrm{g/cm^2}}{2{,}1\,\mathrm{g/cm^3}} = 0{,}3\,\mu\mathrm{m}.$$

The particle density in the target is then

$$\begin{aligned}n_2 &= \varrho\left(^{12}\mathrm{C}\right) \cdot \frac{N_{\mathrm{A}}}{M_{\mathrm{mol}}\left(^{12}\mathrm{C}\right)}\\ &= \frac{64 \cdot 10^{-6}}{0{,}3 \cdot 10^{-4}} \frac{6 \cdot 10^{23}}{12}\,\mathrm{cm}^{-3}\\ &= 1{,}06 \cdot 10^{23}\,\mathrm{cm}^{-3}\end{aligned}$$

The particle flux is

$$\begin{aligned}\Phi_1 &= I_1/Z \cdot e\\ &= \frac{10^{-6}\mathrm{A}}{2 \cdot 1{,}6 \cdot 10^{-19}\mathrm{As}}\\ &= 3{,}1 \cdot 10^{12}\,\mathrm{s}^{-1}\end{aligned}$$

The luminosity is

$$\begin{aligned}L &= \Phi_1 \cdot N_2 = \Phi_1 \cdot n_2 \cdot d\\ &= 10^{31}\,\mathrm{cm}^{-2}\,\mathrm{s}^{-1}.\end{aligned}$$

(b) Particle flux in beam 1 (Fig. 9.5)

$$\Phi_1 = n_1 \cdot b_1 \cdot h_1 \cdot v_1 = \frac{I_1}{Z_1 \cdot e}$$

and in the beam 2

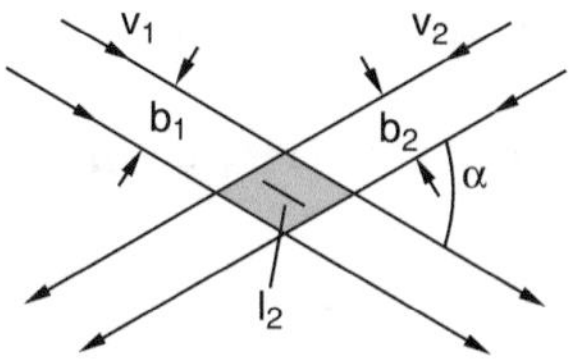

Fig. 9.5 To Solution 4.8b

$$\Phi_2 = n_2 \cdot b_2 \cdot h_2 \cdot v_2 = \frac{I_2}{Z_2 \cdot e}$$
$$v_1 = v_2 \approx c; \quad \sin\alpha = b_1/l_1 = b_2/l_2$$
$$L = \Phi_1 \cdot N_2 = \Phi_1 \cdot n_2 \cdot l_2$$
$$= \frac{\Phi_1 \Phi_2 l_2}{b_2 h v_2} = \frac{\Phi_1 \Phi_2}{hc \cdot \sin\alpha}$$

$$\Rightarrow L = \frac{I_1 \cdot I_2}{Z_1 Z_2 e^2 h \cdot c \cdot \sin\alpha} \quad \text{with} \quad h = \min(h_1, h_2)$$
$$= \frac{4A^2}{1{,}6^2 \cdot 10^{-38} A^2 s^2 \cdot 4 \cdot 10^{-3}\,\text{m} \cdot 3 \cdot 10^8\,\text{m/s} \cdot 0{,}087}$$
$$= 1{,}5 \cdot 10^{29}\,\text{m}^{-2}\,\text{s}^{-1},$$

The luminosity is therefore smaller by 2 orders of magnitude than in example (a)

(c) A particle in beam 1 meets per roundtrip N_2 particles in beam 2. This results in

$$R_1 = N_2 \frac{\sigma}{b \cdot h}$$

reactions within the beam cross section $b \cdot h$.
N_1 particles in beam 1 cause during f round trips per second the reaction rate

$$\dot{R}_{\text{total}} = \frac{N_1 \cdot N_2 \cdot f \cdot \sigma}{b \cdot h}$$

for a continuous operation. With k anti-collinear particle bunches there are $2k$ collision zones. At each of these zones the reaction rate

$$\dot{R} = \frac{N_1 N_2 f \sigma}{2kbh};$$
$$I = NZef, \quad N = kN_{\text{Paket}}$$
$$= \frac{I_1 I_2 \sigma}{2kbhfZ_1 Z_2 e^2}.$$

takes place. The luminosity L is then for $Z_1 = Z_2 = \pm 1$

$$L = \frac{I_1 I_2}{e^2 \cdot 2kbhf}$$
$$= \frac{3{,}2^2 A^2}{1{,}6^2 \cdot 10^{-38} A^2 s^2 \cdot 16 \cdot 10^{-2}\,\text{cm}^2 10^7\,\text{s}^{-1}}$$
$$= 2{,}5 \cdot 10^{32}\,\text{cm}^{-2}\,\text{s}^{-1},$$

This is 3 orders of magnitude higher than in (b). The main reason is the better focusing ($b = h = 1$ mm) and the compression of the particles in bunches.

4.9 The transmitted intensity is

$$I_t = I_0 \cdot e^{-\mu x}$$

$$\Rightarrow x = \frac{1}{\mu} \cdot \ln I_0/I_t$$
$$= \frac{1}{5 \cdot 10^3\,\text{m}^{-1}} \cdot \ln 100$$
$$= 0{,}92\,\text{mm}.$$

A cylindrical rod with diameter $d = 0{,}92$ mm attenuates X-rays with 100 keV energy down to 1%.

4.10 (author Sauerland). The Cerenkov radiation is emitted under the angle ϑ against the flight direction of the particle (Fig. 9.6)

$$\cos\vartheta = \frac{c/n}{v} = \frac{1}{\beta \cdot n} \quad \text{with} \quad \beta = v/c$$

The energy of the particles is

$$E^2 = m_0^2 c^4 + p^2 c^2 = m_0^2 \gamma^2 c^4,$$

because p = $m_0 \cdot \gamma \cdot v$ = 367 MeV/c

$$\Rightarrow m_0 = \sqrt{\frac{p^2}{v^2} - \frac{p^2}{c^2}} = \frac{p}{c}\sqrt{n^2 \cos^2\vartheta - 1}$$
$$= \frac{367\,\text{MeV}}{c^2}\left(1{,}7^2 \cdot \cos^2 51^\circ - 1\right)^{1/2}$$
$$= 139\,\text{MeV}/c^2 = m_{\pi^\pm}.$$

The particle is therefore a π-meson.
The radiated energy W is proportional to the surface of the cone with the aperture angle (90° − ϑ) and the side length $L \cdot \sin\vartheta = v \cdot t \cdot \sin\vartheta$ with L = detector length

$$\Rightarrow W \propto \sin^2\vartheta$$

For $\vartheta = 0$ the radiated energy W becomes zero.

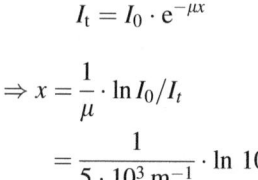

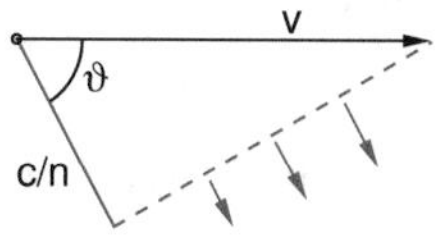

Fig. 9.6 To Solution 4.10

The minimum energy of the particles is

$$E = m_0c^2\gamma \geq m_0c^2 \frac{n}{\sqrt{n^2-1}}$$

with

$$\gamma = \frac{1}{\sqrt{1-\beta^2}}; \quad \beta \geq \frac{1}{n}.$$

For $n = 1.7$ is $E \geq 1.23 m_0 c^2 = 172.5$ MeV

$$\Rightarrow E_{kin} = 33\,\text{MeV}$$

At a path length $v \cdot t = 1$m the wave front on the cone is $\sin\vartheta$ [m]. The radiated power is proportional to the cone surface i.e. $\propto \sin^2\vartheta$.

4.11 The energy of the proton is

$$E = E_{kin} + m_0c^2 = m_0\gamma c^2$$

$$\Rightarrow \gamma = \frac{E_{kin}+m_0c^2}{m_0c^2} = \frac{1877+938{,}5}{938{,}5} = 3$$

$$\Rightarrow \sqrt{1-\beta^2} = \frac{1}{3} \Rightarrow \beta^2 = \frac{8}{9}$$

$$\Rightarrow \beta = \frac{1}{3}\sqrt{8}.$$

The Bethe-formula (4.63) can be written with

$$\frac{1}{\varrho}\frac{dE}{dx} = \frac{dE}{dx}\frac{V_{mol}}{M_{mol}}$$

and

$$n_e = \frac{N_A \cdot Z_2}{V_{mol}}$$

as

$$-\frac{1}{\varrho}\frac{dE}{dx} = \frac{Z_1^2 e^4 N_A Z_2}{4\pi\varepsilon_0^2 v^2 m_e M_{mol}} \cdot \left[\ln\frac{2m_e v^2}{\langle E_B\rangle \cdot (1-\beta^2)} - \beta^2\right].$$

Introducing the classical electron radius

$$r_e = \frac{e^2}{4\pi\varepsilon_0 m_e c^2}$$

we obtain with the mean binding energy of the target electrons $\langle E_B\rangle = 10\,Z_2$ eV and the molar mass M_{mol}

$$-\frac{1}{\varrho}\frac{dE}{dx} = 4\pi(N_A/M_{mol})r_e^2 m_e c^2 \cdot \frac{Z_1^2 Z_2}{\beta^2}\left(\ln\frac{2m_e c^2\beta^2\gamma^2}{\langle E_B\rangle} - \beta^2\right)$$
$$= 0{,}307\,\text{MeV}\cdot\text{cm/g} \cdot \frac{Z_2}{A}\frac{Z_1^2}{\beta^2}\left(\ln\frac{1{,}022\,\text{MeV}}{10Z_2\,\text{eV}}\beta^2\gamma^2 - \beta^2\right).$$

Aluminum: $Z_2 = 13$, $A = 27$, $\langle E_B\rangle = 130$ eV, $\rho = 2.7$ g/cm^3

$$\Rightarrow -\frac{1}{\varrho}\frac{dE}{dx} = 0{,}307\,\text{MeV}\cdot\text{cm}^2/\text{g}\cdot\frac{13}{27}\cdot\frac{9}{8} \cdot \left(\ln\frac{8.176.000}{130} - \frac{8}{9}\right)$$
$$= 1{,}69 \cdot \frac{\text{MeV/cm}}{\text{g/cm}^3}$$

$$\Rightarrow \frac{dE}{dx} = 2{,}7 \cdot 1{,}69\frac{\text{MeV}}{\text{cm}} = 4{,}56\,\text{MeV/cm}.$$

Lead: $Z = 82$, $A = 207.5$, $\langle E_B\rangle = 820$ eV, $\rho = 11.34$ g/cm^3.

$$\Rightarrow -\frac{1}{\varrho}\frac{dE}{dx} = 1{,}136\frac{\text{MeV/cm}}{\text{g/cm}^3}$$

$$\Rightarrow -\frac{dE}{dx} = 12{,}88\,\text{MeV/cm}.$$

If the mass area density is equal for both substances, the light Aluminum stops the particle better by the factor 1.5 than the heavy lead.

4.12 The deBroglie wavelength for neutrons is

$$\lambda_{dB} = \frac{h}{\sqrt{2m \cdot E_{kin}}}.$$

For $E_{kin} = 25$ MeV $= 4 \cdot 10^{-21}$ J is

$$\lambda_{dB} = 1{,}8 \cdot 10^{-10}\,\text{m}.$$

The absorption cross section for neutrons is

$$\sigma_{abs} \approx \lambda_{dB}^2 \cdot W_a,$$

where W_a is the probability that a neutron is captured if it approaches a nucleus closer than the minimum distance $d_{min} = \lambda_{dB}/2$.

While λ depends only on the kinetic energy, W_a depends on the internal structure of the nucleus as well as on the neutron energy (resonances!!) It is approximately

$$W_a \propto \sqrt{E_{kin}} \propto 1/\lambda,$$

The total dependence of σ_a is then

$$\sigma \propto \lambda \propto 1/v$$

4.13 For $E_{kin} < 1$ MeV the deBroglie wavelength is

$$\lambda_{dB} > \frac{6{,}62 \cdot 10^{-34}}{\sqrt{2 \cdot 1{,}667 \cdot 10^{-27} \cdot 1{,}6 \cdot 10^{-13}}}\,\text{m} = 2{,}8 \cdot 10^{-14}\,\text{m} = 28\text{fm}$$

and therefore much larger than the diameter

$$d = 2r_0\sqrt[3]{A} = 2{,}6\sqrt[3]{20}\text{fm} \approx 7\text{fm}.$$

of the nucleus. This means that essentially only the spherically symmetric S-scattering occurs. This can be compared with light diffracted by a slit with width $b < \lambda$, where the central diffraction order of the transmitted light is distributed over the whole angular range and fills the whole half space behind the slit without any structure (since $\sin\vartheta = \lambda/b > 1$ is not defined).

(b) In Vol. 1 Sect. 4.2.3 it was shown, that the ratio of the kinetic energies before and after the collision is

$$\frac{E'_{kin}}{E_{kin}} = \frac{A^2 + 2A\cos\vartheta_1 + 1}{(1+A)^2}.$$

With $\Delta E = E_{kin} - E'_{kin}$ we obtain

$$\frac{\Delta E}{E_{kin}} = 1 - \frac{A^2 + 2A\cos\vartheta_1 + 1}{(1+A)^2}.$$

Averaging over all deflection angles ϑ_1 of the neutron gives with $\langle\cos\vartheta\rangle = 0$

$$\Delta E = \frac{2A}{(1+A)^2} \cdot E_{kin}.$$

4.14 In every inertial system the velocity of light is defined by $c \cdot t = r$ (this is the basis of the Lorentz-Transformations). Therefore with the four-vector R we can form the invariant

$$|\mathcal{R}|^2 = c^2t^2 - r^2$$

For the four-momentum $\mathcal{P}$ we get

$$|\mathcal{P}|^2 = E^2/c^2 - p^2.$$

In an inertial system S_1 where the particle moves with the velocity v_1 is

$$E^2/c^2 = m^2c^2 = \frac{m_0^2c^2}{1 - v_1^2/c^2}$$

$$p^2 = (mv_1)^2 = \frac{m_0^2v_1^2}{1 - v_1^2/c^2}$$

$$\Rightarrow |\mathcal{P}|^2 = \frac{m_0^2(c^2 - v_1^2)}{1 - v_1^2/c^2} = m_0^2c^2,$$

independent of v_1! Therefore $|\mathcal{P}|^2$ must be invariant when changing to another inertial system.

Chapter 5

5.1 The spin of the deuteron is

$$\boldsymbol{s} = \boldsymbol{s}_n + \boldsymbol{s}_p = 0 \quad \text{or} \quad \mathbf{1} \cdot \hbar$$

The total angular momentum in the state $^{2S+1}L_J$ is

$$\boldsymbol{J} = \boldsymbol{L} + \boldsymbol{S}$$

(a) The following levels with $J = 1$ are possible:

$$L = 0, S = 0 \Rightarrow J = 0 :{}^1S_0$$
$$S = 1 \Rightarrow J = 1 :{}^3S_1$$
$$L = 1, S = 0 \Rightarrow J = 1 :{}^1P_1$$
$$S = 1 \Rightarrow J = 0, 1, 2 :{}^3P_{0,1,2}$$
$$L = 2, S = 0 \Rightarrow J = 2 :{}^1D_2$$
$$S = 1 \Rightarrow J = 1, 2, 3 :{}^3D_{1,2,3}.$$

(b) Parity $P = (-1)^L$. Only states with $L = 0$ or $L = 2$ have even parity.
because of the spin-dependent nuclear forces the triplet states (parallel spins) have a lower energy than the singlet states.
For the bound states with $L = 0$ or $L = 2$ only the states ^3S and ^3D are possible. Since the magnetic dipole moment is smaller than the sum of the proton- and neutron moments angular momentum L and nuclear spin S in the D-state must be anti-parallel, i.e. it must be a $^3D_1$-state.

(c) The magnetic moment of the deuteron is

$$\boldsymbol{\mu}_d = \frac{\mu_N}{\hbar}\left(g_{l_p}\boldsymbol{l}_p + g_{l_n}\boldsymbol{l}_n + g_{s_p}\boldsymbol{s}_p + g_{s_n}\boldsymbol{s}_n\right).$$

it is: $\boldsymbol{l}_\mathrm{p} \approx \boldsymbol{l}_\mathrm{n} = \frac{1}{2}\boldsymbol{L}, s_\mathrm{p} \approx s_\mathrm{n} = \frac{1}{2}\boldsymbol{S}$. Da $g_{l_\mathrm{p}} = 1, \quad g_{l_\mathrm{n}} = 0$, $\mu_\mathrm{N}/\hbar \cdot g_{s_\mathrm{p}} \cdot s_\mathrm{p} = 2\mu_\mathrm{p} \cdot \boldsymbol{\sigma}_\mathrm{p}/2$

$$\begin{aligned}\Rightarrow \boldsymbol{\mu}_\mathrm{d} &= \frac{1}{2}\frac{\mu_\mathrm{N}}{\hbar}\boldsymbol{L} + \mu_\mathrm{p}\boldsymbol{\sigma}_\mathrm{p} + \mu_\mathrm{n}\boldsymbol{\sigma}_\mathrm{n} \\ &= \frac{\mu_\mathrm{N}}{2\hbar}\boldsymbol{L} + \frac{1}{2}(\mu_\mathrm{p} + \mu_\mathrm{n})(\boldsymbol{\sigma}_\mathrm{p} + \boldsymbol{\sigma}_\mathrm{n}) \\ &\quad + \frac{1}{2}(\mu_\mathrm{p} - \mu_\mathrm{n})(\boldsymbol{\sigma}_\mathrm{p} - \boldsymbol{\sigma}_\mathrm{n}).\end{aligned}$$

The last term is zero for S = 1, because $\boldsymbol{\sigma}_\mathrm{p} = \boldsymbol{\sigma}_\mathrm{n}$. Replacing $\boldsymbol{L}$ and $\boldsymbol{S}$ by their vector sum $\boldsymbol{J} = \boldsymbol{L} + \boldsymbol{S}$ resp. $\boldsymbol{J} = \boldsymbol{L} - \boldsymbol{S}$, we obtain

$$\begin{aligned}\boldsymbol{\mu}_\mathrm{d}(L,S) &= \frac{1}{2}\left(\mu_\mathrm{p} + \mu_\mathrm{n} + \frac{1}{2}\mu_\mathrm{N}\right)\frac{\boldsymbol{J}}{\hbar} \\ &\quad - \frac{1}{2}\left(\mu_\mathrm{p} + \mu_\mathrm{n} - \frac{1}{2}\mu_\mathrm{N}\right)\frac{\boldsymbol{L} - \boldsymbol{S}}{\hbar}.\end{aligned}$$

The expectation value of th e magnetic moment is therefore

$$\begin{aligned}\mu_\mathrm{d}(L,S) &= \frac{\langle JLS|\boldsymbol{\mu}_\mathrm{d} \cdot \boldsymbol{J}|JLS\rangle}{\langle JLS|\boldsymbol{J} \cdot \boldsymbol{J}|JLS\rangle} \cdot \hbar \cdot J \\ &= \frac{1}{\hbar(J+1)}\langle JLS|\boldsymbol{\mu}_\mathrm{d} \cdot \boldsymbol{J}|JLS\rangle,\end{aligned}$$

With $\langle JLS|\boldsymbol{J}^2|JLS\rangle = \hbar^2 \cdot J(J+1)$

$$\begin{aligned}\Rightarrow \mu_\mathrm{d} &= \frac{1}{2}\left(\mu_\mathrm{p} + \mu_\mathrm{n} + \frac{1}{2}\mu_\mathrm{N}\right)J \\ &\quad - \frac{1}{2}\left(\mu_\mathrm{p} + \mu_\mathrm{n} - \frac{1}{2}\mu_\mathrm{N}\right) \\ &\quad \cdot \frac{L(L+1) - S(S+1)}{J+1} \\ &= \mu_\mathrm{p} + \mu_\mathrm{n} \\ &\quad - \left(\mu_\mathrm{p} + \mu_\mathrm{n} - \frac{1}{2}\mu_\mathrm{N}\right)\frac{L(L+1)}{4},\end{aligned}$$

because $J = S = 1$; $L = 0$ or 2.
We denote with P_D the probability, that the nucleus is in the state with $L = 2$. We then get

$$\begin{aligned}\mu_\mathrm{d} &= (1 - P_D)\mu_\mathrm{d}(L=0) + P_D\mu_\mathrm{d}(L=2) \\ &= (1 - P_D)(\mu_\mathrm{p} + \mu_\mathrm{n}) \\ &\quad + P_D\left[(\mu_\mathrm{p} + \mu_\mathrm{n}) - \left(\mu_\mathrm{p} + \mu_\mathrm{n} - \frac{1}{2}\mu_\mathrm{N}\right)\frac{3}{2}\right] \\ &= \mu_\mathrm{p} + \mu_\mathrm{n} - P_D\left(\mu_\mathrm{p} + \mu_\mathrm{n} - \frac{1}{2}\mu_\mathrm{N}\right) \cdot \frac{3}{2}\end{aligned}$$

$$\begin{aligned}\Rightarrow P_D &= \frac{2}{3}\frac{\mu_\mathrm{p} + \mu_\mathrm{n} - \mu_\mathrm{d}}{\mu_\mathrm{v} + \mu_\mathrm{n} - \frac{1}{2}\mu_\mathrm{N}} \\ &= \frac{2}{3}\frac{0{,}87890 - 0{,}85744}{0{,}37980} \approx 0{,}039.\end{aligned}$$

The admixture of the D share with $s = 2$ to the total wavefunction ψ^2 is only about 4%.

5.2 (a) The mass excess is

$$\Delta = m - A \cdot m_\mathrm{u},$$

with

$$m_\mathrm{u} = 1\,\mathrm{u} = 931{,}4943\,\mathrm{MeV}/c^2.$$

The molecular binding-energy of the molecule ${}^2\mathrm{H}_3^+ = \mathrm{D}_3^+$ is about 1 eV and therefore completely negligible. With $\Delta({}^{12}\mathrm{C}) = 0 \Rightarrow$ (per definition) we get

$$\begin{aligned}3\Delta D - 0 &= 0{,}042306\mathrm{u} \\ \Rightarrow \Delta D &= 0{,}014102\mathrm{u}\end{aligned}$$

$$\begin{aligned}2\Delta H - \Delta D &= 0{,}001548\mathrm{u} \\ \Rightarrow \Delta H &= 0{,}007825\mathrm{u}\end{aligned}$$

$$\begin{aligned}m({}^1\mathrm{H}) &= 1{,}007825\,\mathrm{u} = 938{,}78325\,\mathrm{MeV}/c^2 \\ m({}^2\mathrm{H}) &= m(\mathrm{D}) = 2{,}014102\,\mathrm{u} \\ &= 1876{,}1246\,\mathrm{MeV}/c^2.\end{aligned}$$

(b) For the different masses we get

$$\begin{aligned}m_\mathrm{p} &= m({}^1\mathrm{H}) - m_\mathrm{e} + 13{,}6\,\mathrm{eV}/c^2 \\ &= 938{,}27228\,\mathrm{MeV}/c^2\end{aligned}$$

$$\begin{aligned}m_\mathrm{d} &= m({}^2\mathrm{H}) - m_\mathrm{e} + 13{,}6\,\mathrm{eV}/c^2 \\ &= 1875{,}6136\,\mathrm{MeV}/c^2\end{aligned}$$

$$\begin{aligned}m_\mathrm{n} &= m_\mathrm{d} - m_\mathrm{p} + E_\mathrm{B}(\mathrm{D})/c^2 \\ &= 939{,}5657\,\mathrm{MeV}/c^2.\end{aligned}$$

5.3 β^- decay of the tritium ${}^3\mathrm{H}$.

$${}^3_1\mathrm{H} \rightarrow {}^3_2\mathrm{He}^+ + \mathrm{e}^- + \bar{\nu}_\mathrm{e}.$$

The mass of the atom is

$$m(Z,A) = (A - Z)m_n + Z \cdot m_H - E_B/c^2.$$

The maximum energy of the electron is then

$$\begin{aligned} Q_{\beta^-} &= \left[m(^3_1\mathrm{H}) - m(^3_2\mathrm{He})\right]c^2 \\ &= (m_n - m_H)c^2 - \left(E_B(^3\mathrm{H}) - E_B(^3\mathrm{He})\right) \\ &= (0{,}78235 - 8{,}48182 + 7{,}71806)\,\mathrm{MeV} \\ &= 18{,}60\,\mathrm{keV}, \end{aligned}$$

where the numerical values of $m_n\,c^2$ and $m_H\,c^2$ are taken from "American Handbook of Physics". if the neutrino mass is assumed as $m_\nu = 0$ the maximum energy of the electron is reached for $E_{kin}(\overline{\nu}_e) = 0$, because the recoil energy of the ^{3}He-nucleus can be neglected.
The maximum recoil energy of the ^{3}He-nucleus occurs for $E_{kin}(\overline{\nu}_e) = 0$.

$$\Rightarrow \frac{m_e}{2}\upsilon_e^2 + \frac{m(^3\mathrm{He})}{2}\upsilon_{He}^2 = Q_{\beta^-}.$$

Conservation of momentum

$$\begin{aligned} m_e\upsilon_e &= m(^3\mathrm{He})\upsilon_{He} \\ \Rightarrow \upsilon_{He} &= \frac{m_e}{m(^3\mathrm{He})} \cdot \upsilon_e \\ \Rightarrow E_{kin}^{max}(^3\mathrm{He}) &= \frac{m_e}{m_e + m(^3\mathrm{He})} \cdot Q_{\beta^-} \\ &\approx \frac{m_e}{m(^3\mathrm{He})} Q_{\beta^-} = 3{,}4\,\mathrm{eV} \end{aligned}$$

5.4 The configuration of the $_3{}^6$Li nucleus can be described by

$$3\mathrm{p} + 3\mathrm{n} \rightarrow \alpha + \mathrm{p} + \mathrm{n}.$$

The angular momenta of the 4 nucleons in the α-particle couple to

$$L_\alpha = S_\alpha = T_\alpha = 0.$$

The coupling of the angular momenta of the residual 2 nucleons follow the LS-coupling scheme

$$\begin{aligned} j_p &= l_p \pm \frac{1}{2} = 1 \pm \frac{1}{2}, \\ j_n &= l_n \pm \frac{1}{2} = 1 \pm \frac{1}{2}. \end{aligned}$$

As for the deuteron also here the energy of parallel spins of p and n is lower than for antiparallel spins.

$$\left.\begin{array}{ll} s_p \text{ and } s_n \text{ parallel:} & S = 1 \\ l_p \text{ and } l_n \text{ antiparallel:} & L = 0 \end{array}\right\} J^P$$
$$= 1^+ = {}^3S_1 - \text{state} = \text{Groundstate}.$$

For the excited states the following coupling possibilities exist:

$$\left.\begin{array}{ll} s_p \text{ and } s_n \text{ parallel:} & S = 1 \\ l_p \text{ and } l_n \text{ parallel:} & L = 2 \end{array}\right\} J^P$$

$J^P = 1^+, 2^+, 3^+ = {}^3D_{1,2,3}$ states (Fig. 9.7).
The spin–orbit coupling can be obtained as follows:

$$\begin{aligned} \boldsymbol{L} \cdot \boldsymbol{S} &= (\boldsymbol{l}_p + \boldsymbol{l}_n)(\boldsymbol{s}_p + \boldsymbol{s}_n) = 2\boldsymbol{l} \cdot 2\boldsymbol{s} = 4(\boldsymbol{l} \cdot \boldsymbol{s}) \\ &= \frac{1}{2}(\boldsymbol{J}^2 - \boldsymbol{L}^2 - \boldsymbol{S}^2). \end{aligned}$$

The expectation value is then

$$\begin{aligned} \langle \cdots |\boldsymbol{L} \cdot \boldsymbol{S}| \cdots \rangle &= 4\langle \cdots |\boldsymbol{l} \cdot \boldsymbol{s}| \cdots \rangle \\ &= \frac{1}{2}[J(J+1) - 2 \cdot 3 - 1 \cdot 2] \\ &= \begin{cases} -3 & \text{for } J = 1 \\ -1 & \text{for } J = 2 \\ +2 & \text{for } J = 3 \end{cases} \\ V_{LS} &= \sum a_i \boldsymbol{l}_i \boldsymbol{s}_i = 2a(\boldsymbol{l} \cdot \boldsymbol{s}). \end{aligned}$$

For the energy distances between the ^3D_J-states one obtains

$$\begin{aligned} \langle 1^+|V_{LS}|1^+\rangle - \langle 2^+|V_{LS}|2^+\rangle &= \frac{2a}{4}(-3+1) \\ &= -a \approx 1{,}4\,\mathrm{MeV}, \\ \langle 2^+|V_{LS}|2^+\rangle - \langle 3^+|V_{LS}|3^+\rangle &= \frac{2a}{4}(-1-2) \\ &= -\frac{3}{2}a \approx 2{,}125\,\mathrm{MeV}. \end{aligned}$$

The constant a of the nuclear spin–orbit coupling $a{\cdot}\boldsymbol{L}{\cdot}\boldsymbol{S}$ is negative. The comparison with the experiment gives $a = -1.41$ MeV. The ratio 2:3 of the splittings is quite well reproduced by this simple model.

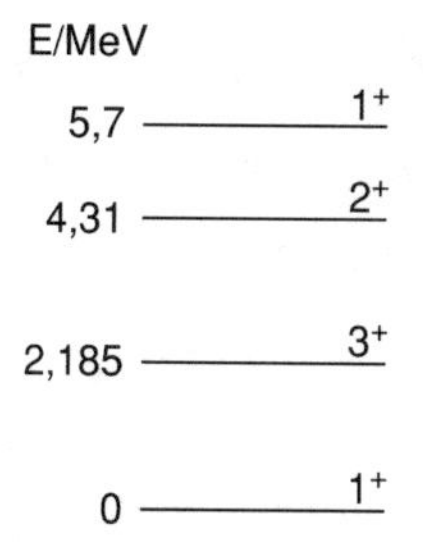

Fig. 9.7 To solution 5.4

5.5 The magnetic moment is

$$\begin{aligned}\boldsymbol{\mu} &= \mu_K \cdot (g_l \boldsymbol{l} + g_s \boldsymbol{s})/\hbar \\ &= \frac{\mu_K}{2\hbar}[(g_l+g_s)(\boldsymbol{l}+\boldsymbol{s}) + (g_l - g_s)(\boldsymbol{l}-\boldsymbol{s})] \\ \Rightarrow \boldsymbol{\mu}\cdot\boldsymbol{J} &= \frac{\mu_K}{2\hbar}\left[(g_l+g_s)\boldsymbol{J}^2 + (g_l-g_s)\left(\boldsymbol{l}^2-\boldsymbol{s}^2\right)\right] \\ \Rightarrow \langle\boldsymbol{\mu}\rangle &= \frac{\mu}{\hbar\cdot J}\langle\boldsymbol{J}\rangle \\ \Rightarrow \mu &= \hbar J\cdot\frac{\langle\mu\cdot\boldsymbol{J}\rangle}{\langle\boldsymbol{J}\cdot\boldsymbol{J}\rangle} = \frac{\langle\mu\cdot\boldsymbol{J}\rangle}{\hbar(J+1)} \\ &= \frac{\mu_K}{2(J+1)}[(g_l+g_s)J(J+1) \\ &\quad + (g_l-g_s)(l(l+1)-s(s+1))] \\ &= \mu_N\cdot J\cdot g(l,s,J),\end{aligned}$$

where μ_N is the nuclear magneton and

$$g(l,s,J) = \frac{[(g_l+g_s)J(J+1) + (g_l-g_s)(l(l+1)-s(s+1))]}{2J(J+1)}$$

is the nuclear Landé-factor.
We therefore obtain:

$$\mu = \begin{cases} \mu_K\left(l\cdot g_l + \frac{1}{2}g_s\right) & \text{for } J = l+1/2, \\ \mu_K\left((l+1)g_l - \frac{1}{2}g_s\right)\frac{2l-1}{2l+1} & \text{for } J = l-1/2, \end{cases}$$

This can be condensed as

$$\mu = \mu_K\cdot J\cdot\left[g_l \pm \frac{1}{2l+1}(g_s-g_l)\right].$$

The curves $\mu(J)$ are named "*Schmidt Curves*".
With the numerical values

$$\begin{aligned} g_l &= 1, \quad g_s = 5{,}586 \quad \text{für Protonen}, \\ g_l &= 0, \quad g_s = -3{,}826 \quad \text{für Neutronen}\end{aligned}$$

one obtains the magnetic moments for

(a) ${}^7_3$Li (1 α-paritcle, 1 proton with $j\,l = 1; j = 3/2$; 2 neutrons with $\boldsymbol{s}_{1n} = -\boldsymbol{s}_{2n} \Rightarrow s_n = 0, \boldsymbol{l}_{1n} + \boldsymbol{l}_{2n} = 0 \Rightarrow J^P = (3/2)^-$

$$\Rightarrow \mu = \mu_K\cdot\frac{3}{2}\left[1+\frac{1}{3}(5{,}586-1)\right]$$

$$\Rightarrow \mu_{theor} = 3{,}793\mu_K, \quad \mu_{exp} = 3{,}2569\mu_K.$$

(b) ${}^{13}_6$C: 3α-particles, 1 neutron with $l = 1, j = 1/2$,→

state: $J^P = (1/2)^-$

$$\begin{aligned}\mu &= \mu_N\cdot\frac{1}{2}\left[0-\frac{1}{3}(-3{,}826-0)\right] \\ &= 0{,}638\mu_K, \quad \mu_{exp} = 0{,}702\mu_K.\end{aligned}$$

(c) ${}^{17}_8$O: 4 α-particles, 1 neutron $l = 2$ $j = 5/2$,→

state: $J^P = (5/2)^+$

$$\begin{aligned}\mu &= \mu_K\cdot\frac{5}{2}\left[0+\frac{1}{5}(-3{,}826-0)\right] \\ &= -1{,}91\mu_K, \quad \mu_{exp} = -1{,}894\mu_K.\end{aligned}$$

5.6 The energy states with quantum number n are

$$E_n = \hbar\omega(n+3/2)$$

with $n = n_x + n_y + n_z = 0, 1, 2, \ldots$
if $n_x = x$ is a natural number, then the sum $n_y + n_z = n - n_x$ must be also a natural number. There are $(n - x + 1)$ possibilities. The total number of combinations of $n_x + n_y + n_z$ is

$$\begin{aligned} N &= \sum_{x=0}^{n}(n-x+1) \\ &= (n+1)^2 - \frac{n(n+1)}{2} \\ &= (n+1)\left[n+1-\frac{n}{2}\right] = (n+1)\left(\frac{n}{2}+1\right).\end{aligned}$$

5.7 If in a gg-nucleus (even number of protons and neutrons) a further proton or neutron is added, it has to occupy a higher not occupied level in the potential model of Fig. 5.24. if this level for neutrons is above the unoccupied levels for protons a β^- decay is possible where according to

$$\mathrm{n} \rightarrow \mathrm{p} + \mathrm{e}^- + \bar{\nu}_\mathrm{e}$$

a neutron is converted into a proton. If the highest occupied proton level lies above an unoccupied neutron level, positron emission according to

$$\mathrm{p} \rightarrow \mathrm{n} + \beta^+ + \nu_e$$

is energetically possible if $\Delta E > \left(m_n - m_p + m_{e^+}\right)c^2$. Note, that in these processes pairs of neutrons or protons are formed and therefore the pair energy in the Bethe-Weizsäcker formula of the nuclear binding energy leads to an increase of ΔE.

5.8 $^{14}_{7}$N is one of the few stable uu-nuclei with 7 protons and 7 neutrons. The lowest energy states are, according to the nuclear shell model (Fig. 5.34), $1s_{3/2}$; $1p_{1/2}$; which can be occupied by 6 protons and 6 neutrons. The angular momenta $\boldsymbol{j} = \boldsymbol{l} + \boldsymbol{s}$ are coupled to a total angular moment of J = 0.

The two shells $1s_{1/2}$ and $1p_{3/2}$ with $\boldsymbol{J} = \sum \boldsymbol{j}_i = \boldsymbol{0}$ are closed shells. The two residual nucleons occupy the next higher $1p_{1/2}$-state. They have the angular moment $j_i = 1/2$, $l_i = 1$, which can couple to the total moment $J = 0$ or 1. The coupling $\boldsymbol{J} = \boldsymbol{j}_1 + \boldsymbol{j}_2$ with parallel vectors $\boldsymbol{j}_i$ results in a lower energy than the coupling with antiparallel moments. Therefore the ground state of $^{14}_{7}$N is a state with $I = J = 1$.

Chapter 6

6.1 The reaction heat Q of the reaction is

$$Q = E_2 + E_3 - E_1.$$

The conservation of momentum demands (Fig. 9.8)

$$\boldsymbol{M}_1 \boldsymbol{v}_1 = \boldsymbol{M}_0 \boldsymbol{v}_0$$
$$\boldsymbol{M}_2 \boldsymbol{v}_r = \boldsymbol{m}_3 \boldsymbol{v}_3$$

where M_0 is the mass of the compound nucleus. The conservation of energy demands

$$E_2 = \frac{M_2}{2} v_2^2 = \frac{M^2}{2}\left(v_0^2 + v_r^2\right) = \frac{M_2}{2}\left[\frac{M_1^2 v_1^2}{M_0^2} + \frac{m_3^2 v_3^2}{M_2^2}\right]$$

$$E_3 = Q + E_1 - E_2 = Q + E_1 - M_2\left[\frac{E_1 M_1}{M_0^2} + \frac{m_3 E_3}{M_2^2}\right]$$

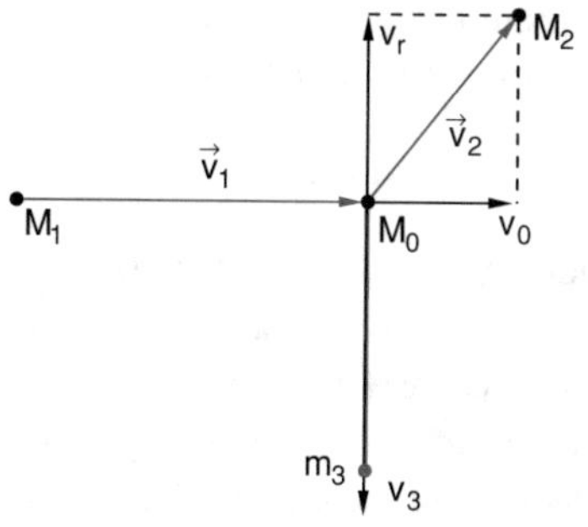

Fig. 9.8 To Solution 6.1

$$E_3\left(1 + \frac{m_3}{M_2}\right) = Q + E_1\left(1 + \frac{M_1 M_2}{M_0^2}\right)$$

$$E_3 = \frac{M_2}{M_2 + m_3}\left[Q + E_1\left(1 - \frac{M_1 M_2}{M_0^2}\right)\right]$$

$$E_2 = Q + E_1 - E_3 = \frac{m_3}{M_2 + m_3}\left[Q + E_1\left(1 - \frac{M_1 M_2^2}{M_0^2 m_3}\right)\right].$$

6.2 For the reaction

$$d + {}^3_1\text{H} \rightarrow {}^4_2\text{He} + n$$

is $M_1 = 2$; $M_0 = 5$; $M_2 = 4$;
$m_3 = 1$; $E_1 = 0.2$ MeV where $M_0 = M_d + M({}^3_1\text{H})$ is the mass of the compound nucleus.
The reaction heat Q is

$$Q = \left(M_d + M({}^3_1\text{H}) - M({}^4_2\text{He}) - M_n\right)c^2 = 17{,}5\,\text{MeV}.$$

The kinetic energy of the neutron is

$$E_3 = \frac{4}{5}\left[Q + 0{,}2\left(1 - \frac{8}{25}\right)\right]\text{MeV} = \frac{4}{5}[17{,}5 + 0{,}14]\,\text{MeV} = 14{,}1\,\text{MeV}.$$

6.3 (a) From Fig. 9.9 we see

$$\sin\varphi_{\min} = \frac{b}{r_1 + r_2}$$
$$\varphi_{\min} = \frac{\pi}{2} - \frac{\vartheta}{2}$$
$$\Rightarrow b(\vartheta) = (r_1 + r_2)\cos\vartheta/2$$
$$\Rightarrow \vartheta = 2 \cdot \arccos\left(\frac{b}{r_1 + r_2}\right).$$

(b) If the neutron collides elastically with a nucleus of mass m_2 the ratio of the kinetic energies E_{kin} before the collision and E'_{kin} after the collision is

$$\frac{E'_{\text{kin}}}{E_{\text{kin}}} = \frac{v_1'^2}{v_1^2} = \frac{A^2 + 2A\cos\vartheta + 1}{(1 + A)^2}$$

$$A = \frac{m_2}{m_1},$$

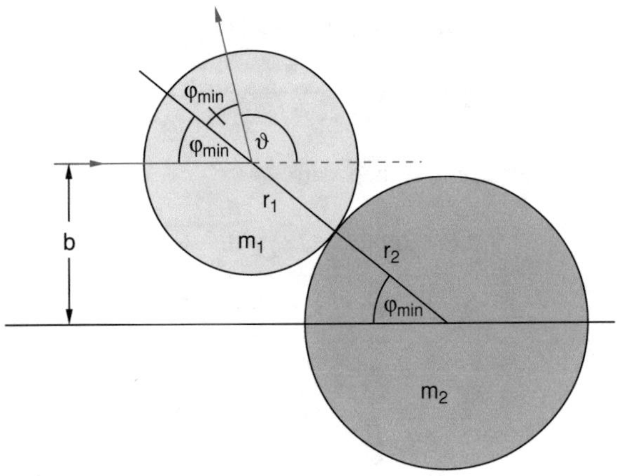

Fig. 9.9 To Solution 6.3

where ϑ is the deflection angle of the neutron in the center of mass system. The energy loss $\Delta E = E_{kin} - E'_{kin}$ is then

$$\Delta E(\vartheta) = \frac{2A(1-\cos\vartheta)}{(1+A)^2}E_{kin}. \quad (9.2)$$

With the relation

$$\cos^2(\vartheta/2) = \frac{1+\cos\vartheta}{2}$$

we obtain

$$\cos(\vartheta/2) = \frac{b(\vartheta)}{r_1+r_2}$$

$$\Delta E = \frac{4A\cdot E_{kin}}{(1+A)^2}\left[1 - \frac{b^2}{(r_1+r_2)^2}\right]. \quad (9.3)$$

(c) The average over all deflection angles ϑ follows from (9.2) with $\langle\cos\vartheta\rangle = 0$

$$\frac{4A\cdot E_{kin}}{(1+A)^2}\left[1 - \frac{1}{\pi(r_1+r_2)^4}\right]\int_0^{r_1+r_2} 2\pi b^3 \mathrm{d}b$$
$$= \frac{4A\cdot E_{kin}}{(1+A)^2}\left(1-\frac{1}{2}\right) = \frac{2A}{(1+A)^2}E_{kin}.$$

For $m_1 = m_2 \Rightarrow A = 1 \Rightarrow \langle\Delta E\rangle = 0{,}5E_{kin}$. This means that the mean energy transfer per collisions of the neutron with H-atoms amounts to ½ of its energy.
For $A = 12 \rightarrow$

$$\langle\Delta E\rangle = \frac{24E_{kin}}{13^2} = 0{,}14E_{kin}.$$

Here only 14% of the neutron energy are transferred per collision from the neutron to the nucleus.
The mean energy transfer can be also calculated from (9.3):

$$\langle\Delta E\rangle = \frac{4A\cdot E_{kin}}{(r_1+r_2)^2\cdot(1+A)^2}\cdot\frac{1}{\pi(r_1+r_2)^2} \cdot \int_0^{r_1+r_2} b^2\cdot 2\pi b\mathrm{d}b$$
$$= \frac{2A}{(1+A)^2}E_{kin}.$$

6.4 We regard the collision A + B → A + B with $v_B = 0$, i.e. B rests in the lab system. The deflection angle is named θ in the lab system and ϑ in the center of mass system. The quantities after the collision are denoted by a single quotation mark. According to Fig. 9.10 is

$$\tan\theta = \frac{a}{b} = \frac{v'_{AS}\cdot\sin\vartheta}{v'_{AS}\cdot\cos\vartheta + v'_{BS}}.$$

With $v'_{BS}/v'_{AS} = m_A/m_B = 1/A$ we obtain

$$\tan\theta = \frac{\sin\vartheta}{\cos\vartheta + m_A/m_B} = \frac{\sin\vartheta}{\cos\vartheta + 1/A},$$

$$\cos\theta = \frac{b}{\sqrt{a^2+b^2}}$$
$$= \frac{1}{\sqrt{1+\tan^2\theta}}$$
$$= \frac{\cos\vartheta + 1/A}{\sqrt{\sin^2\vartheta + (\cos\vartheta + 1/A)^2}}$$
$$= \frac{\cos\vartheta + 1/A}{\sqrt{1 + (2/A)\cos\vartheta + (1/A)^2}}, \quad (9.4)$$

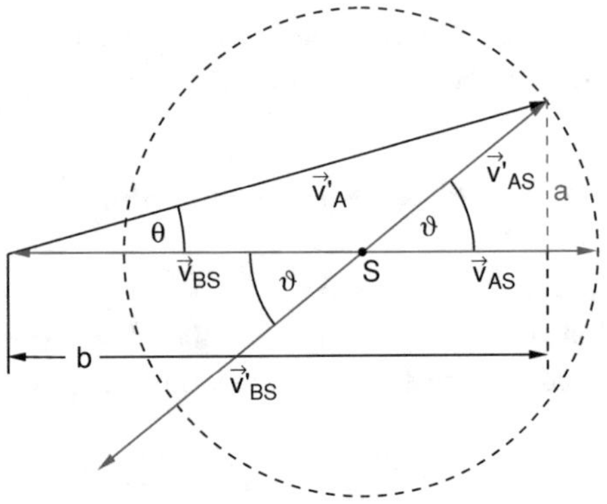

Fig. 9.10 To solution 6.4

$$\sin\theta = \frac{a}{\sqrt{a^2+b^2}} = \frac{1}{\sqrt{1+\cot^2\theta}} = \frac{\sin\vartheta}{\sqrt{1+(2/A)\cos\vartheta+(1/A)^2}}. \qquad (9.5)$$

With the solid angle

$$d\Omega = 2\pi\sin\theta d\theta d\phi$$
$$d\omega = 2\pi\sin\vartheta d\vartheta d\varphi$$
$$d\phi = d\varphi$$

one obtains

$$\frac{d\sigma}{d\omega} = \frac{d\sigma}{d\Omega}\cdot\frac{d\Omega}{d\omega} = \frac{d\sigma}{d\Omega}\cdot\frac{\sin\theta d\theta d\phi}{\sin\vartheta d\vartheta d\varphi}.$$

With $d\cos\theta = -\sin\theta d\theta$ and $d\phi = d\varphi$ this can be written as

$$\frac{d\sigma}{d\omega} = \frac{d\sigma}{d\Omega}\cdot\frac{d\cos\theta}{d\cos\vartheta}.$$

With (9.14) the differentiation with respect to cosϑ gives

$$\frac{d(\cos\theta)}{d(\cos\vartheta)} = \frac{\left[\begin{array}{l}(1+(2/A)\cos\vartheta+1/A^2)^{1/2}\\ -(\cos\vartheta+1/A)\\ \cdot\left(1+(2/A)\cos\vartheta+(1/A)^2\right)^{-1/2}\cdot(1/A)\end{array}\right]}{\left[1+(2/A)\cos\vartheta+(1/A)^2\right]} = \frac{1+(1/A)\cos\vartheta}{[1+(2/A)\cos\vartheta+1/A^2]^{3/2}}$$

$$\Rightarrow \frac{d\sigma}{d\omega} = \frac{1+(1/A)\cos\vartheta}{[1+(2/A)\cos\vartheta+1/A^2]^{3/2}}\cdot\frac{d\sigma}{d\Omega}.$$

6.5 If the nucleus is deformed from a sphere to an ellipsoid (Fig. 5.38) without changing its volume we get

$$\frac{4}{3}\pi R^3 = \frac{4}{3}\pi ab^2 \Rightarrow R^3 = ab^2.$$

With $a = R(1+\varepsilon) \Rightarrow b = R/\sqrt{1+\varepsilon}$ (ε is the deformation parameter) we can calculate the surface of the ellipsoid starting from the equation of the ellipsoid

$$\frac{x^2+y^2}{b^2} + \frac{z^2}{a^2} = 1, \quad x^2+y^2 = \varrho^2,$$

$$\varrho(z) = \left(x^2+y^2\right)^{1/2} = b\left(1-z^2/a^2\right)^{1/2}$$
$$\Rightarrow \frac{d\varrho}{dz} = -\frac{b}{a^2}\frac{z}{\left(1-z^2/a^2\right)^{1/2}}.$$

The surface S is then with

$$dS = 2\pi\varrho ds = 2\pi\varrho\sqrt{d\varrho^2+dz^2} = 2\pi\varrho\cdot\sqrt{1+(d\varrho/dz)^2}\cdot dz$$

$$S = 2\pi\int_{-a}^{+a}\varrho(z)\sqrt{1+\varrho'(z)^2}dz = 2\pi b\int_{-a}^{+a}\left(1-\frac{z^2}{a^2}\right)^{1/2}\cdot\left[1+\frac{b^2}{a^2}\frac{(z/a)^2}{1-(z/a)^2}\right]^{1/2}dz.$$

With the substitution

$$y = \frac{z}{a^2}\sqrt{a^2-b^2}$$

this changes to

$$S = 4\pi ab\frac{a}{\sqrt{a^2-b^2}}\int_0^{\sqrt{a^2-b^2}/a}\left(1-y^2\right)^{1/2}dy$$
$$= 4\pi ab\frac{a/2}{\sqrt{a^2-b^2}}\left[\frac{\sqrt{a^2-b^2}}{a}\cdot\sqrt{1-\frac{a^2-b^2}{a^2}} + \arcsin\sqrt{\frac{a^2-b^2}{a^2}}\right]$$
$$= 2\pi ab\left[\frac{b}{a}+\frac{1}{x}\cdot\arcsin x\right]$$

with $x = \frac{1}{a}\sqrt{a^2-b^2}$

$$\Rightarrow b^2 = \left(1-x^2\right)\cdot a^2.$$

Inserting here $b = R\cdot(1+\varepsilon)^{-1/2}$ and $a = (1+\varepsilon)$ one obtains

$$\frac{R^2}{1+\varepsilon} = \left(1-x^2\right)\cdot R^2(1+\varepsilon)^2$$
$$\Rightarrow x^2 = 3\varepsilon\cdot\frac{1+\varepsilon+\frac{1}{3}\varepsilon^2}{(1+\varepsilon)^3}.$$

$$S = 2\pi R^2(1+\varepsilon)^{1/2}\left[(1+\varepsilon)^{-3/2}+\frac{1}{x}\arcsin x\right].$$

The Taylor expansion of arcsinx is

$$\arcsin x = x + \frac{1}{6}x^3 + \frac{3}{40}x^5 + \frac{5}{7\cdot 16}x^7 + \cdots$$

$$\begin{aligned}\Rightarrow S &= 2\pi R^2\left[\frac{1}{1+\varepsilon} + (1+\varepsilon)^{1/2} + \frac{\varepsilon}{2}\frac{1+\varepsilon+\frac{1}{3}\varepsilon^2}{(1+\varepsilon)^{5/2}}\right.\\
&\left.+\frac{27}{40}\frac{\varepsilon^2(1+\varepsilon+\frac{1}{3}\varepsilon^2)^2}{(1+\varepsilon)^{11/2}} + \frac{5\cdot 27}{7\cdot 16}\varepsilon^3 + \cdots\right]\\
&= 2\pi R^2\big[1-\varepsilon+\varepsilon^2-\varepsilon^3+\cdots\\
&+1+\frac{\varepsilon}{2}-\frac{\varepsilon^2}{8}+\frac{\varepsilon^3}{16}-\cdots\\
&+\frac{\varepsilon}{2}\left(1+\varepsilon+\frac{1}{3}\varepsilon^2\right)\cdot\left(1-\frac{5}{8}\varepsilon+\frac{35}{8}\varepsilon^2\right)\\
&+\frac{27}{40}\varepsilon^2(1+2\varepsilon+\cdots)\left(1-\frac{11}{2}\varepsilon+\cdots\right)\\
&\left.+\frac{5\cdot 27}{7\cdot 16}\varepsilon^3-\cdots\right]\\
&= 4\pi R^2\left(1+\frac{2}{5}\varepsilon^2-\frac{52}{105}\varepsilon^3+\cdots\right).\end{aligned}$$

The surface energy is proportional to R^2. For the Coulomb energy, which is proportional to $1/R$ we obtain

$$E_C \propto \frac{1}{R}\left[1-\frac{1}{5}\varepsilon^2+\frac{26}{205}\varepsilon^3-\cdots\right].$$

6.6 The mass number of the fragments is A_1 and A_2

$$A_1 + A_2 = 235 - 3 = 232.$$

The fragment masses have the ratio $A_1/A_2 = 1.25{:}\ 1$

$$\begin{aligned}\Rightarrow A_1 &= 129 \quad (\text{Xe}),\\ A_2 &= 103 \quad (\text{Rh}).\end{aligned}$$

The mass excesses of these nuclei can be found in tables. It is

$$\begin{aligned}\Delta M(A_1) &= -95{,}216\cdot 10^{-3}\text{u},\\ \Delta M(A_2) &= -94{,}49\cdot 10^{-3}\text{u},\\ \Delta M(^{235}\text{U}) &= +43{,}94\cdot 10^{-3}\text{u}\end{aligned}$$

$$\begin{aligned}&\Rightarrow \Delta M(^{235}\text{U}) - (\Delta M(A_1)+\Delta M(A_2))\\ &= 233{,}65\cdot 10^{-3}\text{u} = \Delta\end{aligned}$$

$$\begin{aligned}\Rightarrow \Delta\cdot c^2 &= 233{,}65\cdot 0{,}9315\,\text{MeV}\\ &= 217{,}64\,\text{MeV}.\end{aligned}$$

One has to subtract

$$\begin{aligned}E_{\text{kin}}(n) &= 3\cdot 2\,\text{MeV} = 6\,\text{MeV}\\ E_\gamma &= 4{,}6\,\text{MeV}.\end{aligned}$$

For the kinetic energy of the fragments remains 207.04 MeV.
This energy is split according to the ratio of the fragment masses

$$\begin{aligned}E_{\text{kin}}(A_1) &= \frac{103}{129}E_{\text{kin}}(A_2)\\ \Rightarrow E_{\text{kin}}(A_2) &= \frac{A_2}{A_1+A_2}\cdot 207\,\text{MeV}\\ &= 91{,}9\,\text{MeV},\\ \Rightarrow E_{\text{kin}}(A_1) &= \frac{A_1}{A_1+A_2}\cdot 207\,\text{MeV}\\ &= 115{,}1\,\text{MeV}.\end{aligned}$$

For a mass ratio $m_1/m_2 = 1.4$ the energy is divided onto the fragments as

$$E_{\text{kin}}(A_1) = \frac{1\cdot 4}{2\cdot 4}\cdot 207\,\text{MeV} = 120{,}75\,\text{MeV},$$

$$E_{\text{kin}}(A_2) = \frac{1}{2\cdot 4}\cdot 207\,\text{MeV} = 86{,}25\,\text{MeV}.$$

6.7 At the fission the repulsive Coulomb energy is transferred into kinetic energy

$$\begin{aligned}\Rightarrow E_{\text{pot}} &= \frac{Z_1 Z_2 e^2}{4\pi\varepsilon_0 r_S} = E_{\text{kin}}\\ \Rightarrow r_S &= \frac{Z_1 Z_2 e^2}{4\pi\varepsilon_0\cdot E_{\text{kin}}}\\ &= \frac{35\cdot 57\cdot 1{,}6^2\cdot 10^{-38}}{4\pi\cdot 8{,}85\cdot 10^{-12}\cdot 200\cdot 1{,}6\cdot 10^{-13}}\,\text{m}\\ &= 1{,}44\cdot 10^{-14}\,\text{m} = 14{,}4\,\text{fm}.\end{aligned}$$

6.8 The maximum appears fort $E_{\text{kin}}({}^{18}_{8}\text{O}) = 95\,\text{MeV}$. The ^{18}O. nuclei are shot onto uranium nuclei at rest. Since $E_{\text{kin}} << m_0\,c^2$ we can use a non-relativistic calculation.: in the center of mass system is

$$\Rightarrow v_S = \frac{m_1 v_1 + m_2 \cdot 0}{m_1 + m_2} = \frac{m_1}{m_1 + m_2}\sqrt{2E_{kin}/m_1}$$
$$= \frac{18}{18+238} \cdot \sqrt{\frac{190 \cdot 1{,}6 \cdot 10^{-13}}{18 \cdot 1{,}67 \cdot 10^{-27}}} \frac{m}{s}$$
$$= 3{,}18 \cdot 10^7 \text{ m/s}.$$

$$E_S = \frac{m_1 + m_2}{2} v_S^2 = \frac{m_1}{m_1 + m_2} E_{kin}^{(L)}$$
$$= \frac{18}{18+238} E_{kin}^{(L)}$$
$$= 0{,}07 E_{kin}^{(L)} = 6{,}68 \text{ MeV}.$$

$$\Rightarrow \Delta E = (95 - 6{,}68) \text{ MeV} = 88{,}32 \text{ MeV}.$$

6.9 An α-particle (mass m_1, velocity v_1) collides with the kinetic energy $E_{kin} = m_1 v_1^2/2 = 17$ MeV onto a tritium nucleus ${}_1{}^3$H (m_2, $v_2 = 0$) at rest. The center of mass velocity is

$$v_S = \frac{m_1 v_1 + m_2 v_2}{m_1 + m_2} = \frac{m_1 v_1}{m_1 + m_2}.$$

The center of mass velocity v_s is not changed by the collision and therefore also not the translational energy

$$E_t = \frac{m_1 + m_2}{2} v_S^2$$
$$= \frac{1}{2} \frac{m_1^2 v_1^2}{m_1 + m_2}$$
$$= \frac{m_1}{m_1 + m_2} E_{kin}(\alpha).$$

Therefore only the energy

$$E_r = E_{kin} - E_t = \frac{m_2}{m_1 + m_2} E_{kin}(\alpha)$$
$$= \frac{3}{7} E_{kin}(\alpha) = 7{,}286 \text{ MeV}.$$

is available for reactions.
The energy balance of the reaction

$${}_2^4\text{He} + {}_1^3\text{T} \rightarrow {}_3^7\text{Li} \rightarrow \text{Reaction products.}$$

is equal to the difference

$$\Delta E = E_e - E_a$$

of the energies in the entrance and the exit channel. it can be calculated from the mass defect

$$\Delta m = M({}_Z^A X) - A \cdot \text{u}$$

(1 AMU = 1.6605 · 10^{-27} kg)
it is therefore sufficient energy available for the production of the nuclide ${}_3^7$Li of the different nuclides (see Table 9.1).

Input channel:

^{4}He + T + E_{kin}	
${}_2^4$He :	2,425 MeV
T :	14,950 MeV
E_r :	7,286 MeV
E_e :	24,661 MeV

It is therefore sufficient energy available for the production of the nuclide ${}_3^7$Li.

Output channel:

^{4}He + D + n	
^{4}He :	2,425 MeV
D :	13,136 MeV
n :	8,071 MeV
	23,632 MeV

⇒ Channel is open

^{5}He + H + n	
^{5}He :	11,390 MeV
H :	7,289 MeV
n :	8,071 MeV
	26,750 MeV

⇒ Channel is closed

^{4}He + H + 2n	
^{4}He :	2,425 MeV
H :	7,289 MeV
2n :	16,143 MeV
	25,857 MeV

⇒ Channel is closed

^{6}He + H	
^{6}He :	17,594 MeV
H :	7,289 MeV
	24,883 MeV

⇒ Just not anymore accessible

^{5}He + D	
^{5}He :	11,390 MeV
D:	13,136 MeV
	24,526 MeV

⇒ Just reachable

^{6}Li + n	
^{6}Li :	14,086 MeV
n :	8,071 MeV
	22,157 MeV

⇒ Channel is open

Table 9.1 .

Nuclide	δm(MeV/c²)
n	8,0713
H	7,2889
D	13,136
T	14,950
^{4}He	2,425
^{5}He	11,390
^{6}He	17,594
^{6}Li	14,086
^{7}Li	14,908

6.10. (a) In problem 6.4 it was shown, that the scattering angle θ in the lab system for the elastic collision of a particle A with a particle B at rest is related to the scattering angle ϑ in the center-of-mass system by

$$\tan\theta = \frac{\sin\vartheta}{\cos\vartheta + 1/A} \quad \text{with} \quad A = \upsilon_{bS}/\upsilon_S.$$

This can be also seen as follows: From Fig. 9.10 (where we replace A by B) we infer:

$$\upsilon_b \cdot \sin\theta = \upsilon_{bS} \cdot \sin\vartheta$$

where v_{Bs} = velocity of B in the CM-system,

$$\upsilon_b \cdot \cos\theta = \upsilon_{bS} \cdot \cos\vartheta + \upsilon_S$$

(v_S = velocity of CM in the lab system).

$$\Rightarrow \tan\theta = \frac{\sin\vartheta}{\cos\vartheta + \upsilon_S/\upsilon_{bS}} = \frac{\sin\vartheta}{\cos\vartheta + 1/A}.$$

For $v_{BS}/v_S < 1$ the denominator cannot become zero. The maximum value $\sin\theta_{max}$ is obtained when $d(\tan\theta)/d\vartheta = 0$. We get with $v_{BS}/v_S = A$

$$\frac{d(\tan\theta)}{d\vartheta} = \frac{\cos\vartheta(\cos\vartheta + 1/A) - \sin\vartheta(-\sin\vartheta)}{(\cos\vartheta + 1/A)^2} = \frac{1 + \cos\vartheta/A}{(\cos\vartheta + 1/A)^2} = 0 \quad \text{for} \quad \cos\vartheta_0 = -A$$

which becomes zero for $\cos\vartheta_0 = -A$.

$$\Rightarrow \tan\theta_{max} = \frac{\sqrt{1-A^2}}{-A+1/A} = \frac{A}{\sqrt{1-A^2}} = \frac{\sin\theta_{max}}{\cos\theta_{max}}$$

$$\Rightarrow \sin\theta_{max} = A < 1.$$

(b) With $M = m_a + m_A$ we get

$$A^2 = \frac{v_{bS}^2}{v_S^2} = \frac{2E_{kin}^S(b)}{m_b} \cdot \frac{m_a + m_A}{2E_{kin}^S(M)} = \frac{(m_a + m_A)^2}{m_a \cdot m_b} \cdot \frac{E_{kin}^S(b)}{E_{kin}(a)},$$

because

$$E_{kin}^S(M) = \frac{m_a}{m_a + m_A} E_{kin}(a)$$

Since $E_{kin}(A) = 0$ the energy balance in the lab system and in the CM system is

$$Q - E_X = E_{kin}(b) + E_{kin}(B) - E_{kin}(a) = E_{kin}^S(b) + E_{kin}^S(B) - \left(E_{kin}^S(a) + E_{kin}^S(A)\right).$$

With

$$E_{kin}^S(B) = \frac{m_b}{m_B} E_{kin}^S(b)$$

$$E_{kin}^S(A) = \frac{m_a}{m_A} E_{kin}^S(a)$$

$$\upsilon_a^S = \frac{m_A}{m_a + m_A} \upsilon_a$$

$$\Rightarrow E_{kin}^S(a) = \frac{m_A^2}{m_a + m_A} E_{kin}(a)$$

$$\Rightarrow Q - E_X = \frac{m_b + m_B}{m_B} E_{kin}^S(b) - \frac{m_A}{m_a + m_A} E_{kin}(a)$$

$$\Rightarrow E_{kin}^S(b) = \frac{m_A m_B}{(m_a + m_A)(m_b + m_B)} E_{kin}(a) + \frac{m_B}{m_b + m_B}(Q - E_X)$$

$$\Rightarrow A^2 = \frac{m_a + m_A}{m_b + m_B}\left[\frac{m_A m_B}{m_a m_b} + \frac{(m_a + m_A) m_B}{m_a m_b} \frac{Q - E_X}{E_{kin}(a)}\right].$$

For $Q - E_x < 0$ and $A^2 > 0$ it is

$$E_{kin}(a) > \frac{m_a + m_A}{m_A} |Q - E_X|.$$

(c) elastic Scattering

$$Q = E_X = 0, \quad m_a = m_b, \quad m_A = m_B$$

$$\Rightarrow A^2 = \frac{m_A m_B}{m_a m_B} = \left(\frac{m_A}{m_a}\right)^2 \Rightarrow A = \frac{m_A}{m_a}.$$

For the elastic scattering of protons or neutrons (m_a = 1 AMU) on nuclei with mass m_A the quantity $A = \upsilon_a'/\upsilon_S$ is equal to the mass number $A = m_A/m_a$ of the target nucleus.

Chapter 7

7.1 The possible orientations of a particle with spin I are given by all possible values $-I \leq m_I \leq +I$ of the projection quantum number m_I. For the reaction products there are $(2I_\pi + 1)$ possible orientations of the pion and $(2I_D + 1)$ orientations of the deuteron. The statistical weight of the reaction products is then $g = (2I_\pi + 1) \cdot (2I_d + 1)$.

The number of possible states in the momentum space per unit volume is

$$n_s(p) = g \cdot \frac{4}{3}\pi \frac{p^3}{h^3}.$$

The density of states is then

$$\varrho = \frac{dn_s}{dE} = \frac{dn_s}{dp} \cdot \frac{dp}{dE} = \frac{4\pi p^2}{h^3} \cdot g \cdot \frac{d}{dE}\sqrt{\frac{1}{c^2}\left(E^2 - (mc^2)^2\right)} = \frac{4\pi p^2 \cdot g}{h^3} \cdot \frac{E}{c^2 \cdot p} = \frac{4\pi p^2 \cdot g}{h^3 \cdot v}.$$

The number of states per unit energy interval is therefore proportional to the square of the particle momentum, but inversely proportional to its velocity. The reaction cross section σ is proportional to the number of final states accessible to the reaction products.
With the statistical weight $g = (2I\pi{+}1)\cdot(2I_d{+}1)$ we et for a given kinetic energy of the protons the cross section

$$\sigma = C \cdot \frac{(2I_\pi + 1)(2I_d + 1)}{v_{pp} \cdot v_{\pi d}} p_\pi^2.$$

For a given relative velocity v_{pp} and a given momentum p_π is the momentum p_d determined. Therefore the factor p_d^2 does not appear in the formal for σ.

7.2 For the decay

$$\mu^- \to e^- + \nu_\mu + \overline{\nu}_e$$

the conservation of momentum is valid:

$$\boldsymbol{p}_e + \boldsymbol{p}_{\nu_\mu} + \boldsymbol{p}_{\overline{\nu}_e} = 0.$$

The electron gets the maximum momentum, if the two neutrinos fly parallel to each other into the opposite direction as the electron. In this case we get

$$\Rightarrow |\boldsymbol{p}_e| = \left|\boldsymbol{p}_{\nu_\mu}\right| + \left|\boldsymbol{p}_{\overline{\nu}_e}\right|$$

Energy conservation demands for $m_\nu = 0$

$$\sum E_{kin} = (m_\mu - m_e)c^2$$
$$\Rightarrow \sqrt{(m_e c^2)^2 + (cp_e)^2} - m_e c^2 + cp_{\nu_\mu} + cp_{\nu_e} = (m_\mu - m_e)c^2$$
$$\Rightarrow \sqrt{(m_e c^2)^2 + (cp_e)^2} + cp_e = m_\mu c^2$$
$$\Rightarrow p_e^{max} = \frac{m_\mu^2 - m_e^2}{2m_\mu} c = \frac{1}{2} m_\mu c \left(1 - \frac{m_e^2}{m_\mu^2}\right) \approx \frac{1}{2} m_\mu c = 2{,}5 \cdot 10^{-3}\,\text{kg m/s}.$$

The electron momentum becomes minimum, if the three momenta forms an equilateral triangle:

$$\Rightarrow p_e^2 = p_{\nu_\mu}^2 = p_{\nu_e}^2$$
$$\Rightarrow \sqrt{(m_e c^2)^2 + (cp_e)^2} + 2cp_e = m_\mu c^2$$
$$\Rightarrow p_e = \frac{2}{3} m_\mu c \pm \sqrt{\frac{4}{9} m_\mu^2 c^2 + \frac{\left(m_e^2 - m_\mu^2\right)}{9} c^2} = \frac{2}{3} m_\mu c \pm \frac{1}{3}\sqrt{m_e^2 c^2 + 3 m_\mu^2 c^2}.$$

Since $m_e << m_\mu$ it follows

$$p_e^{min} = \frac{2 - \sqrt{3}}{3} m_\mu c = 0{,}45 \cdot 10^{-3}\,\text{kg m/s}.$$

7.3 (a) The cross section per nucleon is approximately

$$\sigma \approx \alpha_w^2 \cdot \pi \cdot \lambda^2,$$

where $\lambda = h/p = h \cdot c/E$ is the de-Broglie wavelength of the neutrino. Inserting the numerical values gives:

$$\sigma = 10^{-12} \cdot \pi \cdot (h \cdot c/E)^2 \approx 5 \cdot 10^{-42}\,\text{m}^2.$$

(b) The absorption coefficient is

$$\alpha = \sigma \cdot N_n,$$

where N_n is the number of nucleons per m³. It can be expressed by the mass density ρ and the mass of the proton as $N_n = \rho/m_p$.

With $\rho = 4.5 \cdot 10^3$ kg/m^3 = mean mass density and m_p = $1.67 \cdot 10^{-27}$ kg we get

$$\Rightarrow N_n = 2{,}7 \cdot 10^{30}\,\mathrm{m}^{-3}$$
$$\Rightarrow \alpha = 5 \cdot 10^{-42} \cdot 2{,}7 \cdot 10^{30}\,\mathrm{m}^{-1} = 1{,}35 \cdot 10^{-11}\,\mathrm{m}^{-1}.$$

The mean free path length $\bar{s}$ is then

$$\bar{s} = \frac{1}{\alpha} = 4{,}4 \cdot 10^{10}\,\mathrm{m}.$$

With the earth diameter D $\approx 1.2 \cdot 10^7$ m the fraction

$$\delta = \frac{1{,}2 \cdot 10^7}{7{,}4 \cdot 10^{10}} \approx 1{,}6 \cdot 10^{-4}$$

of all neutrinos traversing the earth is absorbed.

7.4 The quarks inside a nucleon have the orbital angular momentum zero. The intrinsic parities of u-and d-quark must be both positive, because otherwise proton and neutron could not have positive parities.

Proton and neutron have in the deuteron the orbital angular momentum zero (see Sect. 5.1). Therefore the parity of the deuteron is

$$P_d = P_p \cdot P_n(-1)^0 = 1 \cdot 1 \cdot 1 = +1.$$

7.5 The transition $\psi' \rightarrow \psi$ correspond in the charmonium level scheme the transition $2^3s_1 \rightarrow 1^3s_1$ where the quantum number 1 of the orbital angular momentum does not change. Therefore also the parity does not change. No electro-magnetic dipole transitions are allowed.

7.6 According to (5.30) the range r of the interaction between two particles is

$$r \leq \frac{\hbar}{m_A \cdot c},$$

where m_A is the mass of the exchange particle. The mass of the π-meson is

$$m_\pi = 139{,}57\,\mathrm{MeV}/c^2 = 2{,}48 \cdot 10^{-28}\,\mathrm{kg}$$

$$\Rightarrow r \leq \frac{1{,}05 \cdot 10^{-34}}{2{,}48 \cdot 10^{-28} \cdot 3 \cdot 10^8}\,\mathrm{m} = 1{,}4 \cdot 10^{-15}\,\mathrm{m}.$$

$$m_W \approx 80\,\mathrm{GeV}/c^2 \Rightarrow m_W = 1{,}4^2 \cdot 10^{-25}\,\mathrm{kg}$$
$$\Rightarrow r \leq 2{,}4 \cdot 10^{-18}\,\mathrm{m}.$$

7.7 (a) The total energy is

$$E = 2mc^2 = \frac{2m_0c^2}{\sqrt{1 - v^2/c^2}} = Mc^2.$$

With $v/c = 3/5 \Rightarrow$

$$E = \frac{5}{2}m_0c^2 \Rightarrow M = \frac{5}{2}m_0.$$

M is therefore slightly larger than $2m_0$!, because

$$E_{kin} = 2(m - m_0)c^2 = \frac{1}{2}m_0c^2.$$

(b) For the decay the mass of the mass M must be $M > 2\,m_0$. If M was at rest before the decay into two equal masses the total momentum of the decay products must be zero, and for their velocities is:

$$\Rightarrow \frac{2m_0c^2}{\sqrt{1 - v^2/c^2}} = M_0c^2$$
$$\Rightarrow v = \frac{c}{M_0}\sqrt{M_0^2 - 4m_0^2} \quad \text{with} \quad v = |v_1| = |v_2|.$$

If M had before the decay the velocity v_0 in x-direction, energy- and momentum conservation demand: $Mc^2 = 2mc^2$

$$\Rightarrow \frac{M_0c^2}{\sqrt{1 - (v_0/c)^2}} = \frac{2m_0c^2}{\sqrt{1 - v^2/c^2}}; \text{ Energy rate} \qquad (9.6)$$

$$\Rightarrow \frac{M_0v_{0x}}{\sqrt{1 - (v_0/c)^2}} = \frac{2m_0v_x}{\sqrt{1 - (v/c)^2}}; \text{Momentum law} \qquad (9.7)$$

with $v_x^2 + v_y^2 = v^2$. Energy conservation gives the condition

$$M_0\sqrt{1 - v^2/c^2} = 2m_0\sqrt{1 - v_0^2/c^2} \qquad (9.8)$$

$$\Rightarrow v^2 = \left[c^2 - \frac{4m_0^2}{M_0^2}\left(c^2 - v_0^2\right)\right]. \qquad (9.9)$$

Fot $m_0 = \frac{1}{2}M_0 \Rightarrow v = v_0$. For $m_0 < \frac{1}{2}M_0 \Rightarrow v^2 > v_0^2$. Momentum conservation gives

$$\left(1 - v^2/c^2\right) = \frac{4m_0^2 v_x^2}{M_0^2 v_0^2}\left(1 - v_0^2/c^2\right). \qquad (9.10)$$

With (9.8) follows: $v_x^2 = v_0^2$. With $v^2 = v_x^2 + v_y^2 \Rightarrow v^2 = v_0^2 + v_y^2$. With (9.9) then follows:

$$v_y^2 = \left(c^2 - v_0^2\right)\left[1 - \frac{4m_0^2}{M_0^2}\right].$$

The transversal energy of the decay products is

$$E_\perp = 2 \cdot \frac{1}{2} m_0 v_y^2 = m_0\left(c^2 - v_0^2\right)\left[1 - \frac{4m_0^2}{M_0^2}\right].$$

Only for $m_O < ½M_O$ is $E_\perp > 0$.

7.8 Since 14 MeV << $m_O c^2$ one can use the non-relativistic calculation:

$$E_{\mathrm{kin}} = \frac{1}{2} m v^2 \Rightarrow v = \sqrt{2E_{\mathrm{kin}}/m}$$
$$v = \sqrt{2 \cdot 14 \cdot 1{,}6 \cdot 10^{-13}/1{,}67 \cdot 10^{-27}}\,\frac{\mathrm{m}}{\mathrm{s}}$$
$$= 5{,}18 \cdot 10^7\,\mathrm{m/s}.$$

The time of flight = $T = L/v \Rightarrow E_{\mathrm{kin}} = \frac{1}{2} m L^2/T^2$

$$\frac{dE_{\mathrm{kin}}}{dT} = \frac{m \cdot L^2}{T^3} \Rightarrow \Delta E = -\frac{m \cdot L^2}{T^3}\Delta T$$
$$\Rightarrow \Delta E = \frac{m \cdot v^3}{L}\Delta T$$
$$\Rightarrow L = \frac{m \cdot v^3 \cdot \Delta T}{\Delta E}$$
$$= \frac{1{,}6 \cdot 10^{-27} \cdot 5{,}18^3 \cdot 10^{21} \cdot 10^{-9}}{0{,}5 \cdot 1{,}6 \cdot 10^{-13}}\,\mathrm{m}$$
$$= 2{,}9\,\mathrm{m}.$$

Chapter 8

8.1 $H = D \cdot Q = 0.2$ mSv(75 kg. Fot X-rays is $Q = 1$

$\Rightarrow$ the energy dose is

$$D = 0{,}2 \cdot 10^{-3}\,\mathrm{J}/75\,\mathrm{kg} \Rightarrow D = 2{,}7 \cdot 10^{-6}\,\mathrm{J/kg}.$$

The energy of an X-ray quantum is with $h \cdot v$ = 50 keV

$$h \cdot v = 50 \cdot 1{,}6 \cdot 10^{-16}\,\mathrm{J} = 8 \cdot 10^{-15}\,\mathrm{J}$$

$\Rightarrow$ the number of X-ray quanta , absorbed in the body is

$$n_a = \frac{Dm}{hv} = \frac{2{,}7 \cdot 10^{-6} \cdot 75}{8 \cdot 10^{-15}} = 2{,}5 \cdot 10^{10}$$
$$\phi = \frac{d^2N}{dA \cdot dt} = \frac{2{,}5 \cdot 10^{10}}{0{,}1} \cdot 2 \cdot \frac{1}{\mathrm{m^2\,s}}$$
$$= 5 \cdot 10^{11}\,\mathrm{Quantum/m^2\,s}.$$

8.2. The specific weight of aluminum is $\rho = 2.7$ g/cm³. A layer with 20 µm thickness therefore has a mass surface density of

$$\sigma = \frac{2{,}7 \cdot 2 \cdot 10^{-3}}{1}\frac{\mathrm{g}}{\mathrm{cm^2}} = 5{,}4 \cdot 10^{-3}\,\mathrm{g/cm^2}.$$

the energy loss of the α-particles decays (until shortly before the complete stop) according to Fig. 4.54 proportional to the square of the transversal distance x.

$$-\frac{dE}{dx} \approx a_1 x^2 \propto a_2 \sigma^2$$
$$\Rightarrow E = E_0 - \frac{1}{3} a_1 x^3 \propto E_0 - \frac{1}{3} a_2 \sigma^3.$$

For $\sigma = \sigma_1 = 8 \cdot 10^{-3}$ g/cm² is $E = 0$

$$\Rightarrow a_2 = 3E_0/\sigma_1^3 = 3 \cdot 6\,\mathrm{MeV}/8^3 \cdot 10^{-9}\mathrm{g}^{-3}/\mathrm{cm}^6$$
$$= 3{,}5 \cdot 10^7\,\frac{\mathrm{cm^6\,MeV}}{\mathrm{g^3}}.$$

For $\sigma = 5.4 \cdot 10^{-3}$ g/cm² if follows:

$$E = E_0 - \frac{1}{3} \cdot 3{,}5 \cdot 10^7\left(5{,}4 \cdot 10^{-3}\right)^3\,\mathrm{MeV}$$
$$= (6 - 1{,}83)\,\mathrm{MeV} = 4{,}17\,\mathrm{MeV}.$$

8.3 The number of electrons traversing an absorbing or scattering layer with thicknes x is

$$N(x) = N(0) \cdot e^{-x/\langle R\rangle},$$

where $\langle R\rangle$ is the mean range. With $\langle R\rangle$ = 1.5 mm we get

$$N(x = 1\,\mathrm{mm}) = N_0 \cdot e^{-2/3} \approx 0{,}5N_0,$$
$$N(x = 1{,}5\,\mathrm{mm}) = N_0/e \approx 0{,}37N_0,$$
$$N(x = 2\,\mathrm{mm}) = 0{,}26N_0.$$

8.4 $$-\frac{dN(t)}{dt} = A(t) = +\lambda N(t) \rightarrow N(t) = N_0 \cdot e^{-\lambda t}$$
$$\frac{1}{2}N_0 = N_0 e^{-\lambda T} \rightarrow \lambda = \frac{\ln 2}{T} \equiv \tau^{-1}$$

The activity is

$$-\frac{dN(t)}{dt} \equiv A(t)\lambda N(t) = \lambda N_0 e^{-\lambda t} = \frac{\ln 2}{T} N_0 e^{-\lambda t} = A_0 e^{-\lambda t}$$

with

(a)
$$A_0 = \lambda N_0 = \frac{\ln 2}{T} N_0 = \frac{N_0}{\tau}.$$

$$A_0 = \frac{\ln 2}{T} N_0 = 0{,}255 \frac{\text{Bq}}{\text{g}} \cdot 2\ \text{g} = 0{,}510\ \text{Bq}$$

$$1\,\text{g}\,\text{C} \triangleq 6{,}023 \cdot 10^{22}\,\text{Atoms}$$

$$A(t) = A_0 e^{-\lambda t} = A_0 e^{-(\ln 2)/T \cdot t} = 0{,}404\ \text{Bq}$$

$$\Rightarrow t = \frac{1}{\lambda} \ln \frac{A_0}{A(t)} = \frac{\ln A_0/A(t)}{\ln 2} \cdot T = \frac{\ln 0{,}510/0{,}404}{\ln 2} \cdot (5730 \pm 30)\ \text{a} = (1926{,}1 \pm 10)\ \text{a}.$$

The time of die back of the organic material is: (1952.5 − 1926,1 ± 10)A.D. = (26,4 ± 10)A.D. ≈ 13 to 37 A.D.
The monastery at Qumram was destroyed by the Romans in 68 in the war between Romans and Jews (66–77 after Christ)

(b) The initial number of ^{14}C atoms in the sample of 2 g was

$$N_0 = \frac{A_0 \cdot T}{\ln 2} = \frac{0{,}510}{\ln 2}(5730 \pm 40) \cdot 365 \cdot 24 \cdot 60^2 = 1{,}33(1) \cdot 10^{11};$$

At a later time was

$$N(t) = \frac{A(t)}{A_0} N_0 = \frac{0{,}404}{0{,}510} \cdot 1{,}33(1) \cdot 10^{11} = 1{,}053(7) \cdot 10^{11}.$$

(c)
$$\frac{N_{C_{14}}}{N_{C_{12}}} = \frac{1{,}33 \cdot 10^{11}}{12{,}026 \cdot 10^{22}} \approx 10^{-12}.$$

The traffic exhaust contains CO_2 from very old mineral oil. This causes a lower activity of plants close to the highway than that farther away.

8.5
$$N_{235}(t) = N_{235}(0) e^{-\lambda_{235} t}$$

with

$$\lambda_{235} = \frac{\ln 2}{T_{235}} \quad \text{and} \quad T_{235} = 7{,}038 \cdot 10^8\text{a}$$

$$N_{238}(t) = N_{238}(0) e^{-\lambda_{238} t}$$

and

$$\lambda_{238} = \frac{\ln 2}{T_{238}} \quad \text{and} \quad T_{238} = 4{,}468 \cdot 10^9\text{a}.$$

$$N_{207}(t) = N_{235}(0) - N_{235}(t) = N_{235}(0)\left(1 - e^{-\lambda_{235} t}\right)$$

Actinium chain

$$N_{206}(t) = N_{238}(0)\left(1 - e^{-\lambda_{238} t}\right)$$

uranium-radium chain

$$\frac{N_{235}(t)}{N_{235}(t) + N_{238}(t)} = 0{,}72\%,$$

$$\frac{N_{238}(t)}{N_{235}(t) + N_{238}(t)} = 99{,}28\%$$

For $t = 6 \cdot 10^8$ years

$$\Rightarrow \frac{N_{235}(t)}{N_{238}(t)} = \frac{N_{235}(0)}{N_{238}(0)} \cdot e^{-\ln 2(t/T_{235} - t/T_{238})} = \frac{0{,}72}{99{,}28} = 0{,}0072$$

$$\Rightarrow \frac{N_{235}(0)}{N_{238}(0)} = \frac{N_{235}(t)}{N_{238}(t)} \cdot 2^{(t/T_{235} - t/T_{238})} = \frac{0{,}72}{99{,}28} \cdot 2^{(6/7{,}038 - 6/44{,}68)} = 0{,}011931 = \frac{1}{83{,}815}.$$

$$\frac{207 \cdot N_{207}(t) + 206 \cdot N_{206}(t)}{235 \cdot N_{235}(t) + 238 \cdot N_{238}(t)} = \frac{\left[\begin{array}{c} 207 \cdot (N_{235}(0)/N_{238}(0))\left(1 - e^{-\lambda_{235} t}\right) \\ +206\left(1 - e^{-\lambda_{238} t}\right) \end{array}\right]}{\left[\begin{array}{c} 235 \cdot (N_{235}(0)/N_{238}(0)) \cdot e^{-\lambda_{235} t} \\ +238 \cdot e^{-\lambda_{238} t} \end{array}\right]}$$

$$= \frac{\left[\begin{array}{c} 207 \cdot 0{,}011931\left(1 - 0{,}5^{6/7{,}038}\right) \\ +206\left(1 - 0{,}5^{6/44{,}68}\right) \end{array}\right]}{\left[\begin{array}{c} 235 \cdot 0{,}011931 \cdot 0{,}5^{6/7{,}038} \\ +238 \cdot 0{,}5^{6/44{,}68} \end{array}\right]} = 1/11{,}251 = 8{,}89\%.$$

(b) Abundance ratio ^{207}Pb: ^{206}Pb presently:

$$\frac{N_{207}(t)}{N_{206}(t)} = \frac{N_{235}(0)\left(1 - e^{-\lambda_{235}t}\right)}{N_{238}(0)\left(1 - e^{-\lambda_{238}t}\right)}$$

$$= 0{,}011931 \cdot \frac{1 - 0{,}5^{6/7{,}038}}{1 - 0{,}5^{6/44{,}68}} = 6{,}00\%.$$

(c) Age of the earth in case that at the beginning the two isotopes ^{235}U and ^{238}U had the same abundance.

$$N_{235}(t) = N_{235}(0)e^{-\lambda_{235}t}$$
$$N_{238}(t) = N_{238}(0)e^{-\lambda_{238}t}$$

$$\Rightarrow \frac{N_{235}(t)}{N_{238}(t)} = \frac{0{,}72}{99.28} = e^{-\ln 2\cdot(t/T_{235} - t/T_{238})}$$

$$= 0{,}5^{(44{,}68-7{,}039)/(44{,}68\cdot 7{,}038)\cdot\left(t/10^8 a\right)}$$

$$\Rightarrow t = 59{,}37 \cdot 10^8 a \approx 6 \cdot 10^9 a$$

(Age of the earth).

Note: The real age is with $t = 4.5 \cdot 10^9$ years somewhat smaller. The difference comes from the wrong assumption that the two isotopes ^{235}U and ^{238}U had the same abundance at $t = 0$.

8.6 Yield of $^{131}_{53}$I at the fission of ^{235}U by thermal neutrons is about (2/3)·2.9% ≈ 2%. The released energy per fission is about E_f = 190 MeV. The thermal output power

$$Pth = \dot{N}th \cdot E_f = 1000\,\text{MW} \Rightarrow \text{Fission rate:}$$

$$\dot{N}_f = \frac{Pth}{E_f} = \frac{1000 \cdot 10^6 \text{VA}}{190 \cdot 10^6 \cdot 1{,}6 \cdot 10^{-19}\text{VAs}}$$
$$= 3{,}29 \cdot 10^{19}\,\text{s}^{-1}.$$

The production rate of ^{131}I is

$$\dot{N}_I^{(+)} = 2\% \cdot \dot{N}_f = 6{,}58 \cdot 10^{17}\,\text{s}^{-1}.$$

The radioactive decay of ^{131}I is

$$N_I(t) = N_I e^{-\lambda t} \quad \text{with} \quad \lambda = \ln 2/T.$$

The decay rate is then ^{131}I is

$$\dot{N}_I^{(-)} = -\lambda N_I e^{-\lambda t} = -\lambda N_I(t)$$

always proportional the momentary number $N_I(t)$. For the stationary situation is $N_I(t)$ constant, i.e.

$$\dot{N}_I^{(+)} + \dot{N}_I^{(-)} = 0, \quad \text{in order to}\, N_I(t) = \text{const.}$$

With a half-lifetime & T = 8.05 d saturation is reached within a few months.

$$\Rightarrow N_I = \frac{\dot{N}_I^{(+)}}{\lambda} = \frac{2\% \cdot \dot{N}_f \cdot T}{\ln 2}$$
$$= \frac{6{,}58 \cdot 10^{17}\,\text{s}^{-1} \cdot 8{,}04 \cdot 24 \cdot 60^2\,\text{s}}{\ln 2}$$
$$= 6{,}60 \cdot 10^{23}\ ^{131}\text{I-Atome}$$

$$m_I \approx 131 \cdot u = 131 \cdot 1{,}66 \cdot 10^{-27}\,\text{kg}$$
$$\text{with} \quad u = 931{,}5\,\text{MeV}/c^2$$

$$M_I = N_I \cdot m_I = 6{,}60 \cdot 131 \cdot 1{,}66 \cdot 10^{-4}\,\text{kg}$$
$$= 143{,}2\,\text{g}.$$

Assuming a stationary operation of the reactor with a thermal power of 1000 MW it contains the constant amount of 143.2 g ^{131}I. About 20% of the ethereal fission products as for example I; Cs; They are emitted into the atmosphere. This amounts to a radioactive dose of ^{131}I of 0.2·6.58 · 10^{17} Bq = 1.32 · 10^{27} Bq. Germany received in 1986 about 1–2 g ^{131}I i.e. about 1% of the total emission of 143 g.

Time Table of Nuclear- and High Energy Physics

1895 W. C. Röntgen discovers the X-Rays.

1896 A. H. Bequerel discovers radio-active radiation emitted by uranium ore which blackens photo-plates.

1898 M. Sklodowska-Curie and P. Curie discover and separate Polonium and Radium from radio-active ore.

J. Elster and H. F. Geitel explain the radio-activity as transformation of chemical elements.

1900 A. H. Bequerel: identification of ß-radiation as electrons by deflection in magnetic fields P. U. Villard. Detection of γ-radiation and identification as electro-magnetic radiation. E-Rutherford: Discovery of the element Radon.

1905 *O. Hahn* discovers the element Thorium ^{228}Th.

1907 J. *J. Thomson* develops the parabola mass spectrograph.

1908 *E Rutherford*: Identification of α-particles in the radio-active radiation as helium nuclei.

1909 *H. Geiger, R. Marsden*; Scattering experiments with α-particles by gold foils.

1911 *C. T. R. Wilson*: Development of the Cloud Chamber *J. J. Thomson:* Measurements of atomic masses with mass spectrometers, discovery of isotopes *V. F. Hess, W. Kohlhörster*: Discovery of cosmic rays.

1913 N. *Bohr* presents his planetary atomic model F. W. Aston defines the term "isotope" J. Chadwick: Measurement of the continuous ß-spectrum.

1919 *F. SW. Aston*: Development of a high resolution mass spectro-meter. First proof, that nearly all elements have several isotopes *E. Rutherford*: Discovery of the first artificial nuclear conversion.

1921 O. Hahn: Discovery of nuclear isomers.

1922 A. H. Compton: Discovery of the Compton effect.

1924 Visualization of nuclear reactions in the Cloud Chamber *W. Pauli*: Fisrt theoretical hints to the spin of the proton.

1925 *L. Meitner*: Explanation of the γ-radiation as transitions between nuclear states *W. Pauli*: Formulation of the exclusion principle.

1927 *E. Back, S. A. Goudsmit* Explanation of the atomic hyperfine- structure caused by the nuclear spin *P. A. M. Dirac*: Foundation of quantum electrodynamics, postulate of anti-particles. *W. Heisenberg*: Uncertainty relation.

1928 *H. Geiger, W. Müller*: Geiger-Müller counter. G. *Gamow E. U. Condon:* Explanation of the α-decay on the basis of the tunneling effect *R. Wideroe*: proposal of the Betatron principle.

1930 *E. O. Lawrence*: invention of the cyclotron. J. D. Cockroft, E. T. Walton: Invention of the Cascade Accelerator.

1931 *R. J. van de Graaff*: Construction of the first van de Graaff accelerator. *H. Urey*: Discovery of the deuterium.

1932 J. Chadwick: Discovery of the neutron. *E. O Lawrence S. Livingstone*: First operation of a cyclotron. *D. O. Ivanenko, SW. Heisenberg*: Nuclear model with protons and neutrons, Introduction of isospin. *C. D. Anderson*: Discovery of the positron as the first anti-particle.

1933 J. *Estermann, O. Stern:* Experimental determination of the magnetic moment of the proton. W. Pauli postulates the neutrino to explain the spectrum of the ß-decay.

1934 *E. Fermi:* Theoy of the ß-decay. I. Curie, F. Joliot: Discovery of artificial radio-activity and the ß$^+$.decay. P. A. Cernekov, I. M. Frank I. J. Tamm: Discovery of Cerenkov effect. J. Mattauch, RF. Herzog:. Development of the double-focussing mass spectrometer.

1935 *H. Yukawa*: Meson hypothesis of nuclear force. *H. A. Bethe, C. F. von Weizsäcker*: Droplet model and binding energy of the nucleus.

1937 J. DS. Anderson: Discovery of the muon μ as part of the cosmic rays.

1938 *I. Rabi*: Determination of the nuclear magnetic moment, development of nuclear magnetic resonance. H. A. Bethe, C. F. von Weizsäcker: Explanation of nuclear fusion as energy source of stars.

1939: *O. Hahn, E. Strassmann, L. Meitner O. Frisch N. Bohr, J. A. Wheeler:* Discoverý, explanation and theory of nuclear fission.

W. Demtröder, *Nuclear and Particle Physics*, Undergraduate Lecture Notes in Physics,
https://doi.org/10.1007/978-3-030-58313-2

1942 *E. Fermi* et al. Operation of the first nuclear reactor in Chicago.

1944 V. *J. Veksler, E. M.,McMillan*: Synchrotron principle with phase focusing.

1945 First atomic bomb drop on Hiroshima and Nagasaki after trial explosions in the desert of New Mexico.

1946 *F. Bloch, E. M. Purcell*: Development of nuclear magnetic resonance spectroscopy.

1947 *C. F. Powell*: Discovery of the π-Mesons in cosmic radiation. *H. Kallmann*: Development of Scintillation counter.

1948 *O. Haxel, J. H. D. Jensen, H. E. Suess, M. Goeppert. Mayer*: Nuclear spin-orbit coupling, Shell model of nuclei Development of spark chamber.

1950 J. *Rainwater, A. Bohr, B. Mottelson*: Development of a more detailed nuclear model.

1952 *D. Glaser*: Bubble Chamber.

1955 *R. L. Hofstadter*. Investigation of nuclear structure with high energy electrons, O. Chamberlain, E. Segré et al. Discovery of anti-proton.

1956 *E. Reines, C. L. Cowan*;: First experimental detection of neutrinos T. D. Lee, C. N. Yang, Hypothesis of Parity violation for the weak interaction. *S. G. Nilson*: Collective nuclear model.

1957 C. S. Wu et.al, Experimental proof of parity violation. RF. L. Mößbauer: Discovery of recoil-free nuclear absorption (Mößbauer-effect), Nobel Prize 1961.

1960 First electron-positron storage ring in Stanford.

1961 *S. L. Glashow*: First theoretical models for a unification of weak and el, magn. interaction. *M. Schwarz, L, Ledermann,* J, Steinberger: Discovery of μ-neutrino. *V. Fitch, J. Cronin*. Discovery of CP-violation for the weak interaction.

1964 M.*Gell-Mann, G. Zweig*: Quark hypothesis. Experimental evidence of CP violation.

1967 *S. Weinberg, A. Salam*: Development of electro-weak gauge-theory. *J. Friedman, H. Kendall, RF. Taylor*: Experimental confirmation of quark structure of the proton.

1968 *S. van der Meer*: Stochastic cooling of particles in storage rings.

1969 *S. G. Nilson W. Greiner*: Prediction of stability islands of super- heavy nuclei (trans-uranium elements).

1973 Gargamell team at CERN: Experimental proof of "neutral currents".

1974 *S. Ting*, et al. *and S. Richter* et al. Discovery of J/ψ.particle (charmonium).

1975 M. *Perl* et al.: Discovery of τ-lepton.

1977 Discovery of bottom quark.

1979 Discovery and explanation of gluons.

1981 CERN. first operation of SPS proton-antiproton colliders.

1983 UA1 and UA2 Experiments at CERN. Experimental detection of $W^{\pm}$ and Z^0. Gauge Bosons as exchange particles of electro-weak interaction.

1989 First operation of the large electron-positron collider LEP.

1990 Experimental proof that only 3 lepton families exist.

1994 *S. Hofmann, R. Armbruster* et al.: Discovery and production of the elements Z = 110 and 111.

1995 Discovery of the top quark.

1996 GSI: proof of element Z = 112.

2000 Discovery of the τ-neutrino.

2009 Completion of the LHC at CERN.

2012 Discovery of the Higgs boson.

Useful Conversion Factors

Lengths		
1 Å	1 Ångström	$\hat{=}10^{-10}\,\text{m}\hat{=}100\,\text{pm}$
1 f	1 Fermi	$\hat{=}10^{-15}\,\text{m}\hat{=}1\,\text{fm}$
1 AE	1 Astronomical unit AU	$\hat{=}1.49598\times10^{11}\,\text{m}$
1 ly	1 light year	$\hat{=}9.46\times10^{15}\,\text{m}$
1 pc	1 Parsec	$\hat{=}3.09\times10^{16}\,\text{m}$

Time
1 Year $= 3.156\times10^{7}$ s
1 Day $=8.64\times10^{4}$ s

Energy	
1 eV	1.60218×10^{-19} J
1 kWh	3.6×10^{6} J
1 kcal	4.1868 kJ
1 kcal/mol	4.34×10^{-2} eV per molecule
1 kJ/mol	1.04×10^{-2} eV per molecule
From $E = mc^2$ is follows: $1\,\text{kg}\cdot c^2 = 8{,}98755\times10^{16}\,\text{J}$	
With $k = 1{,}380658\times10^{-23}\,\text{J}\,\text{K}^{-1}$ follows for $1\,\text{eV}\hat{=}k\cdot T$ bei $T = 11604$ K	
With $h\cdot\nu = E$ follows for the frequency ν of electromagnetic radiation $\nu = E/h = 2.418\times10^{14}\,\text{Hz}\,\text{eV}^{-1}$	

Angles	
1 rad	57.2958°
1^0	0.0174 rad

(continued)

Angles	
1'	2.9×10^{-4} rad
1"	4.8×10^{-6} rad

Mathematical constants	
π	3.141592653589
e	2.718281828459
ln2	0.693147180559
$\sqrt{2}$	1.414213562373
$\sqrt{3}$	1.732050807568

Approximation formulas for $\lvert x\rvert \ll 1$	
$(1\pm x)^n \approx 1\pm nx$	$\cos x \approx 1 - x^2/2$
$\sqrt{1\pm x}\approx 1\pm\frac{1}{2}x$	$e^x \approx 1 + x$
$\sin x \approx x$	$\ln(1+x)\approx x$

The Greek Alphabet

Letters	Name	Letters	Name
A, α	Alpha	N, ν	Ny
B, β	Beta	Ξ, ξ	Xi
Γ, γ	Gamma	O, o	Omikron
Δ, δ	Delta	Π, π	Pi
E, ε	Epsilon	P, ϱ	Rho
Z, ζ	Zeta	Σ, σ	Sigma
H, η	Eta	T, τ	Tau
Θ, ϑ	Theta	Υ, υ	Ypsilon
I, ι	Jota	Φ, φ	Phi
K, κ	Kappa	X, χ	Chi
Λ, λ	Lambda	Ψ, ψ	Psi
M, μ	My	Ω, ω	Omega

Values of the Physical Fundamental Constants[a]

Quantity	Symbol	Value	Unit	Relative uncertainty in 10^{-6}
Speed of light in vacuum	c	29,9792,458	$m\,s^{-1}$	Exact
Gravitation constant	G	$6.6730 \cdot 10^{-11}$	$m^3\,kg^{-1}\,s^{-2}$	22
Planck constant	h	$6.62607015 \cdot 10^{-34}$	J s	Exact
Reduced Planck constant	$\hbar$	$1.054571817\ldots \cdot 10^{-34}$	J s	Exact
Molar gas constant	R	8.314462618	$J\,mol^{-1}\,K^{-1}$	Exact
Avogadro constant	N_A	$6.02214076 \cdot 10^{23}$	mol^{-1}	Exact
Lohschmidt constant (T = 273.15 K, p = 100 kPa)	N_L	$2.6516467 \cdot 10^{25}$	m^{-3}	Exact
Boltzmann constant	k	$1.380649 \cdot 10^{-23}$	$J\,K^{-1}$	Exact
Molar volume	V_M	$22.41396454 \cdot 10^{-3}$	$m^3\,mol^{-1}$	Exact
(T = 273.15 K, p = 101,325 pa)				
(T = 273.15 K, p = 100 pa)		$22.71095464 \cdot 10^{-3}$	$m^3\,mol^{-1}$	Exact
Elementary charge	e	$1.602176634 \cdot 10^{-19}$	$As \stackrel{Def}{=} C$	0.0003
Electron mass	m_e	$9.1093837015 \cdot 10^{-31}$	kg	Exact
Proton mass	m_P	$1.67262192369 \cdot 10^{-27}$	kg	0.0003
Magnetic constant	$m\mu_0$	$1.256637062 \cdot 10^{-6}$	$V\,s\,A^{-1}\,m^{-1}$	0.00015
Electric constant $1/(\mu_0 c^2)$	ε	$8.8541878128\ldots \cdot 10^{-12}$	$A\,s\,V^{-1}\,m^{-1}$	0.00015
Fine structure constant $\mu_0 ce^2/2h$	α	$7.2973525693 \cdot 10^{-3}$	–	0.00015
Rydberg constant $m_c c\alpha^2/2h^C$	Ry_∞	$1.0973731568160 \cdot 10^{7}$	m^{-1}	0.0000019
Bohr radius $\alpha/(4\pi Ry_\infty)$	a_0	$5.29177210903 \cdot 10^{-11}$	m	0.00015
Proton-electron mass ratio	m_P/m_c	1836.15267343	–	0.00006
Electron charge-to-mass quotient	$-elm_e$	$-1.75882001076 \cdot 10^{11}$	$C\,kg^{-1}$	0.0003
Proton charge-to-mass quotient	$+elm_p$	$+9.57882001560 \cdot 10^{7}$	$C\,kg^{-1}$	0.00031
Atomic mass unit $\frac{1}{12}m(^{12}C)$	AMU	$1.66053906660 \cdot 10^{-27}$	kg	0.0003

Conversion factor
1 eV = $1.6021765634 \cdot 10^{-19}$ J
1 eV/hc = 8065.541 cm^{-1}
1 Hartree = 27.2113845 eV
1 Hartree/hc = $2.194746313 \cdot 10^{5} cm^{-1}$
[a]CODATA, international recommended values (NIST 2018)

Astronomical Constants

Mass of earth	$M_E = 5.9736 \times 10^{24}$ kg	
Mass of moon	$M_M = 7.35 \times 10^{22}$ kg	$\hat{=}\ 0.0123\ M_E$
Mass of sun	$M_0 = 1.989 \times 10^{30}$ kg	$\hat{=}\ 3.33 \times 10^5 M_E$
Radius of sun	6.96×10^8 m	
Distance earth-moon		
Minimum (Perihel)	3.564×10^8 m	
Maximum (Aphel)	4.067×10^8 m	
Mean distance earth-sun	1.496×10^{11} m	
1AU = astronomical unit	$1.49597870700 \times 10^{11}$ m	

Periodic system of elements

Source: Handbook of Chemistry and Physics 94th ed., CRC Press 2013
International Union of Pure and Applied Chemistry (IUPAC), 2004: Pure and Applied Chemistry 75, 1613 (2003) 76, 2101 (2004)
Internet

26 **Fe**	Element wit proton number
55,85	relative atomic mass (in brackets the isotope with the longest half-life)
7,86	Density (g cm–3) [gases in g l–1 at 0 °C]
1538	Melting point (°C)
2861	boiling point (°C)
Eisen	item name

105 Db = Dubnium (Research Center Dubna near Moscow)
106 Sg = Dubnium (Research Center Dubna near Moscow)
107 Bh = Bohrium (Niels Bohr, Danish physicist 1885–1962)
108 Hs = Hassium (lat. Hassia = Hesse, German federal state)
109 Mt = Meitnerium (Lise Meitner, Austrian physicist, 1878–1968)
110 Ds = Darmstadtium (discovered at GSI in Darmstadt)
111 Rg = Roentgenium (Konrad Roentgen, 1845–1923)

1 **H** 1,008 0,090 –259,3 –252,8 Wasserstoff																	2 **He** 4,003 0,179 – –268,9 Helium
3 **Li** 6,941 0,534 180,5 1342 Lithium	4 **Be** 9,012 1,85 1287 2471 Beryllium											5 **B** 10,81 2,34 2075 4000 Bor	6 **C** 12,01 3,51 (D) 3825 (subl.) – Kohlenstoff	7 **N** 14,01 1,251 –210,0 –195,8 Stickstoff	8 **O** 16,00 1,429 –218,8 –183,0 Sauerstoff	9 **F** 19,00 1,69 –219,6 –188,1 Fluor	10 **Ne** 20,18 0,900 –248,6 –246,1 Neon
11 **Na** 22,99 0,97 97,72 883 Natrium	12 **Mg** 24,31 1,74 650 1090 Magnesium											13 **Al** 26,98 2,702 660,3 2519 Aluminium	14 **Si** 28,09 2,33 1414 3265 Silizium	15 **P** 30,97 1,82 (w.) 44,15 277 Phosphor	16 **S** 32,07 2,07 (α) 115,2 444,6 Schwefel	17 **Cl** 35,45 3,214 –101,5 –34,04 Chlor	18 **Ar** 39,95 1,784 –189,4 –185,9 Argon
19 **K** 39,10 0,86 63,4 759 Kalium	20 **Ca** 40,08 1,54 842 1484 Calcium	21 **Sc** 44,96 2,989 1541 2830 Scandium	22 **Ti** 47,88 4,5 1668 3287 Titan	23 **V** 50,94 5,96 1910 3407 Vanadium	24 **Cr** 52,00 7,20 1907 2671 Chrom	25 **Mn** 54,94 7,20 1246 2061 Mangan	26 **Fe** 55,85 7,86 1538 2861 Eisen	27 **Co** 58,93 8,9 1495 2927 Cobalt	28 **Ni** 58,69 8,90 1455 2913 Nickel	29 **Cu** 63,55 8,92 1085 2562 Kupfer	30 **Zn** 65,39 7,14 419,5 907 Zink	31 **Ga** 69,72 5,90 29,76 2204 Gallium	32 **Ge** 72,61 5,35 938,3 2833 Germanium	33 **As** 74,92 5,73 614 (subl.) – Arsen	34 **Se** 78,96 4,81 221 685 Selen	35 **Br** 79,90 3,119 –7,2 58,8 Brom	36 **Kr** 83,80 3,74 –157,4 –153,2 Krypton
37 **Rb** 85,47 1,53 39,31 688 Rubidium	38 **Sr** 87,62 2,6 777 1382 Strontium	39 **Y** 88,91 4,47 1526 3336 Yttrium	40 **Zr** 91,22 6,49 1855 4409 Zirkonium	41 **Nb** 92,91 8,57 2477 4744 Niob	42 **Mo** 95,94 10,2 2623 4639 Molybdän	43 **Tc** (97,91) 11,49 2157 – Technetium	44 **Ru** 101,1 12,3 2334 4150 Ruthenium	45 **Rh** 102,9 12,4 1964 3695 Rhodium	46 **Pd** 106,4 12,02 1555 2963 Palladium	47 **Ag** 107,9 10,5 961,8 2162 Silber	48 **Cd** 112,4 8,64 321,1 767 Cadmium	49 **In** 114,8 7,30 156,6 2072 Indium	50 **Sn** 118,7 7,28 231,9 2602 Zinn	51 **Sb** 121,8 6,684 630,6 1587 Antimon	52 **Te** 127,6 6,00 449,5 988 Tellur	53 **I** 126,9 4,93 113,7 184,4 Iod	54 **Xe** 131,3 5,887 –111,8 –108,0 Xenon
55 **Cs** 132,9 1,878 28,44 671 Cäsium	56 **Ba** 137,3 3,51 727 1897 Barium	**Lanthaniden** 57 **La** — 71 **Lu**	72 **Hf** 178,5 13,31 2233 4603 Hafnium	73 **Ta** 180,9 16,6 3017 5458 Tantal	74 **W** 183,9 19,35 3422 5555 Wolfram	75 **Re** 186,2 20,5 3186 5596 Rhenium	76 **Os** 190,2 22,48 3033 5012 Osmium	77 **Ir** 192,2 22,42 2446 4428 Iridium	78 **Pt** 195,1 21,45 1768 3825 Platin	79 **Au** 197,0 19,30 1064 2856 Gold	80 **Hg** 200,6 13,55 –38,83 356,7 Quecksilber	81 **Tl** 204,4 11,85 304 1473 Thallium	82 **Pb** 207,2 11,34 327,5 1749 Blei	83 **Bi** 209,0 9,80 271,4 1564 Bismut	84 **Po** (209,0) 9,4 254 962 Polonium	85 **At** (210,0) – 302 – Astatium	86 **Rn** (222,0) 9,73 –71 –61,7 Radon
87 **Fr** (223,0) – 27 – Francium	88 **Ra** (226,0) 5,0 700 – Radium	**Actiniden** 89 **Ac** — 103 **Lr**	104 **Rf** (261,1) – – – Rutherfordium	105 **Db** (262,1) – – – Dubnium	106 **Sg** (263,1) – – – Seaborgium	107 **Bh** (262,1) – – – Bohrium	108 **Hs** (265,1) – – – Hassium	109 **Mt** (266,1) – – – Meitnerium	110 **Ds** (269) – – – Darmstadtium	111 **Rg** (272) – – – Roentgenium	112 277–285 – – Copernicium	113 – – – – Noch nicht gefunden	114 (289) – – – Noch kein Name	116 291 – – – Noch kein Name	118 295 – – – Noch kein Name	119 – – – – Noch kein Name	120 299 – – – Noch kein Name

Lanthaniden	57 **La** 138,9 6,16 920 3455 Lanthan	58 **Ce** 140,1 6,77 799 3424 Cer	59 **Pr** 140,9 6,64 931 3510 Praseodym	60 **Nd** 144,2 7,008 1016 3066 Neodym	61 **Pm** (144,9) 7,264 1042 3000 Promethium	62 **Sm** 150,4 7,520 1072 1790 Samarium	63 **Eu** 152,0 5,244 822 1596 Europium	64 **Gd** 157,3 7,901 1314 3264 Gadolinium	65 **Tb** 158,9 8,230 1359 3221 Terbium	66 **Dy** 162,5 8,551 1411 2561 Dysprosium	67 **Ho** 164,9 8,795 1472 2694 Holmium	68 **Er** 167,3 9,066 1529 2862 Erbium	69 **Tm** 168,9 9,321 1545 1946 Thulium	70 **Yb** 173,0 6,966 824 1194 Ytterbium	71 **Lu** 175,0 9,841 1663 3393 Lutetium
Actiniden	89 **Ac** (227,0) 10,06 1051 3198 Actinium	90 **Th** (232,0) 11,72 1750 4788 Thorium	91 **Pa** (231,0) 15,37 1555 2963 Protactinium	92 **U** (238,1) 19,05 1135 4131 Uran	93 **Np** (237,0) 20,45 644 4079 Neptunium	94 **Pu** (242,1) 19,84 640 3228 Plutonium	95 **Am** (243,1) 13,67 1176 2607 Americum	96 **Cm** (247,1) 13,51 1345 – Curium	97 **Bk** (247,1) 13,25 1050 – Berkelium	98 **Cf** (251,1) 15,1 900 – Californium	99 **Es** (254,1) – 860 – Einsteinium	100 **Fm** (253,1) – 1527 – Fermium	101 **Md** (258,1) – 827 – Mendelevium	102 **No** (255,1) – 827 – Nobelium	103 **Lr** (260,1) – 1627 – Lawrencium

Index

W. Demtröder, *Nuclear and Particle Physics*, Undergraduate Lecture Notes in Physics,
https://doi.org/10.1007/978-3-030-58313-2

Printed in the United States
by Baker & Taylor Publisher Services